M000011691

OPTO-STRUCTURAL ANALYSIS

OPTO-STRUCTURAL ANALYSIS

John W. Pepi

SPIE PRESS
Bellingham, Washington USA

Library of Congress Cataloging-in-Publication Data

Names: Pepi, John W., author.
Title: Opto-structural analysis / John W. Pepi.
Description: Bellingham, Washington, USA : SPIE Press, [2018]
Identifiers: LCCN 2018001895 | ISBN 9781510619333 (hard cover) | ISBN 151061933X (hard cover) | ISBN 9781510619340 (PDF) | ISBN 1510619348 (PDF) | ISBN 9781510619357 (ePub) | ISBN 1510619356 (ePub) | ISBN 9781510619364 (Kindle/Mobi) | ISBN 1510619364 (Kindle/Mobi)
Subjects: LCSH: Optical instruments–Design and construction. | Optical engineering. | Structural analysis (Engineering)
Classification: LCC TS513 .P47 2018 | DDC 621.36–dc23 LC record available at https://lccn.loc.gov/2018001895

Published by

SPIE
P.O. Box 10
Bellingham, Washington 98227-0010 USA
Phone: +1 360.676.3290
Fax: +1 360.647.1445
Email: books@spie.org
Web: http://spie.org

Copyright © 2018 Society of Photo-Optical Instrumentation Engineers (SPIE)

All rights reserved. No part of this publication may be reproduced or distributed in any form or by any means without written permission of the publisher.

The content of this book reflects the work and thought of the author. Every effort has been made to publish reliable and accurate information herein, but the publisher is not responsible for the validity of the information or for any outcomes resulting from reliance thereon.

Printed in the United States of America.
First Printing.
For updates to this book, visit http://spie.org and type "PM288" in the search field.

Contents

Preface

Texts on structural and mechanical analysis are numerous, and indeed, this entire text is based on the pioneering works of others in the field. This book, therefore, draws on those texts and presumes a working knowledge of the strength of materials [see J. W. Pepi, *Strength Properties of Glass and Ceramics*, SPIE Press (2014)]. With that foundation, we apply those engineering principles to opto-structural analysis. In the precision world of optics, we are often concerned with displacements and deformations of very small values, from fractions of a wavelength of light to the micron and nanometer (millionths of an inch) order. Furthermore, optical systems designed for flight are often required to be of very light weight. While the analytical techniques in any case are the same as on the macro level, careful analysis is required when moving the decimal point so far to the left.

In preparing to write this book, some thought went into the title. Before selecting the term "opto-structural analysis," an alternative term "opto-mechanical analysis" was considered for the title. However, several excellent texts under that latter title are available. It is certainly not the intent to replace those worthy sources but rather to supplement them. To this end, the title contains the term "opto-structural," perhaps because the author is a structural engineer, but more so, to point out the "static" nature of the topic. If structural analysis is defined as applying to things that *don't move* once they are deformed, mechanical analysis as applying to things that *move* (such as mechanisms), and dynamic analysis as applying to things that *move slightly*, the title selection becomes more clear (although these latter topics *are* discussed in the book).

This book is written with the intent to understand basic structural deformation and stress analysis as applied to optical systems. It provides the tools for first-order analyses required in the design concept phase before entering into the intricate details of a full-up design. Ever-increasing computer technology has allowed former tedious and unwieldy problems to be solved in a fraction of the time by using finite element analysis. Unfortunately, reliance on such fast methods without hand analysis backup can lead to unsuspected errors. Thus, first-order calculations are an excellent way to complement the current state of the industry that relies more on computational design techniques. These calculations accelerate the design process by allowing an understanding of the critical governing parameters and allowing accelerated design trades and sensitivity studies to be performed that decrease schedule and cost. The insights gained from these techniques can then be used to guide the development of appropriate finite element models, including model fidelity, and details focusing on the critical and most sensitive design parameters. These models, in turn, are more efficient and provide the opto-structural engineer a comprehensive and insightful design approach. This approach can then inform the roadmap for risk reduction and environmental testing.

While finite element analysis is paramount to a successful design, the purpose of this text is not to use finite element analysis to validate the hand analysis but rather to use hand analysis to validate the finite element models. The hand analysis forces a discipline that aids tremendously in the understanding of structural behavior. It is the intent, then, not to forget such techniques.

"Forsan et haec olim meminisse iuvabit."*

---

*From Virgil's *The Aeneid* [translation: "Perhaps, someday, we will look back fondly on these things."]

Acknowledgments

Nothing can be learned or known without the pioneering efforts of others. It is with deep gratitude that I acknowledge the work of Stephen Timoshenko, the father of engineering mechanics, whose technical brilliance and straightforward communication skills have set the groundwork for this book.

I would also like to acknowledge the many teachers, professors, supervisors, and peers who have assisted me through the years, without whom this work would not be possible. A special note of thanks is given to Paul Yoder, Jr., who, sadly, is deceased, for his encouragement to produce this text and to Dan Vukobratovich for his insight and expertise.

I am very much indebted to Stefanos Axios, mechanical engineer, for his preparation, editing, and checking of the multitude of equations herein presented, and for his diligence, suggestions, and critique of the manuscript.

I would particularly like to acknowledge and dedicate this book to Francis G. Bovenzi and Joseph E. Minkle, supervisors at Itek Optical Systems, Lexington, Massachusetts, who taught me much of what I know.

Finally, I thank my wife, Sandy, for both her patience and encouragement in the preparation of this text.

John W. Pepi
October 2018

A Note on Units

This tutorial uses a blend of the US (customary) system of units and the International System of Units (metric, SI). Where practical, equivalent values of each are parenthetically noted. Most of the static analyses examples utilize the US system, while most of the fracture mechanics analyses will use SI units. Indeed, match and mismatch will occur; for example, we use the Celsius (SI) system for temperature in all cases. The reader will readily manage to work with the analyses using the simple conversions noted below.

	Unit	
Measure	**US**	**SI**
Force	1 pound (lb)	4.55 newtons (N)
	0.22 pound (lb)	1 newton (N)
Stress, Pressure	1 kilopound/square inch (ksi)	6.895 megapascals (MPa)
	0.145 kilopounds/square inch (ksi)	1 MPa
Stress Intensity	1 ksi-in$^{1/2}$	1.099 MPa-m$^{1/2}$
	0.91 ksi-in$^{1/2}$	1 MPa-m$^{1/2}$
Length	1 inch (in.)	0.0254 meter (m)
	39.37 inches (in.)	1 meter (m)
	mil (0.001 in.)	25.4 microns (μm)
Mass	1 slug (pound/32.2 feet/sec/sec)	14.594 kilograms (kg)
	1 kg	0.0685 slug
	1 pound (mass)	0.454 kg
Gravity Acceleration	32.17 feet/sec/sec	9.814 m/sec/sec
Temperature	1 °Fahrenheit (°F)	0.556 °Celsius (°C)
	1 °C	1.8 °F
Moment, Torque	1 in.-lb	0.113 N-m
	1 N-m	8.86 in.-lb
Density	1 pound per cubic inch (pci)	27.68 grams/cubic centimeter (g/cc)
	1 g/cc	0.036 pci
Volume	1 cubic inch (in.^{3})	16.387 cc
	1 cc	0.061 in.^{3}
Spring Rate/Linear Force	1 pound/inch	175.1 N/m
	1 N/m	0.0057 lb/in
	1 slug = 32.17 pounds mass	
	1 foot = 12 inches	
	1 ksi = 1000 psi	
	1 Pa = 1 N/m^2	
	1000 MPa = 1 GPa	

Chapter 1
Stress and Strain

1.1 Introduction

The opto-structural analyst is concerned with stress and deflection from externally applied loads, such as those occurring during mounting of optics, and internal loads, such as those initiated by gravity or acceleration. Additionally, the analyst is concerned with temperature change, which causes deflection and often causes stress. For cryogenic and high-temperature extremes, such values are obviously crucial; for more benign environments, temperature, loads, and self-weight deflection are still an issue, since we are concerned with fractional-wavelength-of-light changes. Accordingly, this initial chapter provides the basics of structural analysis, which lay the foundation for the chapters to follow.

1.2 Hooke's Law

Before diving into the structural analysis methods required for high-acuity optical systems, it is useful to review the origins of this analysis. While basic and advanced theories and principles of strength of materials and structural analysis have filled volumes, we review here the basis on which everything else follows. We review, therefore, the simple relation developed by Robert Hooke[1] in 1660, when he wrote *ut tensio sic vis*,[2] which literally means, "as the extension, so the force." This expression simply states that force, or load, is directly proportional to deflection for any system that can be treated as a mechanical spring, including elastic bodies, as long as such deflection is small. Simply stated,

$$F = kx, \tag{1.1}$$

where F is the applied force, x is the resulting deflection, and k is a spring, or stiffness, constant. In this theory, the spring is fully restored to its original length upon removal of the load.

A logical extension to Hooke's law relates stress to strain in a similar fashion. Consider a bar of length L and a cross-sectional area A under an axial load P, as shown in Fig. 1.1. Here, we define stress σ as

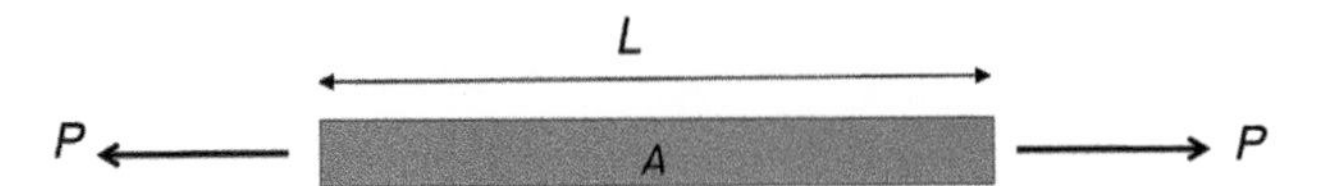

Figure 1.1 Direct tension force application to a one-dimensional (1D) element. Stress is defined as force divided by area and acts normal to the surface of the cross-section.

$$\sigma = \frac{P}{A}. \tag{1.2}$$

Note that this is simply a definition; stress has the units of force divided by area [MN/m^2 (MPa)], or pounds per square inch (psi).

The load in Fig. 1.1 and the resulting stress are considered to be tensile when the object is stretched and are compressive when it is shortened. Tensile and compressive stresses are called direct stresses and act normal to the cross-sectional surface.

Since stress is directly proportional to force divided by area, and strain ε (a dimensionless quantity) is related to deflection as

$$\varepsilon = \frac{x}{L}, \tag{1.3}$$

we can now rewrite Hooke's law as

$$\sigma = E\varepsilon, \tag{1.4}$$

where E is a material stiffness constant; for a solid isotropic material under a unidirectional axial load, E is an inherent property of the material, called its modulus of elasticity, and often referred to as elastic modulus, tensile modulus, or Young's modulus. The modulus of elasticity has the same units as stress (psi) since strain is dimensionless.

Substituting Eq. (1.4) into Eq. (1.2), we now readily compute the axial deflection of the bar of Fig. 1.1 as

$$x = \frac{PL}{AE}. \tag{1.5}$$

While this formulation is quite simplified, computation of stress for 3D solids with loads in multiple directions will be more complex. To illustrate this, and for the sake of completeness, while force is a vector (it has magnitude and direction), i.e., a first-order tensor, stress is a second-order tensor, which is a multidirectional quantity, and follows a different set of rules than the simple laws of vector addition. Further, for anisotropic materials, the stiffness matrix relating stress to strain will, in general, consist of a fourth-order tensor and 21 independent terms, with Hooke's law taking the form of

$$\sigma_{ij} = \sum_{k=1}^{3} \sum_{i=1}^{3} E_{ijkl} \varepsilon_{kl}, \tag{1.6}$$

where subscripts i, j take on values of 1, 2, or 3. Fortunately, in this text, we will not make use of such advanced analyses and will need only to discuss stress and strain in two dimensions, enabling more simplified, yet accurate, analyses. In the case of isotropic loading of 3D solids, the stiffness matrix is reduced to only two quantities, E and G, the latter of which is defined as the shear modulus, or modulus of rigidity. The shear modulus G is related to the elastic modulus E as

$$G = \frac{E}{2(1+\nu)}, \tag{1.7}$$

where ν is the ratio of lateral contraction to axial elongation under axial load and varies between 0 and 0.5 for most common materials. Values of zero are common for cork, for example, and values near 0.5 are common for rubbers, which are essentially incompressible. Another way of saying this is that for a material such as rubber, its volume will be constant under load, while its volume is ever increasing as Poisson's ratio is lowered toward zero. (Theoretical values of Poisson's ratio can be as low as -1, as achieved in certain materials, and are well beyond the scope of this text).

In two dimensions,

$$E_x = \frac{(\sigma_x - \nu\sigma_y)}{E}, \tag{1.7a}$$

$$E_y = \frac{(\sigma_y - \nu\sigma_x)}{E}. \tag{1.7b}$$

Thus, for the purposes of this text, these equations are most useful and preclude the need for unwieldy, 3D constituency matrices. The introduction of the 2D effect gives rise to the additional form of Hooke's law relating to shear stress τ, given as

$$\tau = G\lambda, \tag{1.8}$$

where λ is the (dimensionless) shear strain angle.

Shear stresses act in the plane of the cross-sectional surface. For shear load force V, as depicted in Fig. 1.2, we find the average shear stress as

$$\tau = \frac{V}{A}. \tag{1.9}$$

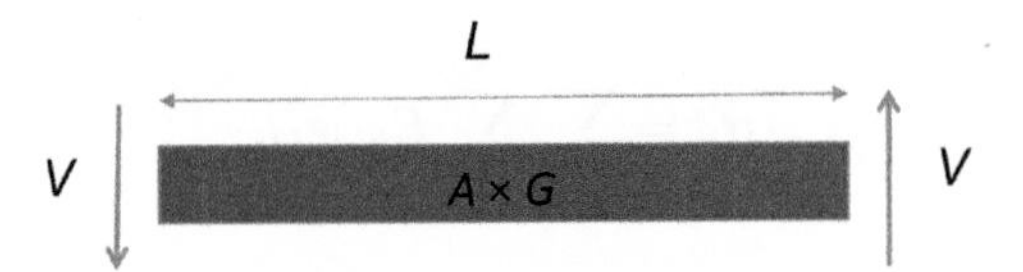

Figure 1.2 Direct shear force application without bending to a 1D element. Stress is defined as force divided by area and acts in the plane of the surface cross-section.

Substituting Eq. (1.8) into Eq. (1.9), we now readily compute the shear deflection (ignoring beam bending for the moment) of the bar of Fig. 1.2 as

$$y = \frac{VL}{AG}. \tag{1.10}$$

1.3 Beyond Tension, Compression, and Shear

Thus far, we have applied Hooke's law in the three translational directions: axial (x, tension/compression) and lateral (y, z, shear). There are also three rotational directions upon which bending and twist moments may act, completing the six possible degrees of freedom. Bending occurs when a moment is applied about either of the orthogonal lateral (y, z) axes, while twisting occurs when a moment [in units of inch-pounds (in.-lb)] is applied about the axial (x) axis. Figure 1.3 depicts these additional degrees of freedom. Again, in these cases, we can use Hooke's law to determine stresses and strains, and, therefore, deformation.

1.3.1 Bending stress

It is worthwhile to illustrate Hooke's law for the case of bending. Consider a beam under pure bending (constant, uniform moment), as shown in Fig. 1.4.

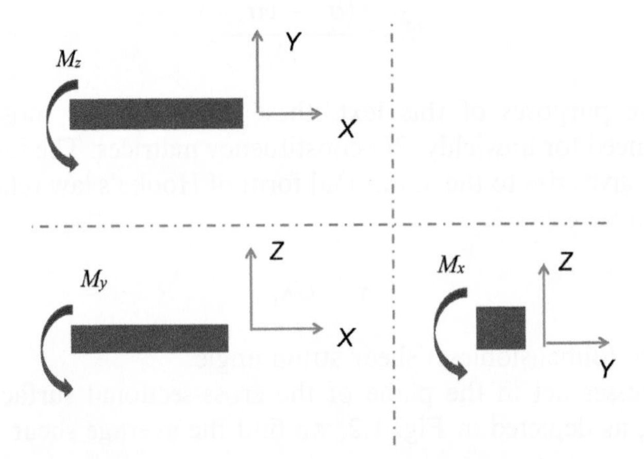

Figure 1.3 1D beam element under bending (about the z and y axes) and twist moments (about the x axis) in rotational degrees of freedom. Bending produces normal stress, while twist produces shear stress.

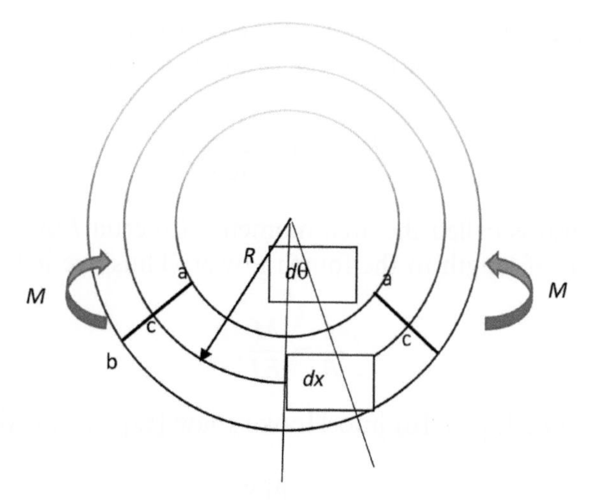

Figure 1.4 Diagram of bending stress showing section curves with radius R under moment loading. Surface a–a shortens, while surface b–b lengthens relative to neutral surface c–c.

The top (concave) surface is shortened, and the bottom (convex) surface is elongated. Somewhere in the middle there is no length change; this is the neutral axis of the beam. Since adjacent planes rotate by an amount $d\theta$, the arc length s at the neutral surface is given as

$$s = dx = Rd\theta, \tag{1.11}$$

where R is the radius of curvature of the beam.

Away from the neutral surface, the beam fibers elongate or shorten by an amount $yd\theta$, and since the original fiber length was dx, the strain is given simply as

$$\varepsilon = \pm\frac{yd\theta}{dx} = \pm\frac{y}{R}, \tag{1.12}$$

where a positive sign indicates tension, and a negative sign indicates compression. We can now apply Hooke's law (Eq. 1.4) and readily compute

$$\sigma = \frac{Ey}{R}. \tag{1.13}$$

These stresses acting over the elemental area give rise to forces that produce the resultant moment. Since there is no net force, from equilibrium it is realized that

$$\frac{E}{R}\int ydA = 0, \tag{1.14}$$

which implies that the neutral axis is at the centroid of the cross-section. The net moment M is the sum of the force distance products, or

$$M = \int y\sigma dA = \frac{E}{R}\int y^2 dA, \tag{1.15}$$

where the integral is called the area moment of inertia I of the cross-section, with dimensions of length to the fourth power. Thus, we have

$$\frac{1}{R} = \frac{M}{EI}, \tag{1.16}$$

and substitution of Eq. (1.16) into Hooke's law [Eq. (1.13)] yields

$$\sigma = \frac{My}{I}. \tag{1.17}$$

The largest value occurs as either tension or compression at the extreme fibers. Denoting the extreme fiber position as $y = c$, the maximum stress is

$$\sigma = \frac{Mc}{I}. \tag{1.18}$$

1.3.1.1 Combined normal stress

If a tensile or compressive axial load exists with a moment load, Eq. (1.18) is added to Eq. (1.2) (normal stresses acting in the same direction can be added):

$$\sigma = \frac{P}{A} + \frac{Mc}{I}. \tag{1.19}$$

It has been said that this equation is 90% of structural engineering; this is an obvious exaggeration, but the equation is, arguably, one of the most commonly used equations in structural analysis.

1.3.2 Bending deflection

For a beam of length L, it is a simple matter to compute the bending deformation y using the approximate parabolic relation

$$y = \frac{L^2}{8R}. \tag{1.20}$$

From Eq. (1.16), for a beam in pure bending,

$$y = \frac{ML^2}{8EI}. \tag{1.21}$$

Most commonly, the moment is not uniform, as is the case when transverse shear loads are introduced. Here, the curvature will vary along the beam, and differential equations—well documented in basic strength of materials literature[3] but not detailed here—give rise to deformations dependent on load and boundary conditions. For the simple case of the cantilever beam shown in Fig. 1.5, the deformation under end load P is

$$y = \frac{PL^3}{3EI}, \tag{1.22}$$

and for the simply supported beam of Fig. 1.6,

$$y = \frac{PL^3}{48EI}. \tag{1.23}$$

Of course, deflection can be accompanied by rotation, which is the slope of the deflection curve. Table 1.1 shows the typical cases of beam deflection and rotation for various loading and support boundary conditions. Support boundary conditions can be *free*, meaning no restraint, and free to translate and rotate; *roller*, meaning free to translate in one direction but restrained in the other, and free to rotate; *pinned*, meaning restrained in translation in both directions but free to rotate; *fixed*, meaning restrained in both translation and rotation; and *guided*, meaning not free to rotate but providing for freedom to translate in one direction.

1.3.3 Shear stress due to bending

Section 1.1 presents the shear stress due to direct shear. When shear is accompanied by bending, the maximum shear stress occurs at the neutral axis and varies to zero at the free boundaries. In this case, the "average" shear

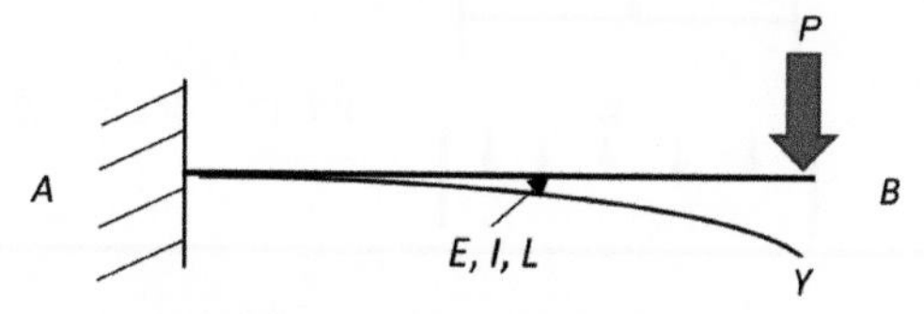

Figure 1.5 Cantilever beam bending under the end load will deflect at end B according to Eq. (1.22). There is no rotation or translation at fixed end A.

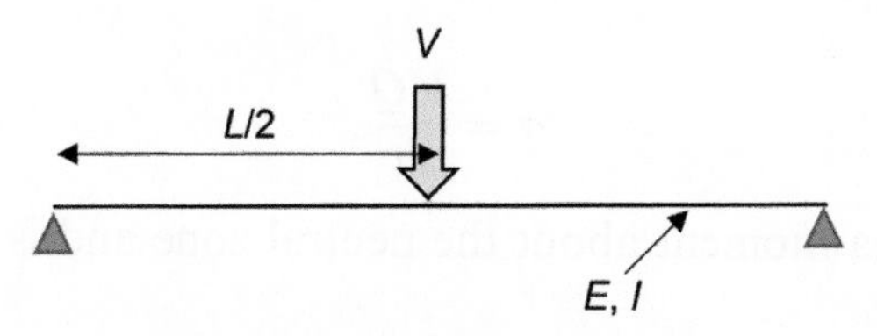

Figure 1.6 Simply supported beam bending under the central load will deflect at the center according to Eq. (1.23). There is no translation at the end points, which are allowed to rotate.

Table 1.1 Moment, deflection, and rotation for various loading and boundary conditions.

	Max. Moment	Max. Deflection	End Rotation A	End Rotation B
Cantilever end load	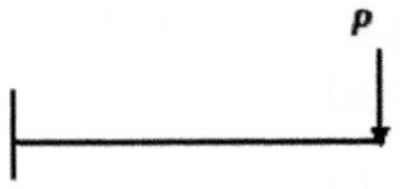 PL	$PL^3/3EI$	0	$PL^2/2EI$
Cantilever end moment	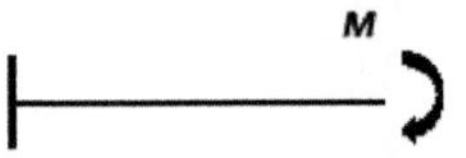 M	$ML^2/2EI$	0	ML/EI
Guided cantilever end load	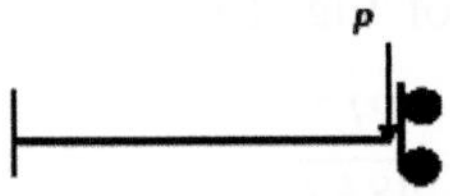 $PL/2$	$PL^3/12EI$	0	0
Cantilever uniform load	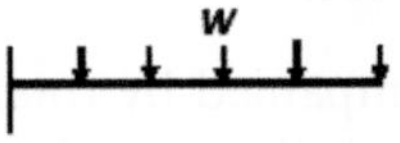 $WL/2$	$WL^3/8EI$	0	$WL^2/6EI$
Propped cantilever end moment load	 M	$ML^2/27EI$	0	$ML/4EI$
Simple support central load	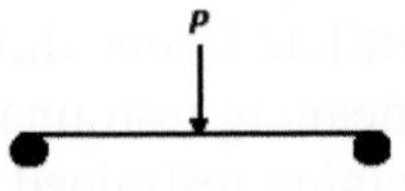 $PL/4$	$PL^3/48EI$	$PL^2/16EI$	$PL^2/16EI$
Simple support end moment	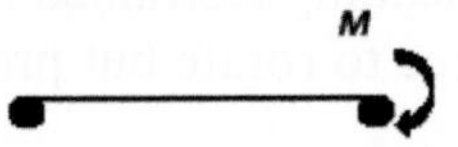 M	$0.0612ML^2/EI$	$ML/6EI$	$ML/3EI$
Simple support end moment	M	$ML^2/8EI$	$ML/2EI$	$ML/2EI$
Simple support uniform load	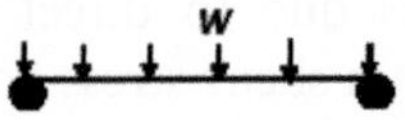 $WL/8$	$5WL^3/38EI$	$WL^2/24EI$	$WL^2/24EI$
Fixed support central load	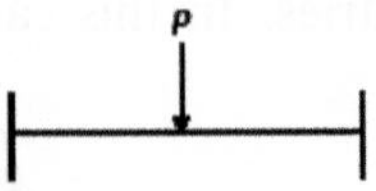 $WL/8$	$WL^3/192EI$	0	0
Fixed supports uniform load	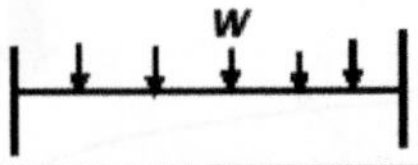 $WL/12$	$WL^3/384EI$	0	0

stress of Eq. (1.9) is exceeded at the neutral zone. The maximum shear stress can be computed as

$$\tau = \frac{VQ}{It}, \tag{1.24}$$

where Q is the area moment about the neutral zone and is given as

$$Q = \int y dA, \tag{1.24a}$$

and t is the thickness of the cross-section at the neutral zone. Equation (1.24) can therefore be rewritten as

$$\tau = \frac{kV}{A}, \tag{1.25}$$

where

$$k = \frac{AQ}{It}. \tag{1.26}$$

For the case of a rectangle,

$$\tau = \frac{3V}{2A}, \tag{1.26a}$$

and for a circular cross-section,

$$\tau = \frac{4V}{3A}. \tag{1.26b}$$

1.3.4 Shear deflection due to bending (detrusion)

Similarly, Section 1.1 presents shear deflection of a beam due to direct shear. When shear is accompanied by bending, shear deflection (sometimes referred to as shear detrusion) depends on both the variation in shear across the beam and the value of Q. In the case of a pure cantilever, we modify Eq. (1.10) and find that

$$y = \frac{kVL}{AG}. \tag{1.27}$$

For other loading and boundaries where the shear varies with beam length, we can use energy methods to compute deflection. For example, for a simply supported beam under a concentrated central load (first row of Table 1.1),

$$y = \frac{kVL}{4AG}. \tag{1.28}$$

The value of k [computed in Eq. (1.26)] assumes that, in computation of shear deflection, the cross-section is free to warp. This is not the case for many conditions of loading where shear changes abruptly, as in the case of the simply supported beam with a concentrated central load. More-complex strain energy formulation shows that, in this case for a rectangular cross-section, we find a modified coefficient as approximately $k = 6/5$, and for a circular cross-section, $k = 7/6$.

Deflection due to shear is generally small compared to deflection due to bending unless the span is short and/or the cross-section is deep. However, for lightweight optics (Chapter 6), shear deflection does have added importance.

1.3.5 Torsion

The final degree of freedom is twist about the axial axis, or torsion. Torque T (in units of pounds) is the torsional moment producing the twist. Again, Hooke's law applies, in this case, for shear [Eq. (1.8)]. Similar to what is done in bending (but not shown here), it is derived that torsional stress τ equals

$$\tau = \frac{\alpha T t}{K}, \tag{1.29}$$

where α is cross-section correction constant; t is the minimum thickness dimension of the cross-section; and K, with units of length to the fourth power, is called the torsional constant. The torsional constant equals the polar moment of inertia J for a circular (solid or hollow) cross-section, where

$$J = 2I. \tag{1.30}$$

In this case, $\alpha = 0.5$ (note that t = diameter), and

$$\tau = \frac{TR}{J}, \tag{1.31}$$

where R is the cross-sectional radius.

For a noncircular cross-section, the torsional constant is not the polar moment of inertia and needs a separate calculation. For a rectangular solid cross-section,

$$K = Bbt^3, \tag{1.32}$$

where b is the long-side width, and t is the short-side thickness of the section. The value of the torsional stiffness constant B is given in the plot of Fig. 1.7 as a function of the width-to-thickness ratio. Note that for thin sections, the value of B approaches 1/3.

The value for the torsional stress constant α is given in the plot of Fig. 1.8. Note that α approaches unity for a thin cross-section.

For hollow, thin-walled (t), closed, rectangular cross-sections,

$$K = \frac{4tA_0^2}{U}, \tag{1.33}$$

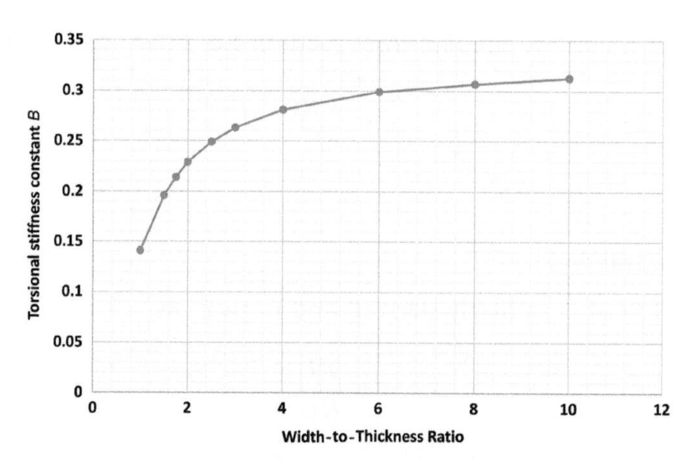

Figure 1.7 Torsional stiffness constant *B* versus width-to-thickness ratio for a rectangular cross-section. Values approach one-third for a thin cross-section.

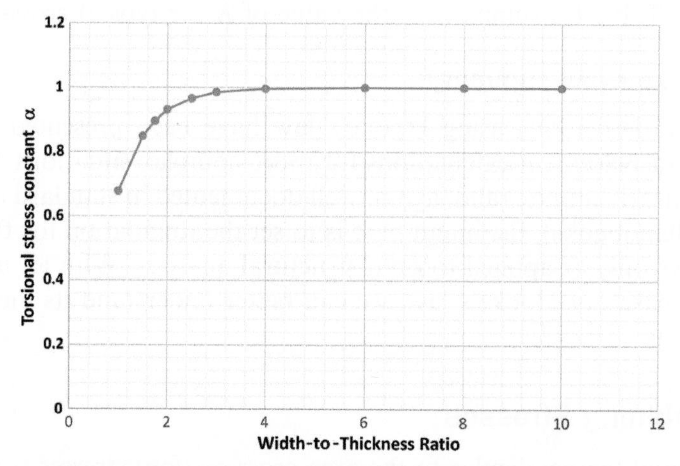

Figure 1.8 Torsional stress constant α versus width-to-thickness ratio for a rectangular cross-section. Values approach unity for a thin cross-section.

where A_0 is the area enclosed by the mean center line of the wall, and U is the perimeter of the mean centerline of the wall.

The value of α for use in Eq. (1.29) is

$$\alpha = \frac{2A_0}{Ut}, \tag{1.34}$$

from which we find that

$$\tau = \frac{T}{2A_0 t}. \tag{1.35}$$

Note that for a hollow, circular cross-section, Eq. (1.35) reduces to

$$\tau = \frac{TR}{J},$$

as it must.

1.3.5.1 Twist rotation

The angle of twist is similarly derived and is given as

$$\theta = \frac{TL}{KG}, \tag{1.36}$$

where, again, K is the torsional constant depending on the cross-section as discussed above. For thin-walled sections such as channel or U shapes, the value of b can be assumed to be the total developed width of the section, to the first order. Table 1.2 summarizes the value of K for typical cross-sections.

1.3.6 Hooke's law summary

Some basic derivations using Hooke's law have been presented. While the stress and displacement calculations for more-complex situations are exhaustive if not nearly infinite (and, again, well documented in standard engineering texts and handbooks), the intent here is to set the foundation for the material that follows only as applied to opto-structural analysis. With an understanding of the basics of Hooke's law, we can better understand its more-detailed formulations.

1.4 Combining Stresses

When normal (perpendicular to the area cross-section) stresses from tension, compression, or bending exist at a point, they can be combined directly. When in-plane shear stresses from torsion or direct shear exist at a point, they can be combined directly. However, as indicated in Section 1.2, stress, unlike force, is not a vector and exists in multiple orientations. Thus, when shear stresses are combined with normal stresses at a given point, they can neither be added algebraically nor vector summed, as the rules of tensor addition will apply. The addition can also be formulated by considerations of equilibrium. At any angle in a plane, the normal and shear stresses are given, respectively, as

$$\sigma = \frac{(\sigma_x + \sigma_y)}{2} + \frac{(\sigma_x - \sigma_y)}{2}\cos 2\theta - \tau_{xy}\sin 2\theta, \tag{1.37}$$

$$\tau = \frac{(\sigma_x - \sigma_y)}{2}\sin 2\theta + \tau_{xy}\cos 2\theta. \tag{1.38}$$

Table 1.2 Torsional constant K for various cross-sections. Dimensions of the constant are in length to the fourth power.

Section		Torsional Constant K
Solid circle	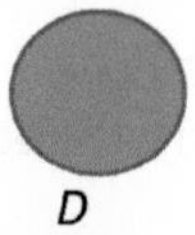 D	$\pi D^4/32$
Solid square	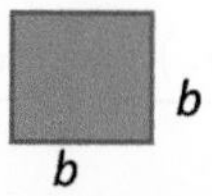 b, b	$0.141b^4$
Solid rectangle	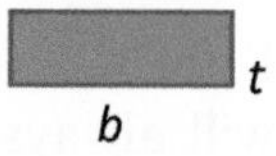 t, b	(see Fig. 1.7)
Hollow square tube	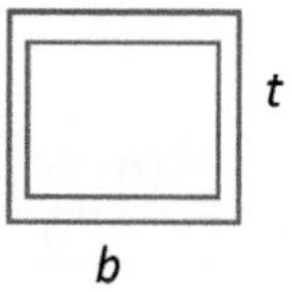 t, b	b^3/t
Round tube	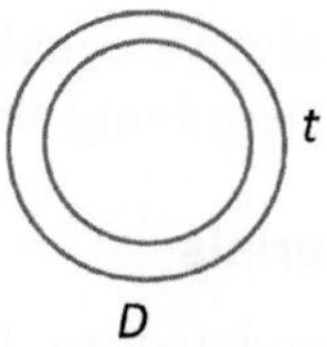 t, D	$\pi D^3 t/4$
Open section (thin wall)	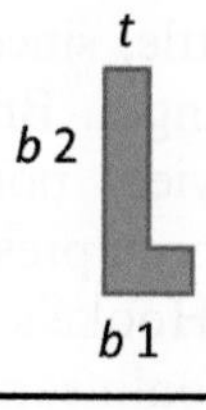 t, $b2$, $b1$	$0.333\ (b1 + b2)t^3$

Because these equations define the normal and shear stresses of a circle, a technique that uses what's called Mohr's circle is very useful in visualizing these stresses through their relationship to each other: At some angle, normal stress will be maximum and will occur where the shear stress is zero.

Differentiating Eq. (1.37) with respect to angle θ, and setting the resulting expression equal to zero (max-minima problem), we can find the angle of the maximum normal stress as

$$\tan 2\theta = \frac{2\tau}{(\sigma_x - \sigma_y)}. \tag{1.39}$$

By substitution, the maximum normal stress is calculated as

$$\sigma_1 = \frac{(\sigma_x + \sigma_y)}{2} + \sqrt{\frac{(\sigma_x - \sigma_y)^2}{4} + \tau^2} \tag{1.40}$$

and is called the major principal stress.

The minimum stress is similarly found as

$$\sigma_2 = \frac{(\sigma_x + \sigma_y)}{2} - \sqrt{\frac{(\sigma_x - \sigma_y)^2}{4} + \tau^2} \tag{1.41}$$

and is called the minor principal stress.

The maximum shear stress will always occur 45 deg from the principal stress angle and is calculated as

$$\tau_{max} = \frac{(\sigma_1 - \sigma_2)}{2}. \tag{1.42}$$

Note that the principal stress always equals or is greater than the applied normal stress and is used for determining strength.

1.4.1 Brittle and ductile materials

Principal stresses are well correlated to test strength data obtained for materials as long as they are brittle, since they generally have higher compressive strength than tensile strength. Brittle materials exhibit a low strain elongation to failure after the yield point is reached. With reference to Fig. 1.4, note that all of what has been presented applies in the linear region of a stress strain diagram, for which Hooke's law applies, i.e., below the material yield point at which it becomes nonlinear.

For ductile materials, the stresses are not conservative, and premature yielding may result. In this case, distortion energy methods are used, resulting in a maximum-stress prediction called von Mises stress. For two dimensions, von Mises stress is given as

$$\sigma_{max} = \sqrt{\sigma_1^2 - \sigma_1\sigma_2 + \sigma_2^2} \tag{1.43}$$

and should be used for materials that have high strain elongation before failure in yield. Note that the von Mises stress is an "equivalent" stress to be compared to the material yield strength and is not a true stress. Based on distortion theory, the premise is that the material fails by distortion, or in shear, as will be shown in the following example.

Using the von Mises criteria for the case of an object in tension (x axis only) and shear, we find from Eqs. (1.40), (1.41), and (1.43) that

$$\sigma_{max} = \sqrt{\sigma_x^2 + 3\tau^2}. \tag{1.44}$$

Under pure shear alone,

$$\sigma_{max} = \sqrt{3}\tau, \tag{1.45}$$

or

$$\tau = \frac{\sigma_{max}}{\sqrt{3}} = 0.577\sigma_{max}. \tag{1.46}$$

Thus, the distortion energy theory predicts that the shear strength is 0.577 times the tensile strength. This relation is common for most metals and other ductile isotropic materials.

A comparison of von Mises and principal stresses for typical, common 2D states of stress is given in Table 1.3.

1.5 Examples for Consideration

It is useful to illustrate the principles we have just discussed with some simple examples. We stress the word simple because the intent of this section is to define the basics and the basis for the material to follow. More-complex calculations will be introduced later as needed.

Example 1. Consider a beam fixed at one end (cantilevered) and loaded at its free end with an axial tensile load (x axis) of $P = 1000$ lbs and a shear Y load of $V = 2000$ lbs. The beam is 5 in. long with a rectangular cross-section of dimensions ½ in. wide by 2 in. deep. It is made of aluminum with an elastic modulus of 1.0×10^7 psi, a Poisson ratio of 0.33, and a yield strength of 35,000 psi.

Compute the following:

a) the normal stress σ_x
b) the shear stress τ
c) the principal stresses σ_1,σ_2
d) the von Mises stress σ_{max}
e) the maximum shear stress τ_{max}
f) the axial displacement x
g) the bending deflection y_b
h) the shear deflection y_s

Table 1.3 Principal and von Mises stresses for various elemental loading types. In general, von Mises stress equals or exceeds principal stress in 2D analysis.

	Principal Stress: Major	Principal Stress: Minor	Von Mises Stress
Uniaxial tension	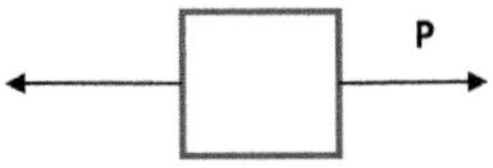 1	0	1
Pure shear	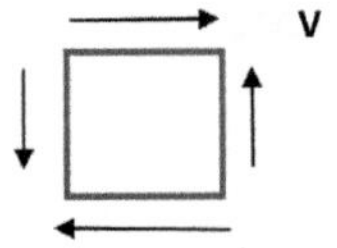 1	−1	1.732
Biaxial tension	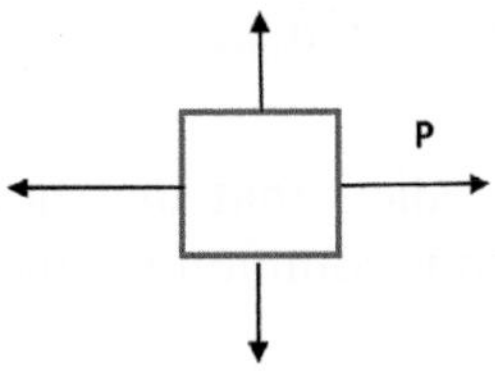 1	1	1
Tension and compression	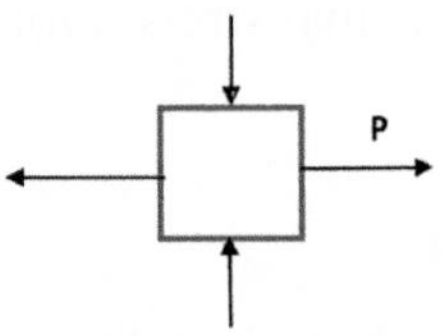 1	−1	1.732
Uniaxial tension and shear	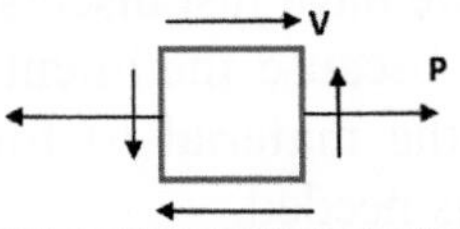 1.618	−0.618	2

Solutions:

a) The normal stress due to the axial load [from Eq. (1.2)] is

$$\sigma = \frac{P}{A} = \frac{1000}{1} = 1000\,\text{psi}.$$

The normal stress due to the shear load results from the maximum bending moment, which is $M = VL$.

The normal bending stress [from Eq. (1.18)] is

$$\sigma = \frac{VLc}{I} = \frac{6VL}{bh^2} = 7500\,\text{psi}.$$

At a particular point at the extreme fiber, we add the normal stresses. The combined normal stress is $\sigma = 1000 + 7500 = 8500\,\text{psi}$.

b) The shear stress [from Eq. (1.26a)] is

$$\tau = \frac{3V}{2A} = 3000\,\text{psi}.$$

c) The major principal stress [from Eq. (1.40)] is

$$\sigma_1 = \frac{\sigma_x}{2} + \sqrt{\left(\frac{\sigma_x}{2}\right)^2 + \tau^2} = 9450\,\text{psi},$$

and the minor principal stress [from Eq. (1.41)] is

$$\sigma_2 = \frac{\sigma_x}{2} - \sqrt{\left(\frac{\sigma_x}{2}\right)^2 + \tau^2} = -950\,\text{psi}.$$

d) The von Mises stress is calculated from Eq. (1.43) as

$$\sigma = \sigma_{\text{max}} = \sqrt{\sigma_1^2 - \sigma_1\sigma_2 + \sigma_2^2} = 9960\,\text{psi}.$$

The von Mises stress is only slightly higher than the principal stress but should be used because the material is ductile.

e) The maximum shear stress is calculated from Eq. (1.42) as

$$\tau_{\text{max}} = \frac{(\sigma_1 - \sigma_2)}{2} = 5200\,\text{psi}.$$

f) The axial displacement [from Eq. (1.5)] is

$$x = \frac{PL}{AE} = 0.0005\,\text{in}.$$

g) The bending deflection is found from Eq. (1.22) or Table 1.1 and is

$$y_{\text{b}} = \frac{VL^3}{3EI} = 0.025\,\text{in}.$$

h) The shear deflection [from Eq. (1.27)] is

$$y_{\text{s}} = \frac{kVL}{AG}, \quad \text{where } k = \frac{6}{5}$$

$$y_{\text{s}} = 0.0032\,\text{in}.$$

The shear deflection can be added directly to the bending deflection. Note that shear deflection is typically small compared to bending deflection unless the beam length is extremely small or the cross-section is very deep.

Example 2. A cantilever beam having properties and dimensions identical to those in Example 1 above is subjected to an end torsional twist of 4000 in-lb.

Compute the following:

a) the shear stress
b) the major principal stress
c) the von Mises stress
d) the angle of twist

Solution:

a) The shear stress [from Eq. (1.29)] is given as

$$\tau = \frac{\mathrm{T}}{\alpha b t^2} = 24800\,\mathrm{psi}.$$

b) The major principal stress [from Eq. (1.40)] equals the shear stress:

$$\sigma_1 = 24800\,\mathrm{psi}.$$

c) The von Mises stress [from Eq. (1.45)] is

$$\sigma_{\max} = \sqrt{3}\mathrm{t} = 43000\,\mathrm{psi}.$$

Note that the von Mises stress is significantly higher than the principal stress, and, in fact, exceeds the yield strength of the material. Since the material is ductile, the von Mises stress should be used; if the principal stress were used, a false sense of security might result, unless the user is aware that shear strength drives the design. In the latter case, if the principal stress were used, the astute analyst would check both the principal and maximum shear stresses, and would see that shear strength drives the design.

d) The angle of twist is computed from Eq. (1.36) as

$$\theta = \frac{TL}{KG} = \frac{TL}{Bbt^3} = 0.066\,\mathrm{rad} = 3.8\ \mathrm{deg}.$$

1.6 Thermal Strain and Stress

As we have seen from Hooke's law [Eqs. (1.1) and (1.4)], when an external force is applied to a member, stress is produced, and that stress is always accompanied by strain. There are cases, however, where strain is applied without producing stress, as occurs under temperature loading. Consider, for example, a beam of length L that is free to expand under a temperature excursion ΔT.

With no restraint, the beam grows an amount

$$y = \alpha L \Delta T, \tag{1.47}$$

where α is the effective thermal expansion coefficient of the material over the temperature range of interest. The beam grows according to the diagram in Fig. 1.9(a). The strain is

$$\varepsilon = \frac{y}{L} = \alpha \Delta T. \tag{1.48}$$

Because this is the natural state in which the beam occurs, there is no stress. Strain without stress is called eigenstrain.

If such a beam were completely restrained from growing [as in Fig. 1.9(b)], the amount it would naturally grow is resisted by a force, which produces stress. Thus, from Eqs. (1.5) and (1.48), we have

$$\begin{aligned} \frac{PL}{AE} &= \alpha L \Delta T; \text{ therefore,} \\ P &= AE\alpha \Delta T, \end{aligned} \tag{1.49}$$

and, therefore,

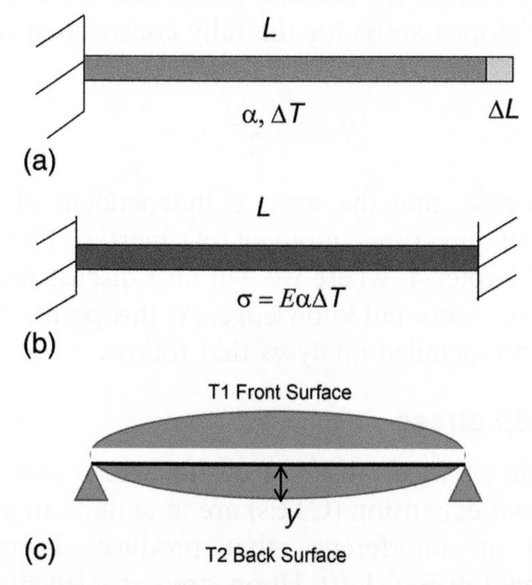

Figure 1.9 Thermal expansion under uniform temperature soak of (a) a stress-free unconstrained beam and (b) a fully constrained beam inducing normal stress σ. (c) Thermal expansion of a stress-free beam simply supported at both ends with a uniform, linear, front-to-back thermal gradient.

$$\sigma = \frac{P}{A} = E\alpha\Delta T. \tag{1.50}$$

This simple equation is very important (because stress developed under thermal strain when constrained will rarely exceed this amount) and serves as an upper bound for first-order calculations. Along with Eq. (1.18), Eq. (1.50) is one of the simplest and most important relationships in opto-structural analysis. (Chapter 4 will expand on this in two dimensions).

Note that the resisting force is independent of length, and the developed stress is independent of both length and cross-sectional area, which is nice. Note further that if a member wants to expand and is not free to do so, the force and stress are in compression; if it wants to shrink and is not free to do so, it is in tension.

Similarly, consider a case in which a thermal gradient is applied through the depth of the cross-section. For a linear gradient, we have again a case of eigenstrain if the beam is unrestrained, and it will bend without stress to the shape shown in Fig. 1.9(c). Here, the radius (of the neutral axis) is

$$R = \frac{t}{\alpha\Delta T}. \tag{1.51}$$

For a positive expansion coefficient and a positive temperature change on the top surface, the top tends to expand and bend the beam in a convex direction. Again, if the beam is fully constrained, stress will develop with the top surface in compression. (In thermal cases, you sometimes have to think backward.) The developed stress for the fully constrained case is

$$\sigma = \frac{E\alpha\Delta T}{2}. \tag{1.52}$$

Note again, in this case, that the stress is independent of length and cross-sectional area or bending (area moment of) inertia. This is nice. We will expand on this in Chapter 4, where we will also discuss nonlinear gradients, which do require cross-sectional knowledge. At this point, we have simply set the stage for the more-detailed analyses that follow.

1.6.1 Thermal hoop stress

A common example of thermal stress occurs when two rings of differing coefficients of thermal expansion (CTEs) are in contact in a thermal environment, resulting in an interference that produces hoop stress in both components, as shown in Fig. 1.10. Hoop stress σ is given as

$$\sigma = \frac{qR}{A}, \tag{1.53}$$

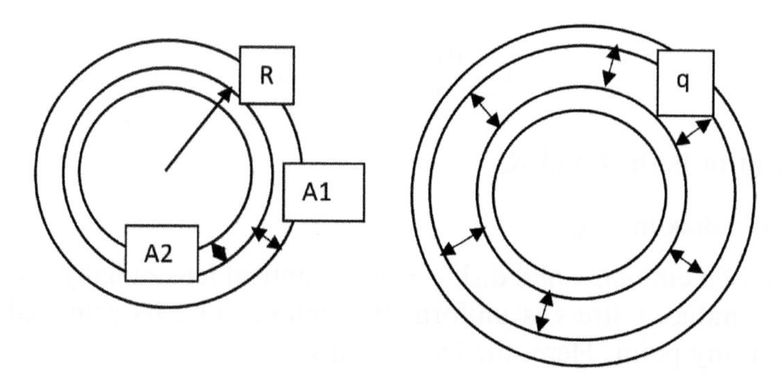

Figure 1.10 Diagram illustrating hoop stress. Two rings of unit width and constant thickness (A1 and A2) produce self-equilibrating pressure q when subjected to a temperature change for given properties of modulus E and differing thermal expansion coefficients α.

where R is the mean ring radius at contact, q is the induced interference pressure in pounds per inch, and A is the cross-sectional area of the individual rings. What remains is to solve for q under a thermal soak condition. From Hooke's law, the induced circumferential strain is

$$\varepsilon = \frac{\sigma}{E} = \frac{qR}{AE}. \tag{1.54}$$

The thermal strain is simply $\Delta\alpha\Delta T$, in which

$$\Delta\alpha = \alpha_1 - \alpha_2,$$

where the subscript numbers denote the outer and inner rings, respectively. From compatibility, we have

$$\alpha_1 \Delta t - \frac{qR}{(AE)_1} = \alpha_2 \Delta t + \frac{qR}{(AE)_2}. \tag{1.55}$$

Solving for q, we obtain

$$q = \frac{\Delta\alpha\Delta T(AE)_2}{\left[1 + \frac{(AE)_2}{(AE)_1}\right] R}. \tag{1.56}$$

The stress is recovered from Eq. (1.53).

Note that if the inner ring is very stiff relative to the outer ring, $(AE)_2$ is set to infinity in Eq. (1.55), and

$$q = \frac{(AE)_1 \Delta\alpha\Delta T}{R}. \tag{1.57}$$

The outer ring stress [from Eq. (1.53)] is

$$\sigma = \frac{qR}{A} = E_1 \Delta\alpha \Delta T \tag{1.58}$$

independent of both A and R.

1.6.1.1 Solid disk in ring

In the case of a circular, solid disk—as in an optical lens radially restrained in a cell—the induced stress is uniform throughout; i.e., its principal stress is identical at any point. Here, the lens strain is

$$\varepsilon = \frac{\sigma}{E} = \frac{q}{Eb}, \tag{1.59}$$

and the induced stress is thus

$$\sigma = E\varepsilon = \frac{q}{b}. \tag{1.60}$$

Therefore, under thermal soak, we can modify Eq. (1.55) to yield

$$\Delta\alpha\Delta T = \frac{qR}{tbE_1} + \frac{q}{E_2 b}, \tag{1.55a}$$

where the subscripts 1 and 2 denote the ring and disk, respectively. Hence, Eq. (1.56) can be written as

$$q = \frac{\Delta\alpha\Delta T}{\left(\frac{R}{tbE_1} + \frac{1}{E_2 b}\right)}. \tag{1.56a}$$

Note from Eq. (1.56a) that if the disk is very rigid relative to the ring, then the ring stress is

$$\sigma = E_1 \Delta\alpha\Delta T \tag{1.58a}$$

independent of the radius, and the disk stress is

$$\sigma = \frac{E_1 t \Delta\alpha\Delta T}{R}, \tag{1.60a}$$

which is *inversely proportional* to the radius.

Note from Eq. (1.56a) that if the ring is very rigid relative to the disk, then the ring stress is

$$\sigma = \frac{E_2 \Delta\alpha \Delta T R}{t} \tag{1.58b}$$

directly proportional to the radius, and the disk stress is

$$\sigma = E_2 \Delta\alpha \Delta T \tag{1.60b}$$

independent of the radius.

Example. We can apply these relationships to the case of a lens cell housing an optical lens. Consider a zinc sulfide lens 1 in. deep b and 4 in. in diameter encased in a 1-in.-deep by 0.10-in.-thick t aluminum lens housing. Over a soak from room temperature to 150 K, compute the stress in the lens and housing. The following effective properties over the range of soak are given:

$$E_1 = 9.9 \times 10^6 \text{psi}$$
$$E_2 = 1.08 \times 10^7 \text{psi}$$
$$\alpha_1 = 2.10 \times 10^{-5}/\text{K}$$
$$\alpha_2 = 5.6 \times 10^{-6}/\text{K}$$

Since the lens is rigid relative to the housing, we find from Eq. (1.57), where $A = bt$, that

$$q = \frac{(1)(0.1)(9.9)(15.4)(143)}{2} = 1100\,\text{lb/in.},$$

and the housing stress from Eq. (1.58) is

$$\sigma = (9.9)(15.4)(143) = 21800\,\text{psi}.$$

The line pressure q on the lens is the same as that on the housing, and the lens stress, under uniform principal stress everywhere throughout, is recovered from Eq. (1.60) as

$$\sigma = \frac{q}{b}; \text{ therefore,}$$

$$\sigma = \frac{1100}{1} = 1100\,\text{psi}.$$

1.6.2 Ring in ring in ring

Similar to the thermal stress induced by the interference of two rings, it is useful to review the case of thermal interference involving three rings. This could occur, for example, when a thin isolation ring is housed between an optic with a central hole and housing, or an insert is bonded to a housing. In this case, we need to consider the strain compatibility relationships between the inner and middle rings, and between the central and outer rings. This is a

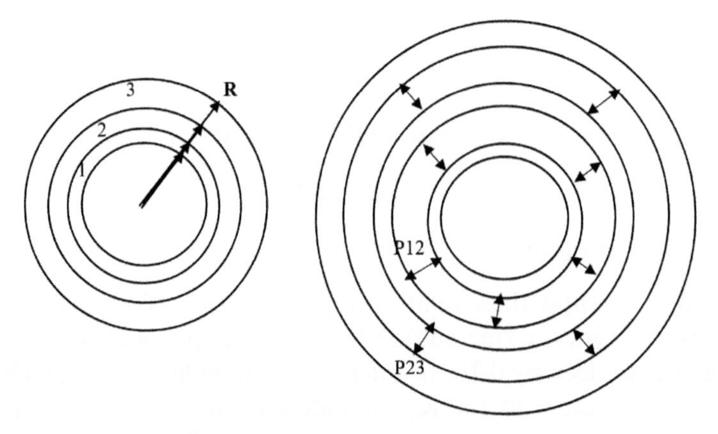

Figure 1.11 Ring-in-ring-in-ring hoop stress. Each ring has an inner and outer radius R, and each can have different modulus, thickness, and expansion characteristics. Expansion leads to self-equilibrating pressure on each surface.

bit more complex than the two-ring problem. Figure 1.11 is a schematic of the three-ring design, along with an equilibrium diagram. In this case, we need not limit thickness to thin rings. Using thick-ring theory,[4] and after tedious calculations, we arrive at the individual ring stresses. We compute both radial and hoop stresses for a total of 12 stresses. (Because the inner and outer radial stresses are always zero, there are ten calculable stresses.) These stresses are given in Table 1.4, and the numerous constants are defined in Table 1.5. These constants are readily programmable to the stress equations of Table 1.4 by use of a spreadsheet. Table 1.6 gives the material property constants used for computation of the hoop and radial stresses defined in Table 1.5.

When the outermost (or innermost) ring is not present and the rings are thin, the problem reduces to the simplified two-ring case of Eqs. (1.53) and (1.56) (believe it or not) with an error difference of less than 5%.

Table 1.4 Tri-ring hoop and radial stresses. Subscripts 1, 2, and 3 refer to inner, middle, and outer rings, respectively; subscripts o and i refer to outer and inner surfaces respectively; r is the radius at the specified interface; other constants are from Table 1.5.

σ_{r1i}	Radial stress inner surface inner ring	0
$\sigma_{\theta 1i}$	Hoop stress inner surface inner ring	$-2p_{12}r_{10}^2/A_{1m}$
σ_{r10}	Radial stress outer surface inner ring	$-p_{12}$
$\sigma_{\theta 10}$	Hoop stress outer surface inner ring	$-p_{12}A_{1p}/A_{1m}$
σ_{r2i}	Radial stress inner surface middle ring	$-p_{12}$
$\sigma_{\theta 2i}$	Hoop stress inner surface middle ring	$(p_{12}\,A_{2p} - 2p_{23}r_{20}^2)/A_{2m}$
σ_{r20}	Radial stress outer surface middle ring	$-p_{23}$
$\sigma_{\theta 20}$	Hoop stress outer surface middle ring	$(2p_{12}r_{2i}^2 - 2p_{23}A_{2p})/A_{2m}$
σ_{r3i}	Radial stress inner surface outer ring	$-p_{23}$
$\sigma_{\theta 3i}$	Hoop stress inner surface outer ring	$p_{23}\,A_{3p}/A_{3m}$
σ_{r30}	Radial stress outer surface outer ring	0
$\sigma_{\theta 30}$	Hoop stress outer surface outer ring	$2p_{23}r_{3i}^2/A_{3m}$

Table 1.5 Constants used for computation of the hoop and radial stresses defined in Table 1.4. ΔT = temperature soak.

A_{1p}	Geometry constant	$r_{1o}^2 + r_{1i}^2$
A_{1m}	Geometry constant	$r_{12o}^2 - r_{12i}^2$
A_{2p}	Geometry constant	$r_{2o}^2 + r_{2i}^2$
A_{2m}	Geometry constant	$r_{2o}^2 - r_{2i}^2$
A_{3p}	Geometry constant	$r_{3o}^2 + r_{3i}^2$
A_{3m}	Geometry constant	$r_{3o}^2 - r_{3i}^2$
c_1	Geometry/material constant	$r_{2i}(A_2p/A_{2m} + \upsilon_2)/E_2 + r_{1o}(A_1p/A_{1m} - \upsilon_1)/E_1$
c_2	Geometry/material constant	$-2r_{2i}r_{2o}^2/(E_2A_{2m})$
c_3	Geometry/material constant	$-2r_{2i}^2r_{2o}/(E_2A_{2m})$
c_4	Geometry/material constant	$r_{3i}(A_3p/A_{3m} + \upsilon_3)/E_3 + r_{2o}(A_2p/A_{2m} - \upsilon_2)/E_2$
d_{12}	Interference inner to middle ring	$\Delta T(-r_{2i}\alpha_2 + r_{1o}\alpha_1)$
d_{23}	Interference middle to outer ring	$\Delta T(-r_{3i}\alpha_3 + r_{2o}\alpha_2)$
p_{12}	Pressure inner to middle ring	$(c_4d_{12} - c_2d_{23})/(c_1c_4 - c_2c_3)$
p_{23}	Pressure middle to outer ring	$d_{12}/c_2 - c_1p_{12}/c_2$

Table 1.6 Material property constants used for computation of the hoop and radial stresses defined in Table 1.5.

E_1	Young's modulus inner ring	Input
υ_1	Poisson's ratio inner ring	Input
α_1	CTE inner ring	Input
E_2	Young's modulus middle ring	Input
υ_2	Poisson's ratio middle ring	Input
α_2	CTE middle ring	Input
E_3	Young's modulus outer ring	Input
υ_3	Poisson's ratio outer ring	Input
α_3	CTE outer ring	Input

1.6.2.1 Case study

Consider a silicon optic with a central hole. The optic is supported by an aluminum hub ring that is rigidly attached by a relatively soft isolation ring of Vespel®, as shown in Fig. 1.12. The assembly is subjected to a soak change of 100 °C. For the dimensions shown, determine the stress in the optic. The effective properties over the thermal range are given in Table 1.7. The maximum hoop stress is obtained from Tables 1.4 through 1.6 as

$$\sigma_{\theta 3i} = \frac{p_{23}A_{3p}}{A_{3m}} = 5360\,\text{psi}.$$

Note that, while this stress level may be well below the allowable value for polished silicon, it would be problematic if excessively deep flaws were present, as discussed in Chapter 12.

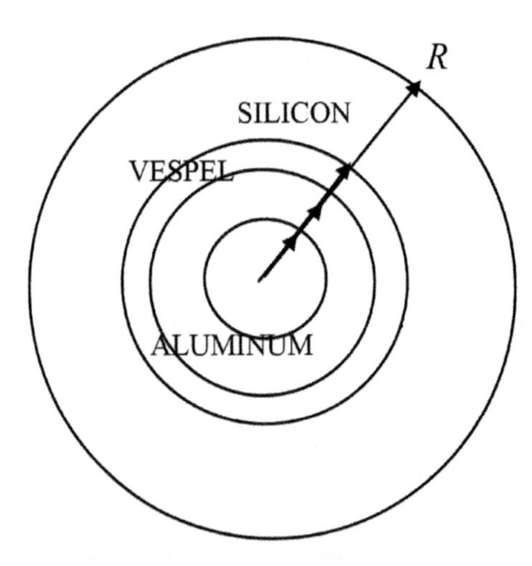

Figure 1.12 Diagram of the case study setup: an aluminum hub is placed in the central hole of a silicon optic that is isolated with Vespel and undergoes a temperature soak to 100 °C. Properties and dimensions are the same as those in the example in Section 1.6.1.1.

Table 1.7 Material and dimensional properties for the case study.

				Radius	
Material	Modulus	Poisson's Ratio (psi)	CTE (ppm/°C)	Inner (inch)	Outer (inch)
Aluminum	1.00E+07	0.33	2.15E-05	5	5.50
Vespel	4.70E+05	0.35	4.00E-05	5.5	5.75
Silicon	1.90E+07	0.2	2.00E-06	5.75	8.00

1.6.3 Nonuniform cross-section

The previous discussion of a uniform cross-section shows thermal stress independent of cross-section or length, approaching the maximum of Eq. (1.50). However, for a nonuniform cross-section, this is not the case.

Consider, for example, the trapped 1D beam of Fig. 1.13 in which the cross-section varies and undergoes a temperature change of ΔT. As it becomes warmer, the beam tends to expand stress free as

$$y = \Sigma \alpha_i L_i \Delta T. \tag{1.61}$$

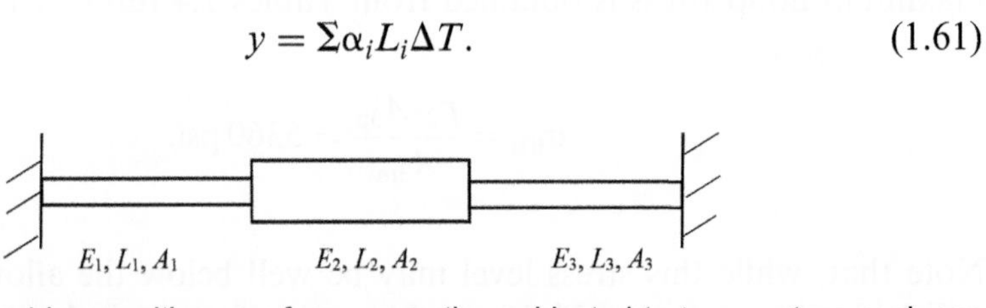

Figure 1.13 A fixed beam with nonunform properties subjected to temperature soak can produce extreme stress conditions.

The rigid wall will not let it expand and pushes back with compressive force P. Since the net deflection is zero, we have, from Hooke's law,

$$P\left[\Sigma\left(\frac{L_i}{A_1E_1}\right)\right] = \Sigma\alpha_i L_i \Delta T \tag{1.62}$$

so that

$$P = \frac{\Sigma\alpha_i L_i \Delta T}{\left(\Sigma\frac{L_i}{A_1E_1}\right)},$$

and

$$\sigma = \frac{P}{A_i}. \tag{1.63}$$

For the case shown in Fig. 1.13, we have

$$P = \frac{(2\alpha_1 L_1 + \alpha_2 L_2)\Delta T A_1 A_2 E_1 E_2}{(2L_1 A_2 E_2 + L_2 A_1 E_1)}, \tag{1.64}$$

and

$$\sigma_1 = \frac{P}{A_1} = \frac{(2\alpha_1 L_1 + \alpha_2 L_2)\Delta T A_2 E_1 E_2}{(2L_1 A_2 E_2 + L_2 A_1 E_1)}, \tag{1.65a}$$

$$\sigma_1 = \frac{P}{A_2} = \frac{(2\alpha_1 L_1 + \alpha_2 L_2)\Delta T A_1 E_1 E_2}{(2L_1 A_2 E_2 + L_2 A_1 E_1)}. \tag{1.65b}$$

Note that, unlike the uniform-cross-section case, stress is now dependent on both area and length.

For the beam of Fig. 1.13, in which the modulus and CTE are constant but the section length and the individual cross-sectional areas vary, we let $\beta = A_2/A_1$ and $\gamma = L_2/L_1$, and substituting into Eq. (1.64), find that

$$\sigma_2 = \frac{2+\gamma}{2\beta+\gamma} E\alpha\Delta T. \tag{1.66}$$

For equal-length sections ($L_1 = L_2$), if the central cross-section is significantly smaller than the end cross-sections, its stress approaches

$$\sigma_2 = \frac{P}{A_2} = 3E\alpha\Delta T, \tag{1.67}$$

which is in considerable excess of that for the uniform case of Eq. (1.50).

Should the thermal stress exceed the yield point of a particular ductile material, the material will not necessarily fail as long as the thermal strain lies under the material elongation capability. It will, however, be in a yielded state, which may require consideration for critical performance criteria. Also, if the load is compressive, buckling can occur, as discussed in the next section.

1.7 Buckling

This introductory chapter concludes with a note on critical buckling. Buckling occurs when a compressive axial load reaches a certain limit, causing instability. It occurs in long, slender beams. We concentrate here on 1D instability, although buckling can certainly occur in plates and shells, which are cases beyond the scope of what is presented here.

Consider the beam of Fig. 1.14 axially loaded along the x axis in compression. If a small load or displacement is applied laterally at the location of the axial load, the beam bends slightly. If the lateral load is removed, the beam returns to its straight position. However, if the axial load is increased, now causing an increased moment due to the lateral eccentricity, the beam becomes unstable and does not return to its straight position when the lateral load is removed. If the axial load increases further, the beam displacement becomes very large, and the beam becomes unstable. This load is called the critical buckling load. Note that this phenomenon will only occur in compression, as tensile loading will serve to straighten any eccentric lateral displacement.

We can compute the critical load by using the bending and curvature relations of Section 1.3 [Eq. (1.16)] and Fig. 1.4 to determine the point of instability. Here, we see that

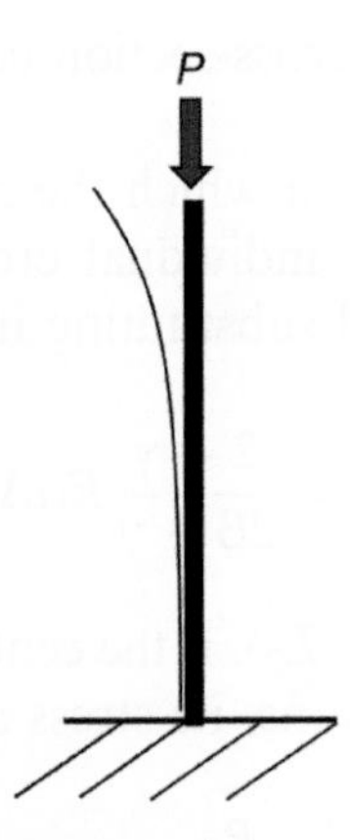

Figure 1.14 Critical buckling instability occurs at a critical load in compression due to small lateral movement. The joint at the application of the load may be free, pinned, or fixed with axial motion allowed. The base can be pinned or fixed. The critical load depends on these boundary conditions.

$$M = \frac{EI}{R} = EI\frac{d^2y}{dx^2} = P(\delta - y). \tag{1.68}$$

This differential equation is readily solved by calculus techniques[3] to produce the critical load value at instability as

$$P_{cr} = \frac{\pi^2 EI}{4L^2}. \tag{1.69}$$

The solution is independent of the material strength and is only a function of its stiffness.

While Eq. (1.69) is solved for the cantilever case, where critical value is the lowest possible, solutions are found for varying boundary conditions. If the beam is simply supported at its ends, the critical load is

$$P_{cr} = \frac{\pi^2 EI}{L^2}. \tag{1.70}$$

If it is fixed at both ends, the load is

$$P_{cr} = \frac{4\pi^2 EI}{L^2}, \tag{1.71}$$

which is the other extreme, so we have bounded the problem.

For many applications in optical structures, buckling needs to be investigated as it may drive the design, even if stress values are below those allowable. We will see an example of this in Chapter 3.

References

1. R. Hooke, *Lectures of Spring*, Martyn, London (1678).
2. R. Hooke, "A Latin (alphabetical) anagram, *ceiiinosssttuv*," originally stated in 1660 and published 18 years later.
3. S. Timoshenko and D. Young, *Strength of Materials*, Fourth Edition, D. Van Nostrand Co., New York (1962).
4. R. J. Roark and W. C. Young, *Formulas for Stress and Strain*, Fifth Edition, McGraw-Hill, New York, p. 504 (1975).

Chapter 2
Material Properties

In all structural analyses, calculations of deflections and allowable stresses are dependent on the characteristic properties of the materials being used in the design. Accordingly, we list here basic material properties that will be useful and utilized throughout the text.

2.1 Properties and Definitions

A complete list of basic material properties would be exhaustively long; here, we concentrate on those that would be of common use in the design of optical systems. These are nominal properties, which will vary somewhat based on the material composition but will serve well for purposes of this text. Definitions are summarized below to aid the reader in understanding.

Density. The weight density of a material is simply the weight of the material divided by its volume. The mass density of a material is the mass of the material divided by its volume, where mass times the acceleration of gravity equals weight, in the particular SI or customary units of interest. When given as grams per cubic centimeter, the mass density is referred to as its specific gravity; a value of 1 is the specific gravity (mass density) of water, such that relative to water, specific gravity is dimensionless.

Elastic modulus. Also called Young's modulus of elasticity, the elastic modulus of a material is the slope of the stress–strain diagram of the material [from Hooke's law (Chapter 1)] in the linear region. Values of the elastic constant can differ somewhat in tension, compression, or bending, depending on the material. Here, we report tensile modulus. Outside of the linear range, elastic modulus may be reported as the tangent modulus (slope of the stress–strain curve at a particular point) or as a secant modulus (slope of the stress–strain curve over a range of values, obtained as a linear value over the range end points).

Shear modulus. Also called the modulus of rigidity, the shear modulus is related to the elastic modulus as function of its Poisson ratio.

Poisson's ratio. Poisson's ratio is the contraction in the lateral (transverse) direction divided by elongation in the direction of the load.

Yield strength. Yield strength is that value of stress at which the material becomes nonlinear, resulting in a permanent deformation after load. For ductile materials, the standard yield point is chosen at 0.2% offset, i.e., that value which, when a specimen is unloaded, results in a permanent strain of 0.002, or 2000 parts per million. While somewhat arbitrary, the standard is based on the point that becomes evident as nonlinear on the measured stress–strain curve.

Micro-yield strength. Sometimes critical for optics and optical supports, where deformations or mount motions must be kept extremely minute, micro-yield strength is defined as that value which, when a specimen is unloaded, results in a permanent strain of 1 part per million. Such a value requires precise measurement and cannot be evidenced or obtained from a typical stress–strain diagram. Micro-yield strength is often referred to as the precision elastic limit (PEL) of a material.

Ultimate strength. The value of stress at which a material fails catastrophically is called its ultimate strength. The ultimate strength may differ in tension and compression, particularly for non-isotropic or brittle materials. For an axially loaded prismatic bar, the ultimate strength is obtained by dividing failure load by the bar's original cross-sectional area (engineering stress). The failure load divided by the bar's smaller neck-down area, which occurs before failure, is called true stress. However, true stress should not be used in calculations for opto-structural analysis.

Shear strength. The value of stress at which a material fails catastrophically when loaded in pure shear is called its shear strength. For ductile materials, shear strength is related to the ultimate tensile strength (as defined in Chapter 1) using von Mises theory.

Coefficient of thermal expansion. The CTE is defined as the value that causes a material to expand or contract when subjected to temperature change, and is given in units per degree of temperature change. When specified at a given temperature, it is called the instantaneous coefficient of thermal expansion. The instantaneous CTE is the slope (tangent point) of the thermal strain diagram at the specific temperature. When specified over a range of temperatures, it is called the secant coefficient of thermal expansion, or, more appropriately, its effective coefficient of thermal expansion. The effective coefficient of thermal expansion is the slope of the thermal strain diagram for the material over the range of values, obtained as a linear value over the range end points.

Coefficient of thermal expansion homogeneity. CTE homogeneity is the variability expected in the material CTE for a given lot, at the three-sigma (3σ) level of probability.

Coefficient of moisture expansion. The coefficient of moisture expansion is defined as the value that causes a material to expand or contract when subjected to moisture (humidity), and is given in units per percent moisture.

Thermal conductivity. Thermal conductivity is the heat-conduction property of a material. Higher values of thermal conductivity mean fewer thermal gradients in the presence of thermal flux, and less local heating due to temperature distribution. For calculations of deformation, thermal conductivity must be used in conjunction with its CTE.

Critical stress intensity. Critical stress intensity is the inherent property of a material to resist failure in the presence of flaws in an inert (dry) environment. Critical stress intensity is also known as the material fracture toughness. This property is measured in both tension (mode I) and shear (modes II and III) but is most commonly applied in mode I (Chapter 11).

Modulus of rupture. The modulus of rupture is the strength property of a material with a known flaw. It has values at which spontaneous fracture occurs in an inert environment. The modulus of rupture is proportional to the material stress intensity factor, and inversely proportional to the square root of the flaw depth (Chapter 11).

Fatigue strength. Fatigue strength is the value of strength at which failure occurs under full or partial stress reversals at a given number of cycles. It is most commonly applied to ductile materials (Chapter 11).

Static fatigue. Static fatigue is the value of strength at which failure occurs under constant stress with time. It is the result of a process called stress corrosion, or slow crack growth, which occurs in a moist environment. Static fatigue is most commonly applied to brittle materials (Chapter 12).

Flaw growth susceptibility. The flaw growth susceptibility exponent is a measured value of a brittle material that relates its crack growth to failure in a moist environment. The higher its value the less susceptible the material is to slow crack growth (Chapter 12).

Glass transition temperature. The glass transition temperature is a property that is specific to materials such as epoxies and is defined as that temperature at which its properties transition between brittle (hard) and rubbery (soft); that is, its elastic properties change somewhat abruptly (Chapter 9).

Data for these properties are shown in Tables 2.1 through 2.4.

2.2 Low-Thermal-Expansion Materials

Since many optical applications require operation over an extreme range of temperatures, and miniscule motions can wreak havoc with performance, materials with a low CTE are important. Such materials, included in Tables 2.1 and 2.2, include glasses and ceramics such as fused silica, ultra-low-expansion

Table 2.1 Mechanical properties of common metals for opto-structural analysis.

General		Mechanical						Thermal	
Materials	Remarks	Density (lb/in^3)	Elastic Modulus (Msi)	Poisson's Ratio	Yield Strength (ksi)	Micro-yield Strength (ksi)	Tensile Ultimate Strength (ksi)	CTE (ppm/K)	CTE Homogeneity (ppm/K)
Al 2024-T4	Bar and rod	0.1	10.5	0.33	42		62	22.4	
Al 6061-T6	Sheet annealed/heat treated	0.098	9.9	0.33	36	18	42	22.5	0.06
Al A356.0-T6	Cast	0.097	10.4	0.33	28		38	20.8	
AlBeMet (AM 162)	Extruded	0.076	29.3	0.17	47		61.9	14.7	
Be O-30	Vacuum hot pressed	0.067	44	0.1	43.5	5	59	11.4	0.04
Be I-70H	Hot isostatic press	0.067	44	0.1	30	3	50	11.4	0.08
Be I-220	Hot isostatic press	0.067	44	0.1				11.4	
Be S200F	Vacuum hot pressed	0.067	44	0.1	35	5	47	11.4	0.1
Be I-400	Hot isostatic press	0.067	44	0.1	60			11.6	
BeCu TH02	Strip	0.298	18.5	0.27	160		185	17.5	
Beralcast 363	Cast	0.078	29.3	.20	31		42	14.2	
Copper	Pure, annealed	0.322	18	0.343	4.8		30.3	16.7	
Gold	Pure	0.697	11.3				14.9	14.2	
INCONEL 718	Bar, Rod, Tube, Plate	0.296	29	0.29	150		180	10.6	
Indium	Cast	0.26	1.83	0.45	0.135		0.232	24.8	
Invar 36	Annealed	0.291	20.5	0.259	40	11.3	71	1.3	
Magnesium AZ31B	Sheet annealed	0.064	6.5	0.35	22	3.2	37	26	
Molybdenum alloy TZM	Arc cast	0.367	46	0.33	90	23.9	120	4.9	
Nickel	Pure	0.322	30	0.31	8.6		46	13.3	
Nickel - Electroless	Coating	0.28	15					12	
Silver	Pure, annealed	0.379	10.3	0.37			18	19	
SS 440C	Bar/annealed	0.28	29	.27	65		110	10.1	
Steel A286	High-strength fasteners	0.287	29.1	.31	120		160		
Steel 18-8 type 300	Standard-strength fasteners	0.29	28	.27	45	10	90	17	
Ti 6Al-4V	Plate annealed/heat treated	0.16	16	0.31	120	70	130	8.9	

Table 2.2 Properties of glass and ceramics. (Strength and crack growth issues are discussed in Chapter 12.)

Materials	Form/Condition	Density (lb/in^3)	Elastic Modulus (Msi)	Poisson's Ratio	CTE (ppm/K)	Homogeneity ΔCTE (ppm/K)
BK-7	As manufactured	0.091	12.27	0.208	8.6	–
Calcium fluoride (CaF_2)	As manufactured	0.115	10.99	0.26	18.7	–
Fused silica	As manufactured	0.0796	10.3	0.17	0.52	0.01
Germanium	Standard grade: optical	0.192	14.9	0.278	5.7	–
Silicon	As manufactured	0.084	19	0.266	2.6	–
Silicon cladding	Cladding	–	–	–	2.93	–
Siliconized silicon carbide	As manufactured	0.105	44.5	0.2	2.43	0.04
Silicon carbide	Chemical vapor deposit	0.115	61	0.2	2.2	0.03
ULE® Standard	Standard	0.0797	9.8	0.17	0 ±0.03	0.012
ULE® Premium	Premium	0.0797	9.8	0.17	0 ±0.03	0.01
ULE® Mirror	Mirror	0.0797	9.8	0.17	0 ±0.03	0.015
ULE® TSG	TSG	0.0797	9.8	0.17	0 ±0.1	0.03
ZERODUR® Class 1	As manufactured: class 1	0.091	13.2	0.24	0 ±0.05	0.02
ZERODUR® Extreme	Extreme	0.091	13.2	0.24	0 ±0.007	0.01
ZERODUR® Special	Special	0.091	13.2	0.24	0 ±0.01	0.01
ZERODUR® Class 0	Class 0	0.091	13.2	0.24	0 ±0.02	0.02
ZERODUR® M	Hysteresis free	0.091	11.8	0.24	–	–
Zn selenide	As manufactured	0.19	9.75	0.28	7.2	–
Zn sulfide (Cleartran™)	As manufactured	0.148	10.8	0.28	5.2	–

Table 2.3 Physical properties of adhesives at room temperature.

	Typical Uncured Physical Properties			Cured Physical Properties						
Materials	**Typical Use**	**Pot Life**	**Cure Time**	**Color**	**Shore D Hardness**	**Elastic Modulus (ksi)**	**Poisson's Ratio**	**Lap Shear Strength Al/Al (psi)**	**Elongation %**	**CTE @ 293 K (ppm/K)**
3M Scotch Weld 2216	Structural adhesive	90 min (75 °F)	7 Days @ 25 °C/ 30 min @ 93 °C	Gray	50–65	150	0.43	3200		100
CT 5047-2	Silver filled electrical Conductive epoxy	60 min	24 hrs @ 25 °C/ 1 hr @ 100 °C		82	150		1000		40
EA9361	Low temp high strain	120 min (77 °F)	5–7 days (77 °F)	Gray	70	155	0.433	3500	40	104
EA 9394	High temperature	90 min. (75 °F)	3–5 days (75 °F)	Gray	88	615		4200		60
Epibond 1210-A/9615-10	Cryogenic adhesive	2–4 hrs.	48 hrs (77 °F)	Blue	80	396	0.38	2500	1.7	80
Epo-tek 301-2	Thin bondline structural adhesive - wicking epoxy	8 hrs (75 °F)	2 days (75 °F)	Clear	82	532	0.358	2000		63
RTV 566	Silicon rubber optic support	90 min. (77 °F)	7 days (77 °F, 50% R.H.)	Dull Orange	Shore A, 61	1		465	120	222
Stycast 2850FT/24LV	Thermal conductivity	30 min.	24 hrs (77 °F)	Black	92	1340		4200	0.73	39

Table 2.4 Properties of select composites and plastics.

	Form/ Condition	Density (lb/in^3)	Poisson's Ratio	0-deg Tensile Strength (ksi)	0-deg Tensile Modulus (Msi)	90-deg Tensile Strength (ksi)	90-deg Tensile Modulus (Msi)	0-deg Compression Strength (ksi)	0-deg Compression Modulus (Msi)	90-deg Compression Strength (ksi)	90-deg Compression Modulus (Msi)	0-deg CTE (ppm/K)	90-deg CTE (ppm/K)
G-10CR Pseudo-isotropic	Fiberglass epoxy	0.0668		34	2.5	–	–			–	–	10.5	–
K13C2U/954-3 Unidirectional	Graphite ester		0.34	199	74.7	2.9	0.71	50.4	74.9	16.3	0.7	−1.26	32.6
K13C2U/954-3 Pseudo-isotropic	Graphite ester		0.34	82	25.5	88	25.6	26.8	24.9	28.1	24.8	−0.88	−0.85
M55J/954-3 Unidirectional	Graphite ester	0.058	0.32	290	45.5	5.5	0.82	134	41.8	27.6	0.82	−0.97	34.6
M55J/954-3 Pseudo-isotropic	Graphite ester	0.058	0.36	108	15.88	65	14.8	48.7	15.3	60.6	14.1	−0.11	−0.18
MC511SN (Norplex - Micarta)	Carbonized fiberglass epoxy	0.0614	0.14	20	1.7	–	–			–	–	18	–
Vespel SP-1	Plastic	0.0517	0.41	11.7	0.36			19.2				44.4	

fused silica (ULE®), and ZERODUR®; silicon; silicon carbide; graphite/resin composites; and the metal Invar. All of these materials exhibit room-temperature expansion below 3 ppm/°C, with some very close to zero expansion. Some comments on these materials are provided next.

2.2.1 Fused silica

Fused silica, or silicon dioxide (SiO_2), is a form of amorphous fused quartz that is typically produced by a synthetic process called flame hydrolysis, resulting in high purity and homogeneity. Fused silica exhibits a CTE on the order of 0.5 ppm/°C at room temperature, and has a CTE inhomogeneity of less than 10 ppb/°C. It is an ideal candidate for cryogenic applications since its expansion coefficient decreases to zero very quickly. The instantaneous CTE is near zero at a temperature of only 150 K (–125 °C). This means that near that operating temperature the presence of thermal gradients in the optic, or small temperature variations, will not affect performance.

Furthermore, the thermal strain over a range to about 100 K is near zero, resulting in an effective, or secant, CTE of near zero over that range, resulting in no change in shape over that range. Figure 2.1 is a plot of a typical

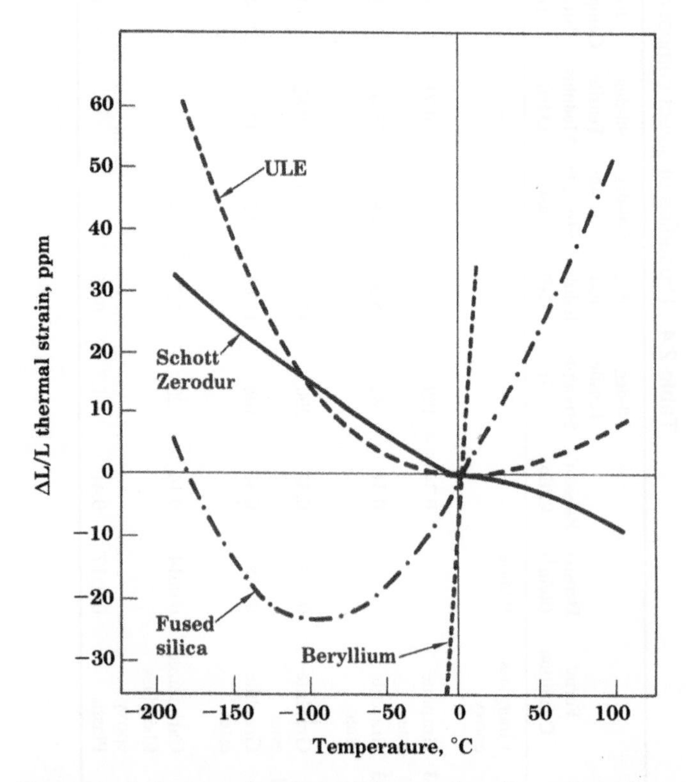

Figure 2.1 Thermal strain as a function of temperature for select glasses and glass ceramics. Comparison to beryllium is provided for illustrative purposes. Note the zero thermal strain for fused silica possible at a select temperature, and negative expansion for select glasses and glass ceramics at cold temperatures.

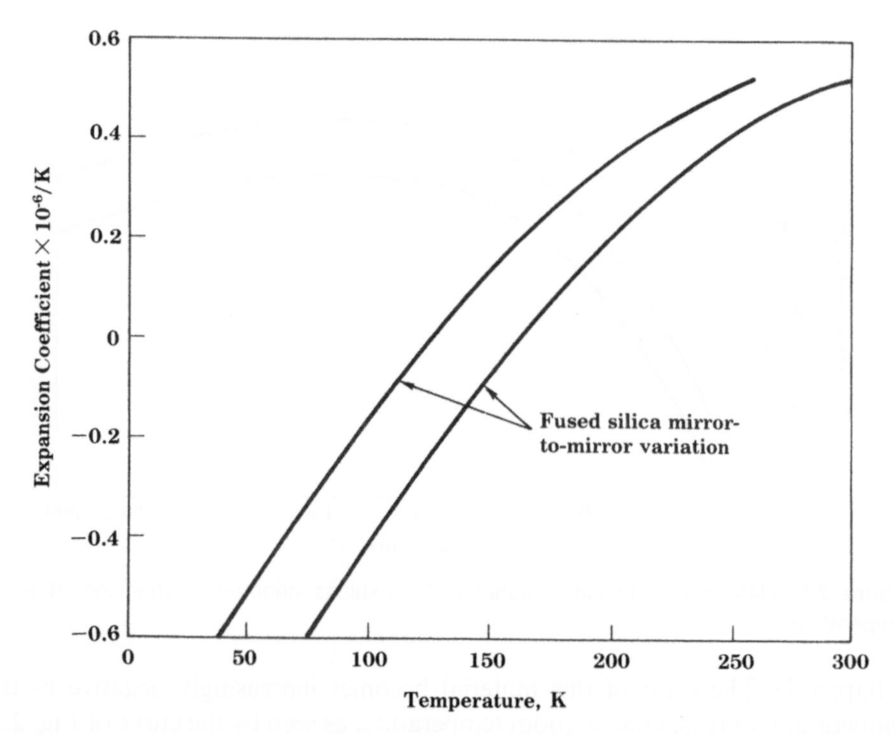

Figure 2.2 Instantaneous CTE for fused silica as a function of temperature. The values are negative below 100 K.

fused-silica thermal strain from room temperature to cryogenic extremes, along with other materials. The instantaneous CTE (tangent of the thermal strain curve at temperature) becomes negative below 100 K, as seen in Fig. 2.2, and eventually near absolute zero, as it does for all materials.

Fused-silica optics can be machine lightweighted using diamond tools or water jets. They can also can be fused, meaning that a facesheet can be attached by bringing the optic to high temperature, resulting in a closed-back design, if desired (Chapter 5). Optics as large as 8 m in diameter have been produced.[1] At near room temperature, fused silica is a good match to the iron-nickel metal Invar, making mounting to it ideal.

2.2.2 ULE® fused silica

Ultra-low-expansion (ULE) fused silica (manufactured by Corning Glass Works[1]) is a fused-silica material doped with precipitated titania (TiO_2), the latter of which with nearly 7.5% content produces nominal zero CTE at room temperature with a guaranteed expansion coefficient of ±0.03 ppm/°C, which, depending on grade quality, can be reduced to ±0.02 ppm/°C. ULE fused silica has a CTE inhomogeneity of less than 15 ppb/°C. Largely due to titania variations, a small residual stress is present that can manifest as a small error for optics if the edges are cut after finishing. Such stress is small (less than 100 psi) and generally inconsequential except for very high-aspect-ratio optics

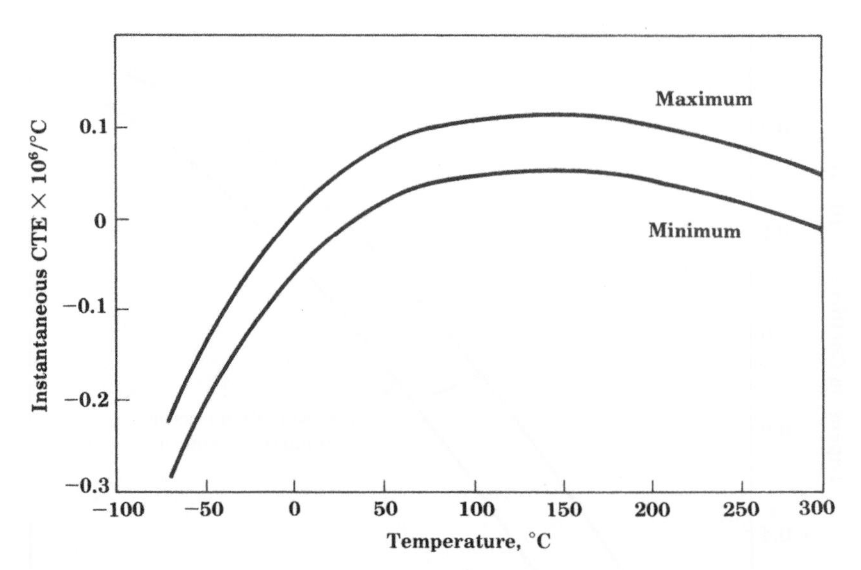

Figure 2.3 Ultra-low-expansion material ULE® exhibits near-zero expansion at room temperature.

(Chapter 7). The CTE of this material becomes increasingly negative as the temperature decreases below room temperature, as seen by the curve of Fig. 2.3.

Like fused silica, ULE can be machine lightweighted and fused. It can also be bonded using a ceramic frit material closely matching in expansion coefficient at a relatively high temperature.

2.2.3 ZERODUR®

Zerodur (manufactured by Schott Glass Works[2]) is a glass ceramic comprising fused-silica material and a host of other oxides. These combine to produce a nominal zero CTE at room temperature with a guaranteed expansion coefficient of ±0.03 ppm/°C, which, depending on grade quality, can be reduced to ±0.01 ppm/°C. Zerodur has an inhomogeneity of less than 15 ppb/°C. Largely due to cooling effects from high-temperature fabrication, a small residual stress is present that can manifest as a small error for optics if the edges are cut after finishing. Such stress is small (less than 30 psi) and generally inconsequential except for very high-aspect-ratio optics (Chapter 7).

Zerodur can be cast in blank sizes up to 4 m in diameter. Its expansion coefficient becomes negative below room temperature, as seen by the curve of Fig. 2.1. Zerodur has been shown to exhibit hysteresis in its expansion coefficient when passed through a temperature above 130 °C and the heating rate is not specifically controlled. This small but observable phenomenon is eliminated in a version designated Zerodur-M in which the manganese dioxide is removed from the formulation. However, demand for such precision is generally not warranted, and the base material itself is almost always sufficient.

Zerodur is subject to a delayed elastic effect (Chapter 7) that manifests slowly when under stress or when stress is relieved. This phenomenon is due to the presence of alkali oxides (notably, lithium oxide) and subsequent stress relaxation under load or load release. Its delayed strain is on the order of 1% but is fully recoverable; unless deformations are high, this does not detract from its excellent properties and machining characteristics.

2.2.4 Silicon

Silicon is a metalloid that is often used as a semiconductor but is ideal for optics due to its low CTE (2.6 ppm/°C at room temperature), excellent polishability, and very good mechanical and thermal figures of merit (Chapter 8). It is grown as a crystal, so its size is limited to less than 18 in. (0.45 m) in diameter without the use of joining techniques. It is particularly suitable in the presence of high-energy lasers or extreme thermal flux, since gradients, and hence deformation (Chapter 4), are minimized. Silicon is very homogenous and ideal for cold applications. Its mechanical and thermal figures of merit in comparison to other materials are very good (Chapter 8).

2.2.5 Silicon carbide

Silicon carbide is a ceramic comprising silicon and carbon that is of significant use in both optics and structures supporting optics. It has a low CTE (~2.4 ppm/°C), high thermal conductivity, and high elastic modulus, making it an ideal and attractive candidate for many lightweight applications. Manufacturing techniques for all silicon carbides require a very high temperature (in excess of 1600 °C). Silicon carbide can be made using several methods, including, for example, manufacture from graphite and conversion in the presence of silicon; chemical vapor deposition (for beta silicon carbide); hot pressing; hot isostatic ally pressing; or casting from silicon carbide powder slip, drying, firing, and postfire siliconizing (for alpha silicon carbide and silicon that is reaction bonded). Casting techniques make the reaction-bonded silicon carbide ideal for lightweight optics; however, because it is a two-phase material, it is not as readily polishable as are glasses and silicon, but it can be polished bare with specialized techniques. Often, a silicon or chemical-vapor-deposited silicon carbide overcoat is made to assist in the polish process. Optics in the 1-m class are achievable without the use of joining methods.

Unlike glasses, silicon carbide can be used as a structural material both for metering and support of optics due to its high fracture toughness. This makes an optical design self-metering with matched silicon carbide optics. Regardless of manufacturing technique, silicon carbide has gained favor as a choice material.

2.2.6 Graphite composites

In general, while not useful as optics due to polishability and moisture effects, graphite composites are ideal lightweight materials for structural support and metering due to their low weight density, low CTE, and high strength. Their resin system (formerly comprising epoxy but now made of cyanate esters or siloxanes) allows for near-zero-CTE layups, described more fully in Chapter 13. This makes for a good match to the glasses and ceramics described in Sections 2.2.1 through 2.2.3. Very large structures can be fabricated from graphite composites.

2.2.7 Invar®

Invar® is a low-expansion metal that exhibits peculiar properties, rendering it worthy of elaboration. It is an iron-nickel alloy invented in 1896 in Paris by a Swiss metallurgist named Charles Guillaume. Invar is a trade name of a French company named Imphy Alloys (currently, ArcelorMittal[3]). Imphy Alloys was involved in developing naval armor using iron-nickel alloys, but Guilllaume, whose specialty was thermography with an interest in metrology, found that Invar had interesting thermal-expansion properties.

Invar® 36, an iron-nickel alloy of approximately 36% nickel and 64% iron ($Fe_{64}Ni_{36}$), was found to have the lowest expansion of all iron-nickel alloys. Guilllaume received a Nobel prize[4] in Physics in 1920 for his discovery—the only Nobel prize ever awarded for metallurgy (he would be in good company, as Albert Einstein won the prize the following year). Invar stands for "invariable" due to its low thermal expansion and constancy over a range near room temperature. The low CTE of Invar is *not* due to any normal rule of mixtures involving iron and nickel mechanical properties, both of which have similar CTE properties, but rather is due to a complex ferromagnetic bond interaction. Guillaume noted (from studies by others) that iron with 22% nickel was nonmagnetic even though iron and nickel themselves were magnetic. He further noted that the alloy exhibited a higher CTE (near that of brass) than either iron or nickel, itself. Then he tested an iron-nickel alloy with 30% nickel and found that, not only was it magnetic, but the alloy exhibited a *lower* CTE than either iron or nickel by themselves. This discovery led to more studies where he showed that 36% nickel exhibited the lowest CTE.

2.2.7.1 CTE and stability

As might be expected, the CTE of iron-nickel alloys is highly dependent on the nickel content: it depends on temperature, chemical composition, and heat treating. Furthermore, its temporal stability (expansion with time in the absence of temperature change) is highly dependent on heat treatment, as will be further discussed in Chapter 16.

Variations in nickel content of 2% will *double* the CTE of Invar at room temperature. For Invar 36 to be certified, its range of nickel must lie between 35.5% and 36.5%. The dramatic effect of the iron-nickel alloy CTE is seen in Fig. 2.4. At 22% nickel, the CTE is quite high—even higher than pure nickel

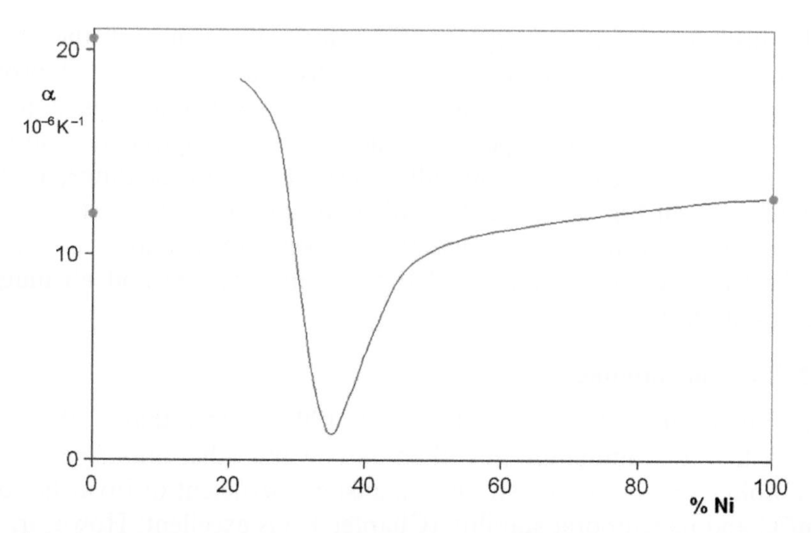

Figure 2.4 Instantaneous CTE for nickel-iron alloys. While iron and nickel have similar CTEs, iron-nickel alloy does not obey the rule of mixtures due to its magnetic interaction.

or pure iron. At 36%, it is at its minimum (invariable, hence, Invar). Beyond 36%, the CTE increases again, eventually approaching that of pure nickel.

Since Guillaume's discovery, we have toyed with his invention: Invar 36 has been annealed, forged, hammered, cold worked, heat treated, rough sawn, furnace cooled, air cooled, slow cooled, fast cooled, quenched in water, quenched in saltwater, quenched in oil, quenched in gas, stress relieved, thermally cycled, decarbonized, purified, unpurified, ceramicized, and supersized. Because of this, the CTE is based on the particular applied process and can vary from –0.3 ppm/°C to +1.7 ppm/°C. Most heat treatments for stability result in values between +0.8 ppm/°C and +1.6 ppm/°C. The *highest* value of CTE for Invar 36 occurs in its annealed state. Heat treatments (quench), machining, and cold work tend to lower the CTE, while re-annealing and stress relieving after machining bring the CTE to a nominal state and result in higher dimensional (temporal) stability. Temporal stability will be discussed in Chapter 16.

2.2.7.2 CTE and temperature

CTE data is obtained from thermal strain curves over a temperature range. As mentioned previously, when specified over a temperature range, CTE is referred to as *effective* CTE, *mean* CTE, *average* CTE, or, more appropriately, *secant* CTE. Since thermal strain is [from Eq. (1.3)] the change in length ΔL divided by the original length, the secant CTE over a temperature range ΔT is simply thermal strain divided by temperature range:

$$\alpha_{\text{sec}} = \frac{\Delta L}{L \Delta T}. \tag{2.1}$$

The instantaneous, or tangent, CTE α_i, is the slope of the thermal strain curve at the given temperature. The instantaneous CTE of properly heat-treated Invar 36 at room temperature (UNS K93603) is nominally 1.3 ppm/°C. Above room temperature, the CTE varies markedly, and below room temperature, the CTE is markedly constant over a wide range, as can be seen in the plots of Fig. 2.5. Note the rather constant slope as the temperature decreases and the dramatic increase the temperature increases, particularly when the Curie temperature (near 270 °C) is approached and all magnetic properties are lost.

2.2.7.3 Invar 36 varieties

Low-carbon Invar. Studies from the early 1990s[5] show that if the carbon content is kept to a minimum (less than 0.01%) and other impurities are kept to an absolute minimum as well, the expansion coefficient of Invar lies below 1 ppm/°C and its temporal stability (Chapter 16) is excellent. However, these specifications require a special billet (bar stock), which is both difficult and costly to obtain.

Free-cut Invar (UNSK-93050). Free-Cut Invar 36 has a small percentage of the element selenium added, which aids in machining, particularly in difficult areas. Its CTE at room temperature is somewhat higher than that of Invar but may be acceptable in many applications. Its stability is fine if properly heat treated after fabrication.

Quenched Invar. After reaching the anneal temperature, Invar can be rapidly cooled by water, salt water, or oil quench. This reduces the CTE to less than 0.6 ppm/°C. However, it is difficult and costly to control the quench process, which can result in uncertainties in the desired value of the CTE. Therefore, quenching is not recommended for general applications unless a specifically low CTE is desired, with the noted caveats.

Elinvar. This variation (ELastically INVARiable) of iron-nickel contains excess chromium to maintain a constant elastic modulus over a wide temperature range. However, its CTE is quite high and, although suitable for watchmakers, it is not a good choice for optical systems.

Silicon-nitride reinforcement. Recently, a ceramicized Invar with a lower density and lower CTE has been introduced in the experimental phase. This variation of Invar is known as silicon-nitride-reinforced iron-nickel. However, additional testing needs to be performed on this material to ascertain its value and dimensional stability with time and temperature.

2.2.8 Iron-nickel varieties

Varieties of Invar with varying nickel content are often referred to as Invar XX, the designation XX being the nickel percentage. Technically, the word Invar (always capitalized) is reserved for Invar 36, so other iron-nickel alloys are more properly designated as alloy XX.

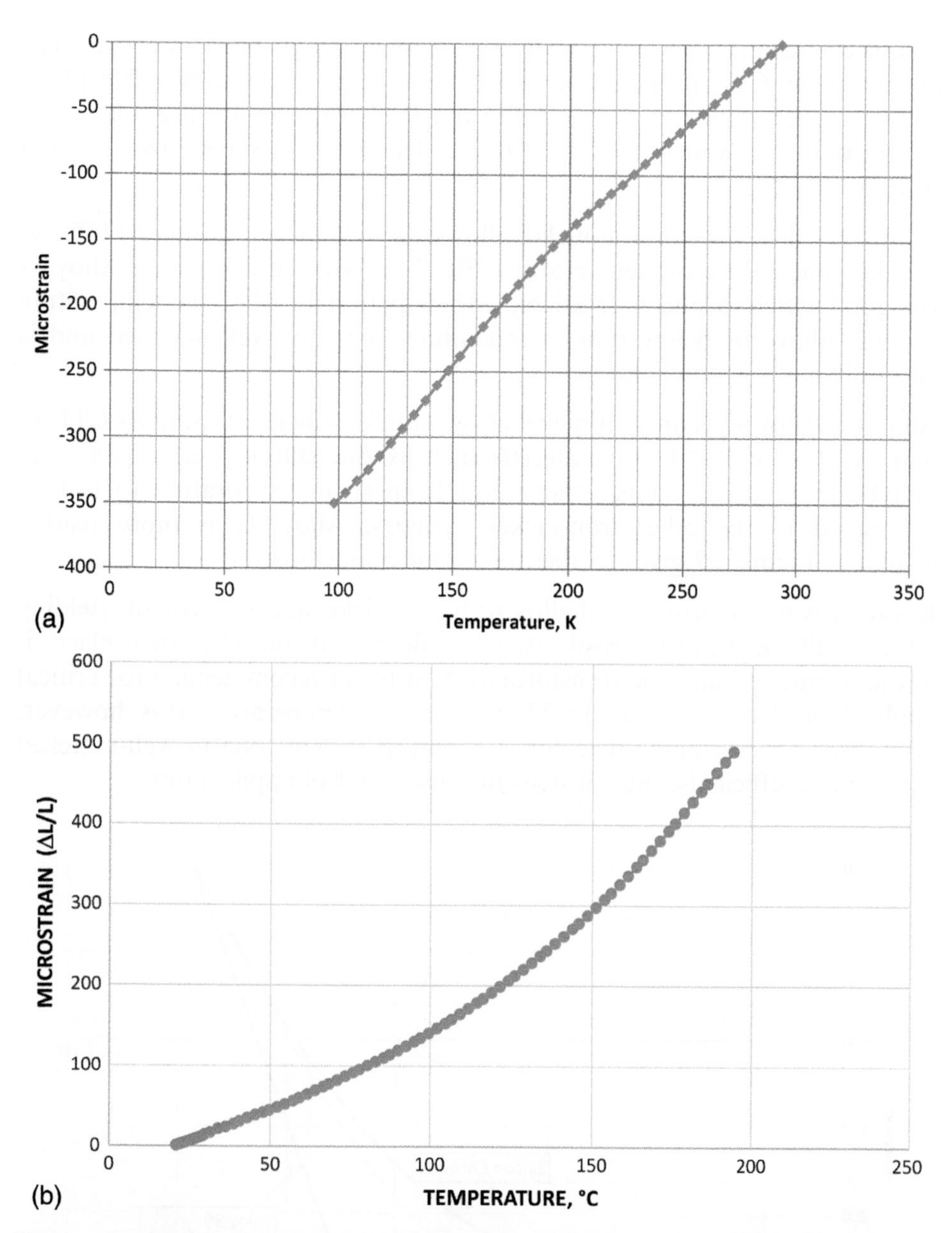

Figure 2.5 (a) Thermal strain of Invar 36 below room temperature. Effective expansion is low over wide temperature extremes. (b) Thermal strain of Invar 36 above room temperature. Instantaneous CTE (slope of thermal strain curve) increases markedly with temperature increase (graphs courtesy of T. Altshuler, Advanced Materials Laboratory, Concord, Massachusetts).

Super Invar (UNSK-93500). Super Invar contains about 32% nickel and 5% cobalt. It exhibits the lowest CTE of the iron-nickel family (less than 0.3 ppm/°C at room temperature) and is very temporally stable at room temperature. However, Super Invar is difficult to fabricate and machine, and is subject to changes in CTE properties if dropped, magnetized, or cooled. The change in its CTE properties when cooled is due to an irreversible phase change, which results in drastic CTE and temporal stability degradation at temperatures as warm as 220–270 K. The use of Super Invar is not recommended.

Alloy 39. Alloy 39 is an iron-nickel alloy with 39% nickel, raising its CTE to 2.3–3.0 ppm/°C at room temperature (Fig. 2.6). While at first glance Alloy 39 is a good match to silicon or silicon carbide materials, when the temperature changes above or below room temperature, it is not well matched and is generally not a good choice.

Alloy 42. Alloy 42 is an iron-nickel alloy with 42% nickel, raising its CTE to make it a good match for materials such as zinc sulfide, germanium, and molybdenum over a temperature range from room temperature to 100 K. Compared to the other iron-nickel varieties, Alloy 42 is more readily available. Figure 2.7 gives its thermal strain characteristics.

Kovar. Kovar is an iron-nickel alloy with 29% nickel and 17% cobalt, yielding a CTE well matched to borosilicate glass. Because its nickel content places it in the regime of magnetic transformation, it is not recommended for critical applications due to phase and CTE changes with temperature. It is, however, often used for packaging detectors for optical systems due to well-matched expansion coefficients; thus, it does find use in select applications.

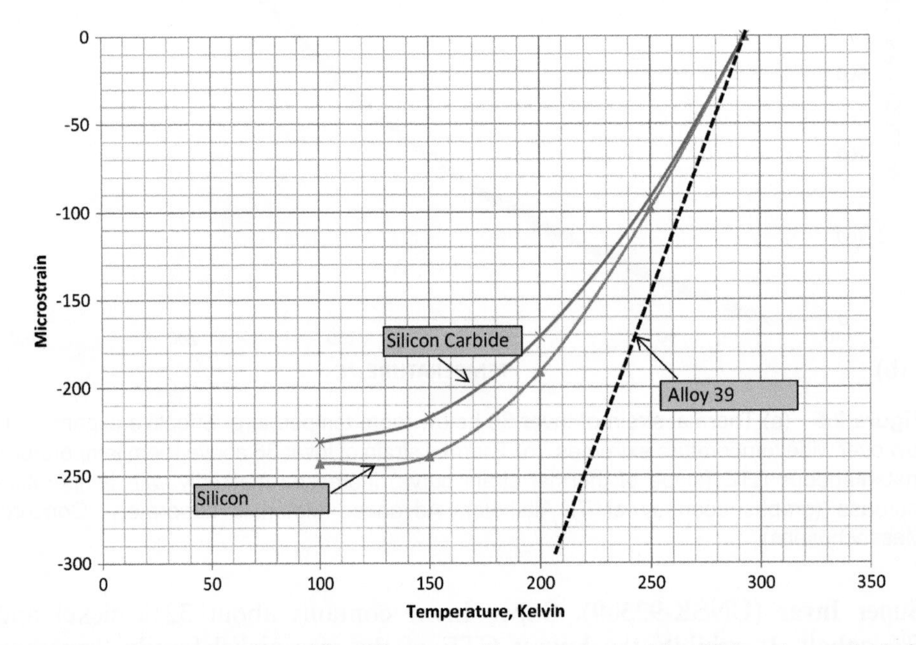

Figure 2.6 Thermal strain of Alloy 39 versus silicon products. Alloy 39 is not a particularly good match to these silicon materials.

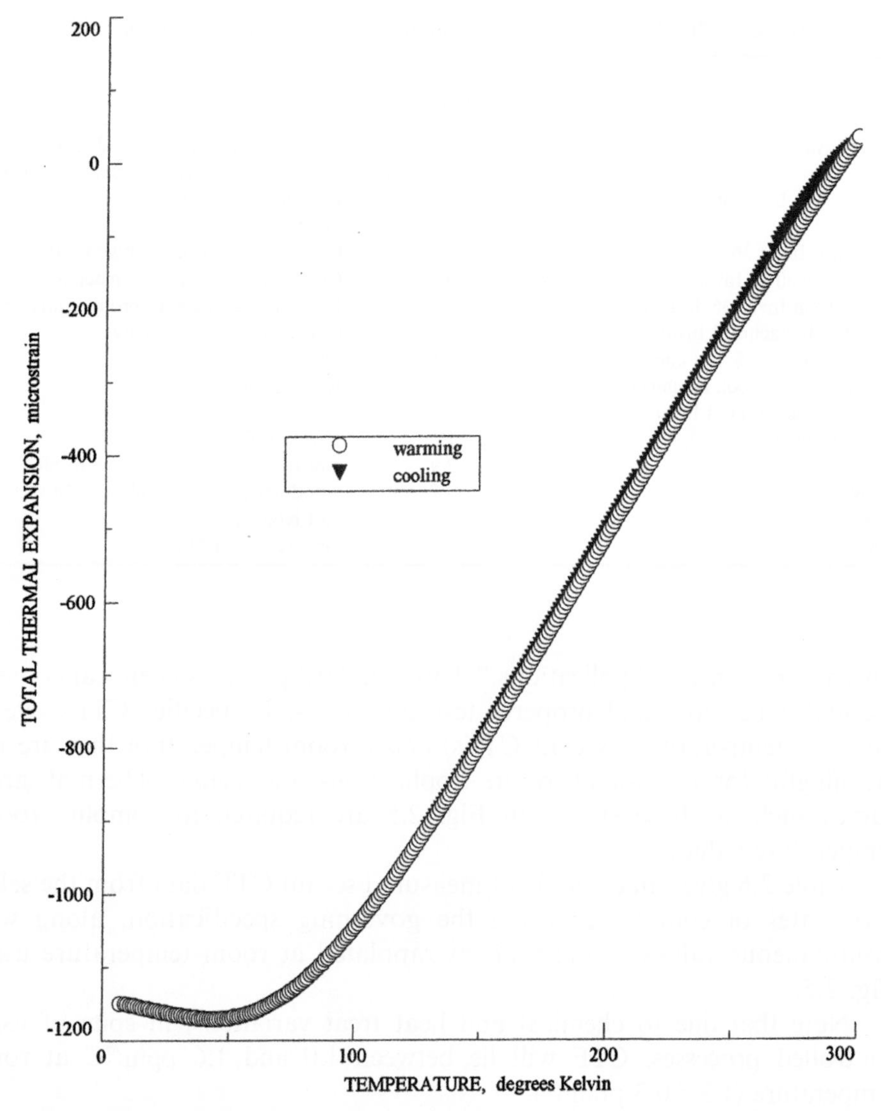

Figure 2.7 Thermal strain characteristics of Alloy 42, which are well matched to ceramics such as zinc sulfide and germanium (test data provided by T. Altshuler, Advanced Materials Laboratory, Concord, Massachusetts).

2.2.9 The iron-nickel family

A summary of the CTE nominal values for the iron-nickel family, including the variations of Invar 36, is provided in Table 2.5.

2.2.10 Governing specifications

The governing specifications for low-thermal-expansion materials is ASTM F 1684, "Standard Specification for Iron-Nickel-Cobalt Alloys for Low

Table 2.5 The Invar family's approximate CTE values at room temperature.

Material	Room Temp CTE ppm/°C	Remarks
Super Invar	0.3	Difficult to fabricate; unstable over temperature; irreversible phase change
Low carbon Invar 36, cold work	0.3	Difficult and costly to produce and obtain; high purity
Quenched Invar 36	0.6	Costly and quench-sensitive CTE
Rough machined Invar 36	0.6	Cuts > 0.050 in. heavily machined
Low carbon Invar 36, heat treat	0.8	Difficult and costly to produce and obtain
Semi-finish machined Invar 36	1.1	Cuts < 0.005 in.; estimate
Annealed Invar 36, fast cool	1.1	estimate
Annealed, slow cool, machined, heat treated, stress relieved Invar 36	**1.3**	**Recommended**
Free cut Invar 36	1.6	Higher CTE, additives
Alloy 39	2.9	Not recommended as match to SiC
Alloy 42	4.7	Good match to Ge, ZnS, and Mo
Kovar	5	Not recommended
Elinvar	7	Not recommended

Thermal Expansion Applications." This standard provides chemical content, thermal expansion, and property test conditions. It specifies CTEs over a range of temperatures (secant CTEs) *above* room temperature that are not meaningful for room-temperature applications and below. Thermal strain curves such as those shown in Fig. 2.5 are required to compute room-temperature values.

Table 2.6 gives an example of measured secant CTE data from the select certificates of compliance using the governing specification, along with instantaneous values of the CTE extrapolated at room temperature using Fig. 2.5.

Note that due to chemical and heat treat variations, in spite of using controlled processes, CTE will lie between 1.0 and 1.6 ppm/°C at room temperature (1.3 ±0.3 ppm/C).

2.2.11 Invar summary

Due to ferromagnetic interactions, Invar 36 exhibits the lowest CTE of all metals at room temperature. Invar 36 is an iron-nickel (FeN) alloy containing nominally 36% nickel and 64% iron. Departures from the 36%/64% nickel/iron content will significantly increase the CTE. Alloy 42 can be useful for matching to zinc sulfide, molybdenum, and germanium. Invar 36's CTE between –40 °C and +40 °C is both low and invariable. Use at temperatures above 50 °C significantly increases the CTE. Impurities (carbon,

Table 2.6 Measured CTE values for Invar 36 versus specification values, and CTE values extrapolated to room temperature using Fig. 2.5.

Sample Lot	Secant CTE 30 °C to 150 °C		CTE Room Temperature (extrapolated)	Comments
	Cert. Data	F- 1684/Spec (ppm/°C)		
1	1.7	1.2–2.7	1.00	Low carbon
2	2.02	1.2–2.7	1.19	
3	2.05	1.2–2.7	1.21	Low carbon
4	2.05	1.2–2.7	1.21	Low carbon
5	2.09	1.2-2.7	1.23	
6	2.15	1.2–2.7	1.26	Low carbon
7	2.16	1.2–2.7	1.27	
8	2.18	1.2–2.7	1.28	
9	2.18	1.2–2.7	1.28	
10	2.19	1.2–2.7	1.29	
11	2.21	1.2–2.7	1.30	
12	2.27	1.2–2.7	1.34	
13	2.33	1.2–2.7	1.37	Free cut
14	2.4	1.2–2.7	1.41	Free cut
15	2.4	1.2–2.7	1.41	
16	2.4	1.2–2.7	1.41	
17	2.4	1.2–2.7	1.41	Free cut
18	2.5	1.2–2.7	1.47	
19	2.6	1.2–2.7	1.53	
20	2.7	1.2–2.7	1.59	
21	2.7	1.2–2.7	1.59	

selenium, etc.) increase the CTE of Invar 36. The highest CTE of Invar 36 occurs in the annealed state for a given composition.

Quenching, cold work, and machining lower Invar's CTE. Invar 36 is temporally unstable without heat treatment. AMS-F-1684 is the governing procurement specification.

Do not use Super Invar. Do not quench without good reason. Finally, process variations for Invar 36 can yield CTEs between 1.0 and 1.6 ppm/°C. Tighter controls on value and range require a special order.

2.3 Not-So-Low-Thermal-Expansion Materials

Some optical materials are attractive despite their high CTEs; their attractiveness lies in high thermal conductivity, light weight, low cost, and high stiffness. Such materials, of course, will demand metering structures of the same material if active focus is to be precluded over temperature excursions. Not-so-low-thermal-expansion materials include aluminum, beryllium, and aluminum-beryllium, which are listed in the next table (see Section 2.3.4). A few comments on these materials are provided here.

2.3.1 Aluminum

Aluminum 6061-T6 has a CTE at room temperature of 22.5 ppm/°C and is one of the more highly expansive materials in the periodic table of elements. As such, it demands an aluminum metering structure. Aluminum has good thermal conductivity and is relatively light weight; it is readily machined and diamond turned. It does require stress relief after machining, resulting in a stable optic. For some applications, aluminum can be polished without cladding, as certain grades allow bare polish while still meeting scatter (BRDF—bidirectional reflectance distribution function) specifications for roughness. Many polishers, however, prefer a polishable cladding, such as nickel, which can cause performance error due to CTE mismatch. Although aluminum is a good thermal conductor, its thermal figure of merit (Chapter 8) is not so good. Its biggest attraction is cost—it is by far the least costly of the options.

2.3.2 Beryllium

Beryllium has a CTE at room temperature of 11.4 ppm/°C and generally demands a beryllium metering structure. It is quite costly and requires special care during machining due to potentially toxic dust inhalation. However, its attractiveness lies in its very good thermal conductivity and, in particular, its extremely low weight density, high stiffness, and subsequent superior mechanical figure of merit. As such, it will result in the lightest design of all possible combinations of materials. While it can be bare polished, it generally requires a polishable cladding, such as nickel. Properly chosen nickel plating with a specific phosphorous content can minimize performance error due to a very good CTE match to the substrate.

Beryllium optical grades, such as O30, which uses fine spherical powder technology, result in very good CTE homogeneity, which shows greatly improved performance over earlier vacuum-hot-pressed (VHP) grades. The VHP grades exhibit both poor CTE homogeneity and poor hysteresis after thermal excursion.

Beryllium can be machined to make good optical surfaces but does require heat treatment to relieve residual stresses, which are particularly present at the machined optical surface due to grain anisotropy.

Residual stress during machining is caused by the machine cut or an abrasive grind. Furthermore, machining the material can cause a CTE increase in the damaged layer due to crystal grain reorientation. Etching before final machining can assist in this regard, and, in theory, a carefully controlled grind procedure can eliminate machining effects using progressively smaller particle sizes until all machining damage is removed. However, temperature relief is often employed as a more promising approach.

Thermal cycling for stress relief is critical if dimensional stability is to be realized over the extremes of temperature excursion.[6] In order to relieve residual stress during machining, temperatures as high as 300 °C are recommended. In reality, one needs even higher temperatures, near 600 °C, to relieve all stresses. Most manufacturers would not attempt that, since both oxidizing and extensive warping could occur. Studies on residual stress relief[7] show minimal stress relief after cycling to under 150 °C, but significant stress relief above 400 °C, and further relief if cycled cold to –200 °C from the 400 °C temperature; cycling below –200 °C provides no additional benefit. Cycling provides more stress relief than does an isothermal treatment. Typically, three cycles are sufficient to stabilize the component. To avoid thermal shock, cycling should typically be done at rates not exceeding 60 °C per hour.

Since full stress relief temperatures are high, it is important to obtain at least partial relief by cycling to just over or under the maximum and minimum survival temperatures expected for the component, respectively. This will stabilize the component as long as the temperature range is not exceeded.

2.3.3 Aluminum-beryllium

Aluminum-beryllium-machined or -cast optics, notably AlBeMet®[8] and Beralcast®,[9] are sometimes used as a happy medium between the use of pure aluminum or pure beryllium, combining the best of both worlds. Aluminum-beryllium is lighter and stiffer, and has a better CTE than aluminum, while being less costly than beryllium. Because it is readily joined, machined, or cast, aluminum-beryllium can be manufactured to produce support and metering structures as well.

2.3.4 Optical metering

Because not all material candidates are amenable to optical fabrication and metering structure fabrication, various combinations of optics and structure can be made. Table 2.7 lists the potential optical candidates described above, alongside the potential structure combination.

2.4 Very High-Thermal-Expansion Materials

Some materials, such as plastics and epoxies, exhibit very high CTEs.

2.4.1 Plastics

Despite their high CTEs, plastics such as Vespel® or Teflon™ are often suitable when a low modulus of elasticity is important, for example, to reduce contact stress (Chapter 16), or as an isolating stress barrier. Thus, while ideal neither for optics nor as a support structure, they do serve a significant purpose in certain applications. These materials exhibit expansion coefficients in excess of 50 ppm/°C at room temperature.

2.4.2 Adhesives

A list of adhesives commonly used in the optics industry for aerospace applications is given in Table 2.8 (specific properties of select adhesives are given in Table 2.3). The list is but a small portion of the more than 1000

Table 2.7 (a)–(f) Select optic and metering candidates for thermal performance. Benefits and regrets of near-athermal design combinations are noted.

(a)

Optic	Metering Structure	Comment
ULE	Graphite composite	Moisture backout
ZERODUR	Graphite composite	Moisture backout
Fused silica	Graphite composite	Moisture backout
Fused silica	Invar	Heavy; temperature limited
Fused silica	Fused silica	Risky
Aluminum	Aluminum	Inexpensive; control gradients
Beryllium	Beryllium	Expensive; ultralite
Silicon	Silicon carbide	Size limited
Silicon carbide	Silicon carbide	Structure joints; structure Size
Silicon carbide	Graphite composite	Focus may be required lightest design (ex all Beryllium)
Silicon carbide	Invar	Focus may be required heavy design

(b)

Optic		Metering Structure	
	CTE (ppm/K)		CTE (ppm/K)
ULE	0.03	Graphite composite	−0.2
ZERODUR	0.03	Graphite composite	−0.2
Fused silica	0.52	Invar	1.3
Aluminum	22.5	Aluminum	22.5
Beryllium	11	Beryllium	11
Silicon	2.6	Silicon carbide	2.4
Silicon carbide	2.4	Silicon carbide	2.4
Silicon carbide	2.4	Graphite composite	−0.2

(c)

Optic		Metering Structure	
	Modulus (Msi)		Modulus (Msi)
ULE	9.8	Graphite composite	15
ZERODUR	13.1	Graphite composite	15
Fused silica	10.6	Invar	20.5
Aluminum	10	Aluminum	10
Beryllium	44	Beryllium	44
Silicon	20	Silicon carbide	44.5
Silicon carbide	44.5	Silicon carbide	44.5
Silicon carbide	44.5	Graphite composite	15

(*continued*)

Table 2.7 Continued

(d)

Optic	Density (lbs/in^3)	Metering Structure	Density (lbs/in^3)
ULE	0.08	Graphite composite	0.06
ZERODUR	0.091	Graphite composite	
Fused silica	0.08	Invar	0.3
Aluminum	0.1	Aluminum	0.1
Beryllium	0.07	Beryllium	0.07
Silicon	0.08	Silicon carbide	0.105
Silicon carbide	0.105	Silicon carbide	0.105
Silicon carbide	0.105	Graphite composite	0.06

(e)

Optic	Conductivity (W/m-K)	Metering Structure	Conductivity (W/m-K)
ULE	1.31	Graphite composite	32
ZERODUR	1.64	Graphite composite	32
Fused silica	1.38	Invar	10
Aluminum	150	Aluminum	150
Beryllium	200	Beryllium	200
Silicon	125	Silicon carbide	150
Silicon carbide	150	Silicon carbide	150
Silicon carbide	150	Graphite composite	32

(f)

Optic	Strength (psi)	Metering Structure	Strength (psi)
ULE	1500	Graphite composite	40000
ZERODUR	1500	Graphite composite	40000
Fused Silica	1500	Invar	71000
Aluminum	42000	Aluminum	45000
Beryllium	35000	Beryllium	35000
Silicon	6500	Silicon carbide	12000
Silicon carbide	12000	Silicon carbide	12000
Silicon carbide	12000	Graphite composite	40000

potential candidates, but it serves as a useful guide for those adhesives most commonly used. Typical uses of these adhesives are also included.

Choice of adhesive depends on many factors, including temperature ranges for operation and survival, material properties, and bond strength. The listed adhesives can be used in bonding most engineering materials, including metals, plastics, ceramics, and glasses, provided the surfaces to be bonded are

Table 2.8 Typical adhesives for structural use in the optics industry for aerospace applications.

Material	Typical Use
3M Scotch weld 2216	Room-to-cryogenic temperatures
EA9361	Low temperature, high strain
EA 9394	High temperature
Epibond 1210-A/9615-10	Cryogenic adhesive
Epo-tek 301-2	Thin bondline structural adhesive: wicking epoxy
RTV 566	Silicon rubber optic support

properly prepared. This issue is discussed in Chapter 9, along with a detailed discussion on selection of the proper candidate for a particular application.

2.5 Strength

The opto-structural analyst is concerned with stiffness. High-elastic-modulus materials exhibit high stiffness for a given cross-section, a property that is required for many optical systems. Additionally, the analyst is concerned with strength; an optical system with high stiffness will perform well only if it survives environmental loading, often demanding not only high strength, but stress levels low enough to preclude permanent deformation, which could negate the benefits of high stiffness. These topics will be discussed in detail in subsequent chapters. Next, some of the basic concepts of stiffness and strength from stress–strain data are presented.

2.5.1 Failure to load

A typical stress–strain plot of a structural steel is shown in Fig. 2.8. It is worthwhile to study the curve, as it illustrates the definitions and concepts used in the analyses that follow.

A tensile load is applied to a bar along its axial dimension to determine its stress–strain characteristics. Typically, a tensile test machine with an integrated load cell and strain extensometer is used to determine stress under load [Eq. (1.2)] and the resulting strain (Eq. 1.3). As the load is increased, the stress-versus-strain curve is initially a straight line, indicating linear elastic behavior, in which case Hooke's law [Eq. (1.4)] applies. Most optical designs use specifications that are concentrated in this region. Here, when the load is removed, the strain returns to zero, its starting point. If the load is continued beyond its linear region, stress is no longer proportional to strain, and less load is required to deform the material. The material may remain elastic for a time, i.e., return to (near) zero strain, until its proportional limit is reached. Beyond that, the material will continue to behave in a nonlinear fashion, taking on a permanent set at the material yield point when the load is removed, and becoming "plastic." Typically, the yield point for many

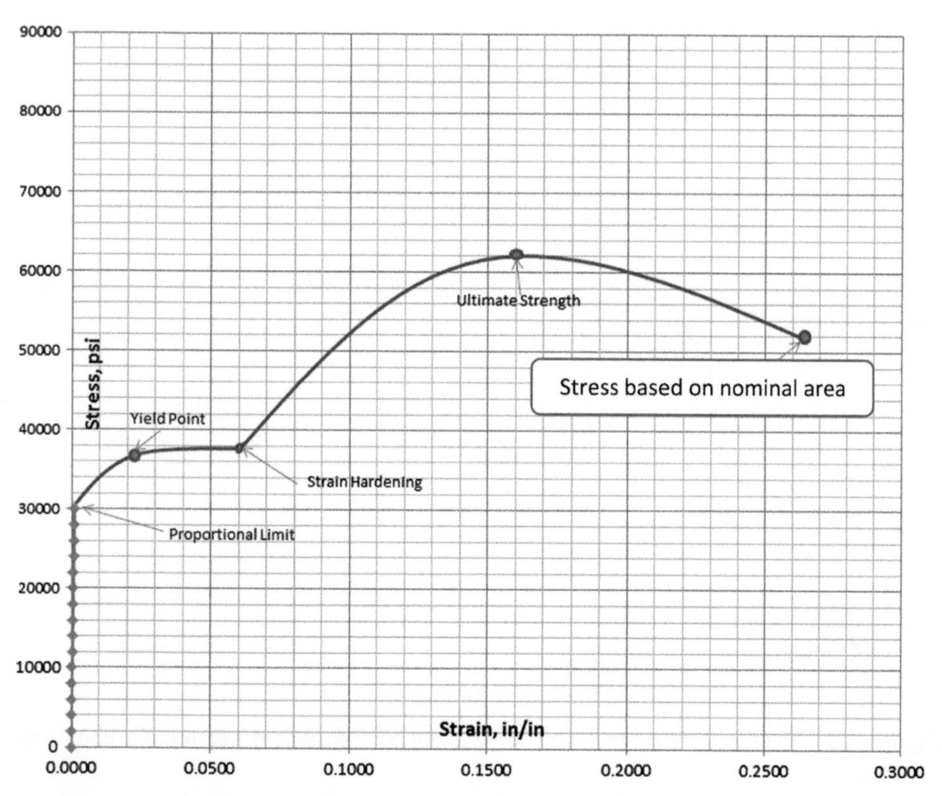

Figure 2.8 Schematic of a typical strain-to-failure diagram for mild structural steel.

materials is considered as the 0.2% offset, or a permanent strain of 0.002 when the load is released.

Load release generally parallels the slope of the elastic region, as shown in Fig. 2.9. As the load is continued, the material stiffens somewhat in a process called strain hardening; here, when the load is removed and then re-applied, the material yield point can be raised significantly, as will its breaking strength. Indeed, such processes are used for many high-strength applications. As the load is increased, eventually the material breaks at a point defined as its ultimate strength. In reality, the material cross-section will reduce in area such that the true strength is higher; however, we use the original area for our analyses, upon which strength reporting (engineering stress) is based.

2.5.2 Yield

When a *ductile* material reaches its yield point, as we discussed, it does not return to zero strain but rather takes on a permanent strain set. For example, consider a typical stress–strain curve for aluminum, as shown in Fig. 2.10. Yield strength can be determined for a particular value of the permanent set. Note that many values of offset can be used; while 0.2% offset is common, in optics sometimes we need to look lower.

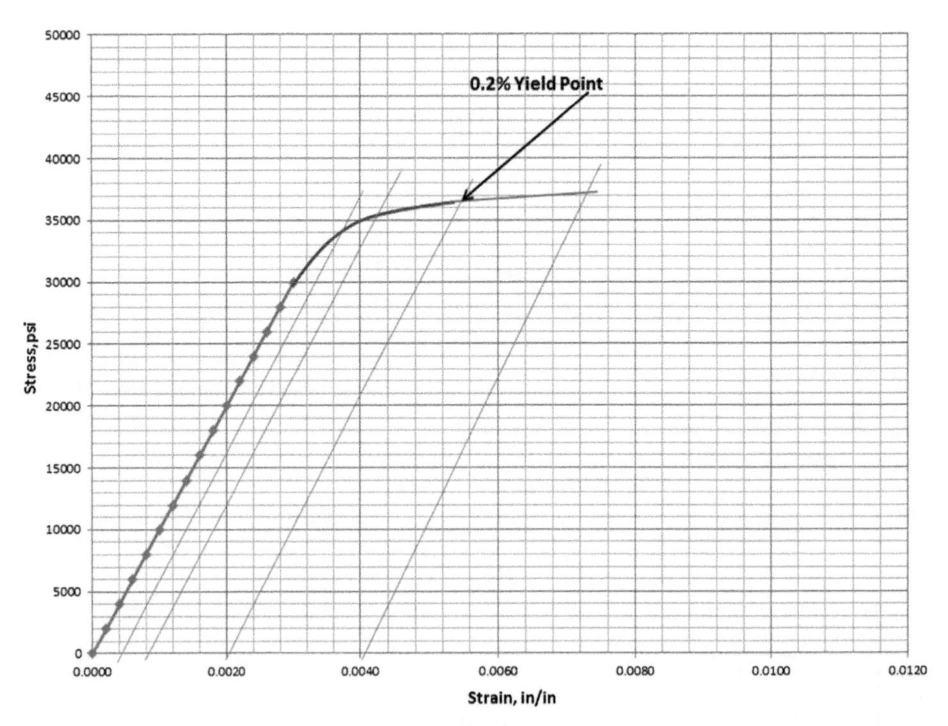

Figure 2.9 Yield point determination for aluminum. Permanent set results upon load release.

2.5.3 Micro-yield

In the optical world, we are sometimes concerned with micro-yield, defined as the stress that results in 1 ppm permanent strain after stress removal, or the 0.0001% offset. This is also referred to as the material precision elastic limit. Since permanent set is so small, 1/2000 of the typical yield point value, it is difficult to measure and requires special techniques. One would be hard pressed to find it on the curve, as it appears linear. Although we imply the word "elastic" and "proportional limit" to mean no permanent set, this is not true at such microscopic levels.

The opto-structural analyst must determine the effects of permanent strain with regard to misalignment and wavefront error, the latter being induced by such misalignment or residual force in the optic. Often, the micro-yield does not pose a problem, so its use needs to be addressed and not blindly applied, which could result in an overly conservative or heavy design. We will see examples of this in Chapter 16.

2.5.4 Brittle materials

Many materials used in optical systems, e.g., glasses and ceramics, are brittle in nature, meaning that there is no clearly defined or existing yield point; the

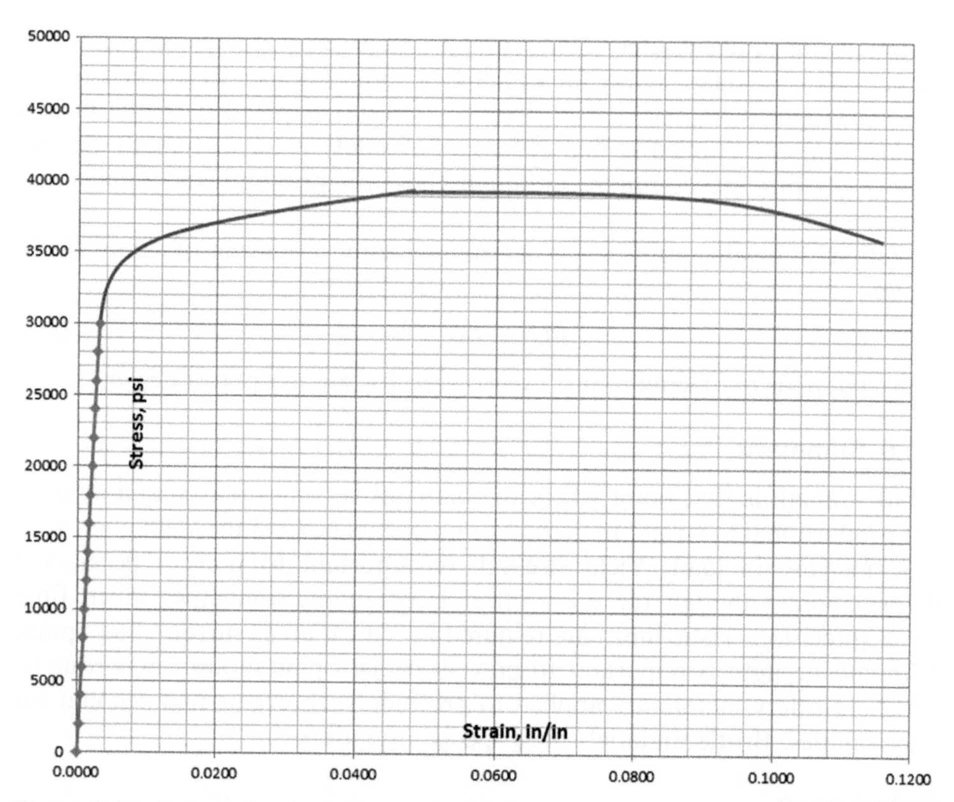

Figure 2.10 Schematic of a typical strain-to-failure diagram for aluminum 6061-T6. The material is elastic below 30,000 psi and has high elongation to failure.

material will fail under load without plastic deformation. Ductile materials, which exhibit nonlinear behavior as stress is increased (Fig. 2.9), can be somewhat forgiving when yield is exceeded, for example, near stress risers or under thermal stress. Brittle materials are not forgiving, so special care needs to be taken, as is more fully discussed in Chapter 12.

2.5.5 Safety factor

To ensure the success of a design, a factor of safety (FS) is generally applied. The safety factor is defined as the material strength (yield, ultimate, etc.) divided by the actual stress. Material strengths for many materials are provided in Tables 2.1 through 2.4. These strengths are generally reported as "A" values (Weibull A), where a 99% probability of survival with 95% confidence is realized. Required safety factors are dependent on customer specification or internal requirements, but for ductile materials are typically specified to be 1.25 for yield strength and 1.4 or 1.5 for ultimate strength. For adhesive bonds, certain composites, and less-characterized materials, a safety factor of 2.0 is commonly applied against ultimate strength. For brittle materials, an applied safety factor of 3 is usual. For lift devices, or where human life is at stake, safety factors of 5 or more are usually applied.

For fatigue (Chapter 11) and stress corrosion (Chapter 12), scatter factors against time are applied in lieu of factors of safety against strength. Scatter factors of four are common, although they are program and application dependent.

Often, allowable stresses are determined using a margin of safety (MS), where

$$MS = FS - 1. \quad (2.2)$$

Here, the margin of safety must exceed zero for an acceptable design.

2.5.6 Summary

Yield strength is that point where the material becomes plastic and stress is not proportional to strain. Increased strain occurs beyond the yield point with little increase in load. Yield strength is important to prevent permanent deformation; although the 0.2% offset is common, sometimes much finer levels (precision elastic limit) are required in critical areas surrounding optics. Ultimate strength is important to prevent catastrophic failure. Accordingly, sufficient safety factors are applied to preclude such events and account for uncertainties. Brittle materials need special attention.

References

1. Corning, Inc., Canton, New York.
2. Schott Glaswerke AG, Mainz, Germany.
3. ArcelorMittal, Stainless and Nickel Alloys, Imphy, Neuvre Department, Bourgogne, France.
4. R. Guillaume, "Invar and Elinvar," *Nobel Lecture* **11** December 1920, from *Nobel Lectures, Physics, 1908–1926*, Elsevier Publishing Co., Amsterdam (1967).
5. S. F. Jacobs, "Variable invariables: Dimensional instability with time and temperature," *Proc. SPIE* **10265**, 102650I, *Optomechanical Design: A Critical Review* (1992) [doi: 10.1117/12.61115].
6. R. A. Paquin, "Dimensional instability of materials: How critical is it in the design of optical instruments?" *Proc. SPIE* **10265**, 1026509, *Optomechanical Design: A Critical Review* (1992) [doi: 10.1117/12.61106].
7. I. Kh. Loskin, "Heat treatment to reduce internal stresses in beryllium," *Metal Sci. Heat Treat* (USSR) p. 426 (1970).
8. Materion Corp., Beryllium and Composites, Elmore, Ohio.
9. Beralcast Corp., a subsidiary of IBC Advanced Allots Corp., Franklin, Indiana.

Chapter 3
Kinematic Mounts

3.1 Kinematics

A kinematic mount, which is ideal for optics and optical structure mounting, is a mount in which all six degrees of freedom (three translations and three rotations) of a 3D object are restrained from moving without overconstraint. Such a mount requires six independently constrained degrees of freedom without redundancy. In other words, the mount is statically determinate as revealed by the six equations of equilibrium in a 3D coordinate system (x,y,z) as

$$\begin{aligned}
\Sigma F_x &= 0, \\
\Sigma F_y &= 0, \\
\Sigma F_z &= 0, \\
\Sigma M_x &= 0, \\
\Sigma M_y &= 0, \\
\Sigma M_z &= 0.
\end{aligned} \tag{3.1}$$

This mount allows for translation or rotation in or about any axis of the reference ground to which it attaches without deformation of the mounted object, which moves as a rigid body.

Consider, for example, the ideally mounted structure of Fig. 3.1, which comprises either three grooves or a ball in cone, ball in groove, and ball on flat. The structure in Fig. 3.1(a) restrains the structure vertically and tangentially. The structure in Fig. 3.1(b) restrains three degrees of freedom in translation (x,y,z); the groove restrains two degrees of freedom in translation (y,z), and the flat restrains one degree of freedom (z) in translation. The equations of equilibrium [Eq. (3.1)] are completely satisfied. Another way of saying this is that we have eliminated three degrees of freedom in translation and three degrees of freedom in rotation (pitch, roll, and yaw) without redundancy.

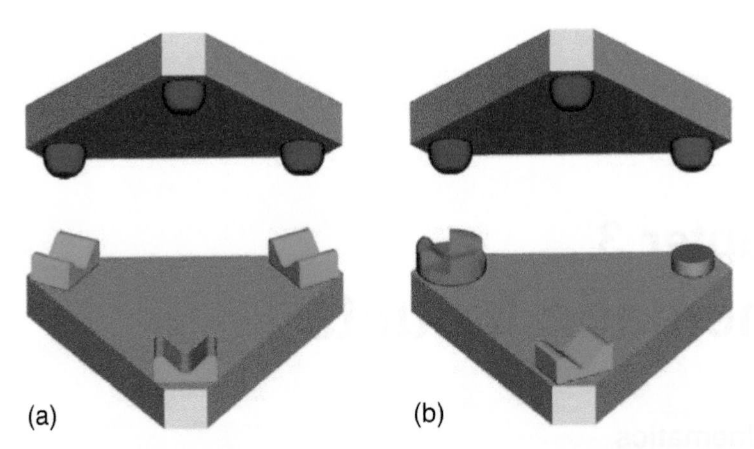

Figure 3.1 Ideal kinematic mounts have a total of six (and only six) determinate restrained degrees of freedom. All other twelve degrees of freedom are unconstrained. (a) Mount restrains structure vertically and tangentially. (b) Mount restrains three degrees of freedom in translation.

The reader might note that not all six-degree-of-freedom constraints are kinematic. The mounts of Figs. 3.2(d) and (e) are not kinematic. In Fig. 3.2(d), the optic cannot be put into equilibrium, as the equations of equilibrium are not satisfied, and it is free to rotate. Note that since constraint is radial, radial thermal expansion would be resisted and impart force, distorting it. A similar argument is made for Fig. 3.2(e). Here, reactions along the x axis are collinear, and again, the equations of equilibrium cannot be satisfied. Figure 3.2(a), on the other hand, is kinematic, as previously noted, as is Fig. 3.2(b), where each mount restrains the optic in the z axis and tangentially, for a total of six degrees of freedom, which satisfies the equilibrium equations. Figure 3.2(c) is kinematic as well, although not as preferable, since all loads in the x axis are resisted at a single point; this will affect stiffness and strength for quasi-kinematic mounting, discussed below. Figure 3.2(f) is another kinematic mount example, where all six degrees of freedom are constrained at a single point.

The kinematic mount allows for distortion-free mounting, which is critical for optics in cases where deformation to a fractional wavelength of light can ruin one's day. This means that the optic (or structure, for that matter, if kinematically mounted) can be aligned, by any motions of the base, as a rigid body without cause for concern. A manager once asked why we use kinematic, statically determinate mounts; after all, we learned to do indeterminate analysis in school, and furthermore, multiple points would support the optic better. While there is a case for multiple-point mounting (Chapter 7), the manager missed the point: kinematics is key to precluding optic distortion. Consider, for example, a nonkinematic mount that is overconstrained (as in Fig. 3.3) by supporting it on four corners. Shown is a thin, square optic that is quite flexible relative to its mount. For symmetry, it might seem like a good idea to support it

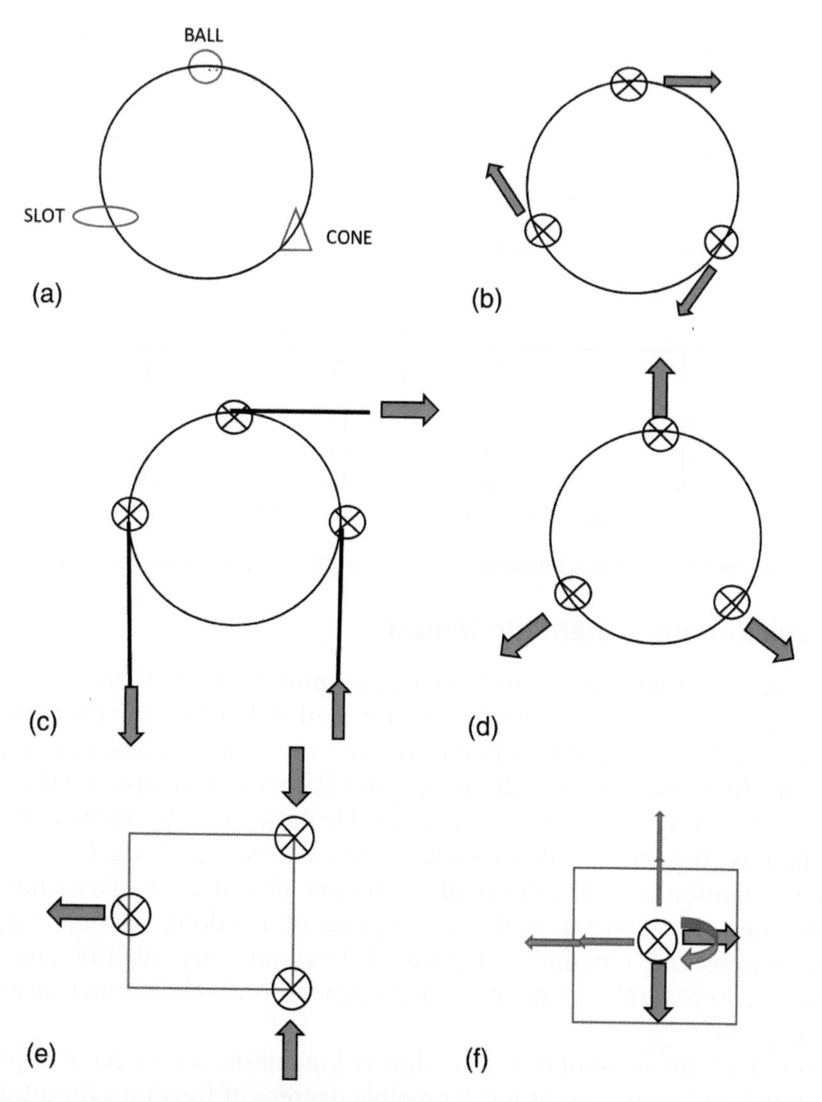

Figure 3.2 Kinematic and nonkinematic arrangements for six constrained degrees of freedom. Cases (a), (b) (vertical and tangential groove restraint), (c), and (f) (double arrow indicates moment) are kinematic, while cases (d) and (e) are unstable.

at its four corner points; the optic would also be capable of carrying more load than if mounted kinematically at only three corners.

However, suppose that three corners are mounted kinematically, and the fourth mount is off by 0.001 in. in the optical axis. The optic is mounted by flexing it 0.001 in. However, 0.001 in. is 40 visible wavelengths of light—well beyond the fractional wavelength required (this is a real-life example that actually occurred)! Needless to say, while we are certainly capable of analyzing statically indeterminate structures, kinematics is the key.

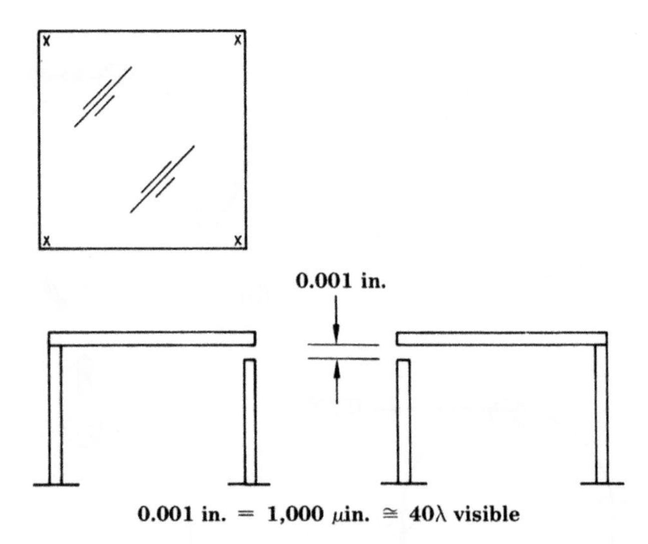

Figure 3.3 Four-point mount; because it is nonkinematic, excessive mount error can occur.

3.2 Quasi-static Kinematic Mount

There are drawbacks to a purely kinematic mount. Since it relies on point contacts, high stresses can develop under load (Chapter 16). Furthermore, since each individual mount is free to rotate (restraining translations only, as required), frictional forces, stiction, and the like will overconstrain the optic. Because friction and stiction are problematic in determination and repeatability, this can wreak havoc with the performance budget.

The situation can be alleviated by proper design of a quasi-kinematic mount—one that constrains the six degrees of freedom, as required, but overconstrains the remaining degrees of freedom only slightly and in a predictable, repeatable, manner. A predictable, repeatable manner means no moving parts.

Note that for a 3-point mount that is kinematic, as we have depicted, each individual mount point has 6 possible degrees of freedom, for a total of 18 degrees of freedom, only 6 of which are restrained, leaving 12 degrees of freedom unrestrained, as required. For quasi-kinematic mounting, these 12 degrees of freedom need to be made very compliant to minimize assembly or thermal forces during mounting or temperature excursion, and hence minimize distortion. This is generally accomplished through a flexurized mounting scheme.

Such compliant mounting is achieved by flexure design, which makes use of bending metal and is quite predictable, as opposed to relying on zero or unpredictable friction (and high contact stress) as demanded by purely kinematic mounts.

Quasi-kinematic mounts generally take the form of a flexurized system of beams, compliant in the required degrees of freedom but stiff in the six required directions, or a system of bipod mounts, also compliant in the required degrees of freedom but stiff in the six required directions. We will illustrate and analyze both types of systems in this chapter.

3.3 Flexure Analysis

A flexure is nothing more than a beam that easily bends and twists in the required degrees of freedom (translation and rotation) but provides sufficient stiffness in the determinate axes, as demanded by kinematics. An example of a three-point flexure design is shown in Fig. 3.4. An optic is supported by mounting the flexures tangent to it and separated by 120 deg. Again, compliancy is not zero, and neither does the required restraint provide zero displacement. Thus, detailed analysis is required to ensure adequacy under mount-induced loading while maintaining appropriate stiffness requirements to meet deflection and fundamental frequency (Chapter 10) criteria.

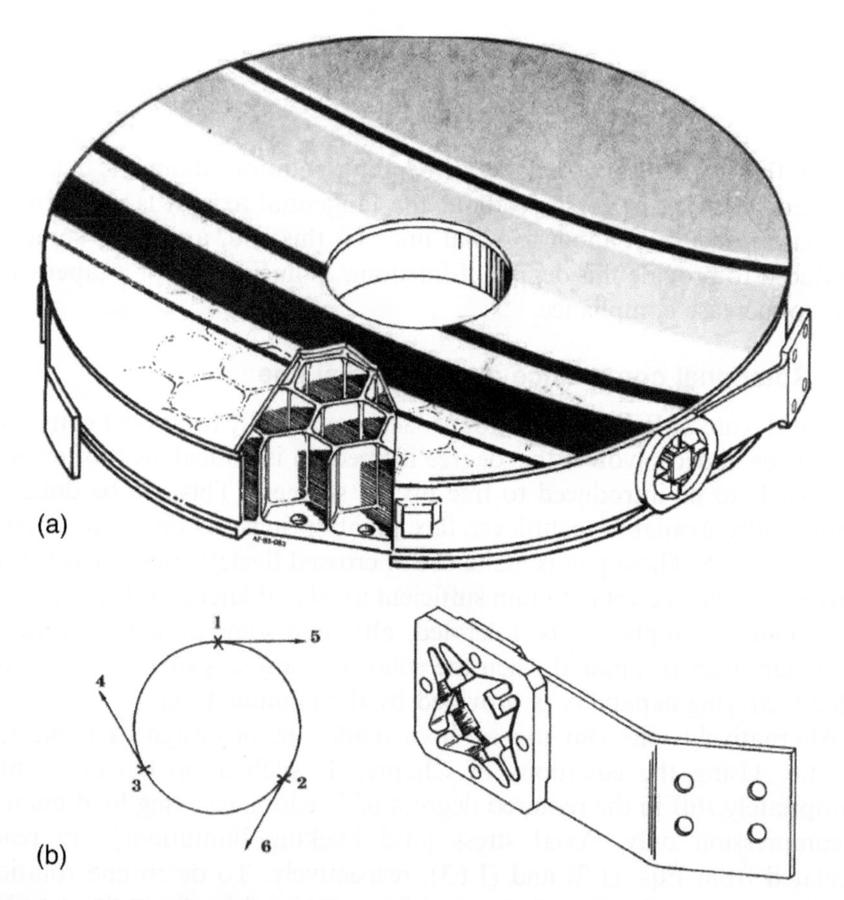

Figure 3.4 (a) Three-point quasi-kinematic mount for a lightweight optic using flexures. (b) Schematic of the flexure mount.

Note that the flexure is relatively stiff in the z axis translation in proportion to its strong axis rigidity in bending EI, where

$$I_s = \frac{tb^3}{12}. \tag{3.2}$$

The flexure is also stiff in the lateral (x and y) axes, where load is resisted in tension and compression in proportion to its area A and modulus, product AE, where

$$A = bt. \tag{3.3}$$

This satisfies the determinate constraint condition.

Note that the flexure is compliant in radial translation in proportion to its weak axis rigidity EI, where

$$I_w = \frac{bt^3}{12}. \tag{3.4}$$

The flexure is also compliant in bending rotation about the z axis and compliant in twist, or torsion, about the tangential axis; it is not, however, compliant in rotation about a radial line. To this end, a pivot is sometimes introduced to provide this degree of freedom, as in Fig. 3.4, or a taper can be made to increase compliance.

3.3.1 Rotational compliance about a radial line

The leaf flexure of Fig. 3.4 does not *per se* have radially rotational compliance without the added pivot. If this degree if freedom is critical, as it often is, the pivot needs to be introduced to free up the stiffness. This can be done with commercially available[1] cantilever flex pivot springs, as depicted schematically in Fig. 3.5. These pivots make use of crossed flexures that provide radial rotation compliance yet maintain sufficient axial and lateral stiffness. A rather large rotational angle can be tolerated, although some decentration occurs. This latter effect is minimal when the rotational angle is small. Tabular data on load-carrying capability is provided by the manufacturer.

Alternatively, one can use a "homemade" pivot design as depicted in Fig. 3.6. Using the equations of Chapter 1, such a pivot can be made appropriately stiff in the required degrees of freedom, carrying load intension or compression only. Axial stress (and bucking limitations) are readily calculated from Eqs. (1.2) and (1.63), respectively. To determine rotational compliance, we make use of compatibility relationships from the cantilever cases of Table 1.1. With reference to Fig. 3.6, for each spoke, we find that

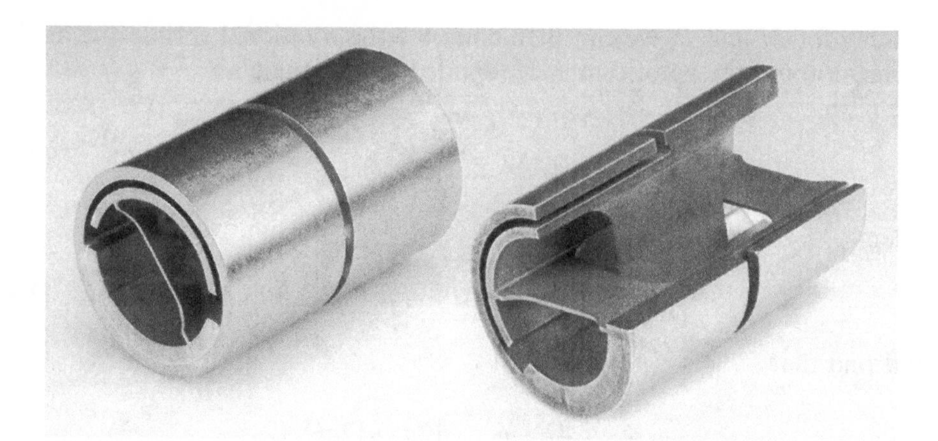

Figure 3.5 (left) Single- and (right) double-ended cross-spring pivot providing radial rotational compliance, yet maintaining load-carrying capability (reprinted from Ref. 1 with permission from Riverhawk Co.).

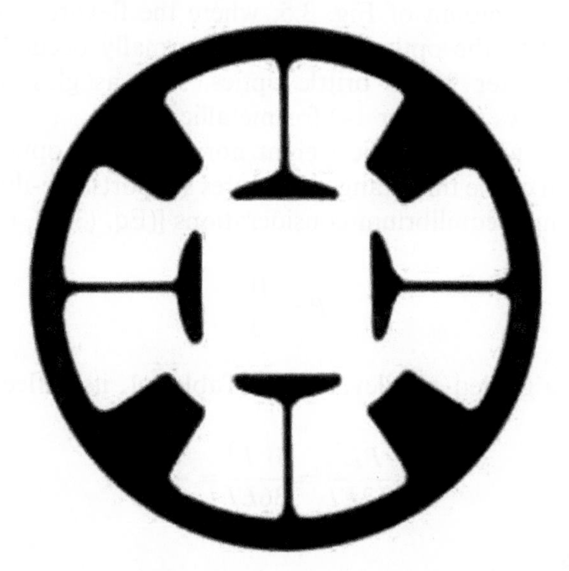

Figure 3.6 Pivot attached to the flexure of Fig. 3.4 provides a high degree of radial rotational compliance. Blades must be sized to resist loading by providing adequate stiffness and strength while precluding buckling.

$$\frac{PL^3}{3EI} - \frac{ML^2}{2EI} = y = a\theta, \tag{3.5}$$

$$\frac{PL^2}{2EI} - \frac{ML}{EI} = -\theta. \tag{3.6}$$

Solving for M and P, we can then compute the rotational spring rate as the moment about the pivot center M_t divided by the angle as

$$K_\theta = \frac{4M_t}{\theta}, \tag{3.7}$$

where

$$M_t = M + Pa, \tag{3.8}$$

and find that

$$K_\theta = \frac{4EI}{L}\left(4 + \frac{12a}{L} + \frac{12a^2}{L^2}\right). \tag{3.9}$$

3.3.2 Analysis: constrained degrees of freedom

Consider, first, the mount of Fig. 3.5, where the flexure is rigidly attached (no radial pivot) to the optic. This would normally occur by means of an epoxy bond (Chapter 9) for brittle optics, such as glass or ceramics, or sometimes by screws (Chapter 14) for metallic optics.

When loaded with the optic weight normal to the optic, i.e., along the optical axis, each of the three tangent flexures supports one-third of the weight according to simple equilibrium considerations [(Eq. (3.1)], or

$$P = \frac{W}{3}. \tag{3.10}$$

The beam acts a guided cantilever; from Table 1.1, its deflection is

$$Y = \frac{PL^3}{12EI} = \frac{WL^3}{36EI} = \frac{WL^3}{3Etb^3}, \tag{3.11}$$

and its stress is

$$\sigma = \frac{Mc}{I} = \frac{6M}{tb^2} = \frac{3PL}{b^2} = \frac{WL}{tb^2}. \tag{3.12}$$

When loaded with the optic weight in the x axis, the compliant top flexure will see little load, while the others will react to most of the weight. From equilibrium,

$$P = \frac{W}{2\cos 30} = 0.577\,W. \tag{3.13}$$

The optic deflection along the x axis is

$$Y = \frac{PL}{AE \cos 30},$$

or

$$Y = \frac{0.577WL}{AE \cos 30}, \tag{3.14}$$

$$Y = \frac{0.666WL}{AE}, \tag{3.15}$$

and the stress is

$$\sigma = \frac{P}{A} = \frac{0.577W}{A}. \tag{3.16}$$

When loaded with the optic weight in the y axis, each flexure shares the load but not equally; from equilibrium, flexure 1 sees the highest reaction as

$$P = \frac{2W}{3} = 0.666W. \tag{3.17}$$

With reference to Fig. 3.4(b), we derive this result as follows: Let the reaction in the Y direction at flexure 1 be A, and let the reaction at flexures 2 and 3, from symmetry, be B. The force in each flexure is F_1, F_2, and F_3, respectively. R is the radius of the inscribed mount circle, with point O as its origin. Under weight W in the Y direction, we have

$$\begin{aligned}
\Sigma M_O &= 0,\\
AR &= 2BR \sin 30 + 2CR \cos 30,\\
W &= A + 2B,\\
AR &= FR + FR,\\
AR &= BR/\sin 30 + BR/\sin 30,\\
AR &= 2BR + 2BR = 4BR,\\
A &= 4B,\\
W &= A + 2B = 6B,\\
B &= W/6,\\
A &= F_1 = 2W/3, \text{ and}\\
C &= W \cos 30/3.
\end{aligned}$$

Noting that $\sin 30 = B/F_2 = B/F_3$, it follows that $F_2 = F_3 = B/\sin 30$, and $F_2 = F_3 = 2B = W/3$. We next find the optic deflection along the y axis as

$$Y = \frac{PL}{AE},$$

or

$$Y = \frac{0.666WL}{AE}, \tag{3.18}$$

which is identical to Eq. (3.15); i.e., stiffness is the same in both lateral directions, although not intuitively obvious to the casual observer.

The stress is

$$\sigma = \frac{P}{A} = \frac{0.666W}{A}, \tag{3.19}$$

which is *higher* than occurs in the x axis [Eq. (3.16)].

3.3.2.1 Example for consideration

Consider the rigid optic mounted on three flexures as in Fig. 3.4(a) (without the shown radial pivot compliance). The optic weighs 10 lbs and experiences 30-g launch load in any direction. For a 6061-T6 aluminum flexure that is 3 in. long, 1 in. deep, and 1/8 in. thick, determine the flexure stress under launch and the self-weight deflection in each orthogonal axis. Use the properties for aluminum found in Table 2.1.

For launch stress along the normal z axis, we find from Eq. (3.12) that

$$\sigma = \frac{WL}{tb^2} = (10)(30)(3)/[(0.125)(1)(1)] = 7200\,\text{psi},$$

which is well below the material allowable stress, as given in Table 2.1.

Along the lateral x axis, we find from Eq. (3.16) that

$$\sigma = \frac{P}{A} = \frac{0.577W}{A} = (0.577)(10)(30)/[(0.125)(1)] = 1385\,\text{psi}.$$

Along the lateral y axis, we find from Eq. (3.19) that

$$\sigma = \frac{P}{A} = \frac{0.666W}{A} = (0.666)(10)(30)/[(0.125)(1)] = 1600\,\text{psi}.$$

For self-weight deflection, along the z axis, we find from Eq. (3.11) that

$$Y = \frac{WL^3}{36EI} = (10)(10)(3)(3)(3)/[(3)(10)^7(0.125)(1)^3] = 0.0007\,\text{in.}$$

For self-weight deflection, along the x or y axis, from the identical Eqs. (3.15) and (3.18), we find that

$$Y = \frac{0.666WL}{AE} = (0.666)(10)(3)/[(0.125)(1)(10)^7] = 0.00016\,\text{in.}$$

3.3.3 Analysis: compliant degrees of freedom

Consider next the required compliance of the tangent flexure system. Suppose that the flexure needs to accommodate a displacement y in the radial direction due to assembly tolerance or CTE mismatch to its supporting bezel.

Here we have a guided cantilever bending about the weak axis. Again, from Table 1.1,

$$Y = \frac{PL^3}{12EI} = \frac{PL^3}{Ebt^3}, \tag{3.20}$$

and the load to the optic is

$$P = \frac{bt^3EY}{L^3}. \tag{3.21}$$

We find the moment to the optic as

$$M = \frac{PL}{2} = \frac{bt^3EY}{2L^2}. \tag{3.22}$$

The bending stress in the flexure is

$$\sigma = \frac{Mc}{I} = \frac{3EtY}{L^2}. \tag{3.23}$$

Note that the stress is independent of the inertia of the beam and dependent only on its thickness, which is rather nice. This is the constant deflection problem: the flexure width does not impact stress, so it may be increased, if needed, which will increase stiffness as well; however, the force to the optic will increase, so a trade is required. This issue will be further discussed.

Suppose, next, that the flexure needs to accommodate a rotation θ in the twist direction, or rotation about the flexure tangential axis due to assembly tolerance.

Here we calculate the twist moment T from Eq. (1.36) as

$$\theta = \frac{TL}{KG} = \frac{3TL}{bt^3G},$$

$$T = \frac{\theta bt^3 G}{3L}, \tag{3.24}$$

which is the torque imparted to the optic.

The shear stress, from Eq. (1.29), is

$$\tau = \frac{3T}{bt^2} = \frac{Gt\theta}{L}, \tag{3.25}$$

again, independent of the flexure width.

We see that the quasi-kinematic flexure mount gives conflicting and opposing requirements; it must be stiff and strong on one hand (and not buckle, either) but weak and compliant on the other hand so as not to impart high loads to the optic while not overstressing the flexure. Methods to solve for an optimum design to meet this challenge are given in Section 3.5. Distortion of the optic and its limits are discussed in Chapter 4.

3.3.3.1 Example for consideration

The rigid optic of Fig. 3.4(a) is mounted to a supportive bezel and due to mount tolerance is subjected to a radial motion of 0.003 in. and a twist about a tangential axis of 0.003 rad. For the flexure of Example 3.3.2.1, determine the forces and moments to the optic.

From Eq. (3.21), we find the radial force as

$$P = \frac{bt^3EY}{L^3} = (1)(0.125)^3(10)^7(0.003)/[(3)(3)(3)] = 2.17\,\text{lbs.}$$

From Eq. (3.22), we find the moment about an axial line as

$$M = \frac{bt^3EY}{2L^2} = (1)(0.125)^3(10)^7(0.003)/[(2)(3)(3)] = 3.26\,\text{in.-lb.}$$

From Eq. (3.24), we find the tangential twist moment as

$$T = \frac{\theta bt^3G}{3L} = (0.003)(1)(0.125)^3(3.8)(10)^6/[(3)(3)] = 2.47\,\text{in.-lb.}$$

In Chapter 4 we will see the effects of such forces on optical performance.

3.4 Bipod

A bipod is an A-frame truss with two legs, as depicted in its simplest form in Fig. 3.7. A bipod, like the flexure beam described in the previous section, bends

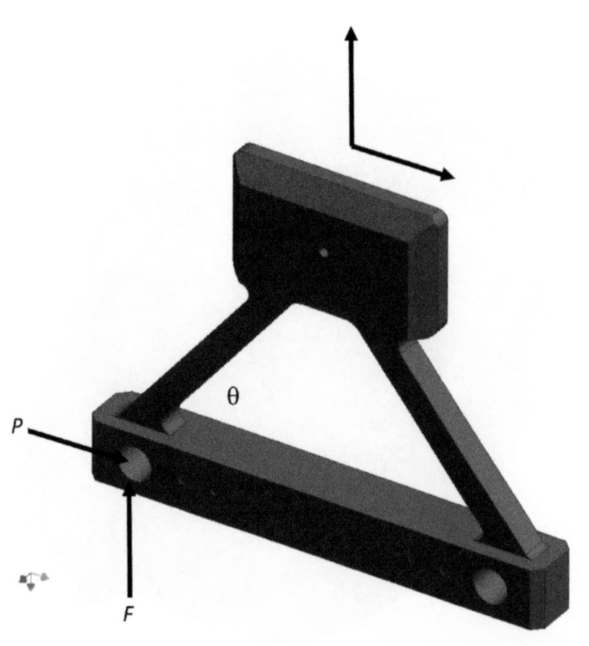

Figure 3.7 An individual bipod is stiff in the required two degrees of freedom and relatively soft (compliant) in the other four.

and twists easily in the required degrees of freedom (translation and rotation) but provides sufficient stiffness in the determinate axes demanded by kinematics. An example of a three-point bipod design is shown in Fig. 3.8. Here, the optic is supported by mounting the bipods normal to it and separated by 120 deg. Again, compliancy is not zero, and neither does the required restraint provide zero displacement. Thus, detailed analysis is required to ensure adequacy under mount-induced loading, while maintaining appropriate stiffness requirements to meet deflection and fundamental frequency criteria.

Note that the bipod is relatively stiff in z axis translation, where load is resisted in tension and compression in proportion to its area A, and modulus, product AE, where, as in Eq. (3.3), $A = bt$.

Generally, such a bipod will be square or circular in cross-section to allow for ease of deformation in both the radial and tangential axes; i.e., $b = t$ for a square section. Square sections are preferable to circular sections from a manufacturing standpoint. This bipod is also stiff in the lateral (x and y) axes, where, again, load is resisted in tension and compression in a similar fashion, satisfying the determinate constraint condition.

Note that for a square cross-section, the bipod is compliant in radial translation in proportion to its rigidity EI, where

$$I = \frac{b^4}{12}.$$

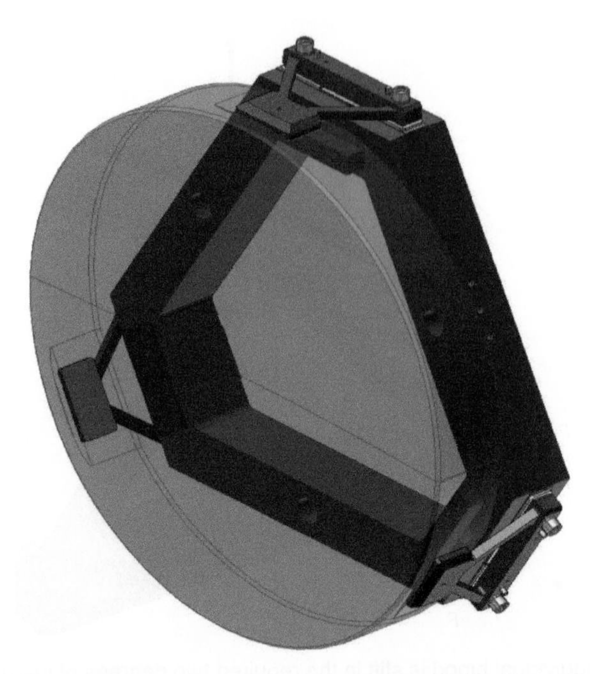

Figure 3.8 A quasi-kinematic bipod series of mounts constrains the optic in six degrees of freedom without overconstraining it.

It also compliant in bending rotation about the z axis, and compliant in twist, or torsion, about the tangential axis; unlike the flexure beam, it is also compliant in rotation about a radial line (a bipod might be considered as a wide tapered flexure with large cutout). The bipod therefore requires no pivot assembly at the apex.

3.4.1 Analysis: constrained degree of freedom

Consider the mount of Fig. 3.8, where the bipod is rigidly attached to the optic. When loaded with the optic weight normal to the optic; i.e., along the optical z axis, each of the three bipods supports one-third of the weight. We can compute its displacement Y and stress σ under the optic weight as follows. Both displacement and stress will be a function of the bipod included angle.

With reference to Fig. 3.7 and truss equilibrium considerations, and neglecting small, secondary bending moments that could develop, each leg reacts with a force of

$$F = \frac{W}{6 \sin \theta}, \tag{3.26}$$

and a resulting stress of

$$\sigma = \frac{F}{A} = \frac{W}{6b^2 \sin\theta}. \tag{3.27}$$

This is the 1-g stress. If a gravity acceleration load of value G is applied (as in a launch environment, aircraft maneuver, shock event, etc.), the force is

$$F = \frac{WG}{6 \sin\theta}, \tag{3.28}$$

and the stress is

$$\sigma = \frac{WG}{6b^2 \sin\theta}, \tag{3.29}$$

where the acceleration is G.

Of course, the leg is not allowed to buckle under acceleration loading. From Eq. (1.63), we find that

$$P_{\mathrm{cr}} = \frac{\pi^2 EI}{L^2} = \frac{\pi^2 E b^4}{12L^2}, \tag{3.30}$$

which needs to be greater than the leg reactive load.

Calculation of displacement is less straightforward for bipods than for flexure beams. Using an energy method such as virtual work, or the dummy load method, both of which are beyond the scope of this text, we find, again along the vertical z axis, that

$$Y = \frac{WL}{6AE \sin^2\theta} = \frac{WL}{6b^2 E \sin^2\theta}. \tag{3.31}$$

When loaded with the optic weight in the x or y axis, the reaction at the bipod apex is, as with the flexure,

$$P = 0.577W \tag{3.32}$$

in the x axis, and

$$P = 0.666W \tag{3.33}$$

in the y axis.

With reference to Fig. 3.9 and truss equilibrium considerations, each leg reacts under acceleration G with a force of

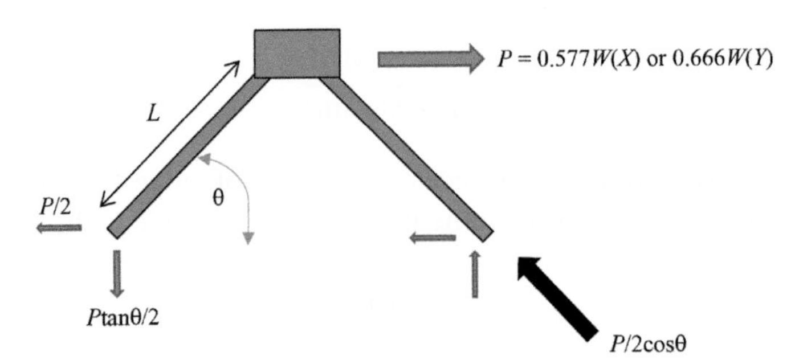

Figure 3.9 Equilibrium diagram for a bipod under lateral load condition. Boundaries are pinned.

$$F = \frac{0.577WG}{2\cos\theta} \tag{3.34}$$

in the x axis, with a resulting stress of

$$\sigma = \frac{F}{A} = \frac{0.577WG}{2b^2\cos\theta}, \tag{3.35}$$

and

$$F = \frac{0.666WG}{2\cos\theta} = \frac{WG}{3\cos\theta} \tag{3.36}$$

in the y axis, with a resulting stress of

$$\sigma = \frac{F}{A} = \frac{WG}{3b^2\cos\theta}. \tag{3.37}$$

Calculation of displacement using the dummy load method gives

$$Y = \frac{WL}{3AE\cos^2\theta}$$

in both the x and y axes, or

$$Y = \frac{WL}{3b^2E\cos^2\theta}. \tag{3.38}$$

3.4.1.1 Optimum base angle

Note that in order to maintain equal stiffness in all orthogonal directions, we can set Eqs. (3.31) and (3.38) equal to one another to find that

$$\sin^2\theta = \frac{\cos^2\theta}{2},$$

or

$$\frac{\sin^2\theta}{\cos^2\theta} = \frac{1}{2},$$
$$\cos\theta = \sqrt{2},$$
$$\theta = \sim 35\,\text{deg}.$$

This is a rather flat angle but is optimum unless the design space envelope is insufficient.

In order to maintain equal strength in the worst-case lateral and vertical axes, presuming equal vertical and lateral load, we set Eqs. (3.28) and (3.36) equal to one another to find that

$$\sin\theta = \frac{\cos\theta}{2},$$

or

$$\frac{\sin\theta}{\cos\theta} = \frac{1}{2},$$
$$\tan\theta = \frac{1}{2},$$
$$\theta = \sim 26\,\text{deg},$$

which is even more flat than required for displacement. Such an angle is less common; often an angle of 45 deg is chosen for practical limitations on envelope, unless requirements for stiffness and/or strength demand otherwise.

3.4.2 Analysis: compliant degrees of freedom

Consider next the required compliance of the bipod system. Suppose that each bipod leg needs to accommodate a displacement Y in the radial direction due to assembly tolerance or CTE mismatch with its supporting bezel.

Here we have a guided cantilever. Again, from Table 1.1, we find that

$$Y = \frac{PL^3}{12EI} = \frac{PL^3}{Eb^4},$$

and hence the load to the optic is

$$P = \frac{b^4 EY}{L^3}. \tag{3.39}$$

We find the moment to the optic as

$$M = \frac{PL}{2} = \frac{b^4 EY}{2L^2}. \tag{3.40}$$

The bending stress in the leg is

$$\sigma = \frac{Mc}{I} = \frac{3EbY}{L^2}. \tag{3.41}$$

3.4.2.1 Example for consideration

Consider the bipod mount design of Fig. 3.8. Requirements dictate that it must support a 5-lb optic under a launch acceleration load of 50 g independently in all orthogonal directions. The maximum self-weight deflection to meet a fundamental frequency requirement (Chapter 10) is specified at 0.0003 in. The maximum bending moment to the optic is to be limited to 1 in.-lb when thermal expansion of the bezel induces a 0.020 in. radial motion. The bipod is moved radially and is made of Invar 36.

Determine the following:

a) the leg stress under launch,
b) the worst axis 1-g deflection,
c) the bending moment under thermal soak,
d) the leg stress under thermal soak, and
e) the critical buckling load.

Solutions:

a) The worst-case stress occurs in the lateral y axis. From Eq. (3.37), the launch stress is

$$\sigma = \frac{F}{A} = \frac{WG}{3b^2 \cos\theta},$$

$$\sigma = \frac{(5)(50)}{3(0.06^2)\cos(45)} = 32{,}700 \text{ psi}.$$

The data in Table 2.1 indicates that this is below the yield stress of Invar 36, but when a factor of safety of 1.25 is considered, it will exceed the allowable yield point and may be too high and unacceptable.

b) From Eqs. (3.31) and (3.38), the vertical deflection is inversely proportional to $1/(\sin^2\theta)$, and the lateral deflection is inversely proportional to $2/(\cos^2\theta)$. For the base angle of 45 deg, we see that the highest deflection is in the lateral (x, y) axes. Hence, from Eq (3.38),

$$Y = \frac{WL}{3b^2 E \cos^2 \theta}.$$

Substituting values, we find that

$$Y = 0.0002\,\text{in.},$$

which is less than the maximum requirement of 0.0003 in. and thus acceptable.

c) The maximum moment under a thermal motion of 0.020 in. is given by

$$M = \frac{b^4 E y}{2L^2}.$$

Substituting values, we find that

$M = 0.106$ in.-lb, which is less that the 1 in.-lb maximum requirement and thus acceptable.

d) The leg stress under a thermal motion of 0.020 in. is given by Eq. (3.41) as

$$\sigma = \frac{3Eby}{L^2}.$$

Substituting values, we find that

$$\sigma = 2950\,\text{psi}.$$

This is acceptably low and thus okay, being well below the allowable yield point value.

e) The critical buckling load is obtained from Eq. (1.63) as

$$P_{cr} = \frac{\pi^2 EI}{L^2},$$

which needs to be greater than the reactive force in the leg of Eq. (3.36). This force is

$$F = \frac{WG}{3 \cos \theta} = 118\,\text{lbs}.$$

$P_{cr} = 8.7$ lbs, which is $\ll 118$ lbs and thus is very unacceptable.

In the above example, we did not satisfy all of the requirements; to do so, we would need to change some parameters, if possible. One can do this by trial and error (varying length L and width b), but there is a better, graphical approach, which is discussed in the next section. In subsequent sections, we will look at a better way to make bipods meet all design criteria.

3.5 Timmy Curves

A graphical representation can be helpful to home in on a design that has conflicting and opposing requirements, as we have seen in the previous chapter. While not a new approach, I call this representation the "Timmy" curves, named after a high school freshman who learned this from his algebra teacher.

The Timmy curves are a series of equations that are plotted using two variables on the x and y axes, where the equations give conflicting requirements; i.e., where a high value of a variable in one set of equations is preferred to be a lower value in another set of equations. We can use the Timmy curves to solve for an optimum solution to the bipod equations of the previous section, where the variables of width and length are in conflict. This is done by solving for one of the variables in terms of the other from the equations, and determining the range of allowable values, with a series of curves and arrows indicating greater-than or less-than acceptable.

We show this now using bipod design examples with requirements to meet acceleration strength, critical buckling, self-weight deflection, assembly or thermal motion moment to the optic, and assembly or thermal motion stress. We take the appropriate equations previously presented and solve for the leg width b as a function of its length L for a given material and base angle, while specifying required displacements, moments, and stresses.

We thus find from Eq. (3.38) that

$$b > \sqrt{\frac{WL}{3EY\cos^2\theta}} \quad \text{(self-weight deflection, lateral axis)}, \tag{3.42}$$

from Eqs. (3.30) and (3.36) that

$$b > \sqrt[4]{\frac{4WGL^2}{p^2E\cos\theta}} \quad \text{(critical buckling)}, \tag{3.43}$$

from Eq. (3.37) that

$$b > \sqrt{\frac{WG}{3\sigma\cos\theta}} \quad \text{(lateral axis launch stress)}, \tag{3.44}$$

from Eq. (3.40) that

$$b < \sqrt[4]{\frac{2ML^2}{EY}} \quad \text{(thermally induced motion moment to optic)}, \tag{3.45}$$

and from Eq. (3.41) that

$$b < \frac{\sigma L^2}{3EY} \quad \text{(thermally or assembly-induced motion stress in leg).} \quad (3.46)$$

We can now plot b versus L for the requirements using a simple spreadsheet program. Once plotted, upward-pointing arrows are used to indicate acceptability where length and width need to be greater, and downward-pointing arrows are used where length and width need to be less. Areas of closure can then be realized for acceptable limits. This is illustrated in the examples below.

3.5.1 Examples

Example 1. Consider again the bipod of Fig. 3.7, where in this case requirements dictate that it must support a 5-lb optic under a launch acceleration load of 50 g independently in all orthogonal directions. The maximum lateral self-weight deflection to meet a fundamental frequency requirement is specified at 0.0003 in. The maximum bending moment to the optic is to be limited to 2 in.-lb when thermal expansion of the bezel induces a 0.010-in. radial motion. If the bipod is made of Invar 36, determine appropriate values of leg width and length to meet all requirements of stress, defection, thermal motion, and buckling. We assume an allowable stress for launch of 24,000 psi and an allowable stress for thermal stress of 8000 psi. Assuming that these two stresses occur simultaneously, the combined stress is still below material yield, with an appropriate 1.25 safety factor. The chosen base angle is 45 deg.

Using Eqs. (3.42) through (3.46), we plot the data in the curves of Fig. 3.10, where the cross-hatched area shows where all requirements have been met. Thus, we can choose a width and length as seen fit, and we have done so without the need for tedious trial and error. This is rather nice.

Example 2. Next, consider the example in the previous section for the bipod of Fig. 3.8, which did not meet all requirements for a chosen width (0.06 in.) and length (5 in.). Let us thus treat these as variables. Again, in this case, the bipod must support a 5-lb optic under a launch acceleration load of 50 g independently in all orthogonal directions. The maximum self-weight lateral deflection to meet a fundamental frequency requirement is specified at 0.0003 in. The maximum bending moment to the optic is to be limited to only 1 in.-lb when thermal expansion of the bezel induces a 0.0200-in. radial motion. If the bipod is made of Invar 36, determine appropriate values of leg width and length to meet all requirements of stress, deflection, thermal motion, and buckling. We assume an allowable stress for launch of 24,000 psi and an allowable stress for thermal stress of 8000 psi. Assuming

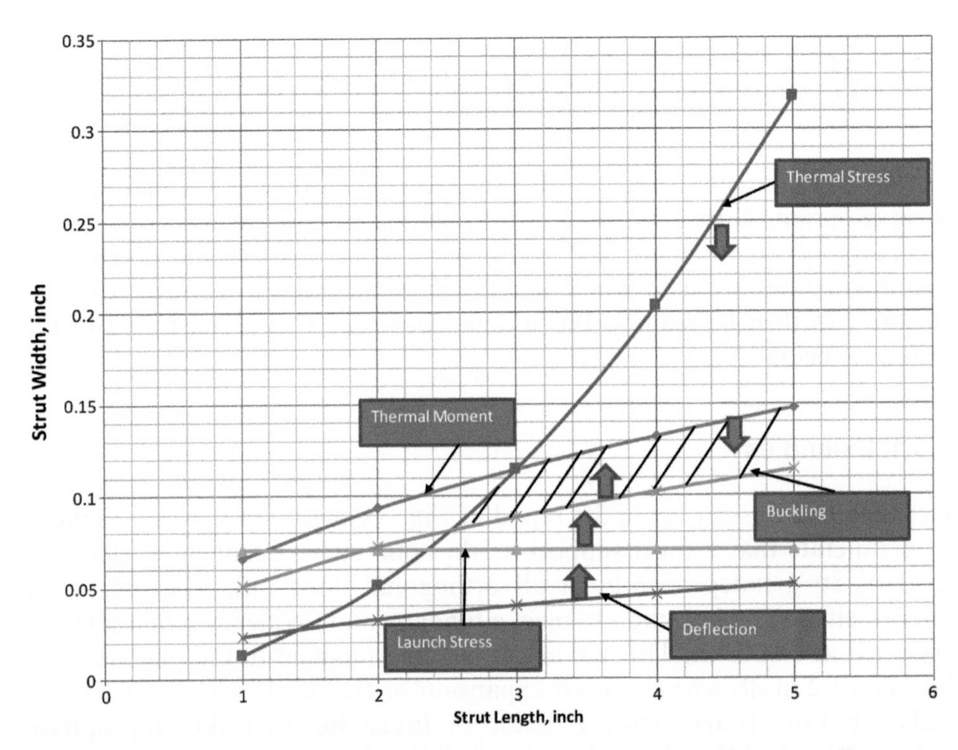

Figure 3.10 Bipod strut width versus length: Timmy curves for Example 1. The cross-hatched area indicates the solution space that meets all requirements concerning buckling, deflection, stress, and performance.

that the two stresses occur simultaneously, the combined stress is still below the material yield, with an appropriate 1.25 safety factor. The chosen base angle is 45 deg.

Using Eqs. (3.42) through (3.46), we plot the data in the curves of Fig. 3.11. Here, unlike Example 1, we find no common area where all requirements are met. While we can meet most requirements, the buckling requirement precludes an appropriate design solution. Thus, we cannot choose a width and length, and appear to be stuck between a rock and hard place. However, we can work around this enigma using a better way to make a bipod, as discussed in the next section.

3.5.2 Other effects

Other parameters besides buckling, deflection, stress, and performance can be plotted on these curves as well. These additional parameters include, for example, assembly tolerance, which can often exceed the thermal strain, and fatigue stress limits under multiple cyclic loading, the latter of which is discussed in Chapter 11.

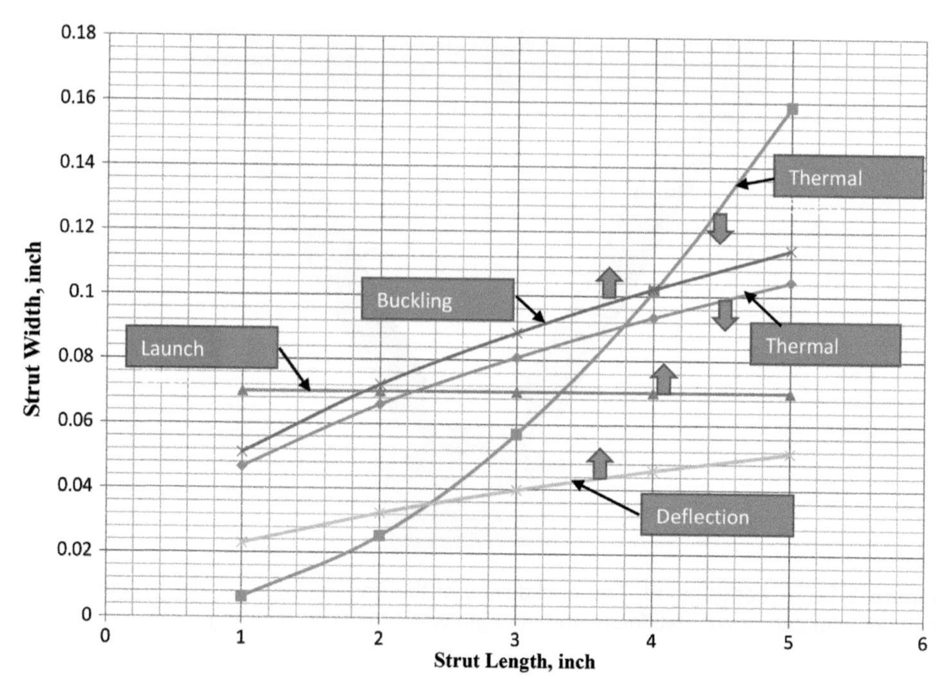

Figure 3.11 Bipod strut width versus length: Timmy curve for Example 2. There is no area of closure for a solution that meets all requirements of buckling, deflection, stress, and performance.

3.6 A Better Bipod

In Example 2 of Section 3.5.1, we saw that the bipod design does not meet all requirements (regardless of the cross-section or length parameters) for the given design criterion. In this case, the buckling requirement drove the design. This can be overcome by the bipod design of Fig. 3.12, where the midsection of the bipod leg length is widened while the end sections remain relatively thin. This has the effect of not only increasing buckling resistance, but also improving self-weight deflection, and hence providing improved stiffness and higher fundamental frequency. If the midsection width is made substantially high, then the bipod length can be considered as the sum of only two short-end lengths, for the purpose of calculating deflection and buckling resistance. Note that the launch stress is unchanged, being a function of only the thin width and independent of length. Making a wider midsection comes at the expense of compliance; however, such compliance is not substantially reduced, since the total length, which includes the midsection length, assists in this regard, as we shall see.

3.6.1 Analysis: constrained degrees of freedom

Considering the midsection to be rigid, which is a good approximation when its width is three to five times the width of the thinner section, we can rewrite

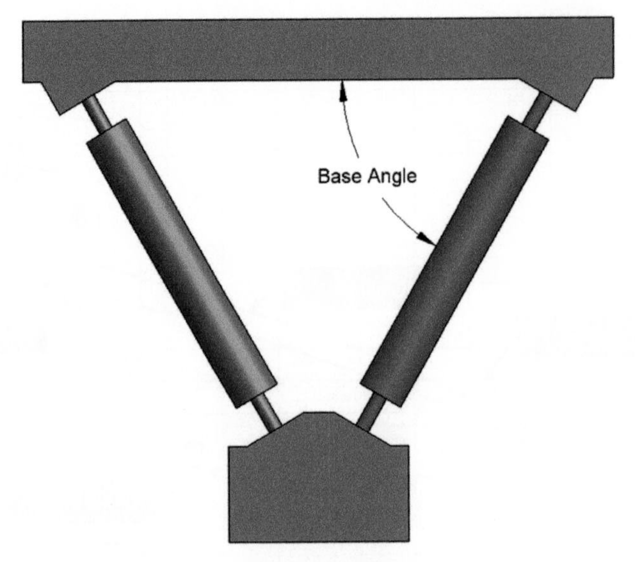

Figure 3.12 A better bipod. The neckdowns of the end sections provide compliance by bending, yet can provide stiffness and strength while resisting buckling.

our parametric equations for the various criteria. We substitute the value of the thin section length sum $2l$ for the total length that we used before.

For stiffness and strength, the equations become (for load in the vertical direction):

$$F = \frac{WG}{6 \sin \theta}, \tag{3.47}$$

$$\sigma = \frac{WG}{6b^2 \sin \theta}, \tag{3.48}$$

$$P_{\text{cr}} = \frac{\pi^2 EI}{L^2} = \frac{\pi^2 Eb^4}{12L^2} > \frac{WG}{6 \sin \theta}, \tag{3.49}$$

$$Y = \frac{WL}{6AE \sin^2 \theta} = \frac{Wl}{6b^2 E \sin^2 \theta}, \tag{3.50}$$

where now $L = 2l$.

When loaded with the optic weight in the x or y axis,

$$F = \frac{0.577WG}{2 \cos \theta} \quad (x\,\text{axis}), \tag{3.51}$$

$$\sigma = \frac{F}{A} = \frac{0.577WG}{2b^2 \cos \theta} \quad (x\,\text{axis}), \tag{3.52}$$

$$F = \frac{WG}{3\cos\theta}(y\,\text{axis}), \tag{3.53}$$

$$\sigma = \frac{F}{A} = \frac{WG}{3b^2\cos\theta}(y\,\text{axis}). \tag{3.54}$$

Equations (3.51) through (3.54) are unchanged from before, as length does not enter. For buckling, we have

$$P_{cr} = \frac{\pi^2 EI}{4l^2} = \frac{\pi^2 Eb^4}{48l^2} > \frac{0.577WG}{2\cos\theta} \quad (x\,\text{axis}), \tag{3.55a}$$

$$P_{cr} = \frac{\pi^2 EI}{4l^2} = \frac{\pi^2 Eb^4}{48l^2} > \frac{WG}{3\cos\theta}(y\,\text{axis}). \tag{3.55b}$$

Note that Eq. (3.55) is derived from Eq. (1.63) using the combined length size $2l$ of the leg strut end. The same answer is realized from Eq. (1.62) (free end) using only one end strut length l.

For deflection,

$$Y = \frac{2WI}{3b^2E\cos^2\theta}\text{in both the } x \text{ and } y \text{ axes.} \tag{3.56}$$

Rewriting in terms of width and length, as before, we have

$$b > \sqrt{\frac{2Wl}{3EY\cos^2\theta}} \quad \text{(self-weight deflection)}, \tag{3.57}$$

$$b > \sqrt[4]{\frac{16WGl^2}{p^2E\cos\theta}} \quad \text{(critical buckling, } y \text{ axis)}, \tag{3.58}$$

$$b > \sqrt{\frac{WG}{3\sigma l\cos\theta}} \quad \text{(launch stress)}. \tag{3.59}$$

3.6.2 Analysis: compliant degrees of freedom

Determination of compliance for the bipod is not as straightforward as before due to the variable cross-section, in spite of the rigidity of the midsection. While it is still a guided cantilever, we need to use superposition principles to determine force P and moment M under assembly or thermal motion. After tedious calculation, the proof of which is readily demonstrated but too large to be contained in the margins of this book, we find that, defining the rigid midsection length as

$$L' = L - 2l, \tag{3.60}$$

$$P = \frac{6EIY}{4l^3 + 6L'l^2 + 3lL'^2}, \tag{3.61}$$

$$M = \frac{P(2l + L')}{2}. \tag{3.62}$$

For a square cross-section of width b in the thin section,

$$P = \frac{Eb^4 y}{(8l^3 + 12L'l^2 + 6lL'^2)}, \tag{3.63}$$

$$M = \frac{Eb^4 Y(2l + L')}{(16l^3 + 24L'l^2 + 12lL'^2)}. \tag{3.64}$$

The bending stress in the leg is

$$\sigma = \frac{Mc}{I} = \frac{3EbY(2l + L')}{(8l^3 + 12L'l + 6lL'^2)}. \tag{3.65}$$

Rewriting Eqs. (3.64) and (3.65) in terms of width and length for use in Timmy curve plots, we have for the thermally or assembly-induced motion moment,

$$b < \sqrt[4]{\frac{M(16l^3 + 24L'l^2 + 12lL'^2)}{EY(2l + L')}}, \tag{3.66}$$

and for the thermally or assembly-induced motion stress,

$$b < \frac{\sigma(8l^3 + 12\mathrm{L}'\mathrm{l}^2 + 6lL'^2)}{3EY(2l + L')}. \tag{3.67}$$

3.6.3 Example for reconsideration

Let us now re-examine the problem of the example of Section 3.4.2.1 (where no solution was found for the straight, thin-leg bipod) in light of the better bipod approach. To reiterate the problem, consider the bipod mount design of Fig. 3.12. Requirements dictate that it must support a 5-lb optic under a launch acceleration load of 50 g independently in all orthogonal directions. The maximum self-weight deflection to meet a fundamental frequency requirement is specified at 0.0003 in. The maximun bending moment to the optic is to be limited to 1 in.-lb when thermal expansion of the bezel induces a 0.020-in. radial motion. If the bipod is made of Invar 36, develop a design chart of thin section width versus length to meet all requirements of acceleration stress, deflection, buckling, and leg thermal moment and leg thermal stress. Compare the results to those of the straight-leg approach of Fig. 3.7. We assume an allowable stress for launch of 24,000 psi and an allowable stress for thermal stress of 8000 psi.

Assuming that these two stresses occur simultaneously, the combined stress is still below the material yield, with an appropriate safety factor of 1.25. Use identical total leg lengths of 5 in.

Rather than using trial and error, we use the Timmy curve approach to home in on a range of solutions. We use Eqs. (3.60) through (3.67) to plot the thin section width b versus its length l. The total leg length is 5 in. such that $L' = 5 - 2l$.

The results of these cases are shown in the Timmy curve of Fig. 3.13. Note that now we do find an area of closure, with quite a large range of options. We have increased the launch strength twofold and buckling resistance sixfold, with only a small (on the order of 25%) increase in compliance, allowing for an appropriate solution. If the envelope is limited, the better bipod approach allows for shorter lengths, which is an added benefit. Note that the cross-section can be made either circular or square to achieve the desired results; however, manufacturing ease dictates the square approach.

Of course, we have not exhausted the trade space; one can trade varying thermal and launch loads, base angle, envelope, and even material, if required, provided that all other requirements presented in this text (some of which are have not been presented yet) are met.

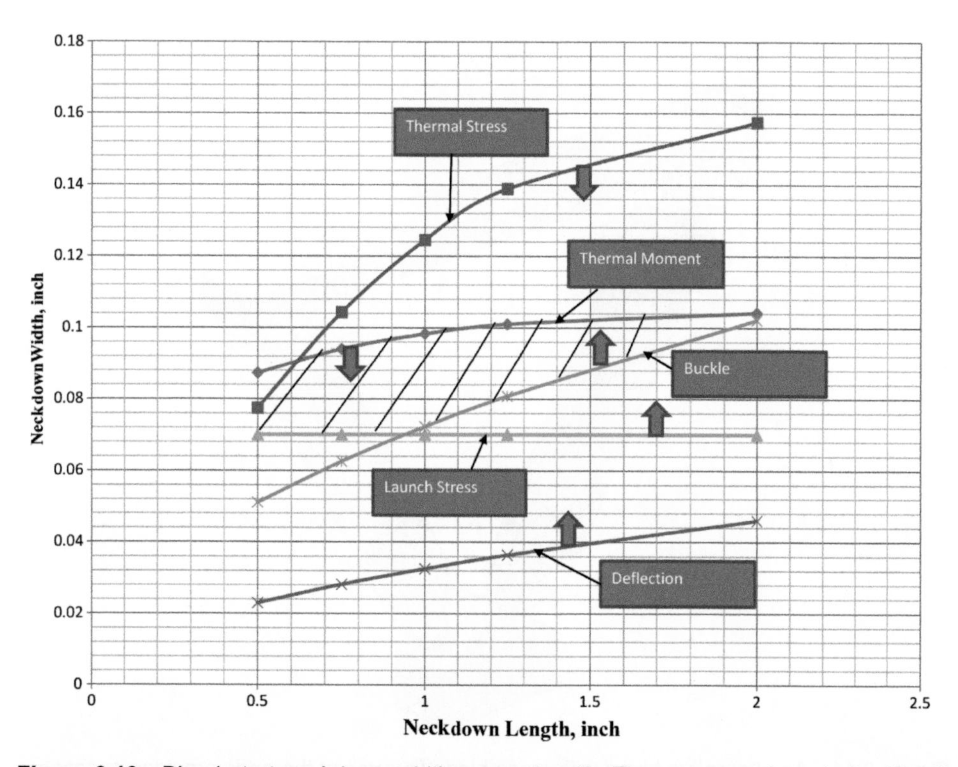

Figure 3.13 Bipod strut neckdown width versus length: Timmy curves for a better bipod than the example in Section 3.4.2.1. All requirements are now met, with a wide trade space indicated by the cross-hatched area.

3.7 An Alternative Bipod

Now that we have a "better" bipod, consider the alternative bipod of Fig. 3.14. Here, we achieve the required flexibility and stiffness at each end by using a set of cross-blade flexures. One blade of each set is compliant in one direction, while the other is stiff in that direction. If we now carry out the calculations previously set forward, the flexibility is improved somewhat over that presented in Section 3.4, with no sacrifice in stiffness. There is, however, a sacrifice in buckling resistance, so, again, a detailed trade will need to be conducted. As is evident in the figure, manufacturing difficulty increases in this approach, warranting a further trade. Sometimes, better is not best.

3.8 Stroke Algorithm

Since bipods are six-degree-of-freedom, near-kinematic systems, they can be employed for phasing and alignment through near-kinematic rigid body motion. This can be done, for example, by incorporating mechanical actuators into the leg struts,[2] as seen in Fig. 3.15. By applying motion along the axes of these struts, degrees-of-freedom motion can be realized without the use of moving parts (the neckdown portion of the strut will bend during such motion).

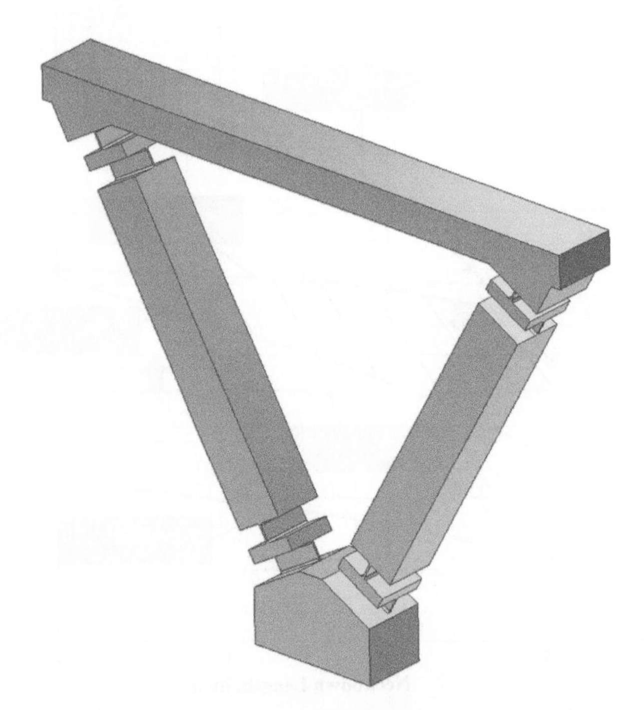

Figure 3.14 An alternative bipod with cross-blade flexures at the ends provides necessary compliance and stiffness and can improve performance; however buckling must be addressed.

Figure 3.15 A bipod mount scheme configured to provide motion with six independent degrees of freedom without inducing significant load to a supported optic or structure, achieved by bending metal without the use of ball joints (reprinted from Ref. 2 with permission).

Consider, for example, the bipod system of Fig. 3.8, where each bipod is symmetrically separated by 120 deg. Let x, y, z, Φ_x, Φ_y, and Φ_z be the desired global motions. Let Δ be an individual actuation extension along a given bipod leg. The maximum independent motions are

$$x = \frac{\Delta}{\sin\phi \cos 30\ \text{deg}},$$

$$y = \frac{\Delta}{\sin\phi},$$

$$z = \frac{\Delta}{\cos\phi},$$

$$\phi_x = \frac{z}{R\cos 30\ \text{deg}} = \frac{\Delta}{R\cos\phi\cos 30\ \text{deg}},$$

$$\phi_y = \frac{z}{R} = \frac{\Delta}{R\cos\phi},$$

$$\phi_z = \frac{\Delta}{R\sin\phi},$$

where Φ is the included half-angle at the bipod vertex.

Table 3.1 Stroke algorithm showing motions required for each bipod strut to achieve rigid body motion in any of six independent degrees of freedom.

Motion Required for In Strut Axis to Achieve Unit Global Displacement or Rotation						
Global Motion Actuator	x	y	z	Φ_x	Φ_y	Φ_z
1	0	+sinΦ	cosΦ	0	−RcosΦ	+RsinΦ
2	0	−sinΦ	cosΦ	0	−RcosΦ	−RsinΦ
3	−CsinΦ	−SsinΦ	cosΦ	+CRcosΦ	+SRcosΦ	+RsinΦ
4	+CsinΦ	+SsinΦ	cosΦ	+CRcosΦ	+SRcosΦ	−RsinΦ
5	+CsinΦ	−SsinΦ	cosΦ	−CRcosΦ	+SRcosΦ	+RsinΦ
6	−CsinΦ	+SsinΦ	cosΦ	−CRcosΦ	+SRcosΦ	−RsinΦ
	C = cos 30 deg		S = sin 30 deg		R = Radius	

Each individual actuator motion produces a coupled global rigid body motion. Judicious motion of individual actuators can produce an uncoupled motion, i.e., unit pure displacement or rotation in each of the six degrees of freedom (including rotation about the z axis, which is not intuitively obvious). This is shown by the complete motion algorithm of Table 3.1. The positive sign indicates an up stroke, while the negative sign is a down stroke. For example, to achieve pure piston z motion, each actuator is moved upward by a value of cos Φ. To achieve pure z rotation, each actuator is moved alternately up and down by an amount R sin Φ.

References

1. Riverhawk Company, New Hartford, New York.
2. Physik Instrumente USA, Auburn, Massachusetts.

Chapter 4
Solid Optics: Performance Analysis

4.1 Wavefront Error and Performance Prediction

The mount studies of Chapter 3 were presented to show the niceties of quasi-kinematic mounts to limit mount-induced forces to the critical structures to which they are attached. For mirrors, particularly reflective mirrors, which are quite sensitive to minute deformation errors, this is of paramount importance.

In many areas of engineering, such as building design, strength is often more critical than deflection. For example, a structural support beam of a warehouse may be allowed to deflect a good portion of one inch over a specified span, as long as it does not crack the plaster in the ceiling below (or cause one to be sea sick!). The beam is strength limited to support high traffic and furniture loads. For optics, deflections are often limited not to one inch but to a millionth of an inch! Strength needs to be checked as well, but deformation is critical.

Designs of high-acuity optics often demand near-diffraction-limited performance, resulting in overall system errors on the order of one-tenth of a wavelength of light. If a system comprises several optics with an associated support structure, as in an optical telescope assembly, each optic and support must be allocated a fraction of the already miniscule system performance budget. The method to fulfill this requirement will be illustrated.

Optical performance deformation is measured as peak-to-valley (P-V) wavefront error (WFE) over the surface, or departure from its desired shape. To compute WFE when deformation is given in inches or meters, we need to divide by the wavelength of light. For the visible spectrum, we choose a wavelength of 0.6328 μm (25 μin.), the helium-neon (HeNe) wavelength used during laser interferometry (optical engineers will adjust for lower and higher ends of the visible spectrum). Mirror deformation results in surface error. For reflective optics, since light will return, the optical path difference is twice the surface error and is known as the wavefront error. Since performance over the full area of the optic, not simply over local peaks, is key, P-V displacements are generally converted to a broad average over the full aperture, referred to as the root-mean-square (RMS) wavefront error. This is given as the square root of the difference between the average of squared wavefront deviations minus the square of the average wavefront deviation. Mathematically, this is expressed as

$$\text{RMS} = \sqrt{\frac{1}{N}\sum_{i}^{N} z_i^2 - \bar{z}^2}, \qquad \bar{z} = \frac{1}{N}\left(\sum_{i}^{N} z_i\right), \tag{4.1}$$

where the subscript denotes the individual displacement of a given point on the surface. The RMS is thus the standard deviation, or the square root of variance, statistically speaking. Since it is derived from squared values, the RMS WFE is independent of the sign of the P-V wavefront deviations and is thus always positive. The RMS WFE must be calculated for a large number of discrete values on the optical surface in order to obtain reliable information.

The RMS WFE is thus related to the P-V WFE error according to Eq. (4.1). The RMS error value is dependent on the shape of the deformation over the optic aperture. It is often convenient to break down the deformed shape into a series of orthogonal polynomials, such as given by a Zernike[1] polynomial representation. While a detailed discussion of these polynomials is beyond the intent of this text, it is sufficient here to say that, because of their orthogonality, one can easily combine shapes and remove certain aberrations without affecting the other aberrations. A particular deformed shape can be broken down into an infinite number of Zernike polynomials, the first dozen or so being all that are required in most applications.

The focus term, also known as power, is a spherical shape and is quite useful, particularly when focus errors can be removed; from an opto-structural standpoint, focus error often occurs under thermal gradients (Section 4.5). The astigmatism terms, known as "potato chip" or "saddle" shape, are often produced by alignment error. From an opto-structural standpoint, astigmatism often occurs in a gravity (Section 4.3) or temperature field, or during fabrication and assembly, as its shape is a common form requiring least energy to obtain. Coma, or "S shape," often occurs when a secondary optic moves laterally (decenter) with respect to the primary optic. Tricorn, also called trefoil, is a three-point shape that often occurs during quasi-kinematic mounting of optics. Finally, spherical aberration, or "sailor hat shape," is often a bi-product of focus change and can wreak havoc with an error budget as it is not readily removed. From an opto-structural standpoint, such spherical aberration often occurs with optics having high curvature and aspect (diameter-to-thickness) ratio (Chapter 7).

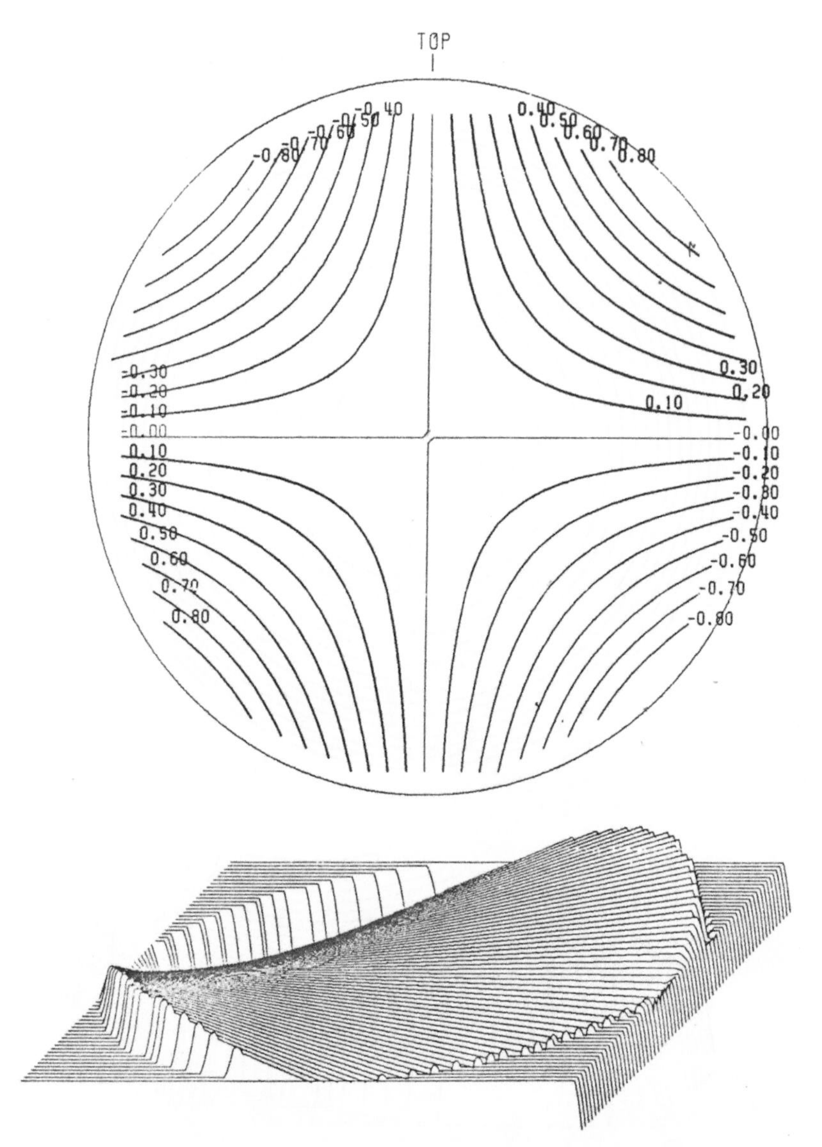

Figure 4.1(a) Zernike contour for astigmatism C(2,−2), $R^2\sin 2\theta$ (saddle shape). A $\cos 2\theta$ C(2,2) variation will produce astigmatism that is 45 deg relative to this contour.

These aberrations are shown pictorially in Figs. 4.1(a) through (e), while Table 4.1 shows the RMS error [obtained from Eq. (4.1)] relative to the P-V deformation. Note that values of the P-V to RMS ratios range from about 3.4 to 5.6 (they will be larger for higher orders). When a shape breakdown is not known *a priori*, a value between 4 and 5 is commonly used.

Figure 4.2 shows a typical performance budget for an optical telescope assembly. The initial error budget is prepared as a "top-down" performance prediction based on experience. However, with the techniques set forth in this

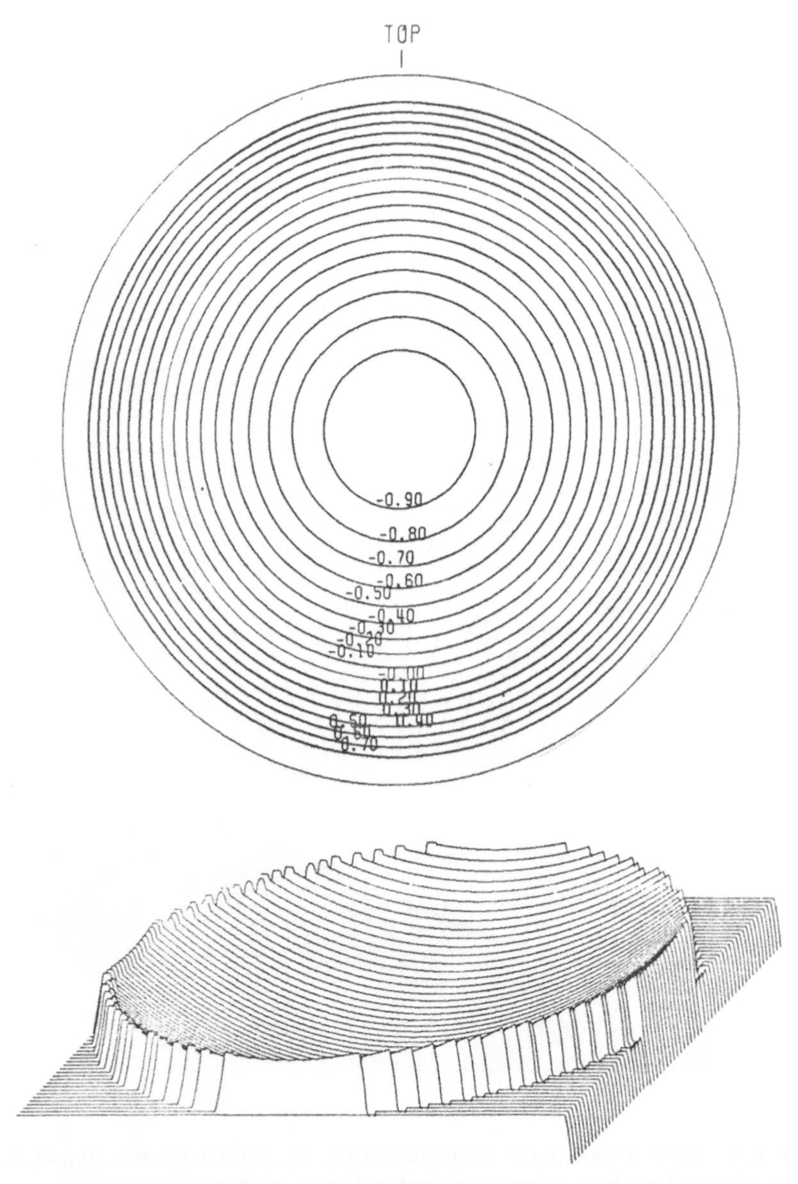

Figure 4.1(b) Zernike contour for focus astigmatism C(2,0), $2R^2 - 1$ (spherical shape).

text, the opto-structural engineer can assist at the beginning with a "bottom-up" approach for the items that can be controlled from an environmental, manufacturing, and assembly standpoint by using first-order principles.

In the sample budget, we assume a top-level error of 0.135 wave RMS in the visible spectrum. Design residual is dictated by the particular design set by the optical engineer and is the minimum error set by the prescription itself. Alignment error is based on optical sensitivities of the system, and

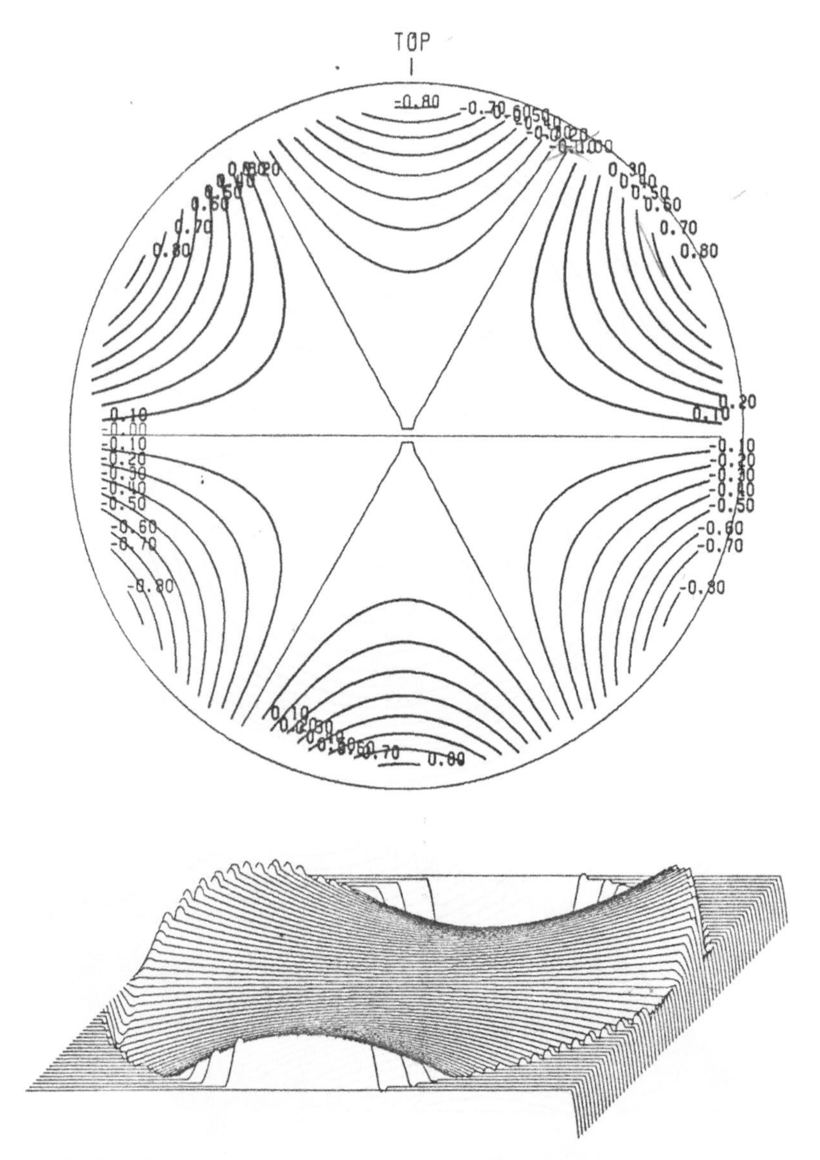

Figure 4.1(c) Zernike contour for tricorn C(3,−3), $R^3\sin 3\theta$ (three-point). A $\cos 3\theta$ C(3,3) variation will produce tricorn that is 60 deg relative to this contour.

manufacturing error in part is set by how well an optic can be figured. The environmental portion is in the hands of the opto-structural engineer, who allocates error from gravity, mounting, polish, thermal soak, thermal gradient (with the aid of thermal engineering), thermal expansion homogeneity, moisture growth, epoxy effects, cladding and coat error, vibration, and so forth. These source calculations are discussed later in the text.

Note that most of the errors are not systematic; i.e., they will exhibit different aberration shapes. As such, errors are not linearly additive but rather

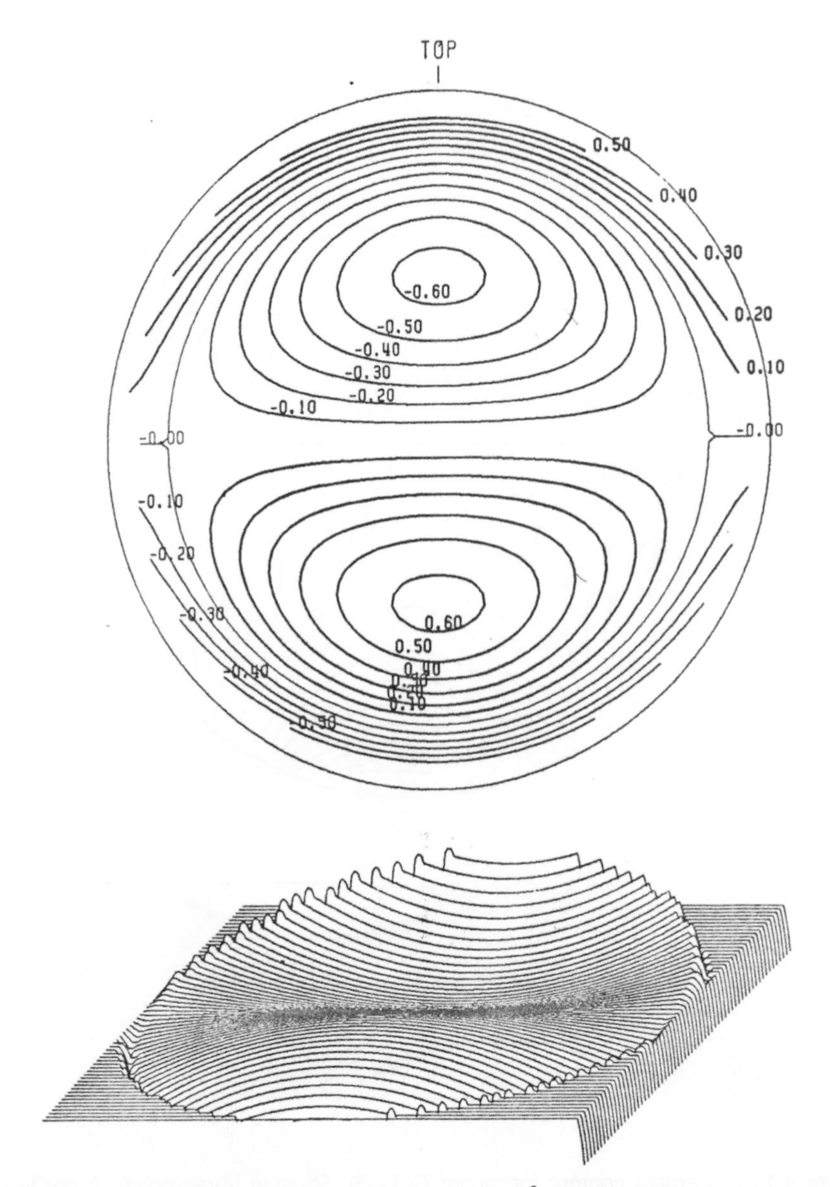

Figure 4.1(d) Zernike contour for coma C(3,−1), ($3R^3 - 2R$)sin2θ (S-shape). A cos2θ C(3,1) variation will produce coma that is 90 deg relative to this coutour.

can be summed using the root-sum-square (RSS) technique. As implied, each error is squared and summed with the other errors, and the square root of the sum of the squares is taken. Mathematically, the top-level error is thus given as

$$\mathrm{RSS} = \sqrt{\sum_{i}^{N} E_i^2}, \tag{4.2}$$

where E is the individual design error of interest.

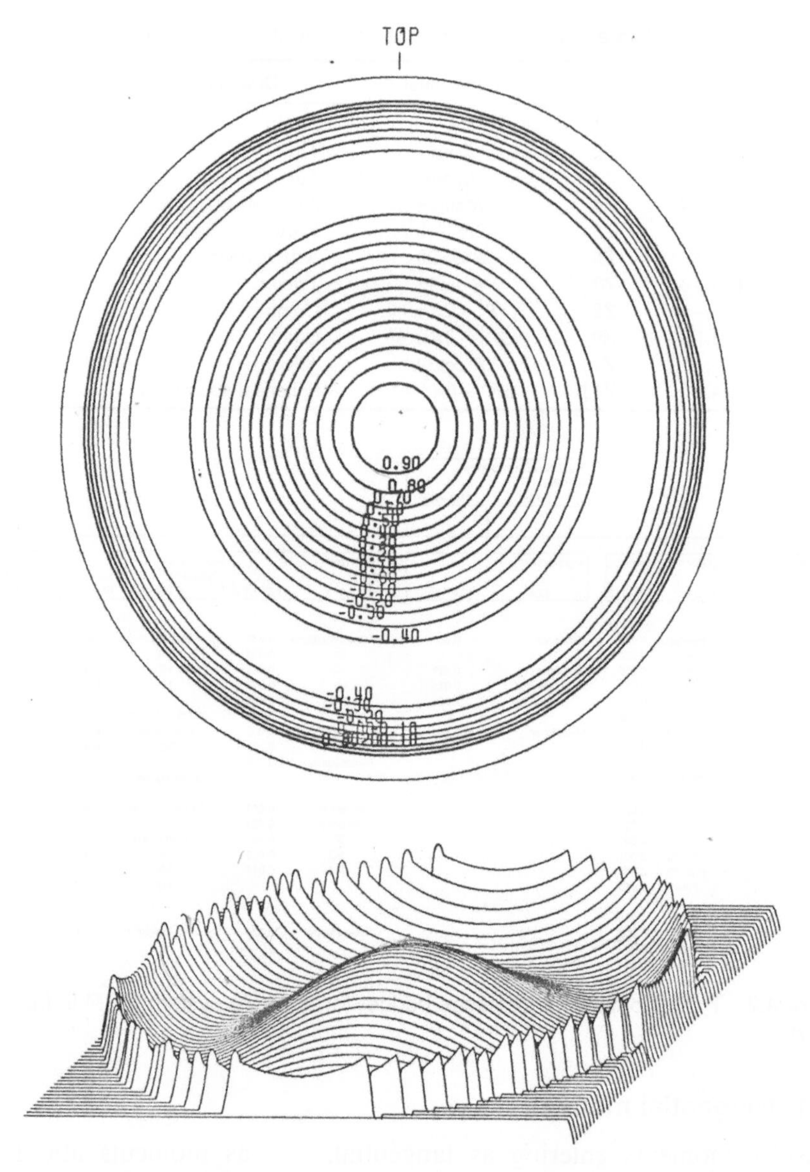

Figure 4.1(e) Zernike contour for spherical aberration C(4,0), $6R^4 - 6R^2 - 1$ (sailor hat shape).

4.2 Mount-Induced Error

In Chapter 3 we saw how quasi-static kinematic mounts can be designed to limit the bending moment to low values for critical optical structures and mirrors both during assembly and under operational environments. We now review the sensitivity of optics to such moments. These moments can be tangential, radial, or axial, depending on the assembly or environmentally induced motion.

Table 4.1 Select Zernike coefficient relationships.

Order	Term		Polynomial	Description	Peak/RMS Ratio
0	C 0,0	Z1	1	Piston	N/A
1	C 1,−1	Z2	$R\sin\theta$	Tip	$2\sqrt{4} = 4$
	C 1,1	Z3	$R\cos\theta$	Tilt	$2\sqrt{4} = 4$
2	C 2,−2	Z4	$R^2\sin 2\theta$	Astigmatism	$2\sqrt{6} = 4.9$
	C 2,0	Z5	$2R^2 - 1$	Focus	$2\sqrt{3} = 3.46$
	C 2,2	Z6	$R^2\cos 2\theta$	Astigmatism	$2\sqrt{6} = 4.9$
3	C 3,−3	Z7	$R^3\sin 3\theta$	Tricorn	$2\sqrt{8} = 5.66$
	C 3,−1	Z8	$(3R^3 - 2R)\sin\theta$	Coma	$2\sqrt{8} = 5.66$
	C 3,1	Z9	$(3R^3 - 2R)\cos\theta$	Coma	$2\sqrt{8} = 5.66$
	C 3,3	Z10	$R^3\cos 3\theta$	Tricorn	$2\sqrt{8} = 5.66$
4	C 4,0	Z13	$3R^4 - 6R^2 - 1$	Spherical aberration	$1.5\sqrt{5} = 3.35$

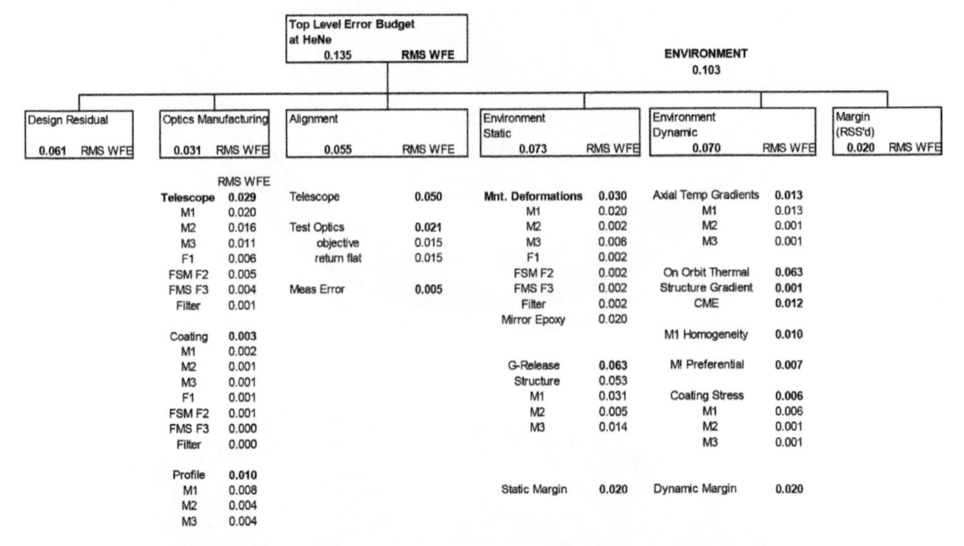

Figure 4.2 Typical optical performance budget for a visible system (FSM is fast-steering mirror).

4.2.1 Tangential moment

Consider moments entering as tangential; i.e., as moments about a line tangential to the circumference. Solutions for three discrete moments are not readily available. However, we can approximate as follows.

Consider the solid, circular optic of Fig. 4.3 with uniform and continuous moment of value M' applied tangentially along its edge, given in inch-pounds per inch of circumference. The deflection solution for a flat, circular plate is readily available and found as[2]

$$Y = \frac{6M'a^2(1 - \nu)}{Et^3}, \tag{4.3}$$

where a is the mirror radius, and t its thickness.

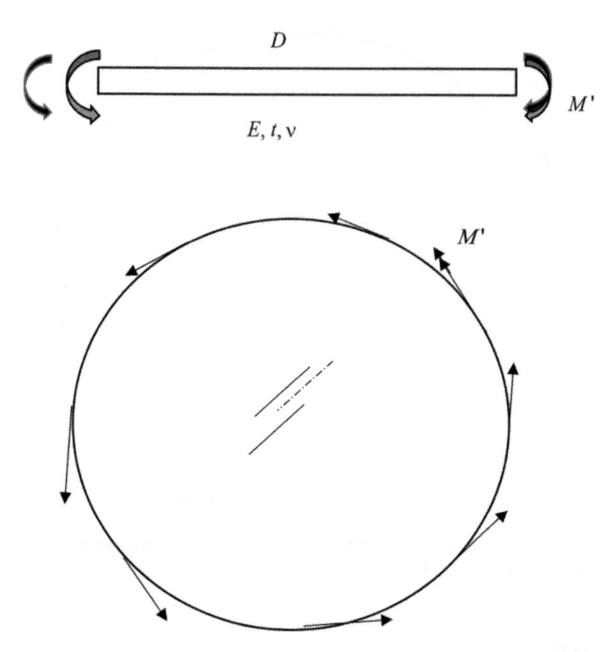

Figure 4.3 Solid optic with continuous-edge moment (inch-pounds per inch) applied tangentially over the circumference.

While the solution applies specifically to a flat plate, it also is a fair approximation for a typical mirror with curvature, as long as the curvature and aspect ratio are reasonably small. Typically, this would apply when the radius of curvature is greater than 2.5 times the diameter of the optic, and for an optic-diameter-to-thickness ratio (aspect ratio) of less than 15:1. Recall that we are looking at first-order analyses here.

For three-point discrete moments of value M each in inch-pounds (as shown in Fig. 4.4), we assume that the local moments are distributed over the circumference. Because the moment is local, the deformation will not be focusable (power) but rather will peak near the moment application, producing trefoil error. The "distributed" moment M' is thus

$$M' = \frac{3M}{2pa}. \tag{4.4}$$

Equation (4.3) becomes

$$Y = \frac{6M'a^2(1-\nu)}{Et^3} = \frac{2.86Ma(1-\nu)}{Et^3}. \tag{4.5}$$

This approximation fares well with finite-element model studies.

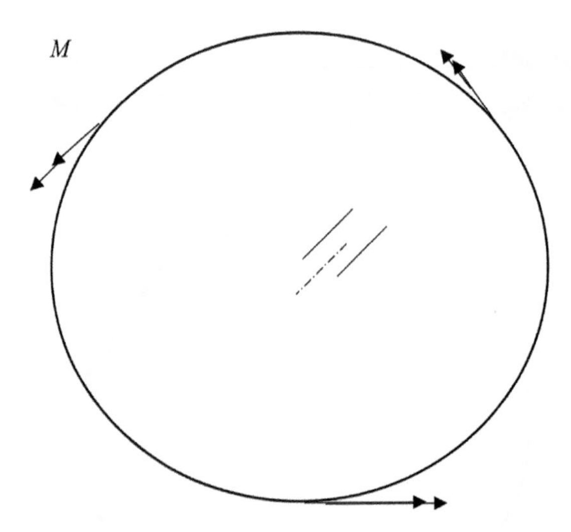

Figure 4.4 Solid optic with discrete tangential edge moments (in inch-pounds) applied at three equally spaced locations.

4.2.2 Radial load

When three-point radial force loads are applied, there is very little deformation for flat optics. For powered optics, a fair approximation can be made by computing the effective tangential moment and using Eq. (4.5). The effective moment is the radial load P_r multiplied by the optic sagitta e, where

$$e = a^2/2R_s, \tag{4.5a}$$

R_s is the mirror radius of curvature, and

$$M = P_r a^2/2R_s. \tag{4.5b}$$

4.2.3 Example for consideration

A circular, fused-silica optical component has a diameter of 20 in. and thickness of 2 in. (aspect ratio 10:1). It is subjected to three equally spaced (120 deg apart) tangentially applied moments of value 2 in.-lb each during bipod mount assembly, as calculated from the equations of Chapter 3 for a given mount design. Compute the optic WFE in visible light.

From Eq. (4.5) and the properties of Table 2.2, $Y = 5.6 \times 10^{-7}$ in. This is P-V surface error. To convert to WFE, multiply by 2 and divide by 2.5×10^{-5} waves per inch: $Y = 0.045\lambda$ P-V WFE visible. Since the shape is part trefoil, part local, part global and other higher-order error, it is reasonable to assume an approximate P-V RMS ratio of 5:1. Thus, $Y = 0.009\lambda$ RMS WFE visible.

Note that this value will likely well accommodate an error budget; if values were significantly higher, an iterative process involving both mount and mirror design is warranted.

4.2.4 Radial and axial moments

When the three-point moment enters about a radial line, a similar distortion results. Here, finite-element analysis indicates an approximate error of

$$Y = \frac{3.08Ma(1+\nu)}{Et^3}. \tag{4.6}$$

When the three-point moment enters about an axial line, there is very little deformation for flat optics, and curved optics deformation is somewhat lower than that resulting from the radial and tangential moments noted above.

Note that if a maximum error induced by loads and moments is desired for input to a performance prediction budget, one can calculate the allowable moment and use the quasi-kinematic mount equations of Chapter 3 to design an appropriate mount.

4.3 Gravity Error

Gravity is the enemy of mirrors. Whether designing high-acuity precision optics for ground, air, or space, the earth's gravity environment is of concern simply due to self-weight deflection.

For ground systems, optical telescopes may have to scan the skies, looking toward zenith and toward the horizon. While the gravity vector direction does not change, of course, the orientation of the optics relative to it does.

Similarly, for optics designed for aircraft use, cameras may be pointing down or toward the horizon for long-range photography or reconnaissance. Again, orientation is changing relative to the gravity field.

Finally, for optics designed for space use, where during orbit there exists, effectively, zero gravity (zero-g), the optic has to be first tested in the 1-g earth environment. In this case, it can be tested in an orientation that minimizes ground test error, to the extent possible, so as to not mask out other important errors to be measured. If this is insufficient, rotational flip tests can be conducted to determine the gravity effects, but such gravity back-out interferometry is costly and limited in range. Alternatively, the optic can be fabricated on a zero-g mount simulator, e.g., a mount simulator with support fluid pistons, or it can be tested with air bag support, both options of which are costly and involve their own errors.

Gravity is the enemy of optics. We now study deformation effects in a variable-orientation attitude relative to the gravitational field for various quasi-kinematic mount zones.

4.3.1 Optical axis vertical

4.3.1.1 Edge mount

With reference to Fig. 4.5(a), when the gravity vector is normal to the optic; i.e., parallel to the optical axis, and the optic is supported at the edge, the optic will deflect. Small optics may be continuously supported at the edge by continuous (or segmented) soft adhesives, such as silicones. In this case, the optic behaves as a simply supported system. For a circular mirror, its P-V error is given[2] as

$$Y = \frac{3(5+\nu)\rho\pi r^4(1-\nu)}{16\pi E t^2}. \tag{4.7}$$

The error is predominantly power [Zernike term C(2,0)].

Returning to our near-kinematic mount approach, relatively large reflective optics are often supported at three points. In this case, when the supports are located at the edge and are equally spaced,[3]

$$Y = \frac{0.434\rho\pi r^4(1-\nu)}{E t^2}. \tag{4.8}$$

Comparing Eqs. (4.7) and (4.8), and using a Poisson ratio of 0.25 by way of example, the three-point deflection is about 40% more than that afforded by the simple support boundary. The error shape is largely three point C(3,3) with some power [C(2,0)]. As such, after converting displacement to waves,

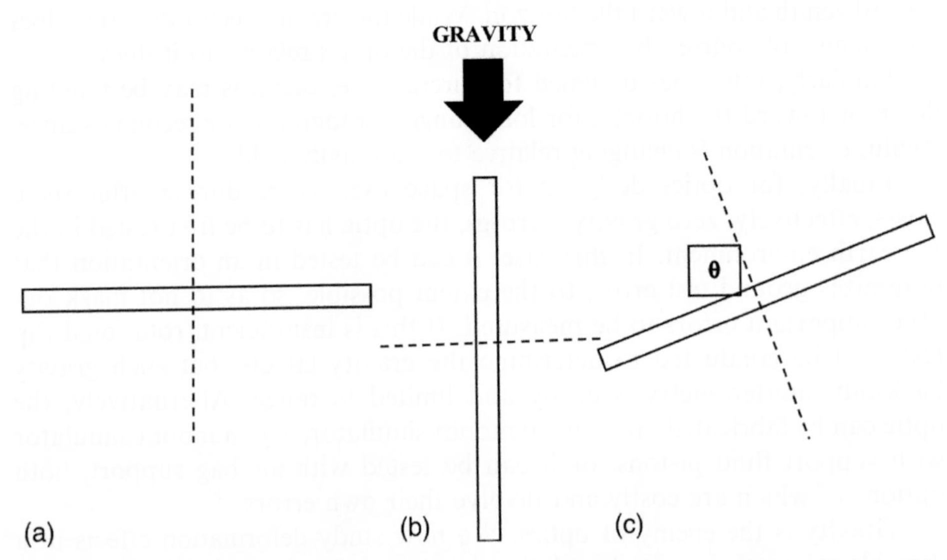

Figure 4.5 Gravity relative to the optical axis: (a) normal to the optic surface, (b) normal to the optical axis, and (c) inclined.

the RMS error is achieved by dividing by about 4.5. While these values apply specifically to flat optics, they are good approximations, again, for curved (powered) optics with the usual low aspect ratios.

4.3.1.2 Internal mount

When real estate and mass budgets allow, and if the optic is mounted at three points internally rather than at the edge, deflection due to gravity can be reduced. If the mount zone is set to near the 0.7 zone of radius, deflection is minimized (the zone of equal area is the 0.707 zone; the optimum support zone is a bit lower in radius, but local mount and area distribution effects render this type of precision unwarranted, and a setting near the 0.7 zone is sufficient. In this case, deflection is reduced nearly fourfold but will not exhibit any focus error, being partially of sailor hat shape [C(4,0)].

As the mirror support zone is moved toward the center, deflection is again increased. This is depicted in Fig. 4.6. The error tends to be more focusable, with an umbrella shape developing at the optimum zone of 0.4 radius. Here, error before focus removal is about twofold higher than that of the 0.7 zone case, but after focus, removal is about twofold lower. Bringing the zone nearer to the center increases error, as depicted. Mounts within the 0.4 zone are uncommon; instead, "mushroom"-mounted mirrors are often employed, as

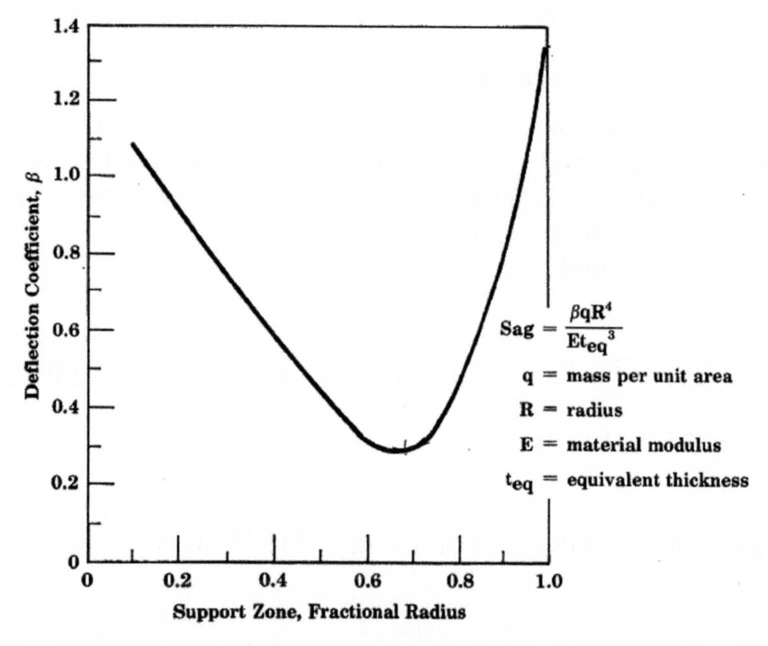

Figure 4.6 Gravitational sag as a function of support zone with a direction normal to the optic on a three-point mount with the optical axis vertical. For solid optics, the equivalent thickness t_{eq} is the thickness t of the optic, while the value of q is pt, its areal density.

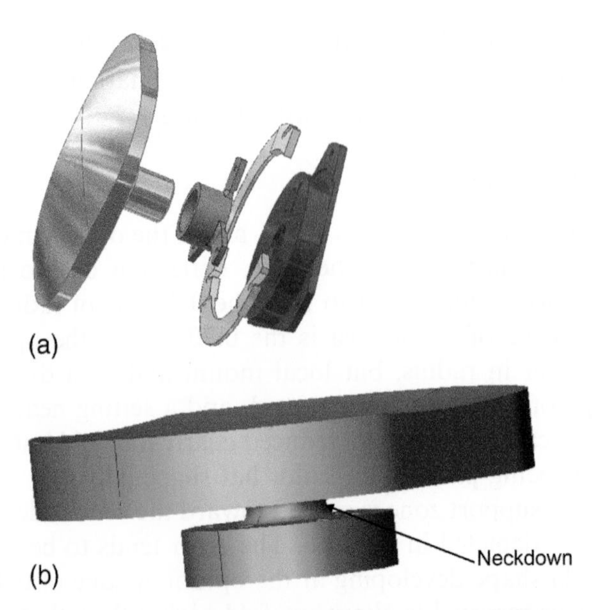

Figure 4.7 A "mushroom" mount: (a) optic bonded to foot, supplying some compliance for mounting, yet adequate stiffness; and (b) integral foot, requiring a neckdown for near-kinematic compliance traded against adequate stiffness and strength.

seen in Fig. 4.7. Note that as the zone approaches zero radius, we have a near-kinematic mount, affording precisely six stable degrees of freedom.

4.3.1.3 Example for consideration

A circular aluminum solid optic of 20-in. diameter and 2-in. depth (aspect ratio 10:1) is supported at three equally spaced kinematic edge points. Compute the RMS visible WFE in a gravity field with the optical axis vertical, i.e., gravity parallel to the optical axis.

From Eq. (4.8) and the properties of Table 2.2,

$$Y = Y = \frac{0.434\rho\pi r^4(1-\nu)}{Et^2} = 0.434(0.1)(\pi)(10)^4(1-0.33)/[(10^7)(2)(2)]$$
$$= 22.8 \times (10)^{-6}\,\text{in.}$$

Using the peak-to-RMS factor of 4.5, the RMS WFE is

$$Y_{\text{RMS}} = \frac{2Y}{4.5(2.5 \times 10^{-5})} = 0.405\lambda_{\text{RMS}}.$$

4.3.2 Optical axis horizontal

4.3.2.1 Edge mount

With reference to Fig. 4.5(b), when the gravity vector is normal to the optical axis (i.e., the optical axis is horizontal) in a gravitational field and the optic is supported at three points at the edge, the optic will deflect out of plane but to a different degree compared to the vertical orientation case.

If the optic is flat, the WFE is minimal (near zero), except for small Poisson effects. For optics with curvature, however, out-of-plane bending and, therefore, WFE occurs. While a theoretical solution to the optical axis horizontal problem is possible, a ready solution was accomplished through detailed finite-element modeling. In this approach, various aspect ratios and curvatures were studied, and a relationship was developed. For the three-point kinematic edge mount, we find the peak error as

$$Y = \frac{0.849\rho\pi R^5}{R_s E t^2}. \tag{4.9}$$

The error produced is largely astigmatic (Fig. 4.1). Thus, to obtain the RMS error, we divide by 4.9, as indicated in Table 4.1. Interestingly, the magnitude of the error is independent of the optic in-plane orientation relative to gravity. The angle of astigmatism, however, will change to a value of one-half of the angle of gravity orientation relative to the mounts. For example, if the optic is rotated 180 deg in plane, the astigmatic error magnitude is the same but opposite in sign.

The astigmatic error is reduced significantly if the mount is located at the mirror centroid. For edge mounting over the mirror thickness, the centroid is not at the edge-mount center for powered optics.

4.3.2.2 Internal mount

As the three-point mount zone is moved radially toward the center, the error decreases, as it did in the vertical axis case. The decrease is due to the mount zone approaching the mirror centroid (not its neutral surface), where error is minimal. The error is no longer astigmatic but is reduced more than fourfold from the edge-mount case. For many optics of typical curvature, this centroidal location is near the 0.7 zone. The ideal minimum error, which is quite sensitive to the centroid offset, is not readily achieved, so using a fourfold factor is reasonable. We thus find that the peak error is

$$Y = \frac{0.212\rho\pi D^5}{R_s E t^2}. \tag{4.10}$$

To obtain the RMS error, we divide by approximately 5.

4.3.2.3 Example for consideration

A circular aluminum solid optic of 20-in. diameter and 2-in. depth (aspect ratio 10:1) with a radius of curvature of 60 in. is supported at three equally spaced kinematic edge points. Compute the RMS visible WFE in a gravity field with the optical axis horizontal, i.e., where gravity is normal to the optical axis.

From Eq. (4.9) and the properties of Table 2.1, we have

$$Y = \frac{0.849\rho\pi R^5}{R_s E t^2} = 11.1 \times 10^{-6}\,\text{in.}$$

Using the 4.5 peak-to-RMS factor, the RMS WFE is

$$Y_{\text{RMS}} = \frac{2Y}{4.5(2.5 \times 10^{-5})} = 0.197\,\lambda_{\text{RMS}}.$$

4.3.3 Zero-gravity test

When the gravity error for orbital g release cannot be readily accommodated during ground test, it is best to test in the angular configuration giving the least error, which is most often with the optical axis horizontal. To determine this, we set Eq. (4.8) equal to Eq. (4.9) and solve for the mirror radius of curvature (with a Poisson ratio of 0.25) as

$$R_s = 2.6R. \tag{4.11}$$

This is a relatively fast system, so for most optics the ideal ground test is performed with the optical axis horizontal. For example, a Cassegrain system exhibits an F-number as

$$F = \frac{R_s}{4R}. \tag{4.12}$$

Substituting Eq. (4.11) into Eq. (4.12) gives

$$F = \frac{2.6R}{4R} = 0.65. \tag{4.13}$$

This is indeed a very fast mirror, so again, optical axis horizontal is the usual and better ground test configuration.

4.3.4 Other angles

We have investigated vertical and horizontal optical axis gravity deformation, and we now review what happens at other angles. As might be expected, the vertical error is reduced by the cosine of the angle off of vertical, and the horizontal error is reduced by the sine of the angle off of vertical. With reference to Fig. 4.5(c) and from Eq. (4.8), for the edge mount,

$$Y_z = Y_z \cos\theta, \tag{4.14}$$

and from Eq. (4.9),

$$Y_x = Y_x \sin\theta, \tag{4.15}$$

where the subscripts denote the axis of orientation.

The two errors will add; however, since the shapes are different (trefoil versus astigmatism), they are not systematic and do not add linearly but rather as root-sum-square (this has been verified by finite-element analysis and optical postprocessing). Thus, the off-angle error is given as

$$Y = \sqrt{(Y_z \cos\theta)^2 + (Y_x \sin\theta)^2}. \tag{4.16}$$

Note, out of curiosity only, that in the special case of curvature where vertical and horizontal errors are equal [$R_s = 2.6$ from Eq. (4.11)], we find Eq. (4.16) to be

$$Y = \sqrt{2Y^2(\cos^2\theta + \sin^2\theta)}, \tag{4.17a}$$

which, from the trigonometric identity, is

$$Y = Y\sqrt{2} \tag{4.17b}$$

and therefore is independent of angle.

4.3.5 Brain teaser

Assume that the optic of Example 4.3.1.3 is fabricated on a three-point edge mount such that, with the optical axis vertical and the optic face up, the gravity error is "polished in" to the optic, ideally resulting in zero error. If the mirror is then supported on three edge points, compute

a) the gravity sag when the mirror is mounted on the ground with optical axis vertical;
b) the gravity sag when the mirror is mounted in an aircraft with optical axis vertical and pointing downward (looking toward the ground directly below);
c) the gravity sag when the mirror is mounted in an aircraft with optical axis horizontal (looking toward the horizon); and
d) the gravity sag in a zero-g space orbit environment.

Solutions:

a) The error polished in to the mirror from Example 4.3.1.3 is 0.405 waves RMS. Thus, on three points, the error is zero because it has been polished in.
b) If the mirror is pointing downward, the error is *double* the original polished-in error, or 0.81 waves RMS.
c) If the mirror is pointing toward the horizon, the polished-in error is relieved according to cos $\theta = 0$ and thus reappears, in addition to the horizontal gravity sag. From Examples 4.3.1.3 and 4.3.2.3, and using Eqs. (4.8) and (4.9),

$$Y_{\text{RMS}} = \sqrt{(Y_z^{\,2} + Y_x^{\,2})} = 0.450 \text{ waves RMS.}$$

d) Since the 1-g error was polished in, it "pops out" in zero gravity. Thus, the error is 0.405 waves RMS.

4.4 Temperature Soak

All materials expand or contract under a uniform temperature change. The degree of expansion is dependent on the material's molecular structure. For ideal kinematically mounted structures, the expansion occurs without stress. For a beam, the expansion is given as

$$\Delta L = \alpha L \Delta T, \tag{4.18}$$

where α is the material CTE, and ΔT is the change in soak temperature. Most materials will expand upon heating and contract upon cooling; however, there are some that do just the opposite. When a material contracts upon heating or expands upon cooling, it is said to have a negative CTE.

We note that the strain is

$$\varepsilon = \frac{\Delta L}{L} = \alpha \Delta T. \tag{4.19}$$

For flat, circular optics, expansion is radial and is given as

$$\Delta r = \alpha r \Delta T. \tag{4.20}$$

For quasi-kinematic mounts, this expansion, relative to its supportive mount (through flexures, bipods, soft adhesive, and the like), needs to be accommodated to preclude excessive WFE.

Additionally, for curved optics, the mirror radius of curvature will change according to

$$\Delta R = \alpha R \Delta T. \tag{4.21}$$

This expansion changes the focal point of the optical assembly and causes focus error; however, such an error can be self-compensating if the metering structure (Chapter 13) to the other optics are of identical thermal expansion characteristics.

In any case, the value of the CTE needs to be properly assessed over the temperature range of interest because, in general, the CTE will not be constant over a given temperature range. At a given temperature, the CTE is called the instantaneous CTE, while over a range of temperatures, it is called the secant CTE, also known as the effective CTE.

Determination of the instantaneous and effective CTEs is made by precision measurements of expansion using probes such as linear variable displacement transducers (LVDTs), or even more-precise laser dilatometers, using interferometry or other methods, such as capacitor dilatometry. Precision to the level of parts per billion can be obtained with these approaches. Measurements are obtained over the temperature range of interest, and thermal strain is determined. From the thermal strain curve, both the instantaneous CTE and effective CTE can be determined.

Determination of the effective CTE is straightforward but is often widely misinterpreted by engineers. Engineers may have access to instantaneous CTE data at different temperatures and may spend a significant amount of time on tedious integration of the data to obtain an effective CTE. In actuality, the effective CTE was already obtained when determining the thermal strain curve, from which both the effective and instantaneous values of CTE are derived.

To illustrate this, consider the thermal strain curve obtained by dilatometry for fused silica, as shown in Fig. 4.8(a). The ordinate axis is strain, which [see Eq. (4.19)] is

$$\varepsilon = \frac{\Delta L}{L} = \alpha \Delta T.$$

The abscissa axis is temperature, so the value of the effective CTE has already been determined over any temperature range. The effective CTE is the difference in strain between any two temperatures obtained from the curve divided by the temperature change. In other words, if the end points of the range are connected by a straight line intersecting the curve (in geometry, this is called the secant), the slope of the line is the effective, or secant, CTE:

$$\frac{\alpha \Delta T}{\Delta T} = \alpha_{\mathrm{eff}}, \tag{4.22}$$

which is independent of the path taken to arrive at the temperature.

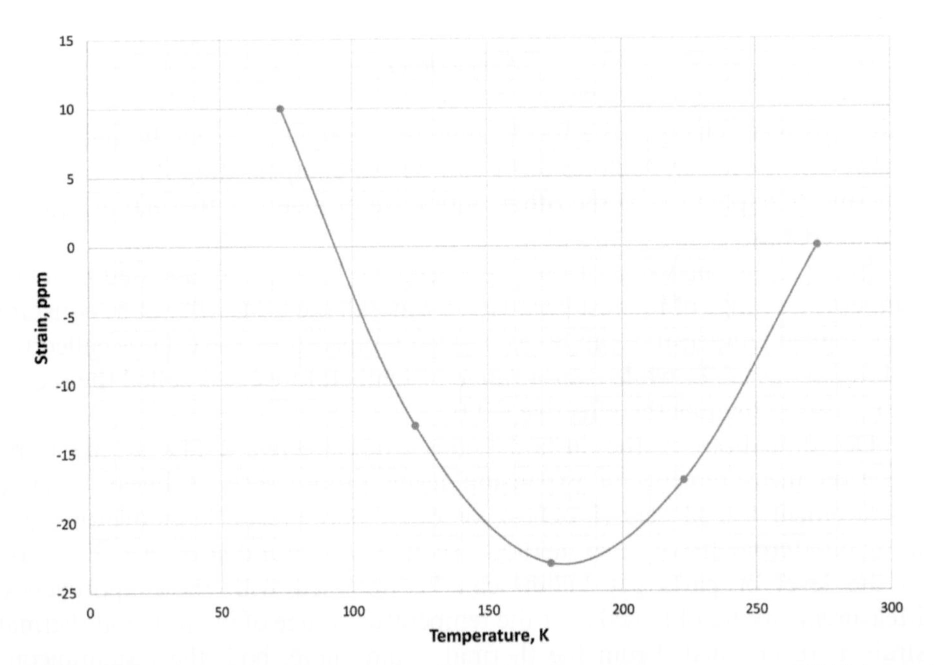

Figure 4.8 (a) Thermal strain of fused silica. Net expansion is zero near 100 K (secant CTE = 0). At 180 K, the local (instantaneous) CTE (slope of the curve) is near zero, precluding gradient concerns.

Note that if the temperature range of interest is small, approaching an infinitesimal amount of change, the secant CTE becomes tangent to the thermal strain curve. At any given temperature, the tangent of the thermal stain curve at that temperature is the instantaneous CTE and is useful for calculations over a very small range of thermal soak. The instantaneous CTE for the fused-silica material is shown in Fig. 4.8(b) as a function of temperature.

The example using fused silica is of particular interest. Consider, further, a fused-silica object subjected to a temperature change from room temperature to 80 K (−193 °C), its operational temperature. From Fig. 4.8(a), the thermal strain is zero and therefore the effective CTE (the slope of the line connecting the end points) is zero, independent of the path taken to get there. That means that there is no net change in shape as the temperature decreases to the operational temperature.

Suppose next that at the operating temperature of 80 K, the temperature is changed by 1 K (1 °C). We can use either the strain curve to determine the effective CTE over this small range, or the tangent (instantaneous) CTE α_i at 80 K from Fig. 4.8(b). We find that $\alpha_i = -0.14$, and hence the change in radius of curvature is

$$\Delta R = \alpha_i R \Delta T. \tag{4.23a}$$

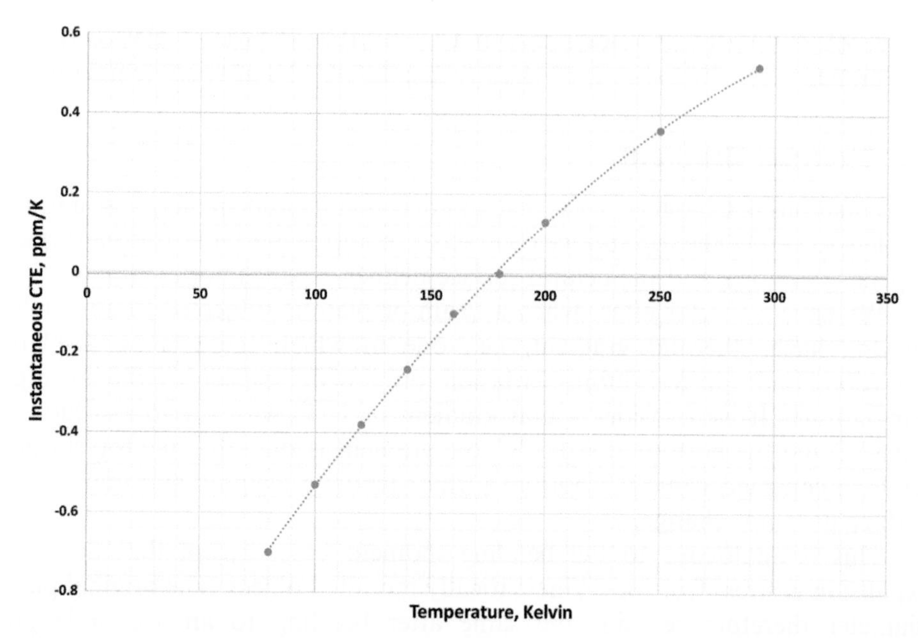

Figure 4.8 (b) Instantaneous CTE of fused silica. At 180 K, the instantaneous CTE is near zero, precluding gradient concerns.

When performance budgets demand, the change in radius of curvature would have to be compensated by the metering structure, if refocus were not possible.

Consider next the same optic subjected to a temperature change from room temperature to 173 K (−100 °C), its operational temperature. From Fig. 4.8(a), the thermal strain is −20 ppm and hence the effective CTE α_e (slope of line connecting end points) is 0.19 ppm/K. The change in radius of curvature is

$$\Delta R = \alpha_e R \Delta T, \tag{4.23b}$$

which means that if performance budgets demand, ΔR would have to be compensated by the metering structure, unless refocus is possible.

Suppose next that at the operating temperature of 100 K, the temperature is changed by 1 K (1 °C). We can use the strain curve in Fig. 4.8(a) to determine the effective CTE over this small range, or the tangent CTE at 100 K from Fig. 4.8(b). Here, $\alpha_i = 0$; hence, there is no change in the radius of curvature, and the system stays in focus.

In the above examples, we see CTEs varying from positive to negative. Negative expansion often occurs for many materials at extremely low cryogenic extremes. Of course, at absolute zero, all materials exhibit zero CTE, as all molecular motion ceases. Some materials, such as ULE® fused silica, ZERODUR® glass-ceramic, and CLEARCERAM®-Z glass-ceramic, exhibit near-zero CTEs, or even slightly negative CTEs at room temperature.

4.5 Thermal Gradient

The equations in Chapter 1 show that a pure bending moment will cause a beam or plate to bend. The top surface contracts z (compression), the bottom surface expands (tension), and the neutral surface neither expands nor contracts.

We have a similar effect when a beam or plate is subjected to an axially varying linear thermal gradient, i.e., one in which there is no in-plane variation, and top-to-bottom variation only with respect to the average temperature. If we consider a unit gradient of 1 °C with the top surface at −0.5 °C and the bottom at +0.5 °C, the gradient is such that the top surface tends to contract, while the bottom surface tends to expand, and the neutral surface does not change.

This is analogous to the bending moment case, except that here the expansion is stress free for a kinematically mounted object. The original plate diameter therefore remains the same after bending to an arc of length s = diameter/r, where r is the plate radius. Referring to Fig. 4.9,

$$s = R\theta, \tag{4.24}$$

where R is the radius of curvature. But with plate thickness t and radius a, and an axial gradient ΔT, the top surface shrinks by an amount

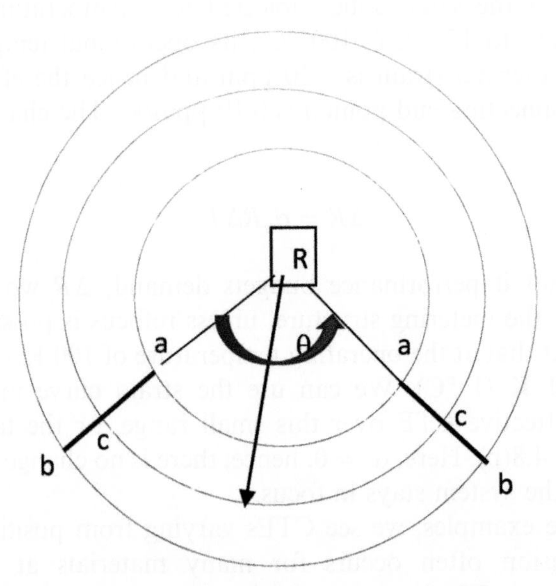

Figure 4.9 Uniform axial (top-to-bottom) thermal gradient. Surface c–c bends to a radius of $t/\alpha\Delta T$; therefore, no stress is produced.

$$\frac{a\alpha\Delta T}{2}, \tag{4.25}$$

and the bottom surface grows by that amount. The curvatures of the top and bottom surfaces are concentric with different radii. Thus, we have a bend without stress and can see from the figure that

$$\frac{\alpha\Delta T}{2} = \frac{t}{2R},$$

or

$$\frac{1}{R} = \frac{aT}{t},$$

where t is the plate thickness;

therefore,

$$R = \frac{t}{\alpha\Delta T}. \tag{4.26}$$

Using the approximate spherical relationship

$$Y = \frac{r^2}{2R}, \tag{4.27}$$

we compute

$$Y = \frac{\alpha\Delta T r^2}{2d}, \tag{4.28}$$

which is the surface displacement of the plate (P-V) under a linear axial gradient. To compute the RMS surface error, we note that the error is all power, so from Table 4.1, we find that

$$Y_{\mathrm{RMS}} = \frac{Y}{\sqrt{12}}.$$

Note that this displacement is *independent* of the area moment of inertia of the cross-section and only a function of its depth. For curved optics with the usual aspect ratio of less than 15:1 and nominal curvature, this displacement serves as a good approximation. In a similar fashion, we can compute the small radius change from basic differential calculus as

$$\frac{dY}{dR} = \frac{-r^2}{2R^2};$$

therefore,

$$dR = \frac{2R^2 dY}{r^2}. \tag{4.29}$$

Unlike the thermal soak case, the change in radius of curvature is not compensated by the metering structure.

4.5.1 Examples for consideration

Example 1. A flat beryllium optic is subjected to a uniformly linear axial gradient of 0.03 °C at room temperature in the absence of any thermal soak change. If the optic is of radius 10 in. and depth 2 in., compute the RMS visible surface error.

From Eq. (4.28) and the instantaneous CTE from Table 2.1, we have

$$Y_{\text{RMS}} = \frac{\alpha \Delta T r^2}{2d\sqrt{12}} = 2.49 \times 10^{-6} \text{ in.}$$

$$Y_{\text{RMS}} = \frac{2.49 \times 10^{-6}}{2.5 \times 10^{-5}} = 0.1\lambda_{\text{RMS surface error}}.$$

Example 2. Consider the same optic as in the previous example but with a radius of curvature of 100 in. Compute the change in radius of curvature.

The P-V sag is approximately the same as for the flat optic due to its low aspect ratio. We can use Eqs. (4.28) and (4.29) to compute

$$\Delta R = \frac{2R^2 \Delta y}{r^2} = \frac{\alpha \Delta T R^2}{d} = 0.001725 \text{ in.}$$

4.5.2 Nonlinear gradients

When the axial gradient is nonlinear, we no longer have a stress-free state. In this case, deflection is a function of the moment of inertia of the cross-section, unlike the previous linear gradient case. For a solid cross-section (to which we limit our discussion for the moment), it can be shown that the effective gradient is given as

$$\Delta T_{\text{eff}} = \int T(z) z dz \left(\frac{h}{I}\right). \tag{4.30}$$

If the integration is formidable, as is often the case, we can use a numerical summation technique by breaking up the section into small parts as

$$\Delta T_{\text{eff}} = \sum_{i}^{n} T_i z_i \Delta z_i \left(\frac{h}{I}\right), \tag{4.31}$$

where:

ΔT_{eff} is the front-to-back effective gradient,
T_i is the temperature at point i in the cross-section,
z_i is the distance from the neutral axis of the cross-section,
Δz_i is the incremental depth of point i in the cross-section,
h is the depth of the cross-section, and
I is the area moment of inertia per unit width of the cross-section; $I = h^3/12$.

Once the effective gradient is calculated, we compute deflection from Eq. (4.28). For kinematic mounting, the shape of the deflection is power, despite its state of stress. For linear gradients, the moment of inertia will cancel out, as is evident in Example 1 of the next subsection.

4.5.3 Examples for consideration

Example 1. Show that the solution for deflection using the effective gradient approach is identical to that produced for the linear axial gradient problem presented in Section 4.5.1.

The linear gradient is 0.03 °C from Example 1 of Section 4.5.1. From Eq. (4.30) we have

$$\Delta T_{\text{eff}} = \int T(z) z dz \left(\frac{h}{I}\right).$$

Over the 2-in. depth, $T(z) = 0.015z$ about the neutral axis of the cross-section shown in Fig. 4.10 (left):

$$\Delta T_{\text{eff}} = 2 \int 0.015z^2 \left(\frac{h}{I}\right).$$

Since $I = \frac{h^3}{12}$, $\Delta T_{\text{eff}} = 0.03\,°\text{C}$, as before.

Alternatively, to avoid the integral, we can break the section into parts and numerically integrate using Eq. (4.31), as has been done for Fig. 4.10 (right). Note that the answer here depends on the number of increments chosen but is nearly identical to the exact solution when ten parts are chosen.

Example 2. A solid optic is subjected to a 1 °C peak nonlinear axial gradient from back to front that follows a parabolic shape, as in Fig. 4.11 (left). Compute the effective gradient and deflection for a 12-in.-diameter solid optic of depth 1 in.

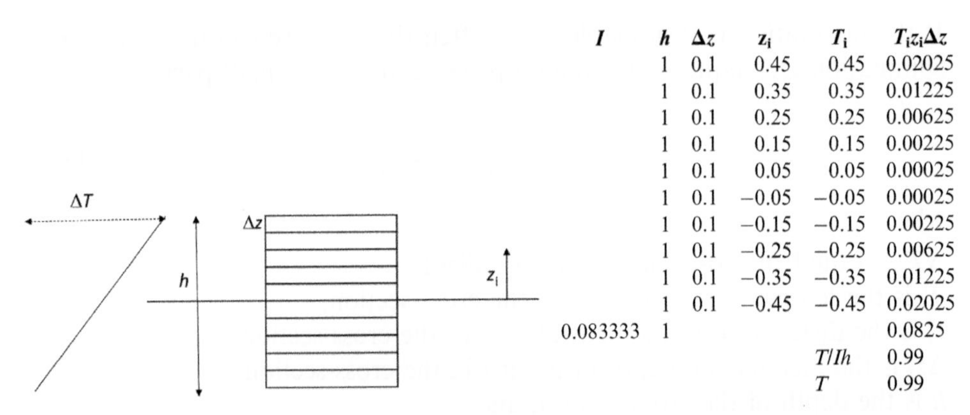

I	h	Δz	z_i	T_i	$T_i z_i \Delta z$
	1	0.1	0.45	0.45	0.02025
	1	0.1	0.35	0.35	0.01225
	1	0.1	0.25	0.25	0.00625
	1	0.1	0.15	0.15	0.00225
	1	0.1	0.05	0.05	0.00025
	1	0.1	−0.05	−0.05	0.00025
	1	0.1	−0.15	−0.15	0.00225
	1	0.1	−0.25	−0.25	0.00625
	1	0.1	−0.35	−0.35	0.01225
	1	0.1	−0.45	−0.45	0.02025
0.083333	1				0.0825
				T/Ih	0.99
				T	0.99

Figure 4.10 (Left) Linear front-to-back gradient section for integration by summation. (Right) Example 1 of Section 4.5.2 computation shows an effective gradient very close to the actual gradient.

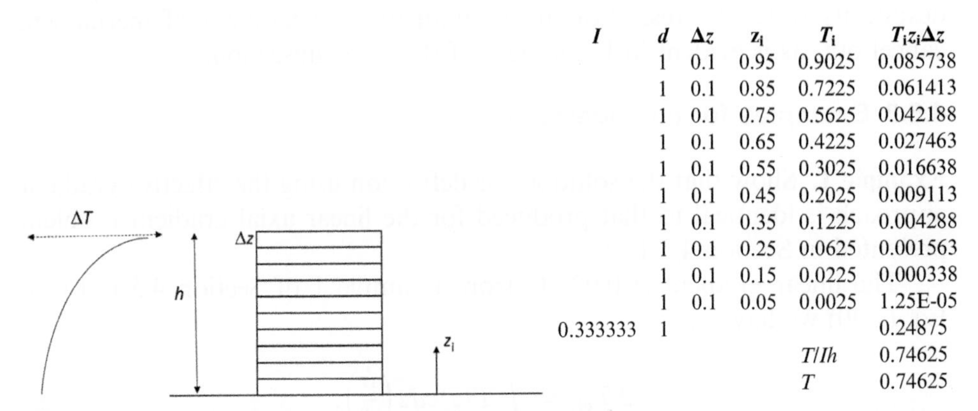

I	d	Δz	z_i	T_i	$T_i z_i \Delta z$
	1	0.1	0.95	0.9025	0.085738
	1	0.1	0.85	0.7225	0.061413
	1	0.1	0.75	0.5625	0.042188
	1	0.1	0.65	0.4225	0.027463
	1	0.1	0.55	0.3025	0.016638
	1	0.1	0.45	0.2025	0.009113
	1	0.1	0.35	0.1225	0.004288
	1	0.1	0.25	0.0625	0.001563
	1	0.1	0.15	0.0225	0.000338
	1	0.1	0.05	0.0025	1.25E-05
0.333333	1				0.24875
				T/Ih	0.74625
				T	0.74625

Figure 4.11 (Left) Parabolic front-to-back gradient section for integration by summation. (Right) Example 2 of Section 4.5.2 computation by integral summation shows an effective gradient that is three-quarters of that in a linear gradient, very close to the actual integral.

To avoid the integral, from Eq. (4.30), we have

$$\Delta T_{\text{eff}} = \sum_{i}^{n} T_i z_i \Delta z_i \left(\frac{h}{I}\right).$$

We break the cross-section into ten parts for numerical integration and compute the temperature at each segment, as shown in Fig. 4.11 (right). Here, it is simpler to integrate from the bottom of the section, where the temperature is zero, to the top, where the temperature is 1 °C, rather than from the neutral axis. In this case, we would use the moment of inertia about the bottom of the section. From the parallel axis theorem, the moment of inertia is

$$I = \frac{h^3}{12} + \frac{h^3}{4} = \frac{h^3}{3}.$$

We find that $\Delta T_{\text{eff}} = 0.75$ °C, or three-quarters of that in the case where the gradient is linear. We can then compute displacement from Eq. (4.28).

Alternatively, with calculus we can integrate using Eq. 4.30. Again, we integrate from the bottom of the section, using inertia as given by the parallel axis theorem:

$$\Delta T_{\text{eff}} = \int T(z)zdz\left(\frac{h}{I}\right), \quad T(z) = z^2, \quad \Delta T_{\text{eff}} = \int z^3dz\left(\frac{h}{I}\right).$$

Using $I = \frac{d^3}{12}$, $\Delta T_{\text{eff}} = 0.75$ °C the exact solution.

Example 3. A solid optic of 12-in. diameter and 3-in. depth is subjected to a 1 °C axial gradient confined to the first 0.2 in. of the top surface. Compute the effective gradient and deflection.

With reference to Fig. 4.12 (left), as before, we use Eq. (4.30), as the integral is not defined by an equation. The results are summarized in Fig. 4.12 (right). Note that the effective gradient (0.96 °C) would be very close to the value of 1 °C if the gradient were purely linear.

If the gradient were confined to the top half of the section only (i.e., uniform over a depth of 0.5 in.), we would find an effective gradient of 1.5 °C, or 50% higher than that produced by a linear front-to-back gradient. This exercise is left to the reader.

4.5.4 Other gradients

When $Y_{\text{RMS}} = \frac{2.49\times 10^{-6}}{2.5\times 10^{-5}} = 0.1\,\lambda_{\text{RMS surface error}}$, a gradient is diametric across an optic, and no error is produced other than tip/tilt and decenter. These

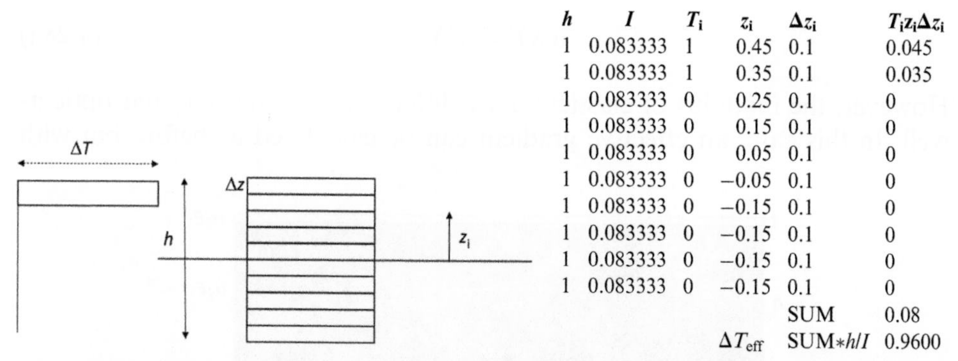

h	I	T_i	z_i	Δz_i	$T_i z_i \Delta z_i$
1	0.083333	1	0.45	0.1	0.045
1	0.083333	1	0.35	0.1	0.035
1	0.083333	0	0.25	0.1	0
1	0.083333	0	0.15	0.1	0
1	0.083333	0	0.05	0.1	0
1	0.083333	0	−0.05	0.1	0
1	0.083333	0	−0.15	0.1	0
1	0.083333	0	−0.15	0.1	0
1	0.083333	0	−0.15	0.1	0
1	0.083333	0	−0.15	0.1	0
				SUM	0.08
			ΔT_{eff}	SUM*h/I	0.9600

Figure 4.12 (Left) Top-section-only (20%) front-to-back gradient section for integration by summation. (Right) Example 3 of Section 4.5.2 computation by integral summation shows an effective gradient that has nearly the same value as a purely linear gradient.

errors need accommodation to a sensitivity error budget but do not cause distortion. If a gradient is radial, linear, or nonlinear, stress, deflection, and therefore WFE are produced. This error is dependent on aspect ratio and curvature, and is beyond the scope of this text; detailed finite-element analysis will bear out such error.

4.6 Coating and Cladding

We turn our attention to determining the stresses and deformations induced by thin-film-stressed surface layers, such as those occurring during deposition or thermal soak conditions for coated or clad optics. An example of these so-called bimetallic strips is illustrated in Fig. 4.13.

All mirrors and lenses require a coating for emissivity, reflectance, transmittance, solar rejection, protection, and the like, depending on the particular requirements of the optical system and the operational wavelength. The coating can range in thickness from 0.1 μm or so (a few micro-inches) to more than 1 μm. A coating can be single layer or multilayer, and bare or protected.

Additionally, for certain materials, such as multiphase silicon carbide, aluminum, and beryllium, for example, optics need to be clad before being coated in order to improve surface roughness and prevent high scatter. These claddings can range in thickness from on the order of several hundred micro-inches (5 μm or so) to a few mils (thousandths of an inch, which converts to 75 μm or so).

The thermal gradient presented in Section 4.5 gives rise to deformation and stresses in the presence of a nonlinear axial gradient in the presence of a constant effective CTE. Similarly, coatings and claddings give rise to deformation and stresses in the presence of nonuniform CTEs in the presence of a constant soak temperature. Equations (4.25), (4.26), and (4.28) are thus valid, using the CTE variation in place of the thermal gradient variation and the thermal soak temperature in place of the CTE, i.e.,

$$\alpha \Delta T = T \Delta \alpha. \tag{4.28a}$$

However, the modulus of elasticity may differ between the coat and optic as well. In this case, an effective gradient can be calculated as before but with

Figure 4.13 Bimetallic strip. Two materials of differing CTEs will cause both stress and deformation under thermal soak conditions.

additional computing of the effective width of the section using the modulus and CTE ratio techniques.

The effective width is the ratio of moduli times the actual width of the section. For an axial gradient, the modification results in

$$\Delta T_{\text{eff}} = \int T(z)\alpha(z)E(z)b(z)z dz\left(\frac{h}{I}\right), \tag{4.32}$$

or, integrating by summation of small increments,

$$\Delta T_{\text{eff}} = \sum_{i}^{n} T_i \alpha_i E_i b_i z_i \Delta z_i \left(\frac{h}{I}\right), \tag{4.33}$$

where:
ΔT_{eff} is the front-to-back effective gradient,
T_i is the temperature at point i in a repeating cross-section,
b_i is the width at point i in the cross-section,
z_i is the distance from the neutral axis of the cross-section to point i,
Δz_i is the incremental depth of point i in the cross-section,
α_i is the CTE ratio (of material 1 to material 2),
E_i is the modulus ratio (of material 1 to material 2),
h is the depth of the cross-section, and
I is the area moment of inertia of a repeating *effective equivalent* cross-section.

In the case of the varying expansion coefficient in the presence of a constant temperature, and for a solid section, the values of b and T are invariant, so

$$\Delta\alpha_{\text{eff}} = \sum_{i}^{n} \alpha_i E_i z_i \Delta z_i \left(\frac{h}{I}\right). \tag{4.34}$$

We then again use Eq. (4.28) in conjunction with Eq. (4.28a) to calculate the deformation, using the CTE of material 1.

Additionally, coat and clad deposition processes can give rise to internal residual stresses, independent of the expansion coefficient. To calculate deformations under such stresses, as well as from temperature change, there is a more elegant way that avoids the tedious integral or summation calculation. We turn to the Timoshenko solution for bimetallic strips.[4]

Since the reference equations are derived for 1D (beam) elements, a modification is required for 2D (plate) elements, as would be the case for optics. (It is noted that edge effects due to shear stresses will develop stresses that are higher than the Timoshenko theory predicts. This will be further discussed in Chapter 9.)

From considerations of equilibrium of forces and bending moments (derivation shown later in chapter), the generalized deformation equation for a beam comprising two materials is given as[4]

$$\frac{1}{R} = \frac{6(\Delta\varepsilon)(1+m)^2}{\left[3(1+m)^2 + (1+mn)\left(m^2 + \frac{1}{mn}\right)\right]}, \tag{4.35}$$

where:
R is the radius of curvature due to deformation,
$\Delta\varepsilon$ is the differential strain between the two materials,
m is the ratio of material thicknesses (T_1/T_2),
n is the ratio of material moduli (E_1/E_2),
T_1 is the thickness of material 1,
T_2 is the thickness of material 2,
E_1 is the modulus of material 1,
E_2 is the modulus of material 2, and
h is total depth of both materials ($T_1 + T_2$).

The generalized stress relation is given as

$$\sigma = \frac{1}{R}\left[\frac{2(E_1 I_1 + E_2 I_2)}{hT} \pm \frac{TE_1}{2}\right],$$

$$I_1 = \frac{{T_1}^3}{12},$$

$$I_2 = \frac{{T_2}^3}{12},$$

$T = T_1$ or T_2 for maximum fiber stress in material 1 or 2.

Note that when the thickness T_1 of one of the materials is very small compared to the other, we can expand the equation algebraically and simplify by eliminating higher-order terms[5] involving T_1 (T_2, T_3, hT) since these tend to zero, as well as by modifying terms involving $h - T_1$, which tend to h.

In this fashion, letting $T_1 = t$ and $T_2 = h$, and using the relationship from Eq. (4.27),

$$Y = \frac{D^2}{8R},$$

where D is a 1D (beam) length. We can readily calculate the deformation. Substituting $m = t/h$,

$$Y = \frac{3E_1 \Delta\varepsilon D^2 t}{4E^2 h^2}. \tag{4.36}$$

The thin-material stress is

$$\sigma_1 = E_1 \Delta\varepsilon. \tag{4.37}$$

The thick-material substrate stress at the interface is

$$\sigma_2 = -E_1 \Delta\varepsilon \left(\frac{4t}{h} \right). \tag{4.38}$$

The thick-material stress on the opposite (free) surface is

$$\sigma_2(\text{free}) = E_1 \Delta\varepsilon \left(\frac{2t}{h} \right). \tag{4.39}$$

Note that the thermally induced stresses in the substrate are substantially lower than those in the thin film, and opposite in sign at the interface. (However, these equations do not apply near the edge of the substrate. In this case, more advanced and generalized theoretical lap bond formulation is required to account for shear restraint. Stresses increase dramatically at the edges and, precluding use of the esoteric and unwieldy general theory, use of detailed finite-element modeling may be mandated. Chapter 9 will discuss the theory and approximations.)

When the stress is thermally induced, under soak temperature T, the strain is given simply [from Eq. (4.19)] as $\Delta\varepsilon = T\Delta\alpha$, and Eqs. (4.37) thru (4.39) become

$$\sigma_1 = E_1 T \Delta\alpha, \tag{4.40}$$

$$\sigma_2 = -E_1 T \Delta\alpha \left(\frac{4t}{h} \right), \tag{4.41}$$

$$\sigma_2(\text{free}) = E_1 T \Delta\alpha \left(\frac{2t}{h} \right). \tag{4.42}$$

The stresses and deformation equations [Eqs. (4.36) through (4.42)] need modification for 2D substrates, as earlier noted. In this case, we can use the 2D Hooke generalized relation [Eq. (1.7a)] discussed in Chapter 1, where

$$\varepsilon_x = \frac{1}{E(\sigma_x - \nu\sigma_y)},$$

$$\varepsilon_y = \frac{1}{E(\sigma_y - \nu\sigma_x)}.$$

For thermal soak or uniform deposition biaxial stress,

$$\varepsilon_x = \varepsilon_y,$$

$$\sigma_x = \sigma_y$$

such that

$$\sigma_x = \frac{\varepsilon_x E}{(1-\nu)}. \tag{4.43}$$

In the case of thermal soak, Eqs. (4.40) thru (4.42) become

$$\sigma_1 = \frac{E_1 T \Delta\alpha}{(1-\nu)}, \tag{4.44}$$

$$\sigma_2 = \frac{-E_1 T \Delta\alpha\left(\frac{4t}{h}\right)}{(1-\nu)}, \tag{4.45}$$

$$\sigma_2(\text{free}) = \frac{E_1 T \Delta\alpha\left(\frac{2t}{h}\right)}{(1-\nu)}, \tag{4.46}$$

and the deformation [Eq. (4.36)] becomes

$$Y = \frac{3E_1 \Delta\varepsilon D^2 t(1-\nu_2)}{4E_2 h^2 (1-\nu_1)}, \tag{4.47}$$

where now D is the major 2D length (diameter, e.g.).

Note that in the case of thin-film intrinsic stress, we have

$$Y = \frac{3\sigma D^2 t(1-\nu_2)}{4E_2 h^2}, \tag{4.48}$$

which recovers the so-called Stoney[6] equation first presented in 1909.

4.6.1 Examples

Simplified first-order calculations are made to determine stresses in coatings, claddings, and substrates along with stress-induced deformation. Substrate stresses are significantly low compared to thin film stresses. Edge effects due to shear stress are not included.

Example 1. Consider a solid mirror made of Zerodur glass-ceramic that is 16 in. in diameter and 2 in. thick with a gold coat of thickness 6 μin. (0.15 μm, or 1500 Å). The mirror undergoes a temperature change from room temperature (293 K) to 200 K. The Zerodur CTE over that range is near zero, while the CTE for the gold is 12.5 ppm/K. Using the modulus properties of Table 2.1, and assuming a Poisson ratio of 0.25 for both materials,

a) compute the RMS visible WFE,
b) compute the stress in the coating, and
c) compute the stress in the optic.

We use Eqs. (4.44) through (4.47) to find that

a) $Y = \frac{3E_1 \Delta\varepsilon D^2 t(1-\nu_2)}{4E_2 h^2 (1-\nu_1)} = 0.294 \times 10^{-6}$ in.

The RMS WFE (noting that the shape is all power) is computed as

$$Y_{\text{RMS WFE}} = 2Y/25(10)^{-6}/3.46 = 0.007\lambda.$$

b) $\sigma_1 = \frac{E_1 \alpha \Delta T}{(1-\nu)} = 18{,}600$ psi.

c) $\sigma_2 = \frac{-E_1 \alpha \Delta T\left(\frac{4t}{h}\right)}{(1-\nu)} = 0.3$ psi (quite low!).

Note that to compute the deformation [computation (a) above], we can alternatively use the method of integration (Eq. 4.32) (with reference to Table 4.2) to find that

a) $Y = 0.294 \times 10^{-6}$ in.,

which is identical to the solution to (a), as it must be, although this method is quite cumbersome.

Example 2. Consider the same mirror as in Example 1 with residual deposition compressive stress in the coating of 4000 psi and compute the RMS visible WFE at room temperature.

We use Eq. (4.48) and the RMS conversion factor to compute $Y_{\text{RMS WFE}} = 0.001\ \lambda$ (quite small).

Example 3. A beryllium optic of 16-in. diameter and 2-in. thickness with a deposited nickel clad layer of thickness 0.003 in. (75 μm) undergoes a

Table 4.2 Tabular data for Example 1 of Section 4.6.1.

E_1	E_2	Ei	α_i	T_i	Δz	z_i	$T_i z_i \Delta z \alpha_i E_i$	h
1.20E+07	1.31E+07	0.916031	1.20E−05	93	0.000006	1	6.13E−09	2
		0.916031	0	93	0.4	0.8	0	
		0.916031	0	93	0.2	0.5	0	
		0.916031	0	93	0.2	0.3	0	
		0.916031	0	93	0.2	0.1	0	
		0.916031	0	93	0.2	0	0	
		0.916031	0	93	0.2	−0.1	0	
		0.916031	0	93	0.2	−0.3	0	
		0.916031	0	93	0.2	−0.5	0	
		0.916031	0	93	0.4	−0.8	0	
						sum	6.13E−09	
						TΔα eff	1.84E−08	
						Y	2.94E−07	

temperature change from room temperature (293 K) to 200 K. The beryllium effective CTE over that range is 10.0 ppm/K, and the nickel effective CTE is 9.5 ppm/K. Using the modulus properties of Table 2.1, compute the RMS visible WFE.

From Eq. (4.47), we compute $Y = 2.28 \times 10^{-6}$ in., and $Y_{\text{RMS WFE}} = 0.053\lambda$, which may significantly degrade optical performance. If the error budget demands, the backside of the optic is often coated as well to negate the effect.

4.7 Rule of Mixtures

Often, as earlier noted, coatings comprise more than a single layer. In this case, we can use the rule of mixtures to compute an equivalent CTE and equivalent modulus, and then use the previous equations to compute deflection.

4.7.1 Two layers

For a two-layer coating, we can use the parallel rule of mixtures to determine the elastic modulus and the CTE. The rule of mixtures is somewhat straightforward; for a material with two layers of different mass density (in which there is no parallel or series rule, as mass is not a directional property), we have, simply,

$$\rho = \rho_1 v_1 + \rho_2 v_2, \tag{4.49}$$

where ρ_1 is the density of material 1, and ρ_2 is the density of material 2.

We introduce the volume relationships because v_1 is the thickness of material 1/total thickness, v_2 is the thickness of material 2/total thickness, and $v_1 + v_2 = 1$.

For the elastic modulus, in parallel, the strain in each layer is the same under load such that, using Hooke's law,

$$E = E_1 v_1 + E_2 v_2, \tag{4.50}$$

where E_1 is the modulus of material 1, and E_2 is the modulus of material 2. This is similar to the rule of mixtures for density. Note that for a modulus in series, which does not apply to coatings, it is the stress in each part that is the same, so in this case, using Hooke's law,

$$E = \frac{E_1 E_2}{E_1 v_2 + E_2 v_1}. \tag{4.51}$$

For our two-layer case, of course, we use the parallel rule to compute the equivalent modulus.

For equivalent CTE, we again invoke the parallel rule. Again, it is the strain compatibility at the boundary that is equal, not the stress compatibility, so we can write

$$\alpha_1 \Delta T - \frac{\sigma_1}{E_1} = \alpha_2 \Delta T + \frac{\sigma_2}{E_2} \tag{4.52}$$

and compute

$$\alpha = \frac{E_1 \alpha_1 v_1 + E_2 \alpha_2 v_2}{E_1 v + E_2 v_2}. \tag{4.53}$$

For the sake of completeness, again not applicable to coatings, the series rule for CTE gives

$$\alpha = \alpha_1 v_1 + \alpha_2 v_2. \tag{4.54}$$

When we now compute the equivalent CTE and modulus, we simply use Eq. (4.47) to compute deflection. Here, the coating equivalent thickness t is the sum of the individual coating thicknesses:

$$t = t_1 + t_2. \tag{4.55}$$

4.7.1.1 Example

The gold-coated optic of Example 1 in Section 4.6.1 has a protective silver overcoat applied that is 12 μin. thick (0.3 μm) with a modulus of 7×10^6 psi and an effective CTE of 7×10^{-6}/K. Compute the deflection of the optic.

Using Eq. (4.50), we compute the effective modulus as

$$E = E_1 t_1 + E_2 t_2 = (7 \times 10^6)(12)/18 + [1.2 \times 10^7 (6)/18] = 8.67 \times 10^6 \text{psi}.$$

Using Eq. (4.53), we compute the effective CTE as

$$\begin{aligned} \alpha &= \frac{E_1 \alpha_1 t_1 + E_2 \alpha_2 t_2}{E_1 t_1 + E_2 t_2} \\ &= [(7)(7)(0.667) + (12)(12.5)(0.333)]/[7 \times 10^6 (0.667) + 1.2 \times 10^7 (0.333)] \\ &= 9.5 \text{ ppm/K}. \end{aligned}$$

We can then enter Eq. (4.47) (using the sum of the thicknesses) to compute the optic deflection.

The stress in the optic and the equivalent stress in the coating can be recovered from Eqs. (4.44) through (4.46), although the individual stress in each of the coating layers cannot be computed.

4.7.2 Multiple layers

For more than two layers in a coating, we can compute the equivalent modulus and CTE in a similar fashion. Here we find, again with v indicating the volume fraction, that

$$\alpha = \frac{\sum E_n \alpha_n v_n}{\sum E_n v_n}, \tag{4.56}$$

and

$$E = \sum E_n t_n. \tag{4.57}$$

For the series rule, which does not apply here but is given for other applications,

$$\alpha = \alpha_1 v_1 + \alpha_2 v_2 + \ldots \alpha_n v_n = \sum \alpha_i v_i, \tag{4.58}$$

and

$$E = \sum_{i=1}^{n} \left[\left(\prod_{i=1}^{n} E_i \right) v_n / E_n \right]. \tag{4.59}$$

While Eq. (4.59) appears formidable, it is simply a mathematical representation of the sum of the sum of the products, and for $n = 2$, reduces to Eq. (4.50). By way of example, if we have four layers, Eq. (4.59) becomes

$$E = \frac{E_1 E_2 E_3 E_4}{E_2 E_3 E_4 v_1 + E_1 E_3 E_4 v_2 + E_1 E_2 E_4 v_3 + E_1 E_2 E_3 v_4}. \tag{4.59a}$$

4.8 Trimetallic Strip

Computing deflection and individual stresses in three layers is commonly required. Two coating layers over an optic, or one layer of material (flexure, for example) bonded (second layer) to a third layer (optic) are some examples. In these cases, we can expand the bimetal case to the trimetal case, from which the bimetal case can be derived.

Consider, then, a plate consisting of three materials rigidly bonded as shown in Fig. 4.14 and uniformly heated over a temperature range T. For small deflections, the curvatures will be equal and constant over the plate, which becomes spherical in form. All forces acting on any cross-section of the plate must be in equilibrium; i.e.,

Figure 4.14 Trimetallic strip. Three materials of differing CTEs will cause both stress and deformation under thermal soak conditions.

$$P_1\left[\frac{t_1+t_2}{2}+\frac{P_2}{P_1}\left(\frac{t_2+t_3}{2}\right)\right]=M_1+M_2+M_3,$$

where $M_1=\frac{E_1'I_1}{R}$; $M_2=\frac{E_2'I_2}{R}$; $M_3=\frac{E_3'I_3}{R}$;

$$E_1'=\frac{E_1}{1-\nu_1};\ E_2'=\frac{E_2}{1-\nu_2};\ E_3'=\frac{E_1}{1-\nu_1};$$
$$I_1=\frac{t_1{}^3}{12};\ I_2=\frac{t_2{}^3}{12};\ I_3=\frac{t_3{}^3}{12}. \tag{4.60a}$$

On each surface, the compatibility of strain must be satisfied; thus,

$$\alpha_1(\Delta T)+\frac{P_1}{E_1't_1}+\frac{t_1}{2R}=\alpha_2(\Delta T)-\frac{P_1-P_2}{E_2't}-\frac{t_2}{2R},$$
$$\alpha_2(\Delta T)-\frac{P_1-P_2}{E_2't}+\frac{t_2}{2R}=\alpha_3(\Delta T)-\frac{P_2}{E_3't_3}-\frac{t_3}{2R}. \tag{4.60b}$$

Eliminating P_1 and P_2 and after tedious calculation and manipulation, we find that

$$\frac{1}{R}=\frac{-\left[\left(\frac{h_{12}+h_{23}}{t_2}+\frac{n_{23}h_{12}}{t_3}\right)(\alpha_1-\alpha_2)+\left(\frac{h_{23}}{n_{12}t_1}+\frac{h_{12}+h_{23}}{t_2}\right)(\alpha_2-\alpha_3)\right](\Delta T)}{\frac{1}{2}\left(\frac{h_{23}{}^2}{n_{12}t_1}+\frac{(h_{12}+h_{23})^2}{t_2}+\frac{n_{23}h_{12}{}^2}{t_3}\right)+2\left(I_1+\frac{I_2}{n_{12}}+\frac{I_3}{n_{13}}\right)\left(\frac{n_{13}t_1+n_{23}t_2+t_3}{t_1t_2t_3}\right)} \tag{4.61}$$

where $h_{12}=t_1+t_2, h_{23}=t_2+t_3$, and $h_{13}=t_1+t_3$ with

$$n_{12}=\frac{E_1'}{E_2'},$$
$$n_{23}=\frac{E_2'}{E_3'},$$
$$n_{13}=\frac{E_1'}{E_3'}.$$

Note that if the bottom layer is eliminated ($t_3 = 0$ and $E_3 = 0$), we find the curvature to be

$$\frac{1}{R} = \frac{(\alpha_2 - \alpha_1)(\Delta T)}{\frac{h_{12}}{2} + \frac{2(n_{12}I_1 + I_2)}{h_{12}}\left(\frac{1}{n_{12}t_1} + \frac{1}{t_2}\right)}, \tag{4.62}$$

which is identical to Eq. (4.35).

Using the curvature for the more-general, three-material case, we can now compute the forces and moments as

$$P_2 = \left[(\alpha_1 - \alpha_2)(\Delta T) + \frac{h_{12}}{2R} + \frac{2M}{h_{12}}\left(\frac{1 - \nu_1}{E_1 t_1} + \frac{1 - \nu_2}{E_2 t_2}\right)\right]\left(\frac{h_{12}}{C_1}\right),$$
$$P_1 = \frac{2M}{h_{12}} - P_2\left(\frac{h_{23}}{h_{12}}\right),$$

where

$$M = M_1 + M_2 + M_3,$$
$$C_1 = \frac{h_{23}(1 - \nu_1)}{E_1 t_1} + \frac{(h_{12} + h_{23})(1 - \nu_2)}{E_2 t_2}, \tag{4.63}$$

and, finally, the normal stresses at the top and bottom of each layer as

$$\sigma_{11} = \frac{P_1}{t_1} - \frac{tE_1}{2R(1 - \nu_1)},$$
$$\sigma_{12} = \frac{P_1}{t} + \frac{t_1 E_1}{2R(1 - \nu_1)},$$
$$\sigma_{21} = \frac{-(P_1 - P_2)}{t_2} - \frac{t_2 E_2}{2R(1 - \nu_2)},$$
$$\sigma_{23} = \frac{-(P_1 - P_2)}{t_2} + \frac{t_2 E_2}{2R(1 - \nu_2)}, \tag{4.64}$$
$$\sigma_{32} = \frac{-P_2}{t_3} - \frac{t_3 E_3}{2R(1 - \nu_3)},$$
$$\sigma_{33} = \frac{-P_2}{t_3} + \frac{t_3 E_3}{2R(1 - \nu_3)}.$$

These quantities are defined in Fig. 4.15.

Then, for a plate of diameter D, it is an easy matter to compute deflection from Eq. (4.27) as

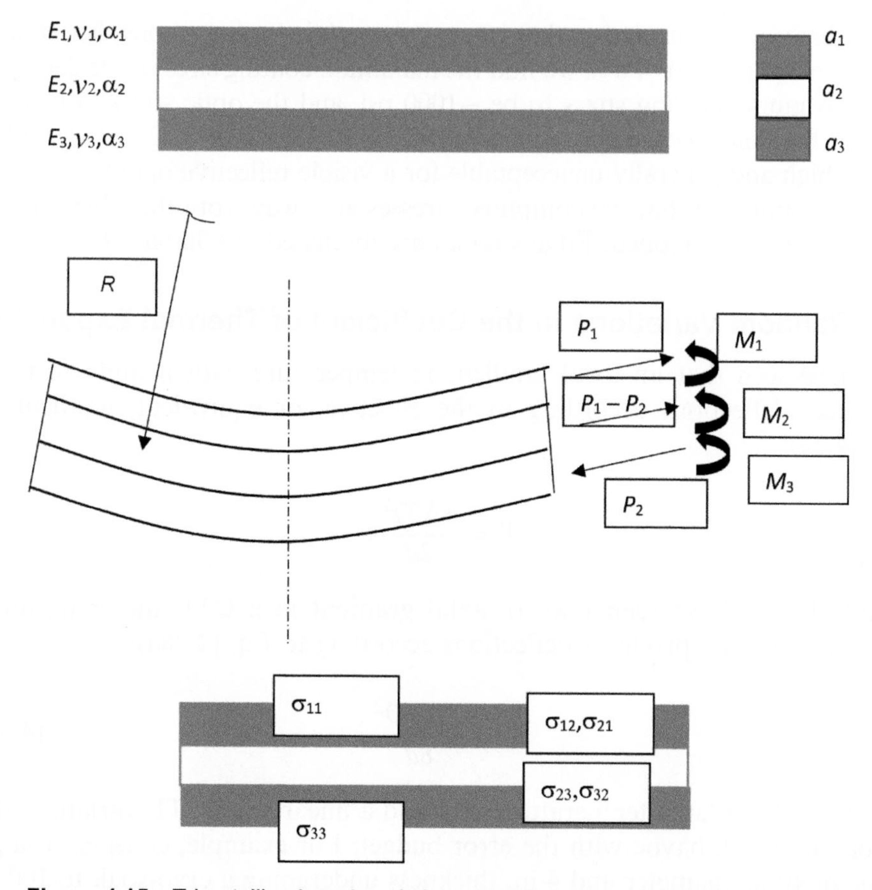

Figure 4.15 Trimetallic plate descriptive analysis terms and interface stresses.

$$Y = \frac{D^2}{8R}.$$

As one can readily see, it is best to program these equations into a spreadsheet to avoid tedious repetition and potential for error, let alone boredom.

4.8.1 Example

A 40-μin. (1 μm)-thick gold coating is deposited over a 0.003-in. (75 μm) nickel cladding that has been deposited on a 1-in. (0.025 m) solid beryllium optic of diameter 16 in. that undergoes a uniform thermal soak change from room temperature (293 K) to 193 K. Using the properties of Table 2.1 and effective expansion coefficients of 12.0 ppm/K for the coating, 10.0 ppm/K for the cladding, and 10.5 ppm/K for the optic, compute the stresses in the coat, clad, and optic, and its deflection. Use a Poisson ratio of 0.25 throughout.

After tedious calculation that is greatly simplified using a spreadsheet, and using Eqs. (4.63) and (4.64), we find the maximum coating stress to be 2500 psi, the maximum cladding stress to be -1000 psi, and the optic stress to be only 10 psi. The deformation is computed as 9.5×10^{-6} inch, or 0.22 λ RMS WFE (quite high and generally unacceptable for a visible reflective optic).

It is cautioned that the computed stresses are away from the edges, where funny things can happen. Edge stresses are discussed in Chapter 9.

4.9 Random Variations in the Coefficient of Thermal Expansion

We have seen that an axial gradient in temperature with a uniform CTE produces deflections according to the effective, or equivalent, gradient of Eq. (4.28):

$$Y = \frac{\alpha \Delta T r^2}{2d}.$$

Similarly, we have seen that an axial gradient in a CTE under uniform temperature soak produces deflections according to Eq. (4.28a):

$$Y = \frac{T \Delta\alpha D^2}{8d}. \tag{4.65}$$

Note that for a large temperature soak and a linear axial CTE variation, the error can wreak havoc with the error budget. For example, consider a large optic of 40-in. diameter and 4-in. thickness undergoing a cryo-soak to 100 K from room temperature of 293 K and a front-to-back variation in CTE of only 1 ppb/K (one part per billion/K, or 0.001 ppm/K). The error is [from Eq. (4.65)] $Y = (193)(0.001)(10)^{-6}\ (40)(40)/8/4 = 9.7$-μin. peak surface error. Using the power RMS factor of $\sqrt{12}$ and converting to visible waves gives $Y = 0.22$ waves RMS.

For systems without active focus, this could be an excessive error. Since almost all optical materials will have CTE variations well in excess of 1 ppb/K (in fact, 10 to 50 times higher) one wonders how performance budgets lacking focus are met, particularly for large optics. The answer lies in the fact that most front-to back variations in materials are randomized in the process; i.e., while variations of more than 1 ppb may exist front to back in one place, variations may be reversed (negative) in another place, or nonexistent in yet another place across the diameter.

For the randomized case, no analytical solution can be readily achieved. However, one can randomize CTEs using finite-element models, compute the analytical result, and compare these results to known interferometric test results in the presence of a constant temperature soak change.

Accordingly, during the 1970s, a series of finite-element models[7] was exercised with random CTE distributions. Considered were optics of various diameters, thicknesses, and radii of curvature. Each was assigned ten different random routines of axially varying expansion [with three-sigma (3σ) distribution] and ten different random routines of in-plane, varying expansion coefficients (with 3σ distribution). Combined with the varying aspect ratio and curvatures, this resulted in a total of over 1000 cases. The results were curve fit to render an approximate equation given as the Pepi–Nagle–Lowe (delta alpha) equation:

$$Y_{\mathrm{RMS\,WFE}} = \sqrt{\left(\frac{56\bar{T}\Delta\alpha_{\mathrm{a}}D^2}{d}\right)^2 + \left(\frac{22\bar{T}\Delta\alpha_{\mathrm{p}}D^{7/2}}{d^{3/2}R_{\mathrm{s}}}\right)^2}, \tag{4.66}$$

where:[7]
$\bar{T}$ is the temperature soak [°C],
$\Delta\alpha$ is the CTE variation, total excursion [in./in./°C],
$\Delta\alpha_{\mathrm{a}}$ is the axial variation,
$\Delta\alpha_{\mathrm{p}}$ is the inplane variation,
D is the mirror diameter [in.],
R_{s} is the mirror radius of curvature, and
d is the mirror depth [in.].

Again, the result is given in visible WFE.

Note that the first term in this equation is the axially varying random gradient. When compared to the linearly varying axial gradient, we see that the error is reduced by a factor of 50! The error generally shows little power, instead being astigmatic with higher-order errors.

Note that the second term in this equation is the laterally (in-plane) varying random gradient. This term is dependent on the mirror radius of curvature: the lower the radius the higher the error. The error again generally shows little power, but after focus removal is highly astigmatic with higher-order errors.

Since Eq. (4.66) is based on analytical finite-element tools, it would need correlation to actual test the data. Accordingly, over the subsequent 25 years since the 1970s, data was obtained from coupon tests for expansion inhomogeneity on multiple optical materials, such as silicon carbide, fused silica, ULE fused silica, Zerodur, borosilicate, beryllium, and aluminum. Then, a literature survey was conducted for various optics of differing diameters to review surface error after soak to cryogenic temperatures after removal of focus. The data is markedly well correlated to the equation. Using interferometer test data for new or existing materials, expansion inhomogeneity may be inferred.

4.9.1 Example

A beryllium optic 60 in. in diameter and 2.5 in. deep with a radius of curvature of 240 in. is brought from room temperature (293 K) to a severe cryogenic soak of 100 K that is stabilized with no axial thermal gradient (uniform soak). Compute the RMS WFE after focus if coupon test data indicate a 3σ random variation in a CTE of 30 ppb/K.

Using Eq. (4.66), $\lambda_{\text{RMS WFE}} = 0.52\,\lambda$.

Conversely, if interferometry had indicated an error of 0.52λ, we could infer the material's homogeneity as 30 ppb/K. This is a very high error for high-acuity optics and will need to be cryogenically nulled by polishing in the error at ambient (room) temperature.

References

1. F. Zernike, "Beugungstheorie des Schneidenver-fahrens und seiner verbesserten Form, der Phasenkontrastmethode," *Physica* **1**, 689–704 (1934).
2. R. J. Roark and W. C. Young, *Formulas for Stress and Strain*, Fifth Edition, McGraw-Hill Book Co., New York (1975).
3. S. Timoshenko and S. Woinowsky-Kreiger, *The Theory of Plates and Shells*, Second Edition, McGraw-Hill Book Co., New York (1959).
4. S. Timoshenko, "Analysis of bi-metal thermostats," *J. Optical Society of America* **11**(3), 233–255 (1925).
5. M. J. Pepi, "Generalized distortion of solid mirrors with stressed surface layer," private memorandum communication, July 1992.
6. G. Stoney, "The tension of metallic films deposited by electrolysis," *Proc. Royal Society A* **82**(533), 172 (1909).
7. J. W. Pepi, "Analytical predictions for lightweight optics in a gravitational and thermal environment," *Proc. SPIE* **748**, pp. 172–179 (1987) [doi: 10.1117/12.939829].

Chapter 5
Lightweight Optics: Optimization

The analysis for solid mirrors sets the stage for the analysis of lightweight optics. Weight is extremely critical for airborne and space optical systems to meet the limitations of fuel consumption, load capacity, and cost. Ground-based systems also have weight limits due to transport, lift, assembly, and other requirements, particularly for large-aperture assemblies. The 200-in. (5-m) Hale Telescope, for example, operating at Mount Palomar, was made using one of the early lightweight mirror approaches. The cast borosilicate optic was about 35% lightweighted compared to a solid of equal depth. This is a far cry from today's lightweighting capabilities, which can exceed 90% weight reduction from the parent solid. By comparison, the 5-m Hale Telescope has a mass density of 900 kg/m^2, while the 6-m James Webb Space Telescope primary optic has a mass density of less than 20 kg/m^2.

5.1 Lightweight Optics

There are several ways to obtain lightweight optics; one, of course, is to make use of thin solids, i.e., high-aspect-ratio optics. Another method makes use of sandwich construction, commonly called a closed-back design, where solid, thin facesheets are attached to a lightweight core. A third possibility uses a thick solid machined to form a lightweight core rib structure with no back sheet. This is commonly called an open-back design.

We will quickly review the benefits and drawbacks of these three lightweighting methods. As seen in Fig. 5.1, the various designs all have the same stiffness-to-weight ratio; that is, they will exhibit essentially the same gravitational sag and the same fundamental frequency. The light mirrors, however, are only 10% of the weight of the solid mirror! This is a very large weight savings as diameter increases. On the other hand, the solid mirror has better stiffness by almost an order of magnitude and would be less sensitive to mount-induced error. If mass is the driver, the open- and closed-back designs are the methods of choice.

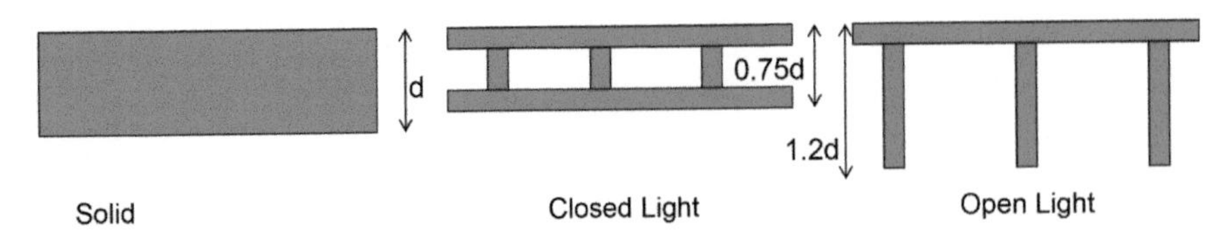

Figure 5.1 Three mirrors having have the same stiffness-to-weight ratio, for which the lightweight options are only 10% of the solid option weight.

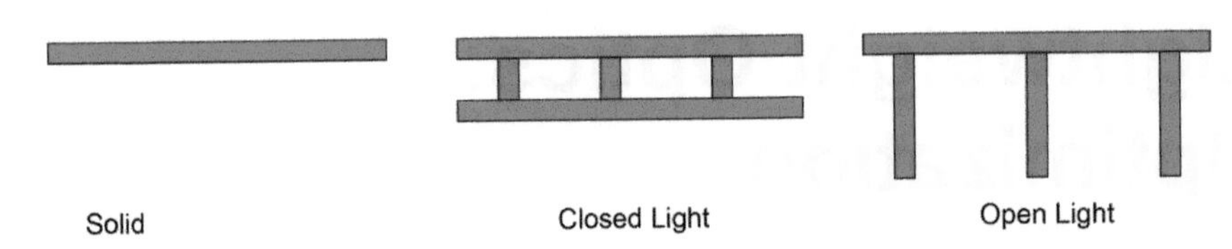

Figure 5.2 Three mirrors having the same weight, for which the lightweight options are 64 times stiffer.

Viewed another way, consider the alternative optics of Fig. 5.2. These mirrors have the same *weight*. However, closed- or open-back lightweighted mirrors have better stiffness by a factor of almost two orders of magnitude and hence much less sensitivity to mount error, and much less sensitivity to gravity sag due to high stiffness-to-weight ratios. This is precisely why we make such lightweight mirrors! For large apertures, the thin solid would likely require a multipoint mount system, as discussed further in Chapter 7.

Note from Figs. 5.1 and 5.2 that, in order to provide the same stiffness and weight, the open-back section is deeper than the closed-back section. As a rule, the equivalence results in a closed-back optic with 75% of the depth of the solid, and an open-back optic of a depth 20% greater than the solid. Calculations provided in the following sections will confirm this.

Key to the lightweight optic design of either open or closed construction is the design of the lightweight core and its attachment to the facesheet(s). For closed-back designs, several attachment methods can be used, depending on the material selection: attachment can be achieved by adhesive bonding, frit bonding, fusing, or direct casting. For open-back designs, the core is generally machined directly into the optic. The open-back approach thus avoids the use of adhesives, which can impact performance due to CTE mismatch. This approach also eliminates the need for more-costly frit bond or fusing at relatively high temperatures, and the need for complex closed-back casting. However, the open-back design does require more envelope for depth, to provide equal stiffness and weight. Furthermore, this design type must be shown to be viable when other effects, such as shear deflection and isotropy of core, are considered. These effects are further discussed later in this chapter. Machining techniques also need consideration, if manufacturing limits to core rib thickness are realized.

5.2 Core Shape

The law of packaging for an equal-size core shape shows that only triangular, square, or hexagonal cores can be packaged on a flat, or nearly flat, surface. The law states that the packaging of identical cells is possible only when the external angle of the shape divided by the internal angle of the shape is an integer (this can be readily proven[1]). With reference to Fig. 5.3, we see that the integer rule (also known as the law of packaging) works only for polygons of 3, 4, or 6 sides; i.e., triangles, squares, or hexagons. At least we have limited the trade space.

5.2.1 Core geometry

It is readily shown that for the triangular, square, and hexagonal shapes, the weight per unit area is identical for equal diameters of an inscribed circle a, as shown in Fig. 5.4. This results in a core side length of a for squares, a spacing across the flats of a for hexagon, and a side length of $\sqrt{3}a$ for the equilateral triangle. We illustrate this by computing the weight of the repeating cell section, as in Fig. 5.5.

Concerning a square, for the core section only, we have

$$\frac{W}{A} = \frac{4bad}{2a^2\rho} = \frac{2bd\rho}{a}, \tag{5.1a}$$

where a is the core spacing, b is the core width, d is the core depth, and ρ is the weight density. For the triangle,

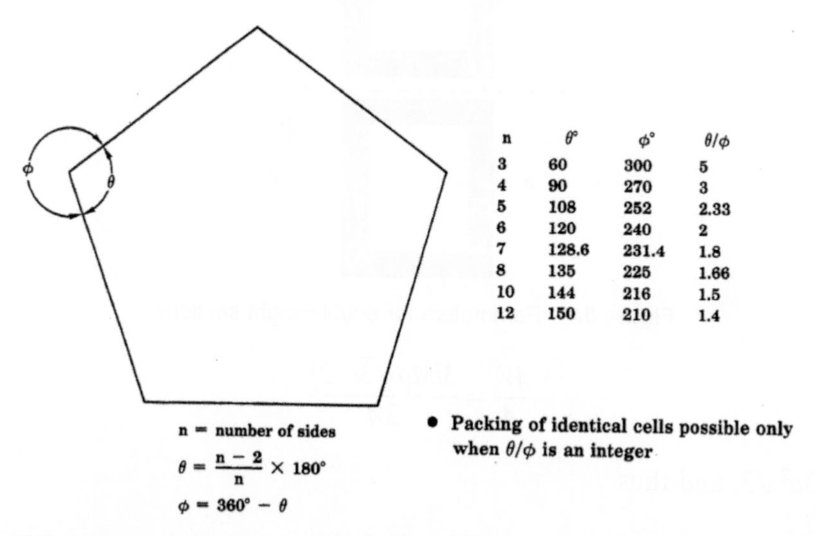

n	θ°	φ°	θ/φ
3	60	300	5
4	90	270	3
5	108	252	2.33
6	120	240	2
7	128.6	231.4	1.8
8	135	225	1.66
10	144	216	1.5
12	150	210	1.4

Figure 5.3 Integer rule for geometrical packaging: Only triangles, squares, and hexagons will package on a flat surface, for equal-size shapes.

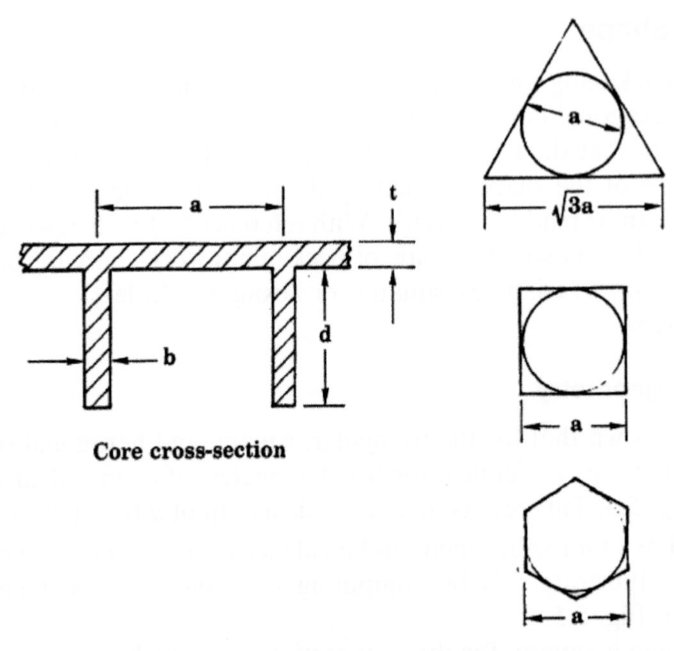

Figure 5.4 Equal weight shapes. These sections all have the same unit weight for an inscribed circle of equal diameter.

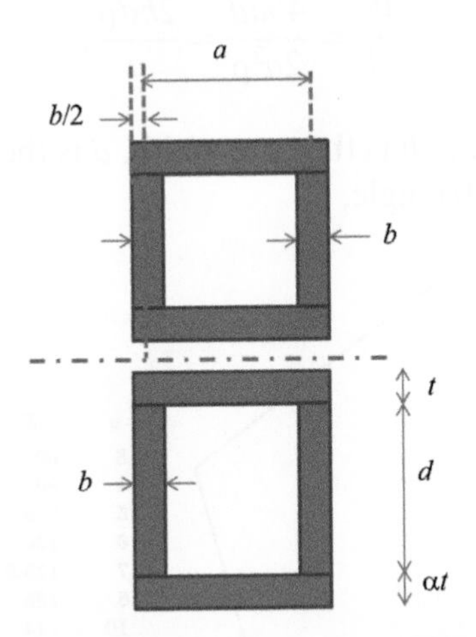

Figure 5.5 Parameters for equal-weight sections.

$$\frac{W}{A} = \frac{3bd\rho\sqrt{3}a(2)}{2A},$$

$A = 3a^2\sqrt{3}$, and thus

$$\frac{W}{A} = \frac{2bd\rho}{a}. \tag{5.1b}$$

Similarly, for the hexagon, we compute

$$\frac{W}{A} = \frac{6bd\rho s}{2A},$$

where s is the side length $= \frac{\sqrt{3}a}{3}$, and $A = \frac{\sqrt{3}a^2}{2}$, and thus

$$\frac{W}{A} = \frac{2bd\rho}{a}. \tag{5.1c}$$

Note that although the core weight in this formulation "double dips" at the ideal interstices, this effect is small for typical core spacing, and negligible for ease of calculation. Note that Eqs. (5.1a), (5.1b), and (5.1c) are identical.

When we combine the facesheet with the core, again with reference to Fig. 5.5, we find the weight per unit area to be

$$\frac{W}{A} = \rho\left[t(1+\alpha) + \frac{2bd}{a}\right], \tag{5.2}$$

where t is the top facesheet thickness, and α is the ratio of the back-facesheet thickness to the front-facesheet thickness. For open-back sections, $\alpha = 0$.

5.2.2 Example

An open-back, 20-in.-diameter optic made of fused silica has a facesheet thickness of 0.20 in., a triangular core spacing of 2 in., a rib width of 0.12 in., and a rib depth of 3 in. Compute the weight per unit area, weight, and percent lightweight from the parent solid.

Using Table 2.2 data and Eq. (5.2),

$$\begin{aligned}\frac{W}{A} &= \rho\left[t(1+\alpha) + \frac{2bd}{a}\right],\\ &= 0.08\left[0.20 + \frac{2(0.12)(3)}{2}\right] = 0.045\,\text{psi}\,(31.7\,\text{kg/m}^2).\end{aligned}$$

$$W = \frac{0.045(\pi)(20^2)}{4} = 14.1\,\text{lbs}\,(6.4\,\text{kg}).$$

The parent solid weighs

$$W_P = \frac{\pi(20)^2(3.2)(.08)}{4} = 80.4\,\text{lbs}.$$

The optic is, therefore, $1 - \left(\frac{14.1}{80.4}\right) = 82\%$ lightweighted.

5.3 Core Stiffness

While all three core types have the same repeating, areal cross-section, they do not necessarily exhibit the same stiffness; indeed, it is the total mirror stiffness with integral facesheet(s) that matters.

For 2D plates, it is evident that square cores provide continuous bending stiffness in two orthogonal directions, while triangular sections exhibit three continuous paths at 60-deg angles. On the other hand, there is no continuous bending path for hexagons. This is critical knowledge for open-back designs, which rely on both the core and the facesheet to provide bending stiffness. Open-back designs with hexagonal cores do not fare well here.

Note that for closed-back sections, most of the bending inertia is achieved for the facesheets and not the core, the latter of which provides the appropriate shear tie. In this case, although there are secondary effects, stiffness is virtually independent of core shape. Typically, a hexagonal core is chosen for ease of machining, since internal core angles are less sharp, resulting in reduced weight at the interstices.

To illustrate this, consider an example of equal-weight, closed-back and open-back sections with the various core geometries. Detailed finite element models were exercised to compare both gravitational deflection and fundamental frequency.

Table 5.1 shows the weight penalty for an open-back section compared to that of a closed-back section with various core geometries, normalized to a weight of 100 lbs. Note that a 10% weight penalty is realized for gravity sag of the open-back section with a square core shape, while a penalty of 60% is paid for frequency of the open-back square section due to a torsional (twist) mode caused by the lack of isotropy. Note that the penalty for the hexagonal-core, open-back section is also 60% for both gravity and frequency due to the lack of bending resistance. Finally, note that the triangular-core, open-back section pays no penalty compared to the closed-back section and is the method of choice.

In a similar fashion, Table 5.2 compares the performance of equal-weight, open-back optics for the alternative core shapes. Again, we see gravity performance degraded by about 15% for the square design versus the triangular

Table 5.1 Open- and closed-back weight penalties (in pounds) for various core configurations.

	Constant Sag Due to Polishing Pressure		Constant Gravity Deflection		Constant Fundamental Frequency	
	Closed	Open	Closed	Open	Closed	Open
Triangle	100	100	100	100	100	100
Square	97	97	100	110	100	160
Hexagon	96	96	100	160	100	160

Table 5.2 Open- and closed-back performance penalties for various core configurations.

Geometric Shape	Weight, lb, Due to Lap Pressure Sag = 3 μin.	Gravity Deflection, μin.	Fundamental Frequency, Hz
Triangle*	100	67	200
Square	97	78	100
Hexagon	96	280	80

*Selected design.

design, and a fourfold degradation (400%!) for the hexagonal core. The fundamental frequency penalty is twofold lower for the square section versus triangular section, and even lower for the hexagonal section. Triangles win!

Figure 5.6(a) shows a glass optic that was manufactured using the closed-back technique[2] with a hexagonal core, while Fig. 5.6(b) shows an open-back glass design with a triangular core. Figure 5.6(c) shows an open-back hexagonal core and is, of course, not the recommended design of choice.

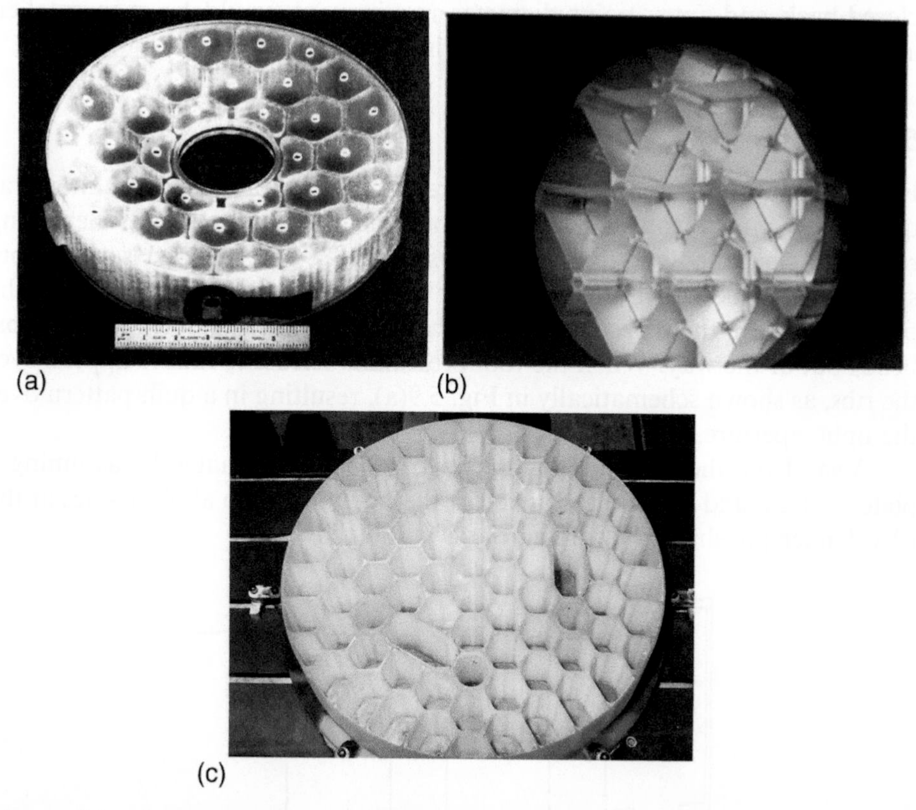

Figure 5.6 Lightweight optics having (a) a closed-back design with a hexagonal core (showing a vented back surface), (b) an open-back triangular-pocket design, and (c) an open-back design with an ill-fated hexagonal core. [Part (a) reprinted from Ref. 2.]

5.4 Partially Closed-Back Optics

As an attempt to avoid closed-back fusion processes of glass ceramics, which is not often possible, manufacturers have machined solid blanks from the back side to produce lightweight pockets of an inscribed diameter "a" but with an entry point of a diameter that is somewhat smaller, leaving a partially closed-back section, as shown schematically in Fig. 5.7. Coring tools designed to open and expand once the entry point is cleared are used for removing material. In this case, hexagonal cores are produced, which maximizes the internal angles for lower-risk machining when compared to triangular sections. As we have seen, hexagonal cores are fine for closed-back mirrors.

However, finite element model studies reveal that unless the entry point diameter is less than three-quarters that of the pocket diameter, there is nothing to be gained over an open-back triangular core section; in fact, the weight is higher for a given depth. This is evident in Fig. 5.8, where we see that the open-back triangle outperforms the partially closed-back hexagon when the entry point is 0.75 times the pocket diameter. If we add a little depth envelope for the open-back triangle to equalize the weight, then the partially closed-back and entry point diameter requirement would be even smaller, approaching the tool limit. Triangles win, again!

5.5 Polish

Polishing and figuring a lightweight core optic can create an effect known as quilting, or print-through, which does not occur in solid optics. When using relatively rigid lap tools that encompass an area greater than several core spacings, the tool tends to polish over the ribs much more so than in the pockets between them because the facesheet, being more flexible than the ribs, bends out of the way. When the tool pressure is released, valleys appear over the ribs, as shown schematically in Fig. 5.9(a), resulting in a quilt pattern over the optic aperture.

Away from the edge, the quilt error can be approximated by assuming a plate of size a and a thickness t fixed at the boundaries on all four sides at the ribs. Under polishing tool pressure q, we find that[3]

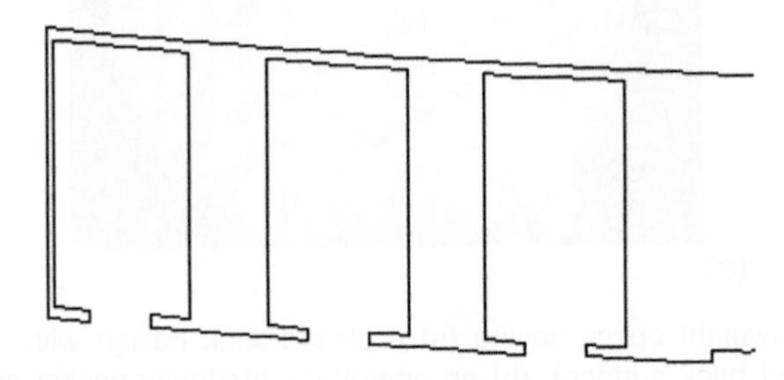

Figure 5.7 Section of a partially closed-back optic.

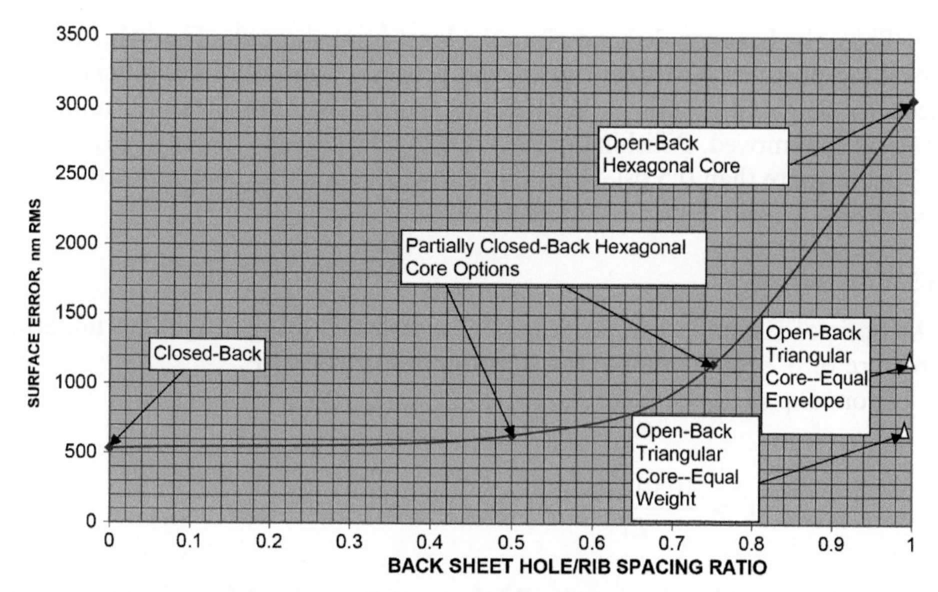

Figure 5.8 Gravity performance for open- and closed-back, or partially closed-back, optics with hexagonal cores and optical axis vertical. Partially closed-back designs cannot exceed the performance of an open-back triangular core design.

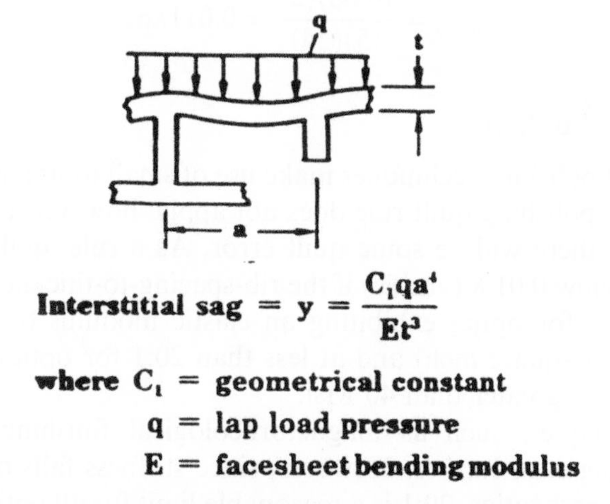

Figure 5.9 Lightweight optics and quilt in which the facesheet sags under pressure, causing more removal over the ribs than between them during polish.

$$Y = \frac{C_1 q a^4}{E t^3}, \tag{5.3}$$

where C_1 is approximately 0.014 for a square core, 0.018 for a triangular core, and 0.011 for a hexagonal core. Here, Y is the P-V quilt surface error. At the edges, the quilt error will be slightly higher.

Thus, the hexagonal shape gives the least print-through. The value of the polishing pressure q will depend on the process. Typically, the optician will need to weight the optic or tool during polish; without pressure, minimal material is removed. Polishing pressures using a rigid lap typically range from 0.2 psi to more than 0.5 psi. With this knowledge and the core geometry, one can readily calculate print-through error.

5.5.1 Example

The optic of Example 1 in Section 5.2.2 is polished using a rigid lap under a pressure of 0.2 psi. Compute the RMS quilt visible WFE.

From Eq. (5.3), we have

$$Y = \frac{(0.018)qa^4}{Et^3} = 0.68\ \mu\text{m peak surface error.}$$

Using a factor of 5 reduction for RMS, multiplying by 2 for wavefront, and converting to waves, we have

$$\lambda_{\text{RMS}} = \frac{(0.68)(2)}{(5)(25)} = 0.011\lambda\alpha.$$

5.5.2 Advanced polish

More-advanced polishing techniques make use of small tools and flexible laps. In this case, the polishing quilt rule does not apply; however, although it will not be as high, there will be some quilt error. As a rule, quilt error can be minimized to below 0.01 λ (visible) if the rib-spacing-to-thickness ratio is kept at less than 15:1 for optics exhibiting an elastic modulus of under 20 Msi (megapounds per square inch) and at less than 20:1 for optics exhibiting an elastic modulus of greater than 40 Msi.

Other techniques, such as magnetorheological finishing (MRF), can eliminate quilting even further. However, since stiffness falls off at high rib-spacing-to-thickness ratios, 20:1 is a reasonable limit for all optics in this case.

If such advanced techniques are not possible, it can be advantageous to minimize quilt error by machining partial depth ribs between the full height ribs. For open-back optics, this approach is sometimes referred to as the "waffle" or "cathedral ceiling" approach, as depicted in Fig. 5.10. Analysis has shown that if such partial ribs are one-third of the full-height rib, sufficient stiffness is provided to keep quilt error near the same limits as if the ribs were full height, thereby reducing mass penalty without significantly increasing the overall bending stiffness of the optic.

The weight per unit area now becomes

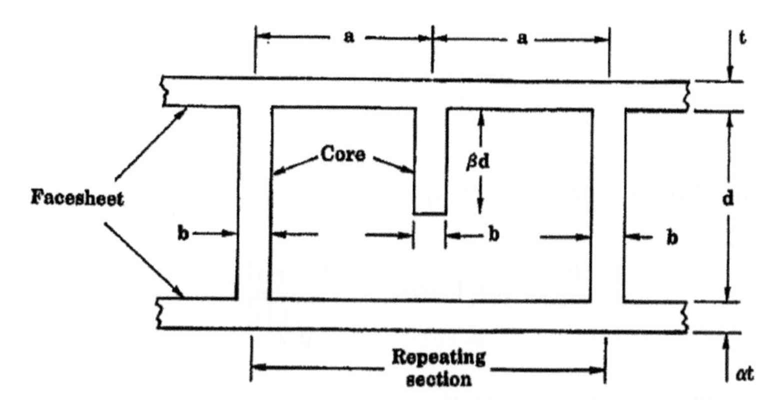

Figure 5.10 Lightweight optics and quilt: partial-depth intermediate ribs ("cathedral ceiling") reduce weight while providing quilt resistance but do not add significant stiffness.

$$\frac{W}{A} = \rho_f t(1 + \alpha) + \frac{bd(1 + \beta)\rho_c}{a}, \tag{5.4}$$

where ρ_f and ρ_c are the weight density of the facesheet and the core, respectively. Here, β is the ratio of partial-to-full rib height. When $\beta = 1$, Eq. (5.4) reduces to Eq. (5.2), as it must. Note that in this formulation, a value of $\beta = 1$ indicates a full-depth rib, i.e., no partial rib.

5.6 Weight Optimization

It is evident from Eq. (5.3) that there are an infinite number of combinations of rib-spacing-to-facesheet thickness ratios possible for a given quilt error. Which combination will yield the lightest-weight design? To answer this, we turn to the maxima-minima problem of differential calculus. Recall that a curve contains a particular point at which the slope is zero, and this point (the first derivative) defines a minimum or maximum. To minimize weight for the rigid lap polish criterion, we solve for the thickness as a function of a in Eq. (5.3), substitute into the weight-per-unit-area equation [Eq. (5.4)], and take the first derivative with respect to thickness to solve for the optimum facesheet thickness. We then substitute this value back into Eq. (5.3) and solve for the optimum spacing. From Eq. (5.4),

$$\frac{W}{A} = \rho_f t(1 + \alpha) + \frac{bd(1 + \beta)\rho_c}{a},$$

and from Eq. (5.3),

$$Y = \frac{C_1 q a^4}{E t^3},$$

we obtain

$$a = \sqrt[4]{\frac{EYt^3}{C_1 q}}. \tag{5.5}$$

Substituting Eq. (5.5) into Eq. (5.4),

$$\frac{W}{A} = \rho_f t(1 + \alpha) + \frac{bd(1 + \beta)\rho_c}{\sqrt[4]{\frac{EYt^3}{C_1 q}}}, \tag{5.6}$$

and thus, differentiating Eq. (5.4) with respect to thickness and setting the derivative equal to zero, we have

$$\begin{aligned} \frac{W}{A} &= \rho_f t(1 + \alpha) + \frac{bd(1 + \beta)\rho_c}{a}, \\ Y &= \frac{C_1 q a^4}{Et^3}, \\ a &= \left(\frac{Ey}{C_1 q}\right)^{1/4} t^{3/4}. \end{aligned} \tag{5.6a}$$

Substituting Eq. (5.6a) into Eq. (5.4), we have

$$\begin{aligned} \frac{W}{A} &= \rho_f t(1 + \alpha) + \frac{bd(1 + \beta)\rho_c}{\left(\frac{Et^3 y}{C_1 q}\right)^{1/4}} \\ &= \rho_f t(1 + \alpha) + \frac{bd(1 + \beta)\rho_c (C_1 q)^{1/4} t^{-3/4}}{(Ey)^{1/4}}. \end{aligned} \tag{5.6b}$$

To optimize, differentiate with respect to t and set the value equal to zero:

$$\frac{d(\frac{w}{a})}{dt} = 0 = \rho_f t(1 + \alpha) - \frac{3}{4} bd(1 + \beta)\, \rho_c (C_1 q)^{1/4} t^{-7/4}, \tag{5.6c}$$

$$\rho_f t(1 + \alpha) = \frac{3}{4} bd(1 + \beta)\, \rho_c (C_1 q)^{1/4} t^{-7/4}, \tag{5.6d}$$

$$\frac{4\rho_f (1 + \alpha)(Ey)^{1/4}}{3bd(1 + \beta)\rho_c (C_1 q)^{1/4}} = t^{-7/4}. \tag{5.6e}$$

Therefore, the optimum facesheet thickness is

$$t = \left[\frac{0.75bd(1+\beta)\rho_c}{\rho_f(1+a)\left(\frac{E_y}{C_{1q}}\right)^{1/4}} \right]^{4/7}, \tag{5.7a}$$

and the optimum core spacing is as given in Eq. (5.6a).

5.6.1 Example

Turn again to the examples of Sections 5.2.2 and 5.5.1, for which in the latter we determined print-through for a given design. We calculated the weight at 14.1 lbs and the quilt error at 0.68 μin. Keeping the quilt error unchanged, we use Eqs. (5.7a) and (5.7b) to find the optimum weight facesheet thickness and rib spacing distance. We find that $t = 0.23$ in. and $a = 2.27$ in., with a resulting weight of $W = 13.9$ lbs. Compared to the 14.1-lb optic, this is a rather modest weight saving; however, there is some benefit due to ease in manufacture, since the facesheet is thicker, lowering the machining risk.

If we now assume an advanced polishing technique in which for the same optic we can tolerate a rib-spacing/facesheet ratio of 15:1, we have

$$K = \frac{a}{t} = 15, \tag{5.8}$$

where $a = Kt$, and K is the ratio of core spacing to facesheet thickness. Here, we substitute into Eq. (5.4) and again take the first derivative:

$$\frac{dW}{dt} = (1+\alpha) - \frac{bd(1+\beta)t^{-2}}{K} = 0, \tag{5.9}$$

$$t^2 = \frac{bd(1+\beta)}{K(1+\alpha)}, \tag{5.9a}$$

$$t = \sqrt{\frac{bd(1+\beta)}{K(1+\alpha)}}, \tag{5.9b}$$

$$a = K\sqrt{\frac{bd(1+\beta)}{K(1+\alpha)}}. \tag{5.9c}$$

For this example, we find that $t = 0.219$ in. and $a = 3.29$ in. with a resulting weight of $W = 11.0$ lbs. Compared to the 14.1-lb optic, this is a rather significant (20%) weight saving.

5.7 Stiffness Criteria

In the above discussions, we have optimized weight to maintain a low polish error; however, we have said nothing about stiffness and stiffness-to-weight optimization. While we have minimized weight for quilt error, we may not have maximized stiffness, nor hence the stiffness-to-weight ratio. We review this next.

For solid sections, we have seen that the bending stiffness, or inertia I, is proportional to its thickness cubed, or

$$
\begin{aligned}
I &= \frac{at^3}{12} \text{ for width } a \text{, and for unit spacing,} \\
\frac{I}{a} &= \frac{t^3}{12} \text{ and} \\
t &= \sqrt[3]{\frac{12I}{a}}.
\end{aligned}
\tag{5.10a}
$$

For lightweight sections, we calculate the equivalent bending stiffness for a repeating section as

$$t_{\text{eq}} = \sqrt[3]{\frac{12I}{a}}, \tag{5.10b}$$

where in this case I is the moment of inertia of the repeating cross-section of Fig. 5.5, i.e., the moment-of-inertia section shaped as the letter I for closed-back optics and that of a section shaped as the letter T for open-back designs.

We can solve for the repeating inertia of an open-back section with full-height ribs by calculating the section's neutral axis and then using the parallel axis theorem. We solve for the neutral plane as

$$¥ = \frac{(a-b)(h-\frac{t}{2})t + (\frac{bh^2}{2})}{(a-b)t + bh},$$

the flange centroidal distance as

$$D_1 = h - \frac{t}{2} - ¥,$$

and the rib centroidal distance as

$$D_2 = ¥ - \frac{h}{2},$$

and compute

$$I = \frac{(a-b)t(h-\frac{t}{2}-¥)^2 + (bh)(\frac{h}{2}-¥)^2 + (\frac{bh^3}{12})}{a}. \tag{5.11}$$

We can solve for the repeating inertia of a closed-back section with full-height ribs and equal top and bottom facesheets by the parallel axis theorem. Since the neutral plane lies at the center of the cross-section,

$$I = \frac{2(a-b)t(\frac{h}{2}-\frac{t}{2})^2 + \frac{2(a-b)t^3}{12} + (\frac{bh^3}{12})}{a}. \tag{5.12}$$

Substitution of either Eq. (5.11) or Eq. (5.12) into Eq. (5.10) gives the equivalent thickness of the section.

We can compute the stiffness-to-weight ratio of the open-back and closed-back sections with full-depth ribs by dividing Eqs. (5.11) and (5.12), respectively, by the weight-per-unit-area relation of Eq. (5.2).

For open-back sections, we have

$$\frac{I}{W} = \frac{(a-b)t(h-\frac{t}{2}-¥)^2 + (bh)(\frac{h}{2}-¥)^2 + (\frac{bh^3}{12})}{\rho\left[t+\frac{2bd}{a}\right]}, \tag{5.13}$$

and for closed-back sections, we have

$$\frac{I}{W} = \frac{2(a-b)t(\frac{h}{2}-\frac{t}{2})^2 + \frac{2(a-b)t^3}{12} + (\frac{bh^3}{12})}{\rho(2t+\frac{2bd}{a})}. \tag{5.14}$$

If we hold all values constant except for the core depth h, and let h_1 be the open-back core depth and h_2 be the closed-back core depth, we can set Eq. (5.13) equal to Eq. (5.14) and solve for the ratio h_1/h_2 that gives equal stiffness-to-weight ratios for open- and closed-back sections. Or, we can first solve for the minimum optimized weight for open- and closed-back sections for a given depth, computing the core spacing in terms of thickness from Eqs. (5.6a) and (5.7a) or Eqs. (5.9b) and (5.9c), and then solve for the value of h_1/h_2 that yields an equal open- and closed-back stiffness.

This computation is best left as an exercise for those readers having some extra time on their hands, since it becomes unwieldy. Alternatively, and more simply, we can use a shotgun approach by spreadsheet formulation of the equations. This approach is demonstrated in the following examples.

5.7.1 Examples

Example 1. Consider a fused-silica-glass, open-back optic that has a core height of 6 in., a core rib width of 0.12 in., and a diameter of 40 in. Under a polishing pressure of 0.2 psi using rigid laps, with an allowable quilt error of 1 μin. peak surface error, determine

a) the facesheet thickness, core spacing, weight, and equivalent solid thickness, with weight optimized for polish;
b) the closed-back facesheet thickness, core spacing, weight, and equivalent solid thickness, with a stiffness-to-weight ratio identical to that of the open-back section in (a);
c) the weight of a solid of the same equivalent stiffness;
d) the stiffness of a solid with the same weight as the lightweight sections; and
e) the solid thickness with the same stiffness-to-weight ratio as the lightweight sections.

Solutions:

a) To find the facesheet thickness, core spacing, weight, and equivalent solid thickness, with weight optimized for polish, we first take Eqs. (5.7a) and (5.7b) to compute the optimum facesheet thickness t and optimum core spacing a to minimize weight, respectively:

$$t = \left[\frac{0.75bd(1+\beta)\rho_c}{\rho_f(1+a)\left(\frac{Ey}{C_1 q}\right)^{1/4}}\right]^{4/7},$$

$$a = \left(\frac{Ey}{C_1 q}\right)^{1/4} t^{3/4}.$$

Here, $b = 0.12$, $d = 6.0$, $\rho = 0.08$, $\beta = 1$, and $\alpha = 0$, and we find that $t = 0.34$ in. and $a = 3.2$ in.

We next compute the weight from Eq. (5.4) as

$$\frac{W}{A} = \rho_f t(1+\alpha) + \frac{bd(1+\beta)\rho_c}{a}.$$

Here, we find that $W/A = 0.064$ psi and hence,

$$W = \frac{\pi D^2}{4(0.064)} = 80\,\text{lbs}.$$

We determine the stiffness using Eq. (5.12) and the equivalent thickness using Eq. (5.10b):

$$¥ = \frac{(a-b)(h-\frac{t}{2})t + (\frac{bh^2}{2})}{(a-b)t + bh},$$

$$D_1 = h - \frac{t}{2} - ¥,$$

and the rib centroidal distance as

$$D_2 = ¥ - \frac{h}{2},$$

$$I = \frac{(a-b)t(h - \frac{t}{2} - ¥)^2 + (bh)\left(\frac{h}{2} - ¥\right)^2 + \left(\frac{bh^3}{12}\right)}{a},$$

and

$$t_{eq} = \sqrt[3]{\frac{12I}{a}}.$$

We find that $t_{eq} = 2.8$ in. The total optic depth is $h_1 + t = 6.34$ in.

b) To find the closed-back facesheet thickness, core spacing, weight, and equivalent solid thickness, with a stiffness-to-weight ratio identical to that of the open-back section in (a), here, we find that a new core height h_2 gives the same stiffness-to-weight ratio as the open-back section. As before, we make use of Eqs. (5.7a) and (5.7b) to optimize the facesheet thickness and core spacing, respectively, for weight, where in this case, $b = 0.12$, $\rho = 0.08$, $\beta = 1$, and $\alpha = 1$.

   We do not have depth h_2 but can use Eq. (5.4) in conjunction with Eq. (5.12) for stiffness of the closed-back section and use a shotgun spreadsheet formulation to find the identical stiffness and stiffness and weight, and hence the identical stiffness-to-weight ratio. This is quickly computed:

$$I = \frac{2(a-b)t\left(\frac{h}{2} - \frac{t}{2}\right)^2 + \frac{2(a-b)t^3}{12} + \left(\frac{bh^3}{12}\right)}{a},$$

$$t_{eq} = \sqrt[3]{\frac{12I}{a}}.$$

We find that $t = 0.18$ in. and $a = 2.6$ in. for the given $W/A = 0.064$ psi, $W = 80$ lbs, and $t_{eq} = 2.8$ in. We find that $h_2 = 3.9$ in. The total optic depth is $d = h_2 + 2t = 4.26$ in.

c) To find the weight of a solid of the same equivalent stiffness, we simply take $W = \rho t_{eq} A = 280$ lbs. Note that this is more than three times heavier than the lightweight section. This is why we make lightweight optics!

d) To find the stiffness of a solid with the same weight as the lightweight sections, we simply set $W = 80 = \rho d A$; $d = \frac{80}{\rho A} = 0.80$ in. Note that this is $\left(\frac{t_{eq}}{d}\right)^3 = \left(\frac{2.8}{0.80}\right)^3 = 40$ times less stiff than the lightweight optic section of the same weight. Again, this is why we make lightweight optics!

e) To find the solid thickness with the same stiffness-to-weight ratio as the lightweight sections, we set the stiffness-to-weight ratios equal as

$$\frac{t_{eq}{}^3}{12\left(\frac{W}{A}\right)} = \frac{d^3}{12\rho d} = \frac{d^2}{12\rho}.$$

Substituting, we find that $d = 5.2$ in. with a weight of 525 lbs!—more than 6 times the weight of the lightweighted cored optic. This is why we make lightweight optics!

From the above examples, we see that the open-back optic is about 1.2 times the depth of the solid optic with the same stiffness-to-weight ratio, and the closed-back optic is about 0.80 times the depth of the solid optic. This is in agreement with the general rule set forth in Section 5.1.

We also note that the open-back optic exhibits the same stiffness and weight as the closed-back optic when only bending deformation is considered. (We will review the effects of core anisotropy and shear deflection in Section 5.9). Meanwhile, several more examples of stiffness and weight comparisons are given below.

Example 2. Consider a fused-silica-glass, open-back optic that has a core height of 4 in., a core rib thickness of 0.12 in., and a diameter of 20 in. Under a polishing pressure of 0.2 psi using rigid laps, with an allowable quilt error of 1 μin. peak surface error, determine

a) the facesheet thickness, core spacing, weight, and equivalent solid thickness, with weight optimized for polish,
b) the closed-back facesheet thickness, core spacing, core height, weight, and equivalent solid thickness, with a stiffness-to-weight ratio identical to that of the open-back section in (a),
c) the weight of a solid of the same equivalent stiffness,
d) the stiffness of a solid with the same weight as the lightweight sections, and
e) the solid thickness with the same stiffness-to-weight ratio as the lightweight sections.

Solutions:
As we did in Example 1 of Section 5.7, we find the appropriate values from substitution of the parameters into the same equations.

a) We find the facesheet thickness, core spacing, weight, and equivalent solid thickness, with weight optimized for polish, as $t = 0.27$ in. and $a = 2.7$ in. We compute the weight as $W/A = 0.051$ psi; hence,

$$W = \frac{\pi D^2}{4(0.051)} = 16\,\text{lbs}.$$

We determine stiffness (equivalent thickness) as $t_{eq} = 2.0$ in. The total optic depth is $h_1 + t = 4.27$ in.

b) To find the closed-back facesheet thickness, core spacing, weight, and equivalent solid thickness, with a stiffness-to-weight ratio identical to that for the open-back section in (a), here, we find a new core height h_2 to give the same stiffness-to-weight ratio as in the open-back section. In this case, we find that $t = 0.14$ in. and $a = 1.7$ in. for the given $W/A = 0.051$ psi, $W = 16$ lbs, and $t_{eq} = 2.0$ in. We find that $h_2 = 2.6$ in. The total optic depth is $d = h_2 + 2t = 2.88$ in.

c) To find the weight of a solid of the same equivalent stiffness, we find that $W = \rho t_{eq} A = 50$ lbs. Note again that this is more than three times heavier than the lightweight section.

d) To find the stiffness of a solid with the same weight as the lightweight sections, we again set $W = 16 = \rho d A$, $d = \frac{16}{\rho A} = 0.64$ in. Note that this is $\left(\frac{t_{eq}}{d}\right)^3 = \left(\frac{2}{0.64}\right)^3 = 30$ times less stiff than the lightweight optic section of the same weight.

e) To find the solid thickness with the same stiffness-to-weight ratio as the lightweight sections, we set the stiffness-to-weight ratios equal to each other:

$$\frac{t_{eq}{}^3}{12\left(\frac{W}{A}\right)} = \frac{d^3}{12\rho d} = \frac{d^2}{12\rho}.$$

Substituting, we find that $d = 3.5$ in. with a weight of 88 pounds! This is almost 6 times the weight of the lightweight cored optic.

From the above example, we see that the open-back optic is about 1.2 times the depth of the solid optic with same stiffness-to-weight ratio, and the closed-back optic is about 0.80 times the depth of the solid optic. This is again in agreement with the general rule set forth in Section 5.1.

Example 3. Consider a fused-silica-glass, open-back optic that has a core height of 6 in. Under an advanced-polishing flexible-lap technique ($a/t = 15$), determine

a) the facesheet thickness, core spacing, weight, and equivalent solid thickness, with weight optimized for polish,

b) the closed-back facesheet thickness, core spacing, weight, and equivalent solid thickness with a stiffness-to-weight ratio that is identical to that of the open-back section in (a),
c) the weight of a solid of the same equivalent stiffness,
d) the stiffness of a solid with the same weight as the lightweight sections, and
e) the solid thickness with the same stiffness-to-weight ratio as the lightweight sections.

Solutions:
As we did in the above examples, we find the appropriate values from the substitution of the parameters into the same equations.

a) We find the facesheet thickness, core spacing, weight, and equivalent solid thickness, with weight optimized for polish, as $t = 0.31$ in. and $a = 4.65$ in. We compute the weight as $W/A = 0.052$ psi and hence,

$$W = \frac{\pi D^2}{4(0.052)} = 65\,\text{lbs}.$$

We determine stiffness (equivalent thickness) as $t_{eq} = 2.6$ in. The total optic depth is $h_1 + t = 6.31$ in.
b) To find the closed-back facesheet thickness, core spacing, weight, and equivalent solid thickness, with a stiffness-to-weight ratio identical to that of the open-back section in (a), here, we find a new core height h_2 to give the same stiffness-to-weight ratio as the open-back section. In this case, we find that $t = 0.17$ in. and $a = 2.6$ in. for the given $W/A = 0.052$ psi, $W = 65$ lbs, and $t_{eq} = 2.6$ in. We find that $h_2 = 3.7$ in. The total optic depth is $d = h_2 + 2t = 4.04$ in.
c) To find the weight of a solid of the same equivalent stiffness, we find that $W = \rho t_{eq} A = 260$ lbs. Note again that this is now more than four times heavier than the lightweight section.
d) To find the stiffness of a solid with the same weight as the lightweight sections, we again set $W = 65 = \rho d A$, $d = \frac{65}{\rho A} = 0.64$ in. Note that this is $\left(\frac{t_{eq}}{d}\right)^3 = \left(\frac{2.6}{0.64}\right)^3 = 60$ times less stiff than the lightweight optic section of the same weight.
e) To find the solid thickness with the same stiffness-to-weight ratio as the lightweight sections, we set the stiffness-to-weight ratios equal to each other as

$$\frac{{t_{eq}}^3}{12\left(\frac{W}{A}\right)} = \frac{d^3}{12\rho d} = \frac{d^2}{12\rho}.$$

Substituting, we find that $d = 5.2$ in. with a weight of 525 lbs! This is eight times the weight of the lightweight, cored optic.

From the above example, we see again that the open-back optic is about 1.2 times the depth of the solid optic with same stiffness-to-weight ratio, and the closed-back optic is about 0.78 times the depth of the solid optic. This is again in agreement with the general rule set forth in Section 5.1.

Example 4. Consider a fused-silica-glass, open-back optic that has a core height of 4 in. Under an advanced-polishing flexible-lap technique ($a/t = 15$), determine

a) the facesheet thickness, core spacing, weight, and equivalent solid thickness, with weight optimized for polish,
b) the closed-back facesheet thickness, core spacing, weight, and equivalent solid thickness, with a stiffness-to-weight ratio that is identical to that of the open-back section in (a),
c) the weight of a solid of the same equivalent stiffness,
d) the stiffness of a solid with the same weight as the lightweight sections, and
e) the solid thickness with the same stiffness-to-weight ratio as the lightweight sections.

Solutions:
We again find the appropriate values from the substitution of the parameters into the same equations.

a) We find the facesheet thickness, core spacing, weight, and equivalent solid thickness, with weight optimized for polish, as $t = 0.25$ in. and $a = 3.8$ in. We compute the weight as $W/A = 0.042$ psi and hence,

$$W = \frac{\pi D^2}{4(0.042)} = 13.2\,\text{lbs}.$$

We determine stiffness (equivalent thickness) as $t_{eq} = 1.85$ in. The total optic depth is $h_1 + t = 4.25$ in.
b) To find the closed-back facesheet thickness, core spacing, weight, and equivalent solid thickness, with a stiffness-to-weight ratio that is identical to that of the open-back section in (a), here, we find a new core height h_2 to give the same stiffness-to-weight ratio as the open-back section. In this case, we find that $t = 0.14$ in. and $a = 2.1$ in. for the given $W/A = 0.042$ psi, $W = 13.2$ lbs, and $t_{eq} = 1.85$ in. We find that $h_2 = 2.4$ in. The total optic depth is $d = h_2 + 2t = 2.68$ in.
c) To find the weight of a solid of the same equivalent stiffness, we find that $W = \rho t_{eq} A = 46$ lbs. Note that this is more than three times heavier than the lightweight section.
d) To find the stiffness of a solid with the same weight as the lightweight sections, we again set $W = 13.2 = \rho d A$, $d = \frac{13.2}{\rho A} = 0.53$ in. Note that

this is $\left(\frac{t_{eq}}{d}\right)^3 = \left(\frac{1.85}{0.53}\right)^3 = 40$ times less stiff than the lightweight optic section of the same weight.

e) To find the solid thickness with the same stiffness-to-weight ratio as the lightweight sections, we set the stiffness-to-weight ratios equal to each other:

$$\frac{t_{eq}{}^3}{12\left(\frac{W}{A}\right)} = \frac{d^3}{12\rho d} = \frac{d^2}{12\rho}.$$

Substituting, we find that $d = 3.5$ in. with a weight of 88 lbs! This is more than six times the weight of the lightweight cored optic.

From the above example, we see once again that the open-back optic is about 1.2 times the depth of the solid optic with same stiffness-to-weight ratio, and the closed-back optic is about 0.75 times the depth of the solid optic. This is again in agreement with the general rule set forth in Section 5.1. Note that using the advanced polishing technique in both Examples 3 and 4, we find a 20% weight savings over that of a rigid-lap polish, with very little loss in stiffness.

5.8 Stiffness Optimization

We have seen how to optimize weight for lightweight core designs in the presence of polishing pressure. This results in the lightest possible design for a optic of a given depth. When we optimize for weight, we do not necessarily optimize for stiffness (which relates to external loading such as mount error, bimetal clad and coat effects, etc.) and stiffness-to-weight ratio (which relates to self-induced gravity error and fundamental frequency). In this case, after satisfying the quilt relationship between the facesheet thickness and core spacing, we can differentiate the stiffness equation with respect to facesheet thickness and solve for the optimum stiffness, and then review the weight penalty over the polishing-optimized design.

With regard to pure stiffness, we use the inertia-per-unit-width relation [(Eq. (5.11) or Eq. (5.12)], and from Eq. (5.7), we have

$$a = 4\sqrt{\frac{Et^3 y}{C_1 q a^4}}$$

for rigid laps or, from Eq. (5.8), $a = Kt$ for flexible laps.

We can differentiate inertia with respect to t to find the optimized thickness for the best stiffness. Here we have [from Eq. (5.11)] for an open-back section,

$$\frac{d\left(\frac{I}{a}\right)}{dt} = \frac{d}{dt}\left[\frac{(a-b)t(h-\frac{t}{2}-¥)^2+(bh)\left(\frac{h}{2}-¥\right)^2+\frac{bh^3}{12}}{a}\right]=0 \tag{5.15}$$

and solve for the optimum thickness to optimize stiffness. In a similar way, for a closed-back section [from Eq. (5.12)], we have

$$\frac{d\left(\frac{I}{a}\right)}{dt} = \frac{d}{dt}\left[\frac{2(a-b)t\left(\frac{h}{2}-\frac{t}{2}\right)^2+\frac{2(a-b)t^3}{12}+\frac{bh^3}{12}}{a}\right]=0 \tag{5.16}$$

and solve for the optimum thickness to optimize stiffness.

In a similar fashion, we can divide the inertia per unit width by the weight per unit area of Eq. (5.2) (for full-depth ribs) and again differentiate with respect to t to find the optimized thickness for the best stiffness-to-weight ratio, or

$$\frac{d}{dt}\left[\frac{I}{a\left[\rho\left[t(1+\alpha)+\frac{2bd}{a}\right]\right]}\right]=0. \tag{5.17}$$

However, a review of these derivatives shows them to be unwieldy at best, if not intractable. It is simpler to plot stiffness and the stiffness-to-weight ratio without differentiation, and to note maxima/minima from the plot data for varying facesheet thicknesses, then to note the maximum or minimum from the curve.

For example, consider Fig. 5.11, which plots stiffness (equivalent thickness), normalized weight, and the stiffness- (normalized equivalent thickness cubed) -to-weight-versus-facesheet-thickness ratio for an advanced-polish, closed-back optic with $K = 20$. The section depth is 1.6 in. Note that the optimum weight occurs at a facesheet thickness near 0.09 in., while the optimum stiffness occurs at very small or very high facesheet thicknesses, where the optic is completely solid, for the given value of K. (A solid optic has the maximum stiffness for a given depth.) The optimum stiffness-to-weight ratio occurs at a higher facesheet thickness, near 0.17 in. Thus, for an optimized stiffness-to-weight ratio, the chart shows a weight penalty of approximately 30%. On the other hand, the stiffness-to-weight ratio is improved by up to 20% for the heavier design; design drivers will determine the best choice, as will core cell packaging to a fixed-diameter optic.

Now consider Fig. 5.12, which plots stiffness (equivalent thickness), normalized weight, and the stiffness- (normalized equivalent thickness cubed) -to-weight-versus-facesheet-thickness ratio for an advanced-polish, open-back optic with $K = 20$. The section depth is 3.0 in. Note that the optimum weight

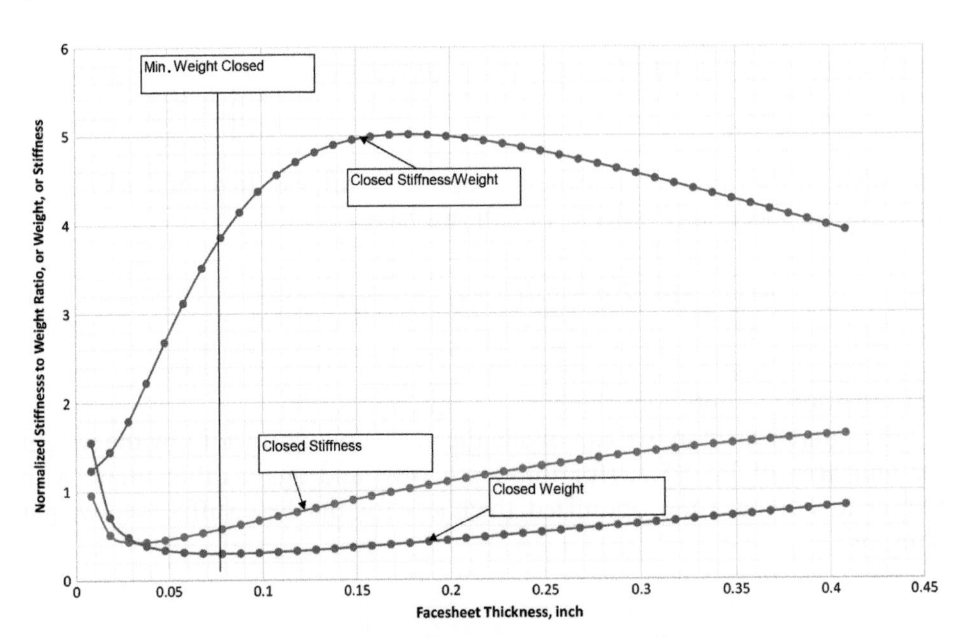

Figure 5.11 Closed-back stiffness and weight optimization. The rib-spacing-to-facesheet-thickness ratio is 20:1 for quilt considerations.

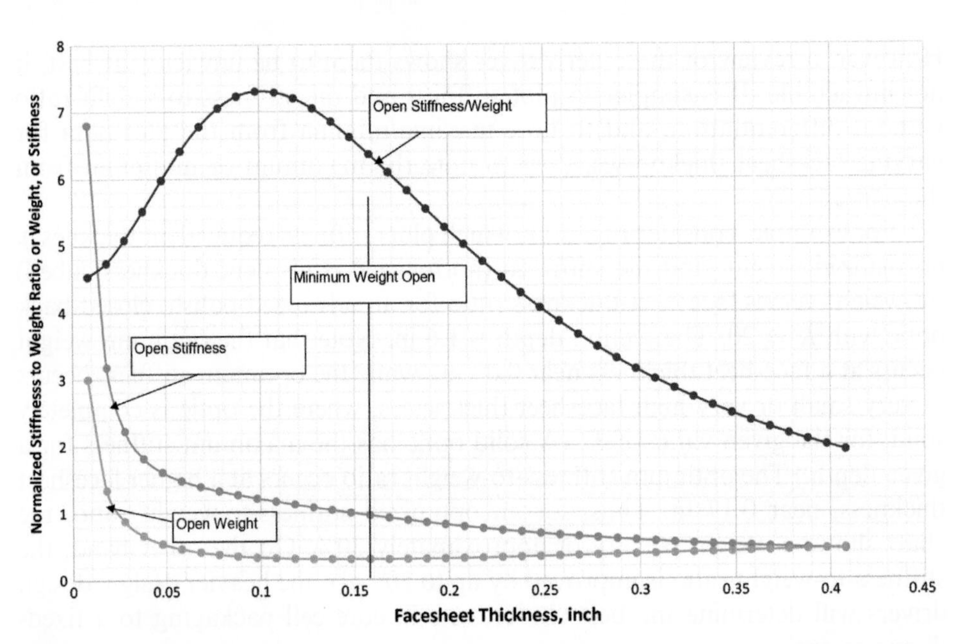

Figure 5.12 Open-back stiffness and weight optimization. $K = 20$ for polish quilt-error weight optimization.

occurs at a facesheet thickness near 0.16 in., while the optimum stiffness occurs at very small facesheet thickness, where the optic is completely solid, for the given value of K. (A solid optic has the maximum stiffness for a given depth.) The optimum stiffness-to-weight ratio occurs at a facesheet thickness near 0.10 in. Thus, in this case, for an optimized stiffness-to-weight ratio, the chart shows an approximate weight penalty of only 10%. Similarly, the stiffness-to-weight ratio is improved by a similar amount; in other words, we have optimized the design for *both* minimum weight and maximum stiffness-to-weight ratio.

5.9 The Great Debate

The previous section presented weight- and stiffness-optimization analysis techniques. It also alluded to the equivalence of open- and closed-back performance for equivalent weight and stiffness. However, we have limited our discussions to isotropic configurations and have also neglected the effects of shear deflection. This leads to a discussion of whether open- and closed-back optics can lead to equivalent performance at a given weight, or equivalent weight at a given performance criterion.

Much has been written about closed- and open-back performance relative to weight, stiffness, and core geometry. The general consensus[4–6] from the literature appears to view closed-back (sandwich) designs for lightweight optics as preferential in this regard. However, caution is advised here, since comparisons are often not made "apples to apples," so to speak. For example, a closed-back optic may be shown to be far superior in stiffness performance at a given depth envelope, whereas an open-back optic may be shown to be far superior in weight performance without the envelope restriction. It is the intent of the following subsections to illustrate comparisons with weight being equal for both closed- and open-back sections. Shear and anisotropy will be duly considered.

5.9.1 Closed-back geometry

Section 5.3 showed that performance is independent of the core geometry, inferring that hexagonal, square, or triangular cores will yield similar results. A study conducted[7] in the 1980s shows analytically, and is confirmed by finite element analysis, that a triangular core may improve performance for a given weight. It is not the intent here to refute this; however, the effects are secondary and hence of only minor importance.[8] Most manufactures do not prefer triangular-core construction in a sandwich design because fabrication techniques favor cores with a greater re-entrant angle, reducing both weight and cost, due to refining the interstitial fillets. The added benefit of the triangular core does not outweigh the fabrication approach. Thus, the opto-structural analyst is not very concerned about core shape.

5.9.2 Open-back geometry

We have already shown in Section 5.3 that open-back cores must be made triangular if we are to avoid the ill effects of stiffness degradation; square cores affect fundamental frequency, while hexagonal cores fair very poorly in both fundamental frequency and self-weight deflection. In this latter regard, the core has no continuous diagonal, horizontal, or vertical path, immensely reducing its bending stiffness. Thus, we limit our discussion on open-back optics to triangular cores.

5.9.3 Open- and closed-back design comparisons

When the envelope of depth is limited, closed-back designs will be stiffer than open-back designs, although significantly heavier and significantly more costly. If the optic is performance limited, one has to pay the price and live with the mass for a closed-back design. However, if the envelope of depth is not a driver, and weight is held constant, how does performance compare?

Figures 5.11 and 5.12 show the effects of optimization on weight, stiffness, and the stiffness-to-weight ratio for open- and closed-back optics, respectively. We now turn our attention to how the two designs compare to one another. We neglect shear deformation, for the moment considering only the bending effects. Comparisons are made for equal-weight optics that meet polishing quilt requirements.

Figures 5.13 through 5.15 show open-back and closed-back comparisons for weight, stiffness, and stiffness-to-weight ratio, respectively, as a function of facesheet thickness for a given depth. As is evident from Fig. 5.13, the two

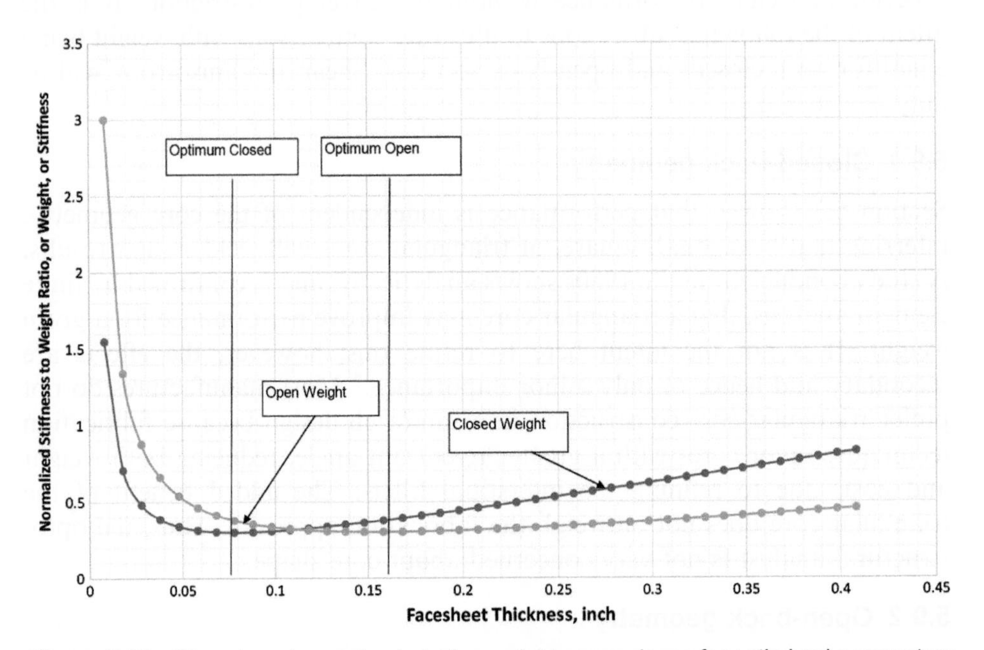

Figure 5.13 Closed- and open-back design weight comparisons for optimized parameters.

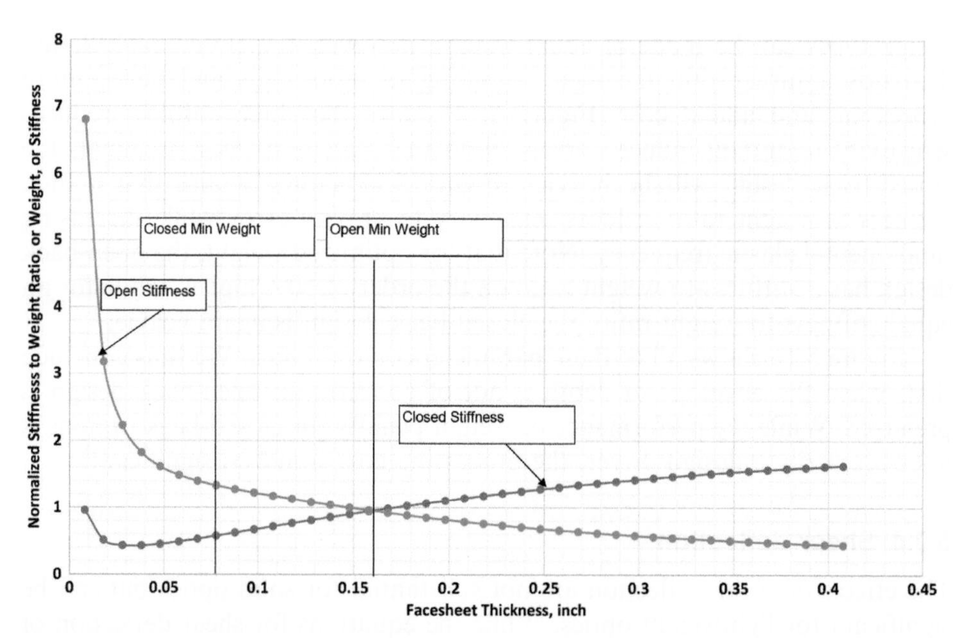

Figure 5.14 Closed- and open-back design stiffness comparisons for optimized parameters.

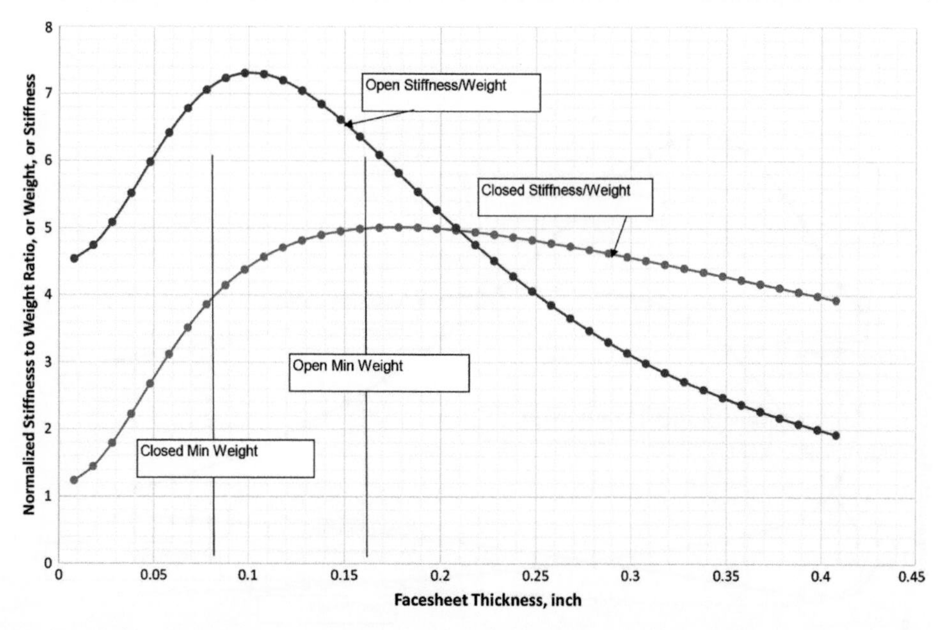

Figure 5.15 Closed- and open-back design stiffness-to-weight ratio comparisons for optimized parameters.

design types can be made of equal weight; however, the closed-back design demands a rather thin facesheet, resulting in more cells and hence more processing and added cost. Figure 5.14 shows that high stiffness reduces sensitivity to mount-induced error. Note that where stiffness is equal, the closed-back design will be heavier. Figure 5.15 shows, again, that a high stiffness-to-weight ratio reduces sensitivity to gravity error while increasing fundamental elastic frequency. Note that for optimized weight, the open-back design has a stiffness-to-weight ratio on the order of 50% higher, while for an equal stiffness-to-weight ratio, the closed-back design is again heavier.

Figure 5.16 shows all of these parameters on one chart. We thus conclude that, when the envelope of depth is not of concern, an open-back design is preferred. While the performance or weight penalty for closed-back options is not necessarily a design driver, the processing cost is indeed significant.

5.9.4 Shear deflection

The effects of shear deflection are not substantial for solid optics but can be significant for lightweight optics. While the equations for shear deflection of beams (Chapter 1) are readily solved, shear deflection of lightweight optics, being rather unwieldy, is not easily formulated, and it is not necessary to expound upon it here. The reader can peruse the references at leisure; however, for simplicity, we can use a modified first-order approximation of self-weight (optical axis vertical) shear deflection first developed by Barnes[9] as

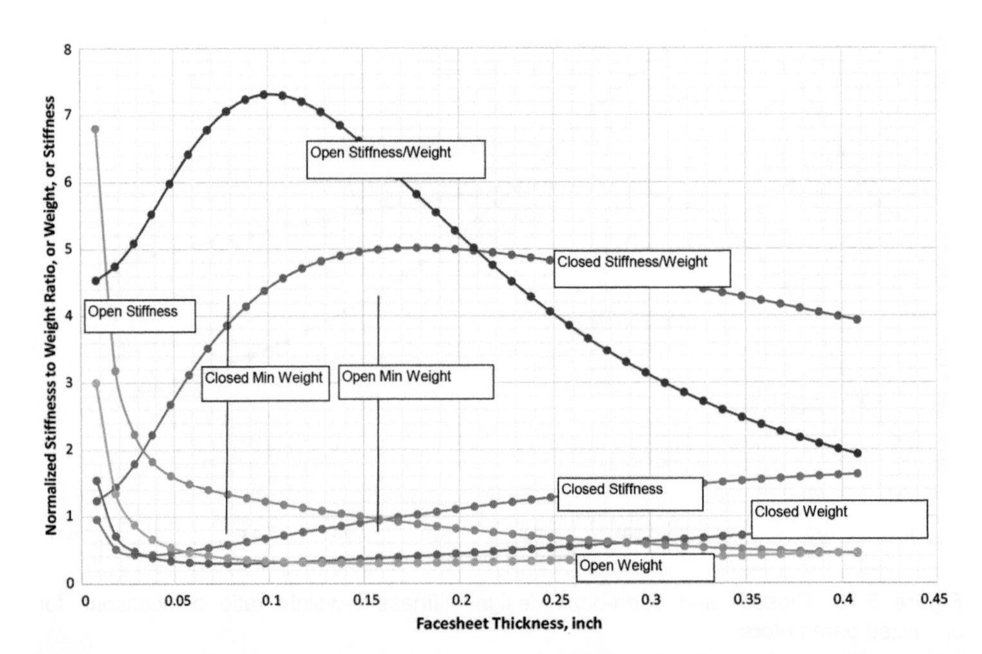

Figure 5.16 Closed- and open-back design comparisons for all criteria of stiffness and weight.

$$Y = \frac{\rho k D^2}{G}, \tag{5.18}$$

where ρ is the weight density; k is a deflection constant, where $k = 1$ for a three-point edge-mounted optics and $k = 0.5$ for an internally mounted optics at the 0.7 zone; and G is the shear modulus of ridigity. This is in line with analysis by Soosar[10] and others, and serves us well here.

Figure 5.17 shows a plot of shear and bending deflection contribution for a 20-in. diameter, 80% lightweighted, open-back glass optic of various depths. Note that even at 5:1 diameter-to-depth ratio, one-half of the deflection is due to shear; however, at aspect ratios of 10:1 and higher, bending deflection dominates.

Note that Eq. (5.18) is somewhat independent of the core depth. This is due to the fact that as depth increases (increasing shear area), weight also increases in direct proportion, negating any benefit or drawback. Recall that the core itself, and not the facesheets, carries most of the shear. This is of interest in that the shear deflection is independent of whether the optic is closed back or open back; a closed-back design is, in essence, equally deformed in shear. Thus, there is no drawback for shear deflection of an open-back optic, regardless of its depth.

5.9.5 Anisotropy

An open-back, triangular-core, lightweight optic is symmetrical along its 60-deg planes. In this sense, it is quasi-isotropic. For circular mirrors, the

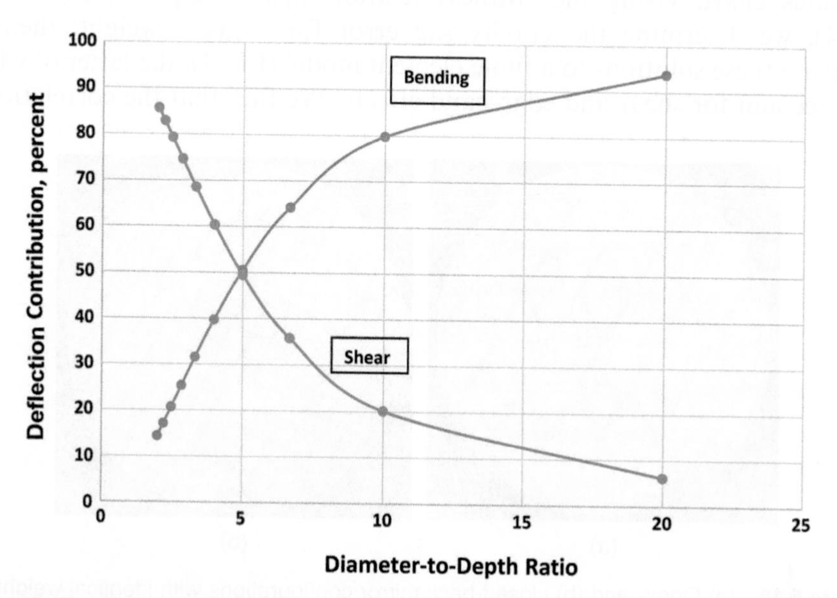

Figure 5.17 Shear and bending deflection contribution versus aspect ratio for a 20-in.-diameter, lightweight, open-back optic.

anisotropy has no significant impact on either self-weight deflection or fundamental mode when compared to a closed-back optic. To illustrate this, consider a 20-in.-diameter optic as shown in Fig. 5.18. Using finite element analysis, this is compared to an equal-weight closed-back optic in Table 5.3 for both self-weight deflection and fundamental frequency. Note that self-weight deflection is identical (Fig. 5.19), as expected, and mode shape is nearly identical with a small penalty for local edge-rib bending, as noted in Fig. 5.20.

5.9.6 Analytical comparison

Using our analytical design equations, we can review again the benefits of an open-back optic. In Fig. 5.21, the bottom-to-top facesheet thickness ratio a from Eq. (5.2) varies; recall that $0 < a < 1$, where $a = 0$ is an open-back optic, and $a = 1$ is a closed-back optic with equal top and bottom facesheets. Using the stiffness relationships of Eqs. (5.11) through (5.14), we determine weight for a given performance criterion of stiffness in terms of a fundamental mode.

Here, we see that the mass of an open-back optic is significantly reduced (15%) compared to a closed-back optic of equal frequency response. Note, however, that if the closed-back optic back-sheet thickness comprises 50% of the thickness of the top facesheet, the mass of the closed design can be less than or equal to that of the open design. On the other hand, as explained in Section 5.9.3, this results in a rather thin back sheet, potentially causing manufacturing or process difficulty.

In Fig. 5.22, the bottom-to-top facesheet thickness ratio varies, as in the previous chart. Using the stiffness relationships of Eqs. (5.11) through (5.14), we determine the gravity sag error for a given weight; then we compare these solutions to a finite element model (FEM), the latter of which will account for shear and edge-band effects. We find that the correlation is

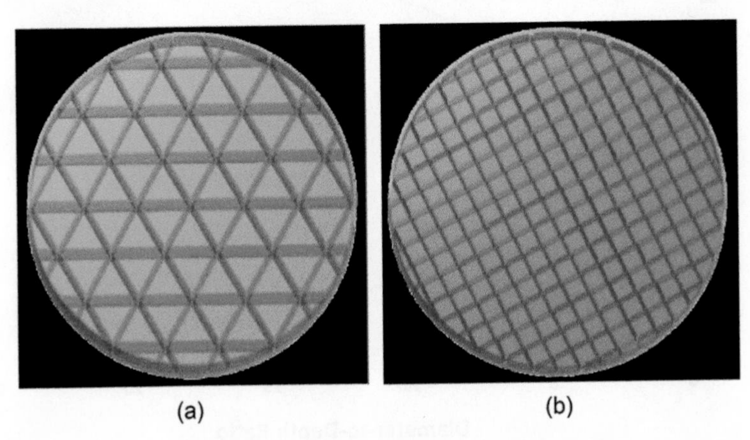

(a) (b)

Figure 5.18 (a) Open- and (b) closed-back mirror configurations with identical weight and quilt error.

Table 5.3 Comparison of open- and closed-back, 20-in.-diameter, ULE™ mirrors optimized for polish quilt. Mirrors are mounted kinematically at three edge points.

Configuration	Rib Width (inch)	Rib Depth (inch)	Facesheet Thickness (inch)	Core Spacing (inch)	Weight (pounds)	Gravity Deflection (micro-inch)	Fundamental Frequency (hertz)
Open	0.12	3	0.29	1.9	19	22.4	705
Closed	0.12	2.12	0.19	1.25	19	22.4	737

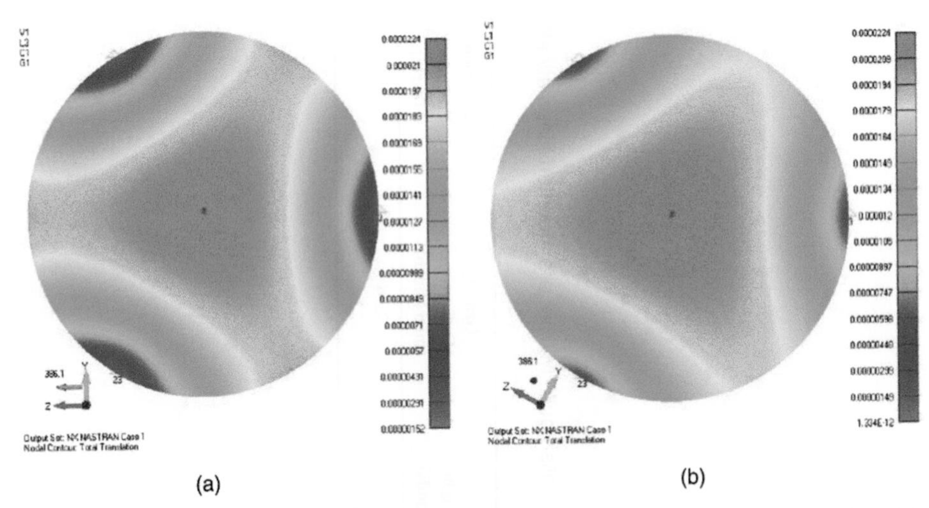

(a) (b)

Figure 5.19 Gravity error for (a) open-back and (b) closed-back optics of equal weight. Both errors are nearly identical in magnitude and shape.

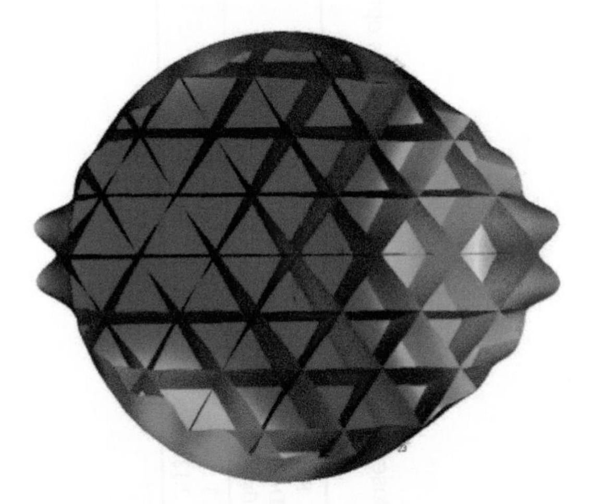

Figure 5.20 The first mode shape of an open-back optic indicates that the edge-rib bending effect reduces the fundamental mode by less than five percent compared to an equal-weight closed-back optic.

good. Note again that for equal weight, the open-back design results in improved gravity performance (up to 30%) unless the closed-back sheet thickness is significantly reduced.

5.9.7 And the winner is...

It is now obvious that an open-back optic with a triangular core will generally equal or exceed the performance of a closed-back optic for a given weight, or

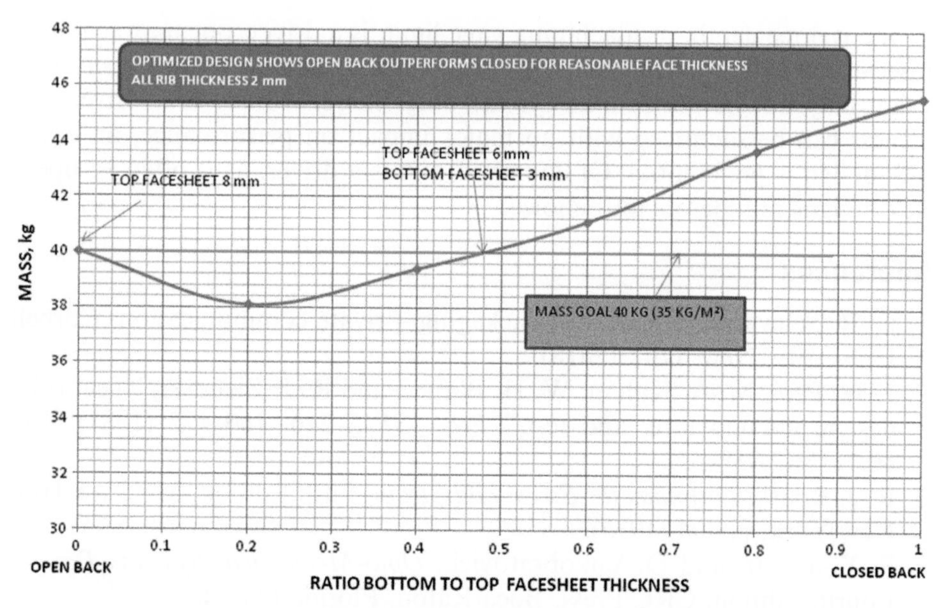

Figure 5.21 Open- and closed-back mass comparison for a 1.2-m glass optic with equal performance (200 Hz) and varying bottom-facesheet thicknesses.

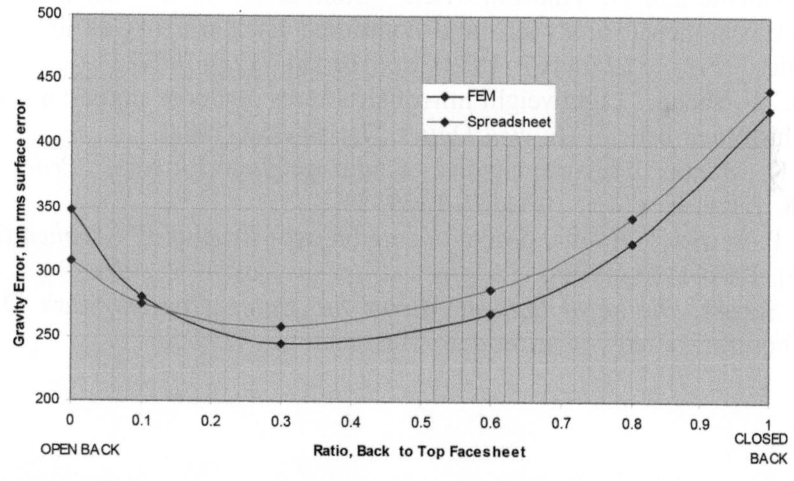

Figure 5.22 Performance for equal-mass (21 kg/m^2) lightweight optics (20:1 cell-size-to-top-facesheet ratio): detailed hand analysis versus finite element model (FEM) results.

improve mass for a given performance criterion as long as envelope of depth is not a driver.

Although the above formulations consider polishing quilt, the above arguments hold even if polishing becomes a nonissue as techniques continue to advance. Due to shear lag, only a portion of the facesheet contributes to stiffness so that the rib-spacing-to-facesheet-thickness ratio should not exceed the 20:1 ratio in any case.

If envelope depth, or machining gage limit, is a driver, one may require a closed-back optic to meet stiffness performance criteria, as the open-back design may not be an option. On the other hand, if mass is a driver and stiffness can be sacrificed with no major impact on performance, an open-back design is preferential. Of course, if cost is a driver, as it often is, open-back designs are the obvious choice.

References

1. C. R. Schwarze, Treatise on packaging, private communication, Cornell University, New York (2018).
2. J. W. Pepi, M. A. Kahan, W. H. Barnes, and R. J. Zielinski, "Teal ruby design, manufacture, and test," *Proc. SPIE* **216**, 160–173 (1981) [doi: 10.1117/12.958459].
3. R. J. Roark, *Formulas for Stress and Strain*, Fifth Edition, McGraw-Hill, New York, p. 404 (1975).
4. P. Yoder, Jr. and D. Vukobratovich, *Opto-Mechanical Systems Design*, Fourth Edition, CRC Press, Boca Raton, Florida (2015).
5. D. Vukobratovich, Ed., *Proc. SPIE* **1167**, *Precision Engineering and Optomechanics* (1989).
6. T. Valente and D. Vukobratovich, "Comparison of the merits of open-back, symmetric sandwich, and contoured back mirrors as lightweight optics," *Proc. SPIE* **1167** (1989) [doi: 10.1117/12.962927].
7. S. C. F. Sheng, "Lightweight mirror structures best core shapes: a reversal of historical belief," *Applied Optics* **27**, 354–359 (1988).
8. P. K. Mehta, "Flexural rigidity of light-weighted mirrors," *Proc. SPIE* **748**, 158 (1987) [doi: 10.1117/12.939828].
9. W. P. Barnes, "Optimal design of cored mirror structures," *Applied Optics* **8**(6), 1191–1196 (1969).
10. K. Soosar, *Design of Optical Mirror Structures*, Charles Stark Draper Laboratories, MIT, Cambridge, Massachusetts (1971).

Chapter 6
Lightweight Optics: Performance Error

We can now compute the displacements for the lightweight, cored optics of Chapter 5 for the various criteria set forth in Chapter 4 for solid optics.

6.1 Mount-Induced Error

The optic is subject to mount loads from assembly or thermal environments, as discussed in Chapter 4. These moments are about either a tangential or radial line. The optic is simultaneously subject to radial loads P from the mounts. The radial loads can be converted to tangential moments to the first order by multiplying by the mount-to-vertex eccentricity e, where

$$e = \frac{r^2}{2R}, \tag{6.1}$$

$$M = Pe. \tag{6.2}$$

6.1.1 Tangential moment

We simply take the equation for solid optic performance from Eq. (4.5),

$$Y = \frac{6M'a^2(1-u)}{Et^3} = \frac{2.86Ma(1-u)}{Et^3},$$

and substitute the equivalent thickness t_{eq} [from Eq. (5.10)] for t:

$$Y = \frac{6M'a^2(1-u)}{Et^3} = \frac{2.86Ma(1-u)}{Et_{\text{eq}}^3}. \tag{6.3}$$

This is the P-V surface error. To convert to RMS wavefront error, multiply by 2 and divide by 2.5×10^{-5} waves per inch and the P-V to RMS ratio of 5:1.

6.1.2 Radial and axial moments

When the three-point moment enters about a radial line, [from Eq. (4.6)] using the equivalent thickness,

$$Y = \frac{3.08Ma(1-u)}{Et_{eq}^3}. \tag{6.4}$$

Note that if a maximum error induced by moments is desired for input to a performance prediction budget, one can calculate the allowable moment and use the quasi-kinematic mount equations of Chapter 3 to design an appropriate mount.

6.2 Gravity

For lightweight-optics gravity performance error, we similarly scale the solid-optics equations using the equivalent thickness approach and account for the reduced mass.

6.2.1 Optical axis vertical

For a three-point edge mount when the optical axis is vertical, we compute the optic weight density from the lightweight equations and modify the solid-optics equation [Eq. (4.8), repeated here] as

$$Y = \frac{0.434\rho\pi r^4(1-u)}{Et^2}.$$

The lightweight optic density and weight can be determined from Eq. (5.4) and compared to the weight of the solid as

$$K_1 = C_2 H, \tag{6.5}$$

where C_2 is the weight of the lightweight optic remaining from its full depth H, and find that

$$Y = \frac{0.434K_1\rho\pi r^4(1-u)}{Et_{eq}^3}. \tag{6.6}$$

For example, for an optic of parent solid height H of 4 in. that is 85% lightweighted and exhibits an equivalent thickness of 1.8 in., we have $K_1 = C_2H = (1-0.85)(4) = 0.6$, and $t_{eq} = 1.8$, which can then be substituted into Eq. (6.6) to determine deflection.

Similarly, for a three-point internal mount, we modify the solid equation from Section 4.3.1.2 and find that

$$Y = \frac{0.109 K_1 \rho \pi r^4 (1 - u)}{E t_{\mathrm{eq}}^3}. \tag{6.7}$$

6.2.2 Optical axis horizontal

When the optical axis is horizontal, Eqs. (4.9) and (4.10) become

$$Y = \frac{0.849 K_1 \rho \pi R^5}{R_s E t_{\mathrm{eq}}^3} \tag{6.8}$$

for the edge mount. As before, this error presumes that the mount is at the mirror edge center, producing astigmatism, and not at the mirror centroid.

For the internal mount,

$$Y = \frac{0.212 K_1 \rho \pi R^5}{R_s E t_{\mathrm{eq}}^3}. \tag{6.9}$$

6.3 Gradients

For the solid-optics case, we saw that under a linear axial gradient, the error produced [Eq. (4.28)] is pure focus and has a value of

$$Y = \frac{\alpha \Delta T r^2}{2d},$$

which is independent of the area moment of inertia of the cross-section and is only a function of its depth.

For lightweight optics, this remains true—the optic remains stress free, and the displacement equation is identical:

$$Y = \frac{\alpha \Delta T r^2}{2H}, \tag{6.10}$$

where, again, H is the full depth of the optic.

6.3.1 Nonlinear temperature gradients

As in the solid-optics case, when the axial gradient is nonlinear, we no longer have a stress-free state. In this case, deflection is a function of the moment of inertia of the cross-section, unlike the previous linear gradient case. For a lightweight-optic cross-section, we use the same equation as was used for the solid optic, but we introduce the width of the cross-section $b(z)$ and find that

$$\Delta T_{\text{eff}} = \int T(z)b(z)zdz(h/I), \tag{6.11}$$

or, integrating by summation of small increments,

$$\Delta T_{\text{eff}} = \sum T_i b_i z_i \Delta z_i (h/I), \tag{6.12}$$

where:
ΔT_{eff} is the front-to-back effective temperature gradient,
T_i is the temperature at point i in a repeating cross-section,
b_i is the width at point i in the cross-section,
z_i is the distance from the cross-section neutral axis to point i in the cross-section,
Δz_i is the incremental depth of point i in the cross-section,
h is the depth of the cross-section, and
I is the moment of inertia of the cross-section.

We then use Eq. (6.10) to calculate the deformation, substituting the effective gradient, i.e.,

$$Y = \frac{\alpha \Delta T_{\text{eff}} r^2}{2H}. \tag{6.13}$$

6.3.2 Example

A lightweight closed-back optic is subjected to a 1 °C temperature change on its top facesheet only, with no change in rib structure or back sheet. If the optic is 4 in. in total height with 0.20-in. facesheets and 0.20 in.-wide ribs spaced 2 in. apart, compute the effective gradient.

We use $\Delta T_{\text{eff}} = \sum T_i\ b_i\ z_i \Delta z_i\ (h/I)$ and break the cross-section into ten parts for numerical integration, then compute the temperature at each segment, as shown in Fig. 6.1. The inertia is

$$I = (2)(2)(0.2)(1.9)^2 + 0.2(3.6)^3/12 = 3.67\,\text{in}^4,$$

$$\Delta T_{\text{eff}} = 0.83\ ^\circ\text{C}.$$

6.4 Coating and Cladding

For thinly coated or clad solid optics [Eq. (4.47)], we saw that

$$Y = \frac{3E_1 \Delta\alpha \Delta T D^2 t(1 - \nu_2)}{4E_2 h^2 (1 - \nu_1)}.$$

For lightweight optics, we substitute t_{eq} for h (the depth of the cross-section) to yield the appropriate solution. While not exact for open-back optics due to

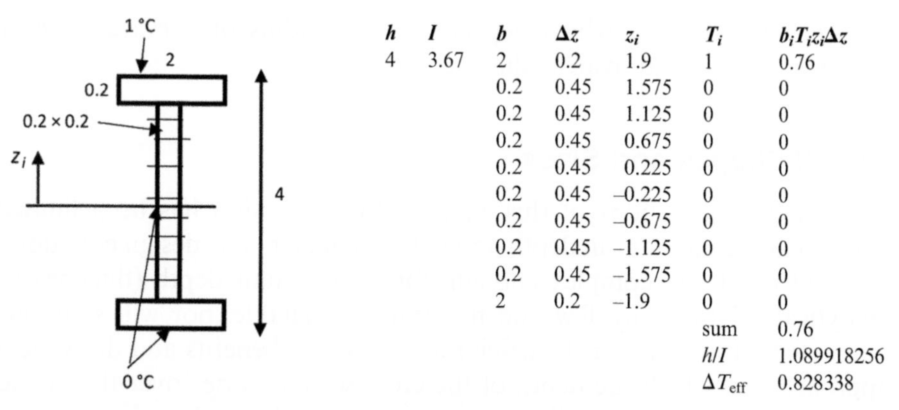

h	I	b	Δz	z_i	T_i	$b_iT_iz_i\Delta z$
4	3.67	2	0.2	1.9	1	0.76
		0.2	0.45	1.575	0	0
		0.2	0.45	1.125	0	0
		0.2	0.45	0.675	0	0
		0.2	0.45	0.225	0	0
		0.2	0.45	−0.225	0	0
		0.2	0.45	−0.675	0	0
		0.2	0.45	−1.125	0	0
		0.2	0.45	−1.575	0	0
		2	0.2	−1.9	0	0
					sum	0.76
					h/I	1.089918256
					ΔT_{eff}	0.828338

Figure 6.1 Data for the example in Section 6.3.2: computation of effective temperature gradient using integration by the summation technique.

the substrate neutral-axis shift, the following equation is nonetheless good enough for a first-order approximation:

$$Y = \frac{3E_1 \Delta\alpha \Delta T D^2 t(1 - \nu_2)}{4E_2 t_{\text{eq}}^2 (1 - \nu_1)}. \tag{6.14}$$

6.4.1 Quilt error

In addition to the power error due to the coat or clad bimetal effect of Eq. (6.14), there can be a quilt error over the lightweight optics core cells. This error is self-canceling at the central pockets, as the compatibility of the slope over the ribs will essentially provide fixity to preclude the error. However, at the edge of the optic, the boundary fixity is reduced, so quilt error will occur. As seen in Eq. (6.14), the error is proportional to the square of the cell spacing and inversely proportional to the square of the facesheet thickness, and the magnitude of the error will depend on the degree of the closeout rib stiffness.

6.5 Random Variations in the Coefficient of Thermal Expansion

We have seen for solid optics that, for a randomly varying CTE, the error under a constant soak condition is taken from Eq. (4.66). For lightweight optics, we have[1]

$$Y_{\text{RMS WFE}} = \sqrt{\left(\frac{56\overline{T}\Delta\alpha_a D^2}{d}\right)^2 + \left(\frac{22\overline{T}\Delta\alpha_p D^{7/2}}{t_{\text{eq}}^{3/2} R_s}\right)^2}. \tag{6.15}$$

The first term under the square root in this equation is the same term as in the solid-optic equation; i.e., it is only a function of the material full height d.

For the second term, which depends on the radius of curvature, we have simply substituted equivalent thickness for depth.

6.6 All Shapes and Sizes

In the previous sections of this chapter, the discussion has been limited to optics that are constant in depth across the diameter. The design equations for constant depth are complex enough; for nonconstant depth (thickness), the analytical solutions are few and not readily available, nor will solutions be attempted. However, it is beneficial to review the benefits and drawbacks of approaches in which the depth of the cross-section varies over the diameter. A more detailed and excellent discussion on sculptured optics is given by Yoder and Vukobratovich.[2,3]

Optics come in many shapes. When viewed from the front (optical) or rear surface, optics can be any combination of shapes that are concave, plano, convex, or sculptured. For refractive lenses, of course, the shape is dictated by the optical prescription. For reflective mirrors, only the front surface is dictated by design; the rear surface shape is dictated by the analyst.

For steering and fold mirrors, the front surface is plano, or flat. For powered optics, which generally take the form of an asphere, the front and rear surfaces can be concave-convex, concave-plano, concave-concave, convex-convex, convex-plano, or convex-concave, where the first word in each term defines the front surface. Further, the rear surface can be sculptured to any desirable shape. The choice of shape will depend on mass, stiffness, stiffness-to-weight ratio, and volume requirements.

Examples of such shapes are shown in Fig. 6.2. Each shape, of course, can be lightweighted by removal of material and mass as determined from geometry; however, mass by itself is not sufficient for system design because wavefront performance is critical under mount, thermal, and gravitational loading. Consider, for example, Fig. 6.2(c), a mirror with a concave front surface and a convex back surface, resulting in a constant thickness (depth) throughout. This has been our baseline for all previous analyses. Here, the back surface curvature radius is concentric (and hence unequal) to that of the front surface. If the back surface is made plano [Fig. 6.2(a)] with central thickness at the vertex the same as in Fig. 6.2(c), stiffness is increased but so is mass. If the optic is mounted with its optical axis horizontal, gravitational error is reduced because (1) the effective curvature is increased and (2) the edge supports render the mounting closer to the mirror centroid. If weight is important, an alternative design would have the radius of the rear surface nearly equal to that of the front surface [Fig. 6.2(d)]. Here, the edges are thinner than the center, so mass is significantly reduced but at the expense of reduced stiffness for mount load considerations. More commonly, the rear surface is sculptured in a single-arch configuration [Fig. 6.2(e)], but again

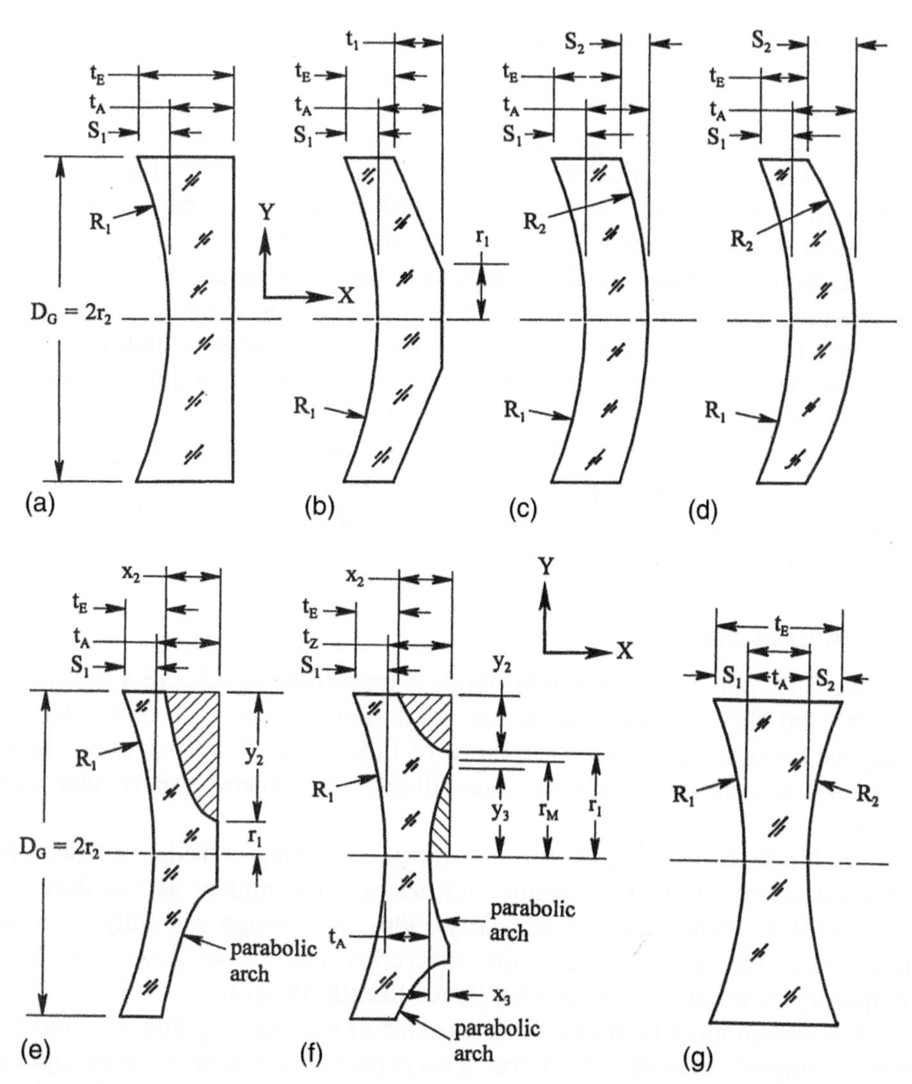

Figure 6.2 Examples of concave mirrors with weight reduced by contouring the rear surface (reprinted from Ref. 2).

stiffness is sacrificed. Even more common is to taper the thickness [Fig. 6.2(b)]. Such a design is readily lightweighted to an open-back configuration and is less costly if stiffness degradation can be tolerated. The tapering to the thin edge often results in self-weight deflection that is focusable, particularly when the optic is mounted near the central zone. This optic is sometimes hub mounted, but another optimum radial zone can be chosen.

A more favorable stiffness-to-weight ratio design is shown in Fig. 6.2(f). This concept is a double-arch design, which, although favorable for gravity error (specifically, if mounted near the six-tenths-diameter zone where depth is maximum), does not readily lend itself to lightweighting and is not a common choice.

For space flight, for which gravity release error needs consideration, testing is usually accomplished with the optical axis horizontal, as discussed in Chapter 4. In some critical cases, a concave-concave design [Fig. 6.2(g)] is preferred; gravity error is greatly reduced because the neutral plane is coincident with the mirror centroid, and the effective curvature is infinite (flat). Only a small Poisson effect needs to be accommodated. A detailed trade for the use of this approach is studied in the next subsection. Such a design, however, is heavier, and impractical for large optics approaching the meter class, as the edges become quite deep. Stiffness is not at optimum since the center portion is the thinnest.

As is evident, the opto-structural analyst has many choices, for which a detailed analysis will substantiate the optimization of the design. The starting point, however, should be a constant-thickness design for first-order analyses, as earlier described.

6.6.1 A case study

Hand calculations for lightweight optics is important to achieve a first-order design, and these calculations make good sanity checks for finite element analysis; however, precise performance of these optics is difficult to achieve without finite element analysis. Accordingly, we review a particular case study.

To illustrate a tradeoff for a sculptured optic, consider a telescope designed to operate in a cryogenic and orbital environment as was done for the Teal Ruby experiment[4] in the early 1980s. The design was fully qualified and tested, but unfortunately, the spacecraft was never flown due to a temporary shutdown of the NASA Space Shuttle Program.

The system must be shown to be capable of maintaining integrity under a rather stringent set of design criteria. This is particularly true when the optical performance of the telescope under all environmental loadings must be held to an extremely precise level. In fact, in order to meet performance in terms of optical resolution, the overall error of the design must be held to one-tenth of one wavelength or less (RMS) of near-infrared light.

The system must also be light in weight, in this case, held to less than 60 lbs. This is necessary in order to meet the spacecraft payload capability for launch from a shuttle to a higher orbit, since other onboard experiments, refrigerants, and the like require substantial weight. Additionally, the system must be inherently stiff to minimize optical element motions caused by the on-orbit gravity release condition. That is, since the system is aligned in the gravitational environment on Earth, such an alignment will spring out or change in the zero-gravity environment of outer space. Such misalignments must be kept within budget for ground test if costly gravity back-out interferometry is to be avoided.

Consider, then, the lightweight mirror design of the primary optic,[4] the largest mirror in the telescope system (approximately 20 in. in diameter). In order to meet its weight balance of less than 16 lbs, a solid fused-silica mirror would have been limited to a depth of about 0.65 in. [Fig. 6.3(A)]. Such a thin meniscus, as we have seen, results in a mirror of low stiffness, low strength, large thermal sag, and excessive distortion due to gravity sag and mount-induced loading. Mirror lightweighting was therefore a must. The first approach was to analyze a mirror that has been cored out from the back to produce the open-back "waffle" design shown in Fig. 6.3(B). Supported at three edge points in a quasi-kinematic fashion with a sufficient depth of about 3 in., a 16-lb mirror was possible—with strength and stiffness being substantially increased. An approach that involved fusing the front and back facesheets to a core central section was also conceived, as shown in Fig. 6.3(C). This optic is 2 in. deep to maintain the same 16-lb weight. In either case, when this mirror was reviewed analytically, unacceptable gravity sag was still found, even if the mirror were to be earth aligned on edge (optical axis horizontal). Gravity back-out interferometry was not an option.

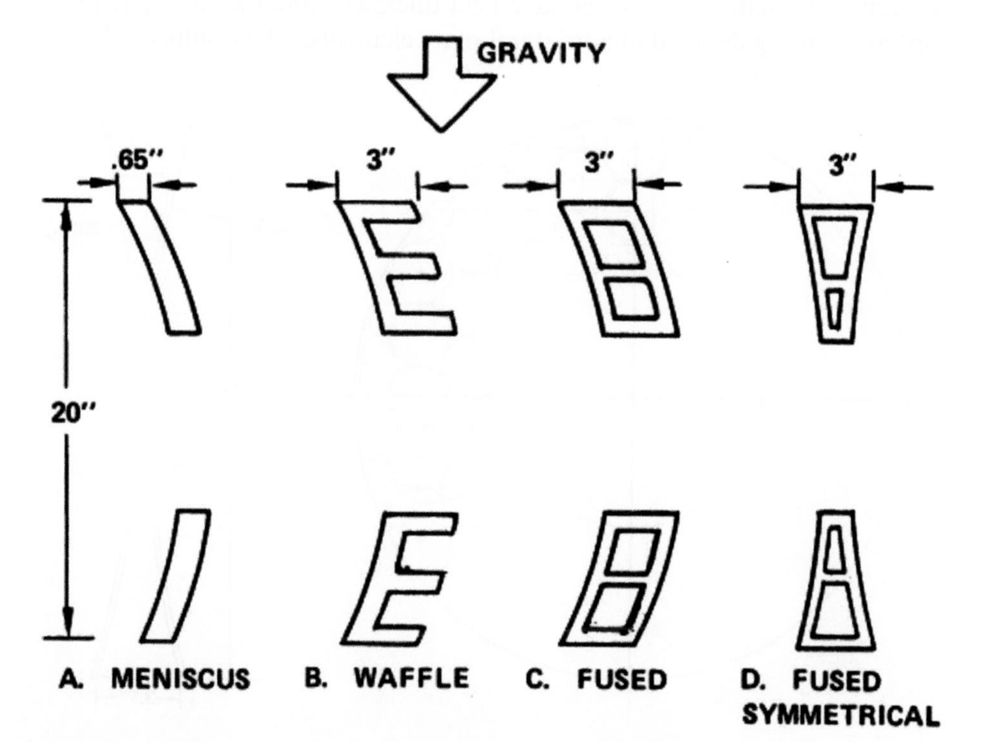

Figure 6.3 Design options to minimize weight. Only the shape in part D can meet the gravity test requirement (reprinted from Ref. 4).

The gravity sag with optical axis horizontal is due to the curvature of the mirror causing a curved-neutral-axis/center-of-gravity offset. The gravity error could be minimized by moving the mount zone internally to be closer to the mirror centroid, but then the required bulkhead back support would be too heavy. Alternatively, if additional support at the edge is introduced, the nonkinematic (more-than-three-point) mount would create more problems than it would solve. The solution was to use the symmetrical, fused-facesheet, bi-concave concept shown in Fig. 6.3(D). While stiffness is not optimum, the error produced by a well-toleranced flexure mount design could be accommodated. Such a design has a straight neutral axis and a center of gravity directly in its line of action. Gravity sag, when aligned on edge, is maintained at a minimal value. The large astigmatic distortion apparent in the cases of Figs. 6.3(A) through (C) is non-existent. To minimize the weight of this design, use of the optimization technique described in Section 5.6 was made. Choosing the proper core spacing and facesheet thickness minimizes local intercellular deflection effects on polishing (i.e., print through) when using a large lap-polishing pressure of 0.2 psi.

To ascertain the validity of the design, including the effects of mirror shear, the detailed design shown in Fig. 6.4 was modeled using the NASTRAN finite element analysis digital routine. Here the elements were configured to match the selected facesheet thickness, and the core spacing was configured using detailed quadrilateral plate elements. The results of all of the

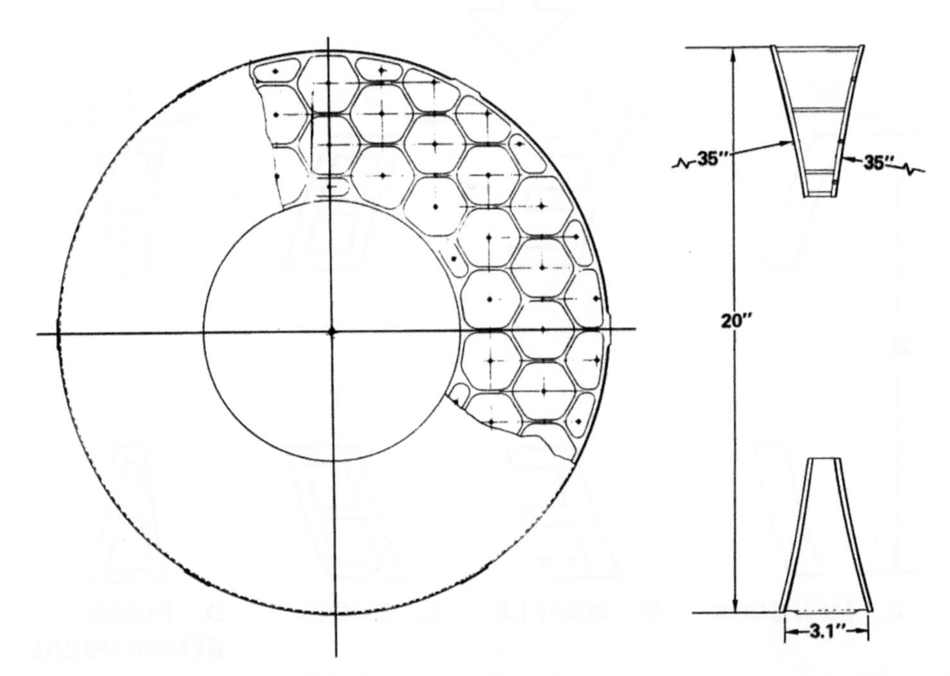

Figure 6.4 Detail of a closed-back bi-concave design modeled using finite element analysis techniques to determine environmental performance (reprinted from Ref. 4).

environmental loading extremes—launch-, gravity-, and mount-induced errors together with thermally induced errors—are verified via the math model, showing design adequacy. Also included are the effects of the vent holes present in the back facesheet at each core pocket. These holes are required for use in vacuum (of space) to release trapped air in the cells and hence preclude the potential for thin-facesheet failure. This minor deviation from symmetry has minimal impact on the overall mirror performance.

6.6.1.1 Manufacturing

Manufacture of closed-back, lightweight, bi-concave optics is expensive. Before lightweighting operations were begun, the core blank of our case study was generated to the double-concave configuration. Core drilling and grinding utilizing bonded diamond and diamond-plated annular or core drills and end mills were used to reduce rib thickness to the 3-mm range, as illustrated in Fig. 6.5.

The preparation of the faceplates utilizes existing fabrication techniques. The plates are fabricated as meniscus blanks approximately three-quarters of an inch thick on conventional, optical-sphere-generating equipment. Extra faceplate thickness is required to provide sufficient stiffness to minimize slumping or sagging of the faceplate into the core pockets during the subsequent fusing operations. Faceplates are then later thinned to the final thickness by a generating machine after the blanks have been fused. Acid etching results from a further need to remove subsurface fracturing after machining of the glassy components. In development of the mirrors, several problems associated with this process were worked out in order to achieve a good fusion. These problems involved abrasive contaminants that became

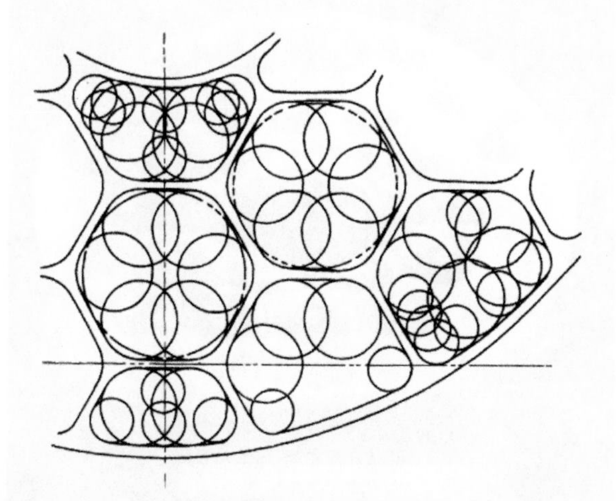

Figure 6.5 Glass-core, lightweight design technique using diamond-plated core drills (reprinted from Ref. 4).

nucleating sites for the growth of cristobalite during the fusion process, preventing a clear fusion at the seal plane.

The above problems were therefore eliminated by regenerating the seal surfaces with a fine-grit bonded diamond wheel (with a hardness near that of the 400 Diamond Series) after acid etching to remove any possible contamination that might interrupt a clear seal. The fusing temperature is quite high, near 1700 °C, compounding problems. Furthermore, the sealing cycle (for sealing each faceplate to the core) is required twice. Once the sealed monolithic mirror blank has been thoroughly annealed, it is ready for optical surfacing and is handled as a conventional glass component in the optical fabrication cycle. The initial operations involve optical sphere generating and loose abrasive grinding of the two faceplates to their final thickness of 0.21 in. At this stage, the mirror blank is very close to its final weight goal of 16 lbs. From this point, the grind/polish cycle is undertaken to put a spherical surface on the rear plate and an $f/1.7$ ellipse on the front mirror surface. Figure 6.6 shows the final mirror product, in this case, the tertiary mirror (12 in. in diameter). Shown on the mirror are the external, integral boss pads used for the flexure mounts that tie it at three points to a graphite composite bezel.

6.6.1.2 Fabrication postscript

Since the fabrication of the Teal Ruby primary mirror, many improvements have been made to facilitate the manufacture of lightweight optics. Corning Inc., for example, has developed a low-temperature fusion technique for egg-crate core bonding in which temperatures are significantly lower than the fused-silica softening point discussed above. Furthermore, frit, a glass powder with a chemically enhanced organic vehicle, is used as an adhesive for joining

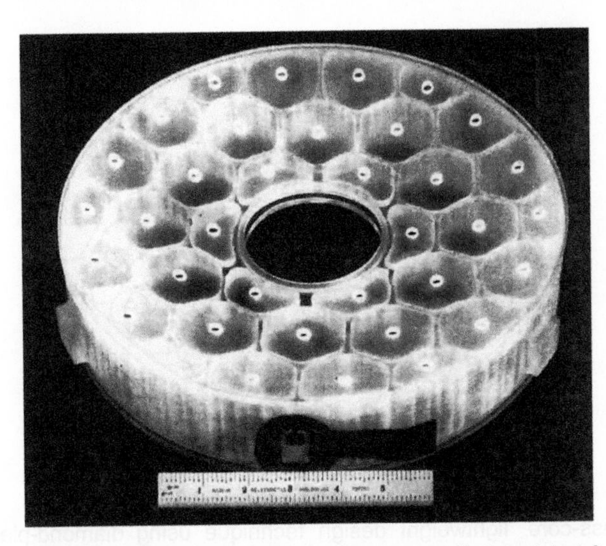

Figure 6.6 Final mounted design of a bi-concave optic (reprinted from Ref. 6).

the facesheet to the core; again, temperatures well below 1000 °C are realized, resulting in a safer process. A proprietary frit CTE by Corning has been well matched to that of ULE™ glass such that performance for both strength and deformation are met without any degradation from the baseline parent material performance.

To further improve lightweight-optics potential, Corning has also developed a water-jet core-cutting technique that leaves core cell walls as thin as 0.08 in. (2 mm). Correspondingly, Schott[5] has developed a proprietary open-back machining technique for its low-expansion ZERODUR® product, leaving wall thicknesses as thin as 0.08 in. (2 mm). In either case, light acid etching removes machining stresses without the need for expensive and dangerous heavy chemical etching.

References

1. J. W. Pepi, "Analytical predictions for lightweight optics in a gravitational and thermal environment," *Proc. SPIE* **748**, pp. 172–179 (1987) [doi: 10.1117/12.939829].
2. P. R. Yoder, Jr., *Mounting Optics in Optical Instruments*, Second Edition, SPIE Press, Bellingham, Washington, p. 306 (2008).
3. P. R. Yoder, Jr. and D. Vukobratovich, *Opto-Mechanical Systems Design*, Vol. **2**, Fourth Edition, CRC Press, Boca Raton, Florida, p. 83 (2015).
4. J. W. Pepi and R. J. Wollensak, "Ultra-lightweight fused silica mirrors for a cryogenic space optical system," *Proc. SPIE* **183**, p. 131–137 (1979) [doi: 10.1117/12.957406].
5. T. Hull, A. Clarkson, G. Gardopee, R. Jedamzik, A. Leys, J. Pepi, F. Piché, M. Schäfer, V. Seibert, A. Thomas, T. Werner, and T. Westerhoff, "Game-changing approaches to affordable advanced lightweight mirrors: extreme Zerodur lightweighting and relief from the classical polishing parameter constraint," *Proc. SPIE* **8125**, 81250U (2011) [doi: 10.1117/12.896571].
6. J. W. Pepi, M. A. Kahan, W. H. Barnes, and R. J. Zielinksi, "Teal-Ruby design, manufacture, and test," *Proc. SPIE* **216**, 160–173 (1981) [doi: 10.1117/12.958459].

Chapter 7
Large Optics

7.1 Multipoint Mounts

Gravity is the enemy of mirrors. When optics become large enough, the ideal three-point mount may produce gravitational performance error in excess of system requirements. While large is a relative term, with diameter-to-thickness (aspect) ratio coming into play, in general, when the size of an optic approaches the 1-m class, more than three points may be required to properly support it. In this case, the opto-structural analyst needs to try to maintain a quasi-kinematic design.

Multipoint mounts can take on numerous forms. Here we summarize only some types. An excellent treatise on multipoint mounts is given by Yoder and Vukobratovitch.[1]

When the optical axis is vertical, back mounts for large optics can consist of pneumatic or hydraulic pistons, air bag support, counterweights, ring mounts, multipoint zonal mounts, or whiffletree mounts. When the optical axis is horizontal, support can be achieved with multiple edge mounts, including radial support, whiffletree supports, tangent rod supports, or a central hub support. For variable-orientation mirrors, a combination of the above is required. Often, due to the nonkinematics of a given situation, active, actuated control may be required.

When a mirror is supported on multiple, evenly spaced points, gravitational sag is driven by the deflection between such supports. When bending alone is considered, we can compute the sag approximately[2] to the first order as

$$Y = \frac{0.07\rho a^4(1 - \nu^2)}{Et^2}, \tag{7.1a}$$

and for lightweight optics as

$$Y = \frac{0.07qa^4(1 - \nu^2)}{Et_{\mathrm{eq}}{}^3}, \tag{7.1b}$$

where q is the weight/area, and a is the mount spacing.

A caution here is that, except for very high-aspect-ratio optics, deflection due to shear will dominate. Unlike three-point mount mirrors, where shear deflection is generally a fraction of the bending deflection, for multipoint mounts, shear defection can be significantly higher than the bending deflection by up to an order of magnitude. Detailed finite element analysis is required to properly assess deformation.

A multipoint optic resting on a bed of fluid pistons[3] is shown in Fig. 7.1. The pistons are weighted according to the load they support, and the optic is "floated" in an open configuration. The only sag due to gravity occurs between the piston supports. The pistons are then closed or locked for fabrication and released again for testing.

7.1.1 Example for consideration

Consider a high-aspect-ratio solid ZERODUR® (Table 2.2) optic of diameter 80 in. and thickness 3 in. floated on a bed of properly weighted fluid pistons. If the P-V surface error due to gravity is limited to 0.005 visible wave, determine the required piston spacing.

We convert waves to inches as

$$Y = 0.005 \times (2.5 \times 10^{-5}) = 1.25 \times 10^{-7}\text{in.}$$

From Eq. (7.1),

$$Y = \frac{0.07\rho a^4(1-\nu^2)}{Et^2} = 1.25 \times 10^{-7}, \text{so}$$

Figure 7.1 Fluid piston mount. Multipoint supports minimize gravitational sag during polish (reprinted from Ref. 3).

$E = 1.31 \times 10^7$, $\nu = 0.25$, $r = 0.091$, and $t = 3$; therefore, $a = 7.0$ in., which is the required spacing.

7.2 Zonal Mount

We saw in Chapter 4 that when a three-point quasi-kinematic mount is used, the optimum support zone to minimize gravitational sag with the optical axis vertical is near the seven-tenths-diameter zone. If we increase the number of mounts at that zone, we can further reduce the error. Equations reported by Williams and Brinson[4] can be used to compute the gravitational sag for any number of discrete points. The error is given as

$$Y = \frac{K_{\mathrm{W}} \rho a^4}{E t^2}, \tag{7.2}$$

where K_{w} is dependent on the support zone, support number, and location of the desired defection. Figure 7.2 shows the sag as a function of the number of supports at the seven-tenths-diameter zone for edge and center locations of the optic. Note that the error is significantly reduced when the support points are increased from three to six, with diminishing returns beyond that.

The key, however, is to maintain near-kinematic properties in these supports; otherwise, the mount error may be significant. This is accomplished by use of a whiffletree, or Hindle mount support.

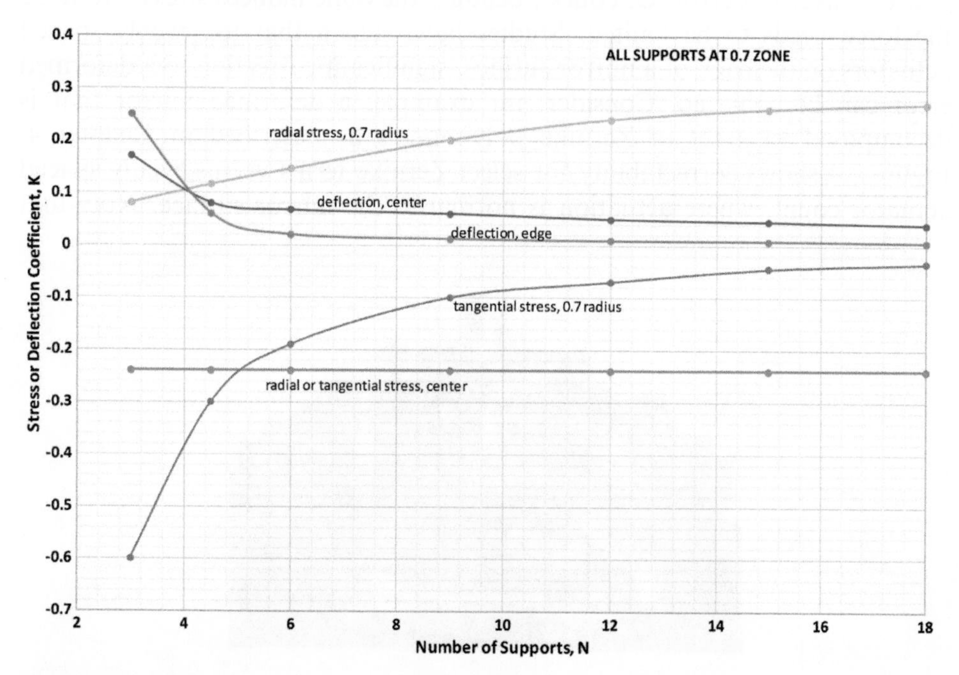

Figure 7.2 Edge and center sag at the internal zone versus the number of support points at angle π/N (adapted from Ref. 4).

7.3 Hindle Mount

A Hindle mount, commonly referred to as a whiffletree, is a multiple-point mounting scheme first described by J. H. Hindle[5] in a classic 1945 paper. The mount relies on a pivot arrangement, or whiffletree, using multiple supports that are precisely arranged and equally loaded from 6, 9, 18, or more points, cascading to 3 supports.

A typical Hindle-mount arrangement[3] is shown in Fig. 7.3. Six groups of three points are connected to a plate, which is in turn connected to three beams mounted at three points, for a total of eighteen support points. Note that each support point has near-zero rotation restraint to maintain the required kinematics. If this were not the case, and the supports were rigidly attached, the mount would behave like a three-point mount, with no added benefit. The Hindle mount supports vertical loads only; for lateral or horizontal loading, an additional support scheme is required in the form of edge or center hub mounts.

7.4 Active Mount

Often, for large, multipoint-support optics, the error may be such that some form of actuated, or active, control is required to get the figure back into shape. In this case, forces are applied to the mirror to bend it to achieve the desired wavefront error. Of course, bending the optic induces stresses to it, so the force needs to be limited. Studies have shown that for evenly spaced actuators on the back of a mirror surface, significant correction to a deformed error can be achieved. Consider, for example, a deformed mirror that is decomposed to a set of Zernike polynomials as described in Chapter 4. Figure 7.4 shows correctability for select Zernike terms versus evenly spaced actuator count, where actuation is normal to the mirror surface. Note that

Figure 7.3 A Hindle mount (whiffletree) for a Keck Telescope primary mirror segment (reprinted from Ref. 3).

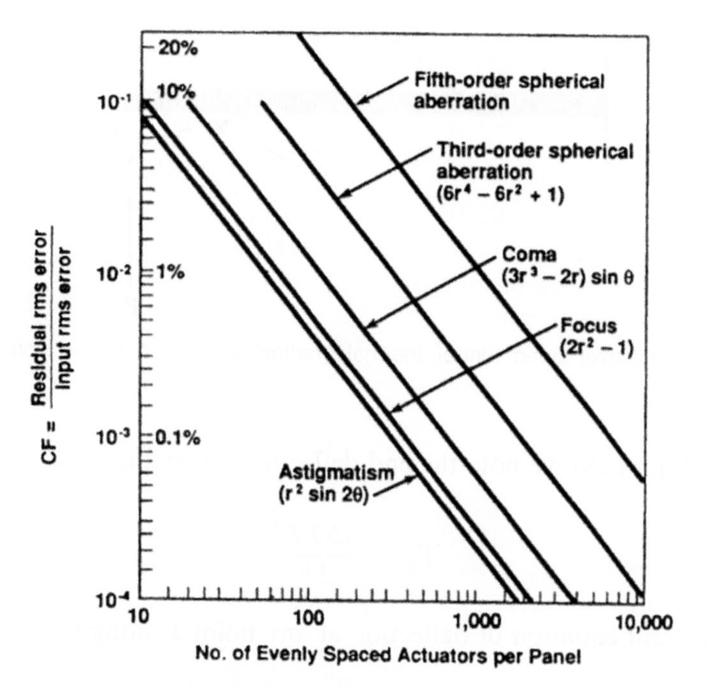

Figure 7.4 Active-mount correctability with as few as 10 evenly spaced actuators can correct more than 90% of astigmatism error.

astigmatic errors are easiest to correct, with an increased actuator count required for higher-order aberrations. The ordinate axis shows the percent error remaining from the given shape; for example, an error of 1 wave can be reduced to 0.01 waves (99% correctability) with only 60 actuators. This number of actuators will correct third-order spherical error by about 90% but will be ineffective for higher-order terms. The figure, again, is for evenly spaced actuators; for rotationally symmetrical error, such as power (focus), closely spaced edge actuators will reduce count, since they can provide a moment in push–pull fashion, which more readily corrects such error.

Note further that correctability for a given displacement amplitude is *independent* of mirror size or thickness, although the force required to achieve the result will depend on these parameters. This relationship may not be initially obvious; however, it is readily shown in the following example.

7.4.1 Active-mount correctability illustration

Consider the cantilever beam shown in Fig. 7.5 of length L, CTE α, rigidity (modulus–inertia product) EI, and thickness h subjected to an axial thermal gradient ΔT. If a single actuator force is applied at the free end, show that the correctability is independent of the material properties, thickness, and length, and determine the correctability.

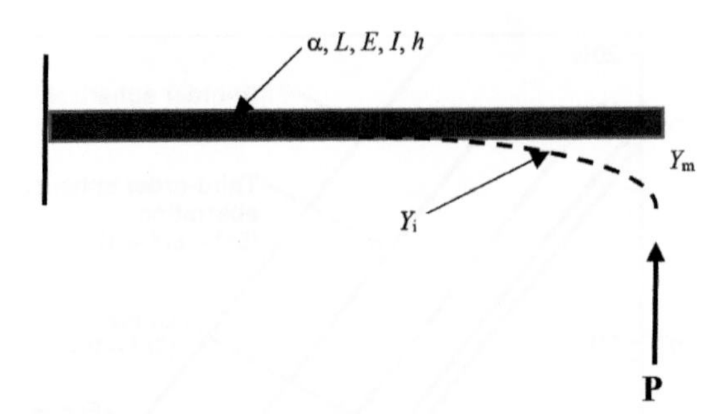

Figure 7.5 Cantilever beam under thermal gradient with an actuated corrective force applied.

From Eq. (4.28) we note the end deflection under the gradient as

$$Y_{\mathrm{m}} = \frac{\alpha \Delta T L^2}{2h},$$

and the general equation of deflection at any point x along the beam as

$$Y_{\mathrm{i}} = \frac{\alpha \Delta T x^2}{2h}.$$

Applying an actuator force P at the end to restore the peak deflection to zero, we have, from Eq. (1.22) and compatibility,

$$\frac{PL^3}{3EI} - Y_{\mathrm{m}} = 0.$$

Solving for the actuator force,

$$P = 3EIY_{\mathrm{m}}/L^3.$$

The deflection equation along the beam length under this restoring load can be obtained from handbooks[6] as

$$Y_{\mathrm{p}} = \frac{P}{6EI}(3x^2L - x^3).$$

Substituting $P = 3EIy_{\mathrm{m}}/L^3$ into the above equation,

$$Y_{\mathrm{p}} = \frac{Y_{\mathrm{m}}}{2L^3}(3x^2L - x^3),$$

independent of rigidity.

The net deflection curve is the difference between the original shape and the corrective force shape:

$$Y_{\mathrm{net}} = Y_{\mathrm{i}} - Y_{\mathrm{p}}.$$

The correctabilty is the peak deflection divided by the original maximum (for optics we would use the RMS displacement):

$$C = Y_{\mathrm{net}}/Y_{\mathrm{m}}.$$

Noting that the peak will occur at some distance $x = kL$ along the mean, where $0 \leq k \leq 1$,

$$C = (k^3 - k^2)/2,$$

independent of rigidity, thickness, and length. Of course, the force (and hence the stress in this case) is dependent of these properties, so this would need to be addressed to ensure capabilities.

We can compute the peak correctability by taking the first derivative of $C = (k^3 - k^2)/2$ and setting it to zero (maxima–minima problem) to find that $k = 0.67$ and correctability $C = 0.074$. That is, the residual displacement is only 7.4% of the original displacement with only one actuator force correction.

7.4.2 An active-mount mechanism

A precision actuation device is required to correct error to fractional-wavelength precision. Such a device is depicted in Fig. 7.6. This novel actuator[7] will precisely control the contour of a mirrored surface supported on a thin mirror faceplate, maintaining wavefront quality under the influence of thermal or pressure effects, or under the influence of an otherwise-aberrated wavefront. The actuator can comprise a magnet and coil (voice coil actuation), stepper motor, or piezo device, operatively connected by way of a web of stabilizing flexures to an actuating rod in contact with the rear surface of a mirror. The position of the actuator rod may be precisely controlled through electrical signals to the coil. A position sensor that uses a linear variable displacement transducer (LVDT) monitors the position of the actuator rod. The actuator set comprising an appropriate number of actuators to provide the desired correction is mounted to a mirror-supporting substrate. A small force is applied in conjunction with restraining flexures to provide millionths-of-an-inch precision.

7.5 Large-Aspect-Ratio Optics

7.5.1 Funny things happen at infinity

A physics professor was once asked by his students about the Doppler effect. The Doppler effect for sound indicates an infinite frequency when the velocity

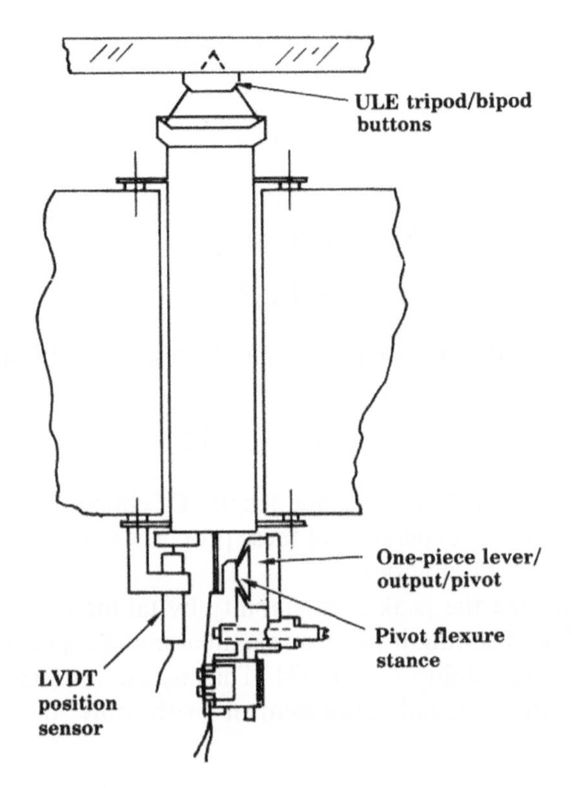

Figure 7.6 Example linear actuator applying corrective force loading normal to the surface (reprinted from Ref. 8).

of a passing object reaches the speed of sound relative to a stationary observer. The student wondered what an infinite frequency meant. The professor paused a moment and then said… "Aha! Funny things happen at infinity!"

So, too, it is with large optics, for which aspect ratios of diameter to thickness approach the infinite. Of course, aspect ratios of infinity are not possible, but once an optic aspect ratio exceeds about 20:1, funny things one would not observe at lower ratios start to become noticeable. This chapter focuses on some of these amusing phenomena and the accompanying analyses.

7.5.1.1 A case study

The infinite brings out the infinitesimal. To make this point for the subject of this chapter, consider the following case study for an airplane engine. Although not an optical topic, it brings home the theme for the optical topics to follow: that funny things do indeed happen at infinity.

Back in the late 1960s, a large jet engine was being designed for the Boeing 747 aircraft, currently one of the largest commercial aircraft ever built. The fan jet engine was on the order of 8 ft in diameter and was designed for a thrust load of 40,000 lbs. During ground test, the engine was observed to "ovalize," i.e., to depart from its required round shape and become egg shaped. This out-of-roundness effect caused two problems: (1) the spinning turbine blades in the aft section, requiring tight tolerance to their housing, would interfere at the minor diameter of the oval, causing blade damage; and (2) at the major diameter of the oval, a large gap would exist, drastically reducing the specific fuel consumption goal for the engine.

The reason for the ovalization problem was unknown; various conditions of thermal gradient variations around the circumference were considered and ruled out, as were other potential causes. Attention was turned to the thrust load; thrust from the engine was reacted at a single point, as it was in previous jet engine designs. The engine for the Boeing 707, its predecessor, had met all of its requirements nicely. The single-thrust point load for that design was 20,000 lbs, and the engine diameter was 3 ft.

Mathematical models were exercised to see whether a high thrust load could be the root cause. Unfortunately, none of the models could predict the ovalization. The models were based on Donnell's shell theory; however, analysis and computer models could not handle point loads, even if represented as a circumferentially distributed Fourier series function (finite element modeling programs were yet to be developed).

While engineers puzzled over the problem, a technician rigged a single jack screw to an empty, 2-lb metal coffee can with its end caps removed. Upon loading, he readily demonstrated the can taking on an egg shape! Why did this not happen to the smaller engine for the Boeing 707? Again, it is a matter of scale. The smaller engine would indeed become egg shaped, but the change in shape would be too small to notice or be of concern. Its thrust was one-half of that of the larger engine, and its diameter was nearly three times smaller. Funny things do indeed happen as design limits are pushed to the infinite.

The solution to the problem, by the way, was to design a yoke to transfer the thrust to two points 90 deg apart, thereby cancelling out the ovalization. As an aside, in the ensuing years, when finite element modeling came into its own, the ovalization was readily predicted by applying an out-of-plane point load to the engine model shell elements—something that standard theory could not readily do.

We now turn to some optical case studies.

7.5.2 How large is large?

Since the term "large" is relative, there are varying definitions for large optics and optics with large diameter-to-thickness ratios. Certainly, a 20-in. optic is large by comparison to a 1-in. optic. For purposes of these discussions, an optic is considered large when it approaches the 1-m (40 in.) -diameter class. At this point, it is difficult to achieve high-acuity performance using standard, three-point, near-kinematic mounting without the use of active or passive correction techniques.

The performance of optics with large aspect ratios is not always dependent on diameter alone. For example, a review of the equations for bimetal analysis shows error that is proportional to a power of the aspect ratio. In this case, optics with a given aspect ratio perform the same whether they are 1 in., 20 in., or 40 in. or more in diameter. On the other hand, other assembly or environmentally induced errors, such as mounting and gravity, are dependent on diameter and thickness power orders that differ. Table 7.1 is a summary of the relationships between diameter and thickness for various environments. These relationships will be clarified in the case studies to follow.

The earliest design of astronomical telescopes used a rule for aspect ratio that stated "when in doubt, make it stout." This implied that the use of aspect ratios of 6:1 or less would mitigate concerns of poor performance under a variety of design criteria. Of course, this results in very heavy mirrors. Advanced fabrication and metrology techniques have raised this conventional aspect ratio to 8:1 or 10:1. Detailed analysis is still required, of course, but for the purposes of these studies, where funny things happen, we define large-aspect-ratio optics as those exhibiting a diameter-to-thickness ratio in excess of 20:1.

The performance criteria for all optics are numerous. The optic must be generated, cut, ground, polished, tested, mounted, and coated. During these stages, the effects of residual stress, temperature, gravity, assembly tolerances, mount force errors, metrology, tool pressure, bimetallic distortion, surface flaws, and humidity must be duly considered to meet the stringent criteria of fraction-of-a-visible-wavelength performance.

The laws of physics that govern the design of mirrors with high aspect ratios are, of course, the same as those governing the design of mirrors with low aspect ratios. The difference, however, lies in the magnitude of distortion induced in the high-aspect-ratio designs by the operational, fabricating, or test environments. Depending on the particular environmental conditions,

Table 7.1 Aspect ratio relationships for various environments.

Environment	Error
Coating	$(D/t)^2$
Cladding	$(D/t)^2$
Gravity	D^4/t^2
Mounting	D/t^3
Humidity	$(D/t)^2$
Residual stress	$(D/t)^2$
Thermal gradient	D^2/t
Thermal soak	D^2/t
Grinding	$(D/t)^2$
Edge cut	D^3/t^2
Lightweighting	$(D/t)^2$
Metrology	D^2/t^3
Temporal	D^2/t

performance errors can be orders of magnitude higher for these optics than for those of more conventional optics. What is "in the noise" or unmeasurable for that latter class suddenly becomes a design driver.

To illustrate these points, we review some actual case studies, both analytical and actual, for large-aspect-ratio optics. The case studies involve aspect ratios between 25:1 and 250:1, and are, again, not simply thought exercises but rather real case studies. These studies will provide particular insight into peculiar effects that might be missed when designing for more conventional systems. Both lightweight (cored) optics and thin-meniscus solid optics are considered.

7.5.3 Cladding

The term cladding generally applies to a thin-layer deposit to an optical surface to enhance its polishability. While techniques are available to polish and figure difficult, high-scatter materials (such as beryllium, multiphase silicon carbide, and aluminum), generally, such optics designed for visible wavefront quality have a clad deposit to assist in both ease of material removal and smoothness. Typically, a nickel/phosphorous cladding is preferred over aluminum and beryllium optics for this reason; the phosphorous content of the cladding can be tailored to match the CTE of beryllium (but not aluminum) to within 500 ppb/°C. For silicon carbide, a cladding generally consists of silicon or vapor-deposited silicon carbide that is CTE-matched to the substrate material to within 500 ppb/°C. This is important in minimizing the bimetallic effect discussed in Chapter 4.

Claddings generally range in thickness from 10 μm (0.0004 in.) to 100 μm (0.004 in.). Demand for larger astronomical telescopes of the 8-m class led to several proposals of a solid optic with low thickness for mass savings. One such proposal for the European Very Large Telescope (VLT) program using a thin-meniscus design suggested aluminum as the candidate material, as a cost savings measure. The design considered an 8-m (320 in.) -diameter primary reflective optic having a thickness of 0.16 m (6.4 in.), with a resulting aspect ratio of 50:1. The mirror is designed to operate in a changing thermal environment with a soak extreme of 20 °C about the nominal set point.

To fabricate the optic to the required low wavefront error (well below 0.1 visible surface wave) and low surface roughness (10 Å), a thin, 0.003-in., clad coat of nickel is to be deposited on the optical surface.

Using the bimetal equation of Chapter 4 [Eq. (4.47)],

$$Y = \frac{3E_1 \Delta\varepsilon D^2 t(1 - \nu_2)}{4E_2 h^2(1 - \nu_1)},$$

we compute the error under thermal soak by referring to the material properties from Table 2.2. We find that $Y = 1250$ μin. $= 50\ \lambda$ peak surface error, some 500 times over budget!

In Section 7.5.8 we will see that this error can be reduced somewhat to the order of 25 waves due to the mirror curvature, but even with active focus compensation, the optic will be hopelessly out of specification.

7.5.4 Coating

The term coating generally applies to a very thin layer deposit to an optical surface to enhance its reflectivity, optimize emissivity or transmission, or provide a protective surface. Unlike claddings, which can be polished to meet surface requirements, coatings are deposited after polish. As such, wavefront performance error needs to be met without postpolish capability; therefore, needless to say, the coating layer needs to be quite thin to preclude the bimetal effect both under thermal soak conditions and in the presence of residual stress that may be induced in the coating process. Typically, coatings range in thickness between 0.1 μm (1000 Å) and 1 μm.

While the design of conventional-aspect-ratio optics generally precludes the need for concern of wavefront degradation caused by the coating, this may not be the case for large-aspect-ratio optics. Consider, for example, the primary mirror design for the Large Active Mirror Program (LAMP)[8] shown in Fig. 7.7. The optic is a point-to-point 2-m (80 in.) ULE™ glass segment of a 4-m-diameter mirror. The optic was fabricated as part the Space Defense Initiative ("Star Wars") program. The mirror is a solid meniscus with a thickness of only 0.17 m (0.67 in.), with a resulting aspect ratio of 116:1.

The optic demanded a very highly reflective coating to resist and reflect a high-energy laser beam in order to prevent excessive substrate heat up. Such a coating, a multidielectric, was achievable with a very thin 0.150-μm (1500 Å) deposit. Test coupons showed the internal residual stress in the coating to be 6,000 psi.

Using the bimetal equation of Chapter 4 [Eq. (4.48) for residual stress],

$$Y = \frac{3\sigma D^2 t(1 - \nu_2)}{4E_2 h^2},$$

we compute the error under the stress by referring to the material properties from Table 2.2. We find that $Y = 38$ μin. $= 1.5\lambda$ peak surface error to the first order. Considering the optical budget of 0.05λ, it becomes obvious that active control is required. As we will see in Section 7.5.8, this error is reduced somewhat due to the optic curvature, and is partly focus and partly spherical aberration, but is still well over the noted budget. Fortunately, using active control, with 40 actuators mounted to the optic rear surface (Fig. 7.4), the mirror can be corrected to well within the prescribed error budget.

7.5.5 Humidity

Neither glass nor ceramic mirrors are hygroscopic; i.e., they do not take on or release water and are dimensionally stable. However, for large, brittle optics,

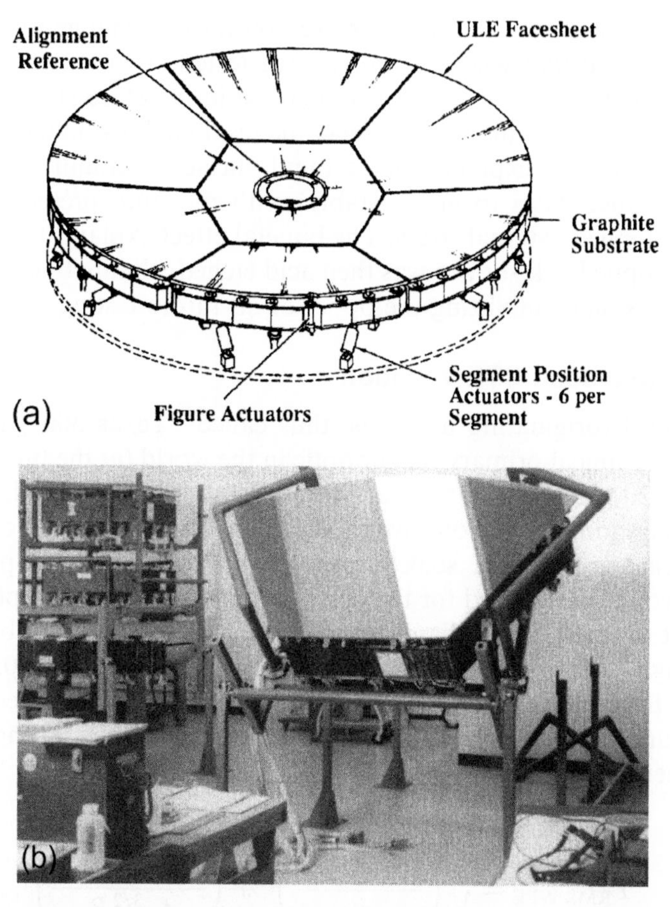

Figure 7.7 Large active segmented primary mirror from the LAMP. (a) Seven segments of the complete 4-m-aperture lightweight active optic and (b) an individual segment of a large-aspect-ratio optic (reprinted from Ref. 8).

funny things can happen in the presence of moisture due to slow crack growth, as described fully in Chapter 12.

Consider, for example, the primary mirror design for the LAMP as discussed in Section 7.5.4 and shown in Fig. 7.7. The mirror aspect ratio is 118:1. The optical surface was control ground and polished, as discussed in Chapter 12, leaving a pristine surface. The back side, on the other hand, was generated and ground, the last step being achieved with 40-μm diamond grit particles, which is more than sufficient for proper radius, roughness, and figure, as it is not the surface of interest.

The mirror reflective surface, which was control ground and polished to better than 1 μm, was tested in the winter months for optical quality. Testing was done at a controlled room temperature, and the relative humidity was near 30% due to low dew points and hence low moisture availability. When summer arrived, the mirror was again tested and found to have changed by a value of 5 μm of power. Although the reason for this change was puzzling at first, it was noted that when the mirror was tested at room temperature, the relative humidity was over 50%. The increase in moisture since the previous winter test allowed the moisture to relax the residual stress in the ground back surface. Chapter 12 explains that ground surfaces contain residual stress, which will cause flaws to grow (stably) with moisture present even in the absence of applied external stress. The bimetal effect explains the power error nicely! The optic back surface was then acid etched, eliminating the flaws and residual stress, and stabilizing the optic for all future testing.

7.5.6 Thermal soak CTE variation

In a proposal (originating in Texas, thus called "Texas 300") to build the largest astronomical primary mirror optic in the world (at the time, the largest was the the 200-in. Hale reflector at Mount Palomar), a 300-in.-diameter thin, solid, primary mirror was conceived. ULE glass was chosen as the baseline for thermal reasons, where a soak change of up to 10 °C was possible. No cladding would be required for the glass. For cost savings, the optic would be flowed out to the required diameter with a thickness of 5 in. to be fabricated from a single boule of glass. This results in an aspect ratio of 60:1.

While the glass is quite homogeneous, from Table 2.2 we see that nonetheless a random peak variation of up to 15 ppb/°C occurs. Using Eq. (4.66), repeated here,

$$Y_{\mathrm{RMS\ WFE}} = \sqrt{\left(\frac{56\overline{T}\Delta\alpha_{\mathrm{a}}D^2}{d}\right)^2 + \left(\frac{22\overline{T}\Delta\alpha_{\mathrm{p}}D^{7/2}}{t_{\mathrm{eq}}{}^{3/2}R_s}\right)^2},$$

we find that $Y = 0.38$ waves from the random axial variation, and $Y = 0.35$ waves from a lateral in-plane variation—errors that are well above the total budget and neither of which is focusable!

7.5.7 Thermal gradient

To meet the demands of a fraction-of-a-wavelength residual error imposed by high-resolution optical systems, large, high-aspect-ratio optics can exhibit a significant error, even in a relatively benign thermal environment with materials exhibiting near-zero CTEs.

This study evaluates a thin, circular, solid, spherical mirror made of Corning ULE fused-silica glass subjected to thermal solar irradiance. We consider the behavior of the optic under a uniform axial gradient, i.e., a gradient that varies linearly from top to bottom along the optical axis through the mirror thickness. The optic has an aspect ratio of 50:1, which is a radical departure from conventional optics in the usual 6:1-to-10:1 regimes.

7.5.7.1 Theoretical analytical solution

Chapter 4 explains that a free or kinematically mounted circular flat plate subjected to a linear thermal gradient through its thickness results in a spherical deformation well approximated by the relation [Eq. (4.28)]

$$Y = \frac{\alpha \Delta T r^2}{2h},$$

where, again, α is the material coefficient of thermal expansion, and ΔT is the difference in temperature between the top and bottom surfaces of the plate.

For circular kinematic mirrors that have a finite radius of curvature, however, the mirror shell under the described gradient no longer remains in the stress-free state afforded by the flat condition such that Eq. (4.28) no longer applies. Although Eq. (4.28) will bode well for mirrors with modest curvature and aspect ratio, the thermal bending of a circular segment of a thin spherical shell subjected to a constant temperature gradient through its thickness needs to be examined.

The deflections have been shown[9] to equal those of a thin shell subjected to an edge moment M, given by

$$M = \frac{\alpha \Delta T D(1+\nu)}{h}. \tag{7.3}$$

Note that while the edges are free, the thermal moments are not self-canceling as is the case for flat plates, and give rise to thermal stresses. For uniform-thickness shell segments of diameters less than one-half of the spherical radius (focal ratio, $F > 1$, optically speaking) the shallow-shell approximation[6] may be used. For the edge-moment-only case, the solution may be put into the form

$$\frac{Y}{\alpha \Delta T r^2 2h} = C_1\{[b_{\mathrm{er}}(x)] - 1\} C_2 b_{\mathrm{ei}}(x), \tag{7.4}$$

where C_1 and C_2 are complex constants, and the values within the square brackets are complex Bessel functions, or Kelvin functions. The solution to this equation is resolved with infinite-series representations[6] that are far too complex for this text and unwieldy at best. In order to correlate the exotic theory, a detailed finite element model was prepared for a high-aspect-ratio optic. Consider such an optic with the parameters given in Table 7.2.

The model was initially run as a flat plate, and the results were found to agree with the theoretical solution to within one-half of one percent. A radius of curvature of 175 in. was then included to represent a typical curved optic, and the thickness of the shell was varied. Utilizing a constant unit gradient through the mirror thickness, the resulting displacements from the math

Table 7.2 Parameters for a thin-shell (high-aspect-ratio) optic.

	Designation	Value (inch)
Radius	r	20
Radius of curvature	R	175
Thickness	h	
Axial temperature difference	ΔT	
Strain gradient	$\alpha\Delta T/h$	1 ppm

model were then input to a Zernike polynomial postprocessor to determine residual error both before and after focus. The results[9] are shown in Fig. 7.8, alongside the theoretical solution afforded in the paragraphs above. The correlation is markedly good.

As evidenced by the curves, the departure from the flat plate error is most pronounced as the diameter-to-thickness ratio increases. For a D/t of 50:1, as in our design, the error is about 2.5 times less than that of the flat plate without correction. After focus, however, about 16% residual error still remains, whereas the flat plate error is entirely focusable. For a D/t of 10:1, the uncorrected error is very close to that of the flat plate, while the residual error remaining after focus is less than 1%. We conclude that the shell effect reduces the residual error before a focus fit under a thermal axial gradient.

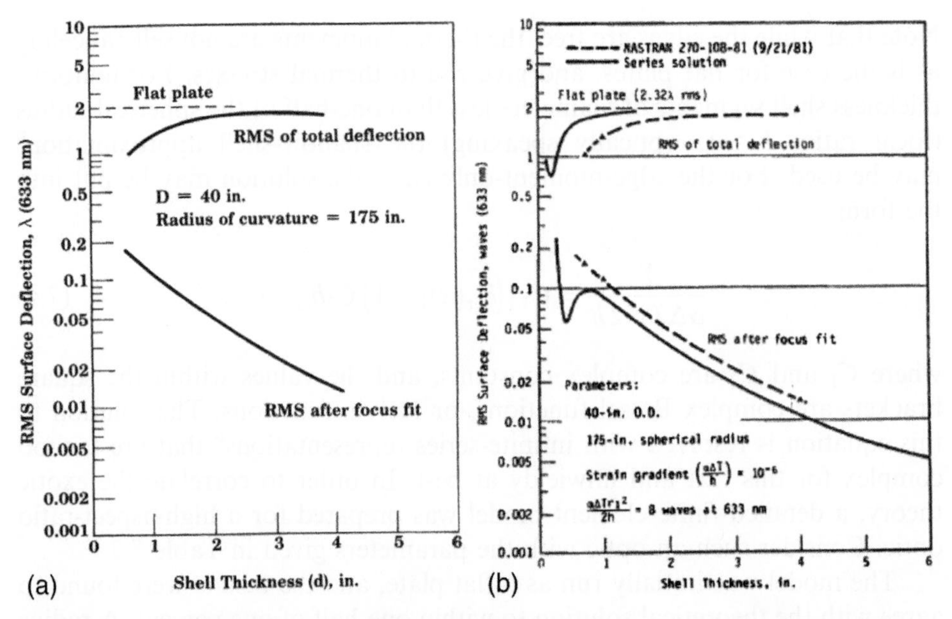

Figure 7.8 Large-optic model correlation between theory and finite element modeling is markedly good. (a) The thin-shell effect under axial gradient reduces error at the expense of focusabilty (reprinted from Ref. 8). (b) Comparison of theory and finite element model (reprinted from Ref. 9).

After focus, the residual error may be significant. This phenomenon, again, is validated by the shell theory that was presented earlier.

7.5.7.2 Thermal performance under various conditions

We turn again to the 300 in. × 5 in. "Texas 300" optic of Section 7.5.6. Various thermal levels were superimposed, as could occur from gradients under solar load and from thermal soak due to diurnal changes. Soak and gradient values were incorporated into a finite element model in conjunction with the measured CTE data.

As before, the CTE varied throughout the model, as is representative of typical ULE data. Variations of $\pm 0.015 \times 10^{-6}$/°C about a nominal value were thus considered after initially imposing a uniform CTE field. The mirror was independently subjected to a 1 °C soak and 1 °C axial gradient, and the results were postprocessed to obtain the RMS error before and after focus correction.

The results from these cases are shown in Table 7.3. For the case of a uniform soak and uniform CTE, the residual error is small before focus and entirely correctable after focus. When the radial variations of the CTE are considered, the uniform soak case yields a tenfold greater error before focus, less than 50% of which is refocusable. For the case of a uniform axial gradient, nearly 30% residual error remains after focus, with no variation in CTE. This is in agreement with the theoretical results earlier presented. Imposition of a variation in the CTE for the gradient case further increases the error residual. While the errors are small, larger soaks and gradients could demand other compensation techniques.

7.5.8 Metrology

Can a mirror be made so large with such a high aspect ratio that it cannot be tested? Possibly. Consider a very large optical segment designed and fabricated for the aforementioned Space Defensive Initiative program. The outer segment has a maximum dimension of 4 m and is 0.017-m thick, for an aspect ratio of 235:1! In order to test such an optic, it is supported on a bed of up to 300 fluid

Table 7.3 Uniform soak and gradient effects on a large, thin optic. High aspect ratio and curvature affects performance after focus (reprinted from Ref. 8).

	1 °C Soak Change		1 °C Axial Gradient	
	Error From Perfect CTE Value	Error From Radial CTE Variation	Error From Perfect CTE Value	Error From Radial CTE Variation
Uncorrected	0.0024	0.032	0.131	0.073
Focus	0.0000	0.019	0.037	0.046

- All values in waves (surface) at 633 nm.
- Nominal perfect CTE 0.015×10^{-6}/°C.
- Radial CTE variation $\pm 0.015 \times 10^{-6}$/°C.

pistons, each carrying their share of the mirror weight of 660 lbs (300 kg). Thus, each piston reacts with 2.2 lbs (1 kg). However, manufacture of the pistons may result in a small hysteresis (due to diaphragm friction and the like) on the order of one gram, which is very small. Nevertheless, for large-aspect-ratio optics nothing can be considered small. Accordingly, a finite element model was developed using quadrilateral plate elements for the optics and fluid support elements for the pistons. A "one-sigma-gram" was applied to the model; i.e., each piston was loaded with a random ±1 gram (0.01 N force) peak error. The results are shown in Fig. 7.9, where a 1-μm error (ten times the budgeted wavefront error) is realized! Such an optic demands active correction in order for these errors to be removed, but testing *per se* is indeed difficult.

7.5.9 Gravity

In Chapter 4 we saw the effect of gravity on performance. Gravity deformation is critical in a ground-based system that is required to operate at different orientations as well as in a space-based system that must be aligned and tested in a 1-g environment but operated in the effective zero-gravity orbital space

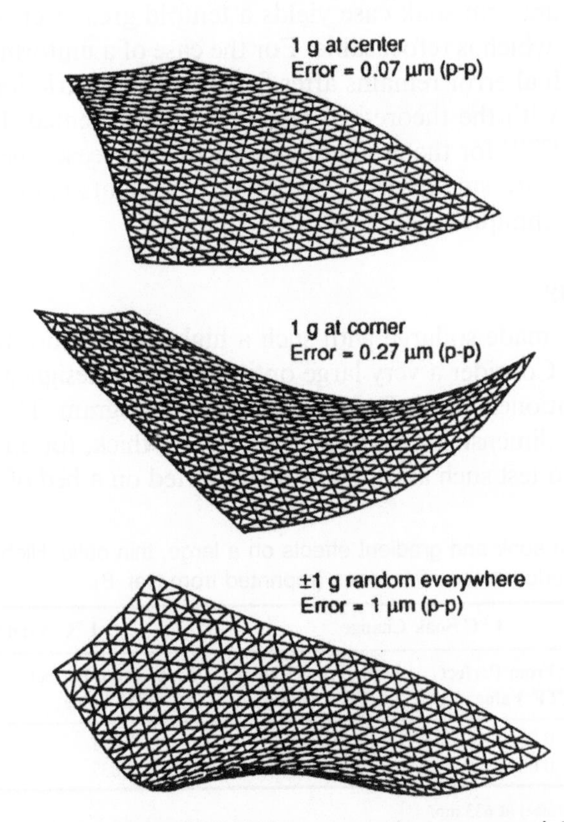

Figure 7.9 Metrology mount sensitivity to force errors (p-p mean peak to peak). Large and high-aspect-ratio optics can be gram sensitive (reprinted from Ref. 8).

environment. Chapter 4 demonstrated that for most meniscus (concave-convex) powered optic systems, the gravitational error is lower when tested with the optical axis horizontal. In this case, Eq. (4.9) gives the displacement as

$$Y = \frac{0.849\rho\pi R^5}{R_s E t^2}.$$

Consider, then, the large primary optic for the James Webb Space Telescope (JWST). The design consists of 18 1.5-m (60 in.) segments, comprising a 6.5-m-diameter mirror. It is a lightweighted, open-back beryllium design, resulting in the lightest optic ever made in this class, in terms of areal density. The total budget (as in many other precision astronomical systems) is on the order of 0.1 wavelength peak surface error in the visible regime. We now compute the gravitational sag under ground test with the optical axis horizontal.

Again, to the first order, from Eq. (4.9), we find that $Y = 24$ μin., or about one wavelength of visible light—some ten times the entire budget. Such an error, if polished in for a ground test, would reappear in zero gravity, defeating its purpose. Thus, clever interferometry is required to back out the ground test error in order to understand other alignment, mounting, or thermally induced errors. Modern techniques and improvements in interferometry have allowed this; by testing the optic in various orientations, gravitational sag can be compared to analytical predictions, and, more importantly, can be determined to well within the required precision, despite the high magnitude of error realized in the ground test.

7.5.10 Edge machining

The 10-m-diameter Keck Telescope, one of the largest of the telescopes in the world, consists of 36 1.8-m-diameter off-axis hyperbolic hexagonal mirror segments. These comprise near-zero-CTE glass ceramic substrates manufactured by Schott Glaswerke of Germany under the trade name of ZERODUR®. The blanks are approximately 1.9 m in diameter and 7.5 cm thick. To produce the aspheric segments (consisting of six different configurations) in a timely fashion, scientists at the University of California have developed the technique of stressed mirror polishing.[10]

This polishing method employs the introduction of shears and moments about the segment's circumference to bend the mirror into the reverse of the desired shape. A true sphere is then ground and subsequently polished into the segment, after which the loads are removed and the desired optical prescription is obtained. Next, as shown in Fig. 7.10, the segment is cut to the hexagonal shape and a central hole with a 0.25-m diameter is core drilled partially through its back. The segment is then mounted to its final support structure.

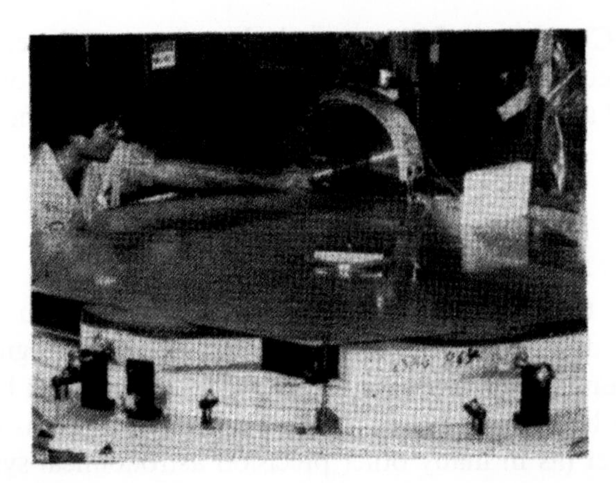

Figure 7.10 Edge-cut machining using a diamond saw for a Keck segment. Release of residual hoop stress bows the optic (reprinted from Ref. 3).

After the segments are cut, their shape may warp slightly due to residual stress levels in the blank itself. While this may be unnoticeable in conventional optics, it can be significant in optics with high aspect ratios.

For the subject case, consider a circular disc subjected to hoop compressive stresses at its outer periphery and tensile stresses toward its center. This is a likely scenario in the forming process for Zerodur glass ceramic, in which the outer periphery is cooled more rapidly than the center due to heat transfer laws. As the center continues to shrink (positive CTE) during cooling (after ceramicizing) and anneals against the already cooled, more-rigid outer edges, tensile stresses are set up near the center with hoop compression near the edge, as shown in Fig. 7.11. The locked-in stresses are a function of the

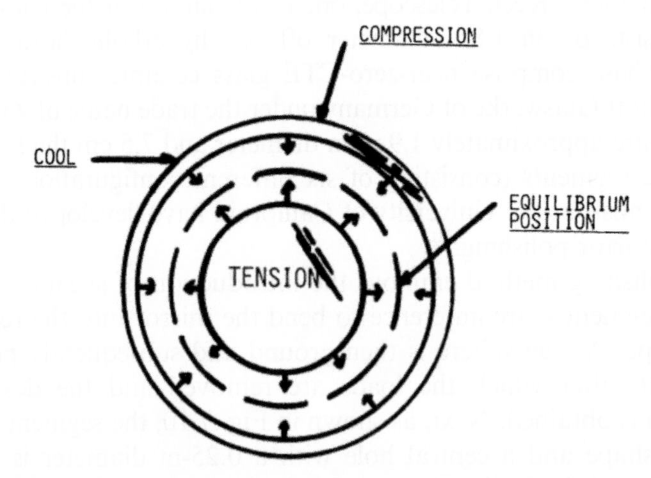

Figure 7.11 Cooling effect after ceramicizing at high temperature will leave residual stress in the optic (reprinted from Ref. 3).

expansion characteristics of the material, and because this CTE value is small, so are the stress levels.

Birefringence tests on the Keck blanks show values of compression near the edge on the order of 10–50 psi, depending on the boule and process. Birefringence measurements are made with a polarimeter manufactured by Strainoptic Technologies, Inc. and are based on the Sénarmont compensation method. Operation of this system and derivation of the polariscope equations for stress determination are well documented in the literature. As with any polarized measurement, only the difference in principal stresses (see Chapter 16) are readily made, with more-complex techniques using oblique-incidence viewing required to isolate the stresses. Thus, an element in pure (bi-directional) tension gives no indication of stress if the usual technique is applied. Nonetheless, measurements near the edge, where radial stress is zero, give a good estimation of the magnitude and direction of internal edge stress.

To simulate the stress condition, a finite element model is again utilized. A sensitivity case is exercised in which only the outer periphery is subjected to a temperature change. This sets up hoop stresses that, if removed, will deform the piece due to the effect of radial load applied to a curved segment (finite radius of curvature). When exercised, the model shows a change in power (second order of Zernike symmetrical), making the segment more concave, and a change in the fourth-order spherical of about 8% of the power[3] term, opposite in sign, with a much smaller effect on sixth-order and higher-order distortions, as depicted in Table 7.4.

A refinement to the math model input was made to account for zonal cooling by removal of the cut edge (simplified as a circular rather than hexagonal cut). A temperature change is imposed on the edge zone, inducing stress in the optic, and the resulting value is scaled to the actual measured stress determined by the birefringence. The resulting deformation is the warping effect. The central, partial-depth hole that is required to mount the segment is further removed from the math model, using the plate element offset technique to simulate the partial depth of the 2D plate model. Subtraction of this result from the edge-cut model will yield the approximate

Table 7.4 Residual error of a thin meniscus due to edge hoop stress (data from Ref. 3).

Sensitivity to Hoop Compression for a 1.9-m Keck Disc (35-m radius of Curvature)		
Values Normalized to 1.0 on Power		
Zernike	**Term**	**Value**
C(2,0)	Power	1.000
C(4,0)	Spherical Aberration	0.077
C(6,0)	6th-Order Spherical	0.002

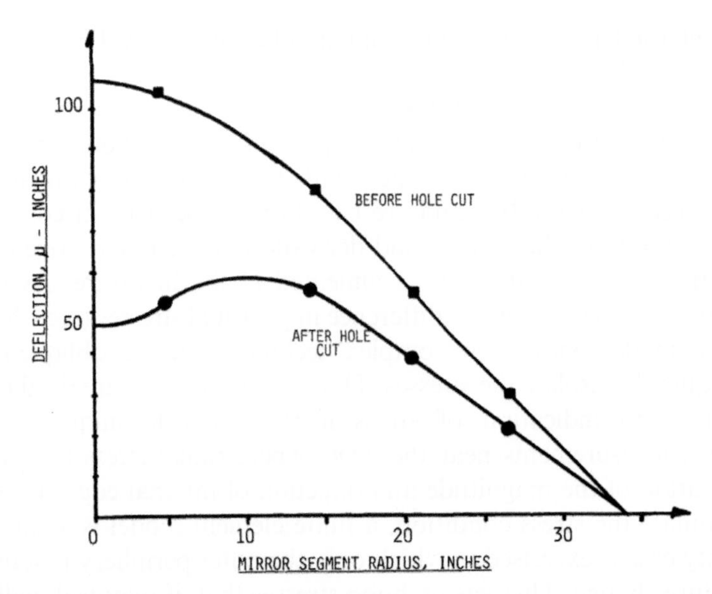

Figure 7.12 Theoretical effects on deflection for a central hole cut in a high-aspect-ratio Keck optic (reprinted from Ref. 3).

effect due to the hole, as can be seen in Fig. 7.12. This reveals a shape that is discernable by symmetrical terms through the eighth order, plus a central dimple over the hole diameter.

Based on pre- and postcut interferometry, the measured warping from cutting on the predominant power term is shown in Fig. 7.13 as a function of birefringence for eight of the cut segments. The birefringence test correlation is good, except for segment 9, whose birefringence and pedigree are quite dubious. Figure 7.14 shows the correlation as a function of the total shape

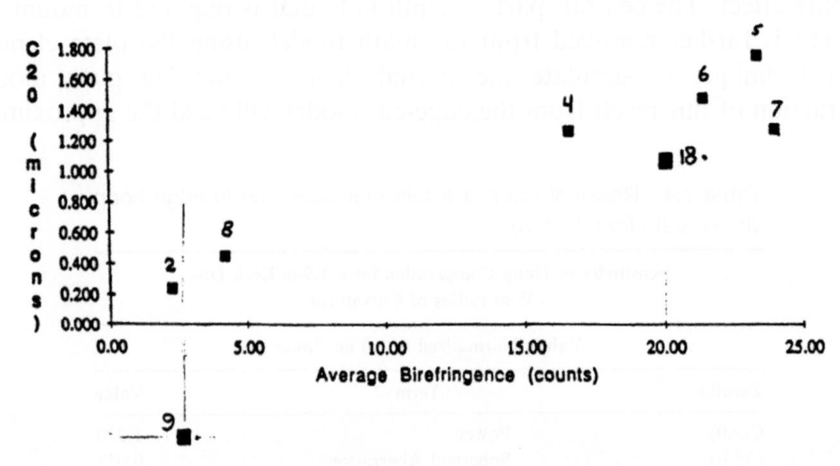

Figure 7.13 Birefringence measurements on Keck segments correlate well to power change after edge cut (reprinted from Ref. 3).

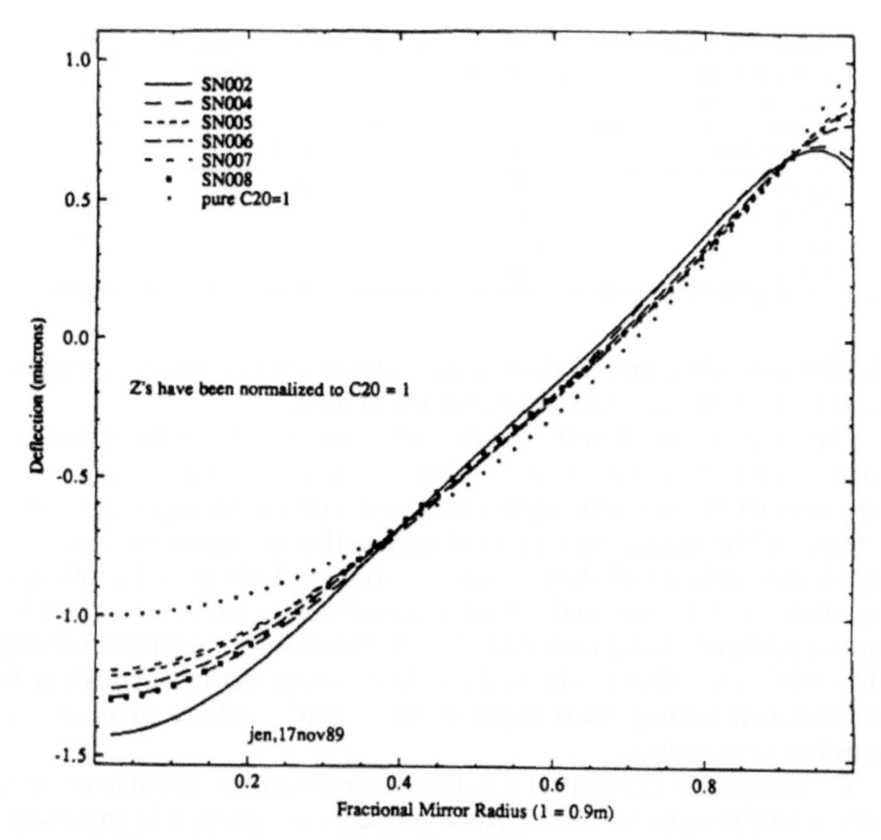

Figure 7.14 Mirror shapes after machining characterized by Zernike fit through the eighth order (reprinted from Ref. 3).

described by Zernike through the eighth order. A systematic universality of shape is noted, although by no means is the warping identical.

Of interest here is the correlation between the aforementioned measurements and the analytical model, which is quite good in magnitude, shape, and sign. Table 7.5 gives the theoretical values of warp for the four symmetrical Zernike terms. While in the theory the effects of edge and hole cut can be separated, test results include both effects simultaneously, since no post–edge-cut/pre–hole-cut data is taken.

7.5.11 Delayed elasticity

As mentioned in the Boeing jet engine story, the infinite brings out the infinitesimal. When the aspect ratio becomes high, unforeseen effects at the atomic level can wreak havoc with performance. This is borne out in the time-dependent phenomenon of delayed elasticity exhibited by certain glass ceramics. Delayed elasticity is a phenomenon in which an elastic material that

Table 7.5 Residual theoretical warping of the Keck meniscus due to edge and central hole machining. Higher-order errors remain after focus removal (data from Ref. 8).

Zernike Term	Edge Cut	Center Cut	Total
C(2,0)	0.7	0.3	1
C(4,0)	−0.06	−0.13	−0.19
C(6,0)	0	0.05	0.05
C(8,0)	−0.01	−0.08	−0.09

is loaded to produce stress and strain in its elastic region does not immediately return to its strain-free state after removal of load.

The effect is related to the alkali oxide content of the glass ceramic and rearrangement of the ion groups within the structure during stress. The effect is apparent under externally applied load, and is elastic and repeatable; i.e., no hysteresis of the permanent set is evidenced within measurement capabilities. Nonetheless, delayed elasticity must be accounted for in determining the magnitude of distortion under load (delayed elastic creep) and upon load removal (delayed elastic recovery). This is particularly important for large, lightweight optics that might undergo large strain during fabrication and environmental loading, as is experienced in gravity release or in dynamic control of active optics.

The delayed elasticity effect was first observed during manufacture of the primary mirror segments for the Keck Telescope program. It is important to note that these effects in no way affect the performance of the telescope, since the delayed strain is fully accommodated in analysis, test, and final assembly. Indeed, low, near-zero-CTE ceramics such as Zerodur are a staple of the optical industry.

The Keck primary optic segments were stress polished, as described in Section 7.5.10. Some of the segments were stressed to an aspheric departure of up to 100 μm from the base sphere. During stressing and unstressing of the Zerodur optical segments, measurements of the surface profile always indicated higher values than the theory indicated, as validated by a detailed finite element mathematic model.

Furthermore, measurements made over the period of time showed a continual increase in value. Finally, measurements after release of stress following several weeks of stressing were always higher than measurements made when the optic was initially moving toward the stressed condition.

As might be expected, several different sources that might explain these observations were possible, since the amount of measurement change with time was on the order of a few tenths of a micron. To measure such effects without the use of interferometry was certainly suspect. Additionally, the fluid mount supporting the optic was subject to concerns of stabilization after load application and removal. Finally, clamping loads and thermal changes could cause errors in the optic as a function of time. All of these sources could give the appearance of creep in the glass and are briefly discussed.

A typical plot of the surface profile after stressing is shown in Fig. 7.15 as a function of time. This curve is generated after removal of the loads for one of the innermost-diameter segments whose deflection magnitude is shown in Table 7.6. As evidenced in the table, the predominant change is in the power (focus) term. The curve is generated by measuring along a diameter of the optic using a multipoint profilometer placed on the surface. The profilometer

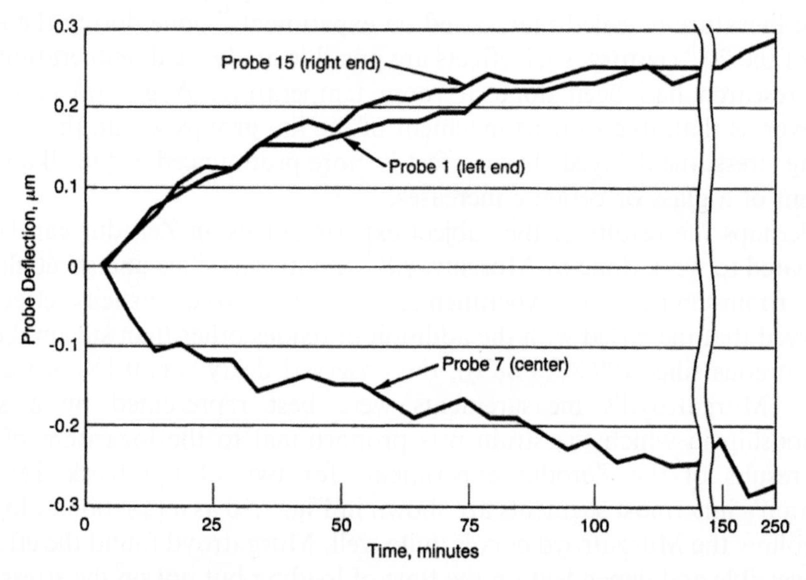

Figure 7.15 Delayed elasticity. Optic changes shape with time due to stress relaxation induced by alkali bonds (reprinted from Ref. 11).

Table 7.6 Tabulation for an inner Keck segment. The power term of 78 μm will leave up to 1% residual error after load removal (data from Ref. 11).

Zernike Coefficient		Distortion Peak (microns)
4	C(2,–2)	0
5	C(2,0)	–78.016
6	C(2,2)	12.81
7	C(3,–3)	0
8	C(3,–1)	0
9	C(3,1)	5.194
10	C(3,3)	–0.008
11	C(4,–4)	0
12	C(4,–2)	0
13	C(4,0)	–0.0525
14	C(4,2)	–0.002
15	C(4,4)	0

consists of a graphite-epoxy beam in which electronic linear variable displacement transducer (LVDT) probes are housed, as shown in Fig. 7.16. The profilometer is connected to a personal computer, and measurements are set to be read every 5 minutes for 4 hours in the presented case. When shown as a function of probe location in Fig. 7.17(a), a systematic power term drift with time is realized. While the profilometer stability was initially suspect, tests show that instability to be small and random, as seen in Fig. 7.17(b).

The recorded observations could be explained by creep of the ceramic at room temperature under load and delayed recovery upon removal. A review of the literature revealed that, based on experiments[9] done during the earlier part of the 20th century, such effects are possible at elevated temperatures, but little research had been done at room temperature. A general consensus, however, is that, due to rearrangement of the ion groups within the structure during stress, the delayed elastic effect is more pronounced as the alkali oxide content of a glass or ceramic increases.

Perhaps the results of the subject experiment using Zerodur can best be compared to work done by Murgatroyd[12] on vitreous silica and sheet glass. In those room-temperature experiments, a small delayed-elasticity effect was observed that increased with the addition of oxides other than silicon dioxide. For vitreous silica (99.86% SiO_2), the observed delay was 0.1% of the total strain. Murgatroyd's measurements were best represented by a simple relationship in which the strain was proportional to the logarithm of time. The results of the Zerodur experiments for two of the Keck Telescope program's innermost segments are shown in Fig. 7.18 as a function of log time and follow the Murgatroyd curves quite well. Murgatroyd found the effects to be reversible and dependent on the time of loading but not on the stress level; that is, the ratio of delayed strain to total strain is a constant. For the Zerodur ceramic materials reported on here, stress levels in the optic are on the order of only 100 psi. When these stress levels are extrapolated to several weeks of loading, a delayed elasticity strain of nearly 1% (10 times that of vitreous silica) is realized. Interestingly, Benjamin Franklin's work on glasses in the 19th century bore out the same results—alkali oxides cause stress relaxation—and Franklin showed this that delayed elasticity at room temperature is proportional to the logarithm of time.

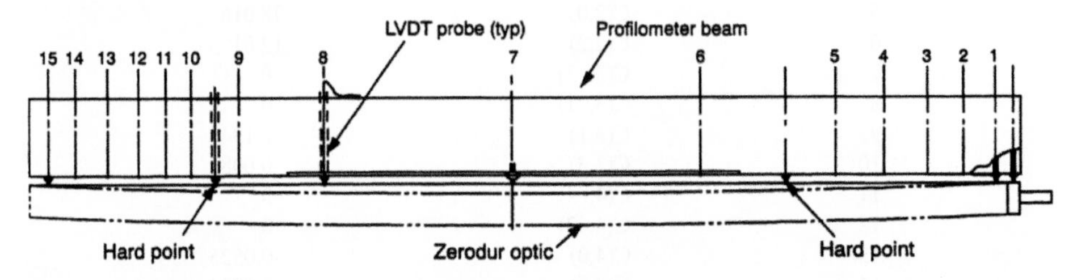

Figure 7.16 Optic profilometer. A near-zero-CTE composite beam has LVDT probes encased to achieve accuracy to better than 0.05 μm (reprinted from Ref. 11).

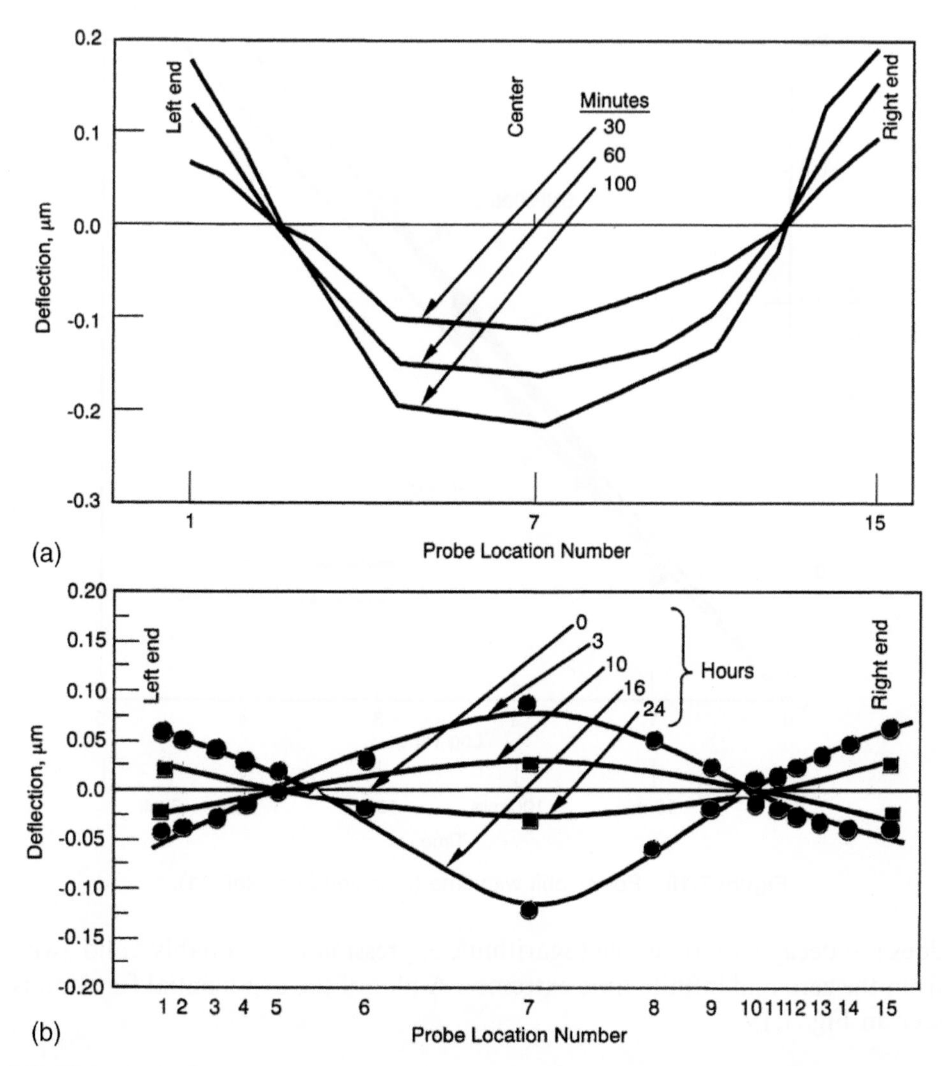

Figure 7.17 (a) Delayed elasticity. Optic changes shape with time due to stress relaxation induced by alkali bonds. (b) Profilometry. The drift with time due to composite/transducer stability is small and random, not systematic (reprinted from Ref. 11).

To further confirm the Keck Telescope data for delayed elasticity, additional experiments were conducted on glasses that do not contain alkali oxides by comparing test samples to those with alkali oxides, most notably, lithium oxide, which exhibits the highest potential for relaxation. While a relaxation of up to 1% in the alkali oxide glass was observed at room temperature after two weeks of loading, no such delayed recovery was noted in the alkali-oxide-free glass, for which full recovery occurred in a few minutes.

Note that, while the actual theory of delayed strain behaves as an exponential function of time, unlike the logarithmic approximation, which

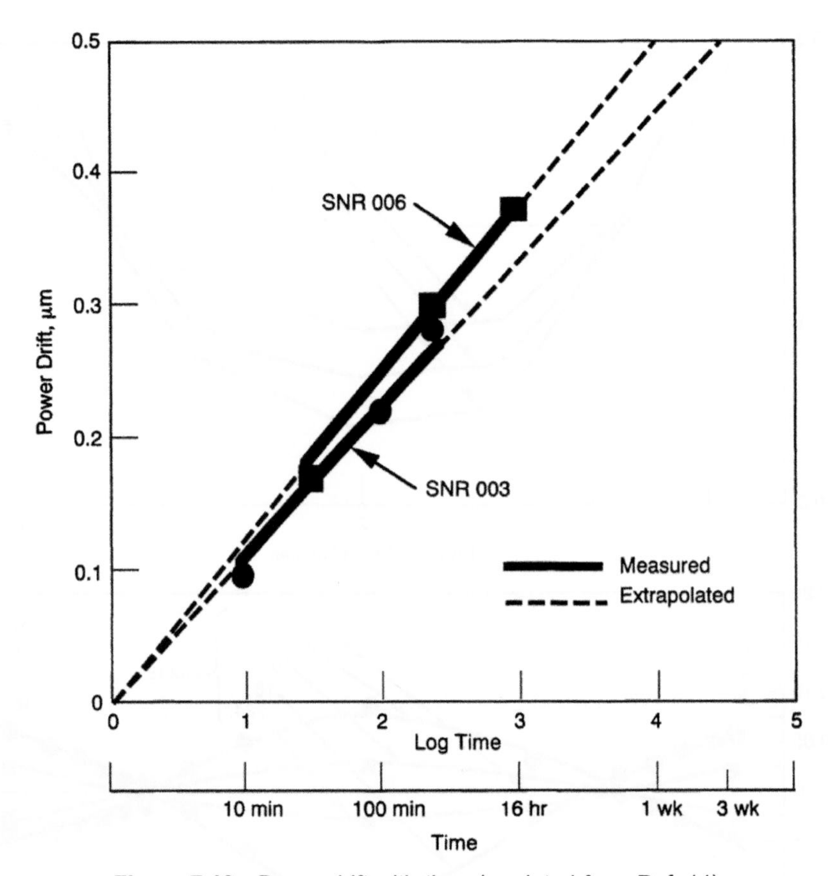

Figure 7.18 Power drift with time (reprinted from Ref. 11).

does not decay with time, the logarithmic expression is reasonably valid away from the zero and infinite time extremes. A plot of the exponential function is seen in Fig. 7.19.

7.6 Performance Comparisons

With reference to the studies of the previous section, it is of interest to compare large and high-aspect-ratio optics performance to conventional optics. This gives the reader a sense of why design considerations for large, lightweight optics are important, since displacements unseen in smaller optics suddenly increase to significant proportions.

In this light, Table 7.7 compares a 1-m-diameter optic with a low aspect ratio of 6:1 to four other optics: (1) a 2-m-diameter optic with an aspect ratio of 15:1; (2) a 4-m-diameter optic with an aspect ratio of 25:1; (3) an 8-m-diameter optic with an aspect ratio of 40:1; and, (4) an 8-m-diameter optic with an aspect ratio of 150:1. The chart is normalized to the 1-m-diameter conventional optic for the purposes of the comparison. All of

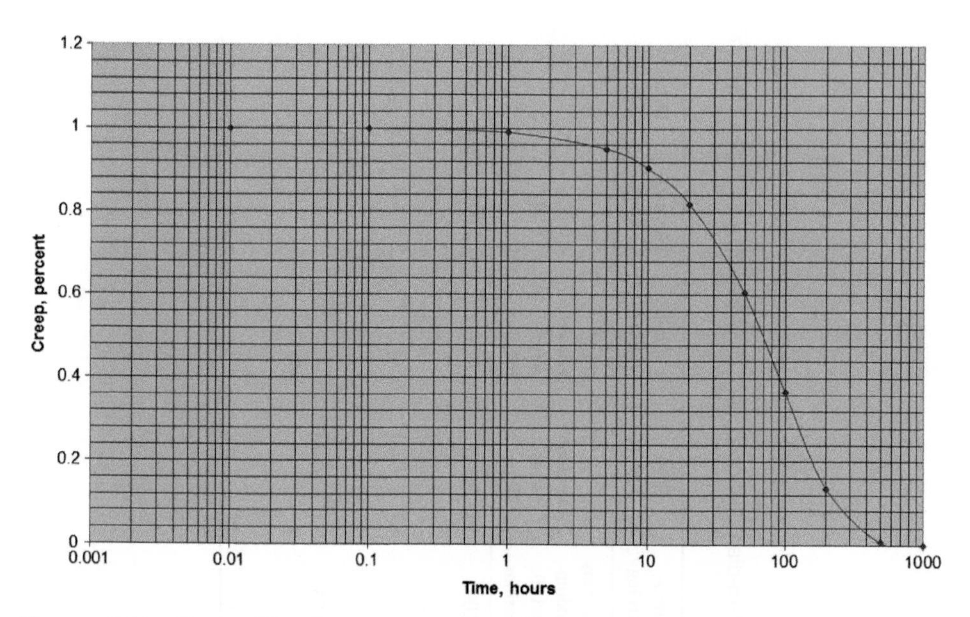

Figure 7.19 Exponential delayed-elasticity effects (in Zerodur) approximated as a logarithmic function over a large time domain.

Table 7.7 Sensitivity of high-aspect-ratio large optics relative to a conventional optic (data from Ref. 8).

Size (meters)	Aspect Ratio	Normalized Performance Error			
		Grind/cut/coat	Metrology	Thermal Soak	Gravity
8	150	580	1760	190	37300
8	40	40	33	51	2620
4	25	16	16	16	256
2	15	6	7	5	24
1	6	1	1	1	1

the designs, incidentally, are in the domain of optics that are already manufactured or under consideration, and in fact are less sensitive than some large optics currently being fabricated.

As evidenced by the table, performance errors are increased up to several orders of magnitude in some cases of large, thin optics. To this end, Table 7.8 shows the expected performance error for select programs highlighted in the above section, when compared to a conventional-meter-class optic. Once again, the infinite brings out the infinitesimal.

7.7 How Low Can You Go?

As we have seen, it is difficult to meet performance requirements for large and lightweight optics without some form of multipoint support, focus correction,

Table 7.8 Large optics with high-aspect-ratio performance error compared to a conventional 1-m-diameter optic with a low aspect ratio. Values are in peak-to-valley surface microns (data from Ref. 8).

Program	Diam	D/T Ratio	Grind	CTE Homog.	Coating	Moisture	Gravity	Force Metrology	Edge Cut	Light-weight	Thermal Gradient	Delayed Elastic
	meter		40-μm grit	15 ppb/K to 100 K	1500 Å @ 6 ksi	50% RH	3-point F1.5 Horizontal	10 grams random @ 100 locations	Residual stress 20 psi	ΔCTE residual stress 200 psi	1 °C (after focus)	1%
Conventional	1	6	0.03	0.06	0.003	0.01	0.035	0.001	0.03	0.09	0.002	0.01
Keck	1.9	25				0.3			1			1
Gemini	8	40				0.6						
VLT	8.2	47				0.8						0.3
Texas 300	7.5	60									0.13	
JWST	1.4	80		0.5			0.75					
Lamp	2	118	10		1	5		0.75				
Alot	2.6	150								60		
Los	4	235	40					3				
Code S	1	500			2	5					0.3	

active control, one-time set, nulling, environmental control, etc. This leads to the question of how light a high-aspect-ratio optic can be made for meter-class optics and beyond. Table 7.9 presents areal mass densities for several large and lightweight optics fabricated between 1950 and 2003. The table is presented in order of mass density and is not necessarily chronological. All optics listed are designed to meet visible-quality wavefront requirements. It is of interest to note that one of the first lightweight large optics (the ground-based Hale Telescope) is not very light at all, nor is one of the early large space optics (the Hubble Telescope). Note, on the other hand, how light the latest-generation large space optic (the JWST) is made. Many controls (gravity back out, focus correction, cryogenic nulling, environmental control, etc.) were required to achieve this while maintaining performance criteria. Is there an ultimate limit to how lightweight such large optics can be made? The analyses presented in this chapter indicate that we may already be at that limit with the state-of-the-art JWST. While there are proposals for thin-meniscus large optics approaching 10 kg/m^2 mass density, these optics will have extreme difficulty meeting performance requirements, even in the presence of active control.

7.8 Extremely Large-Aspect-Ratio Optics

In Section 7.5.8, we discussed apertures with aspect ratios as high as 235:1. The curious reader might inquire about even higher aspect ratios. Such may be the case for very thin, solid-membrane shell optics, which can be made extremely lightweight with aspect ratios as high as 500:1 (to infinity and beyond!). These optics can be made of lightweight materials, such as composites, and coated using a replication process.

As we saw earlier, such thin optics are sensitive to large residual error under an axial thermal gradient after removal of focus because they become less focusable as the aspect ratio increases. Furthermore, for replicated optics in which the replication layer CTE differs from that of its substrate, thin optics are sensitive to large residual error under an axial soak condition due to the bimetallic effect (see Chapter 4). As in the case of a thermal gradient, error is less focusable as aspect ratio increases.

The attractiveness of such an optic is, of course, its mass; exclusive of hardware, the optic itself is less than 5 kg/m^2 and with active control hardware and substrates may still lie below 10 kg/m^2. But is the aspect ratio so high that it cannot be tested or made operational?

For example, while conventional active optics (aspect ratio less than 25:1) can be ground tested and mitigate laboratory gravity error while maintaining good correctability, optics having aspect ratios on the order of 500:1 cannot be readily ground tested without excessive wavefront error. Fabrication error is based on gram-sensitive uncertainties and is high.

Low-order errors demand correction by compensation techniques to a degree where surface actuation corrects remainder and higher-order errors.

Table 7.9 Lightweight optics fabricated between 1950 and 2003 and designed to meet all performance specifications and criteria.

Year	Material	Application	Telescope	Diameter (meters)	D/T_{eq}	Configuration: Type	Configuration: Lightweight (% removal)	Mass/area (kg/m^2)
1950	Borosilicate	Ground	Hale	5	10	Open	35	900
1991	ULE	Ground	Gemini	8	40	Solid	0	440
1980	ULE	Space	Hubble	2.4	14	Closed	70	200
1990	Zerodur	Ground	Keck	1.9	25	Solid	0	165
2003	ULE	Air	ABL	1.5	12	Closed	80	80
1985	ULE	Space	ULM	1.5	10	Open	93	45
2002	ULE	Space	Kepler	1.4	12	Closed	86	45
1993	ULE	Space	Los	4	235	Solid	0	38
1995	Silicon carbide	Space	OAMP	1.4	40	Open	88	30
2003	Beryllium	Space	JWST	1.4	80	Open	85	18

Low-order errors may readily exceed tens of microns before correction. Importantly, as noted above, errors due to temperature changes and gradients are of concern.

We consider the following example. A 40-in.-diameter (1 m) composite membrane mirror has a thickness of 0.080 in. (2 mm) and radius of curvature of 175 in (4.4 m). If a replicated epoxy layer of 0.001 in. is applied at the surface, compute the optic error for a soak condition of 5 °C and axial thermal gradient of 0.33 °C/in. For the soak condition, we start with the usual flat plate bimetallic solution of Eq. (4.47), and for the gradient condition, we use the usual flat plate solution of Eq. (4.28).

Since the aspect ratio of this optic is 500:1, and the optic is curved (thin meniscus), the flat plate approximation does not apply. This was evident in Section 7.5.7.1, where complex Bessel functions were required in the shallow-shell theory. We can solve these (not easily) and modify the equations to yield

$$y = K\frac{\alpha\Delta Tr^2}{2d} \tag{7.5}$$

for the gradient, and

$$y = K\frac{3E_1\Delta\varepsilon D^2t(1-\nu_2)}{4E_2h^2(1-\nu_1)} \tag{7.6}$$

for the soak, where K is a reduction factor computed from the complex theory.

As we have seen, the solution to Eq. (7.6) is unwieldy, so here we make use of finite element analysis to plot K as a function of aspect ratio. The reduction factor K is the same for both the soak and gradient conditions, since the thermal moment is equivalently reduced.

Note from Fig. 7.20 the good news that at a high aspect ratio the error is reduced significantly from a flat plate; the bad news from Fig. 7.21 (soak case and bimetal effect) and Fig. 7.22 (axial gradient effect) is that the error is not focusable whatsoever. The worst news is that the magnitude of the error for the bimetal case is quite high, meaning that some form of active control may be required for correction in this environment. Similar effects can be shown for moisture-induced-replication epoxy resin and composite substrate dry out. These optics are obviously not a panacea, given the current technology.

7.9 Summary

As suggested in this chapter, what is infinitesimal for conventional optics suddenly manifests when both diameter and aspect ratio increase toward the infinite. Perhaps movie director Jack Arnold—who must have had a math background in projective geometry—says it best: "... so close, the infinitesimal and the infinite—two ends of the same concept. The unbelievably small and the

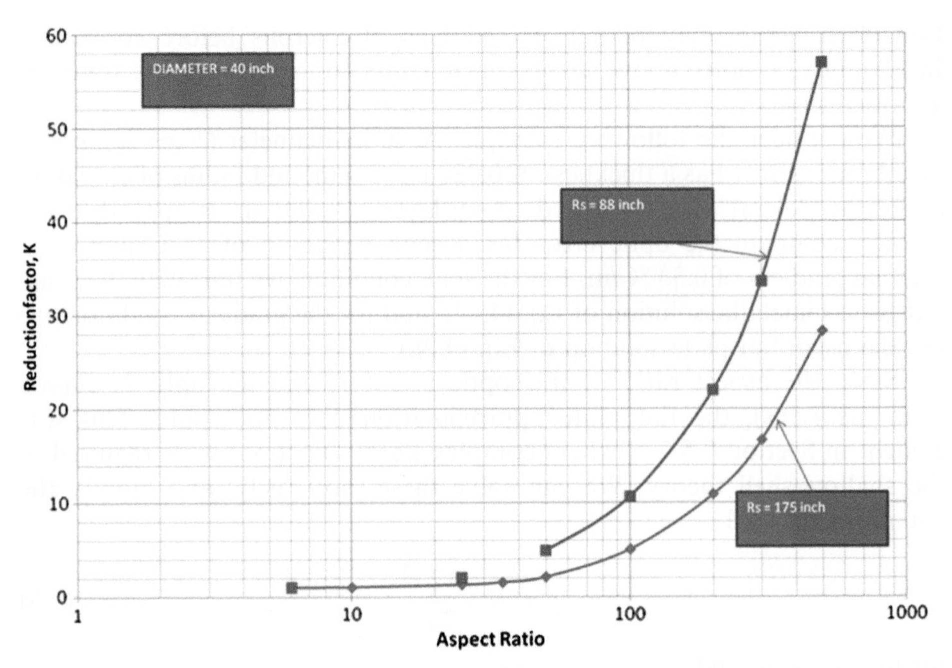

Figure 7.20 Reduction factor for various aspect ratios to be used to correct nominal error from a flat plate under gradient and bimetallic soak.

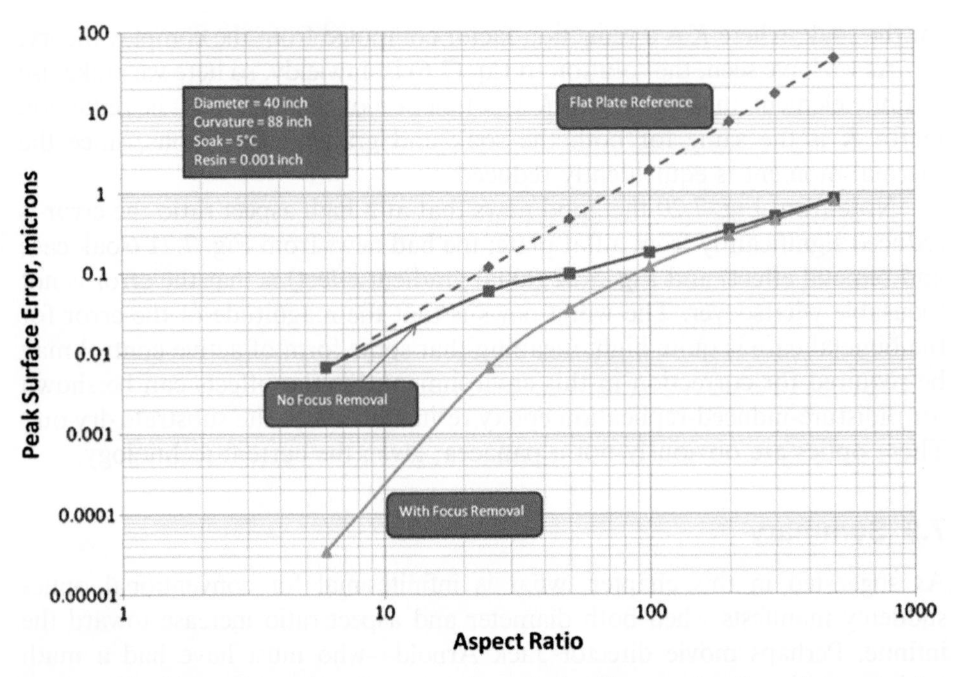

Figure 7.21 Bimetallic performance error for a fast optic as a function of aspect ratio. High-aspect-ratio optics exhibit less magnitude but little focus.

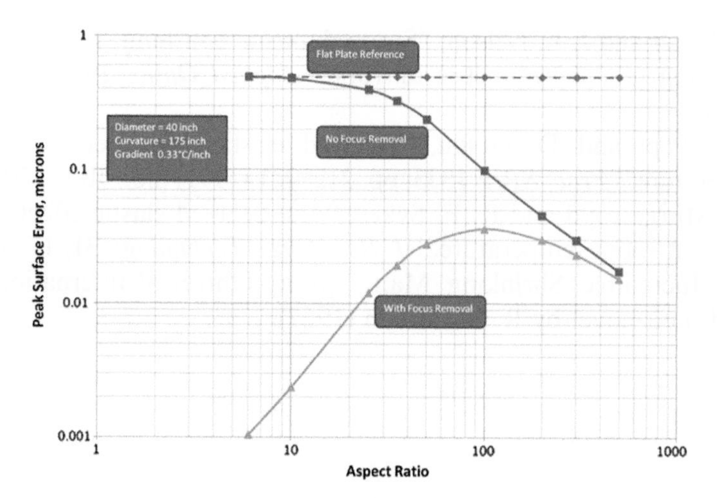

Figure 7.22 Performance error for a slow optic under an axial temperature gradient as a function of aspect ratio. High-aspect-ratio optics exhibit less magnitude but little focus in the presence of an axial thermal gradient.

unbelievably vast eventually meet—like the closing of a giant circle. Yes, smaller than the smallest, I meant something, too... there is no zero—I exist!"[13]

References

1. P. R. Yoder, Jr. and D. Vukobratovich, *Opto-Mechanical Systems Design*, Vol. **2**, Fourth Edition, CRC Press, Boca Raton, Florida, Ch. 4, p. 141 (2015).
2. S. Timoshenko and S. Woinowsky-Krieger, *Theory of Plates and Shells*, Second Edition, McGraw-Hill, New York, p. 249 (1959).
3. J. W. Pepi, "Test and theoretical comparison for bending and springing of the Keck ten-meter telescope, *Proc. SPIE* **1271**, pp. 275–287 (1990) [doi: 10.1117/12.20417].
4. R. Williams and H. Brinson, "Circular plate on multipoint supports," *J. Franklin Inst.* **297**(6), 429–497 (1974).
5. J. H. Hindle, "Mechanical flotation of mirrors," in *Amateur Telescope Making*, Book One, Scientific American, New York (1945).
6. W. D. Pilke, *Formulas for Stress, Strain, and Structural Matrices*, John Wiley & Sons, New York, p. 520 (1994).
7. J. Pepi and C. Finch, "Fine Figuring Actuator," U.S. Patent 4,601,553 (1986).
8. J. W. Pepi, "Design considerations for mirrors with large diameter to thickness ratios," *Proc. SPIE* **10265**, pp. 207–231 *Optomechanical Design: A Critical Review* (1992) [doi: 10.1117/12.61107].
9. J. W. Pepi and W. P. Barnes, "Thermal distortion of a thin fused silica mirror," *Proc. SPIE* **450**, pp. 40–49 (1984) [doi: 10.1117/12.939265].

10. J. Lubliner and J. E. Nelson, "Stressed mirror polishing. 1: A technique for producing non-axisymmetric mirrors," *Applied Optics* **19**(14), 2332–2340 (1980).
11. J. W. Pepi and D. Golini, "Delayed elasticity in Zerodur® at room temperature," *Proc. SPIE* **1533**, pp. 212–221 (1991) [doi: 10.1117/12.48857].
12. J. B. Murgatroyd and R. F. Sykes, "The delayed elastic effect in silicate glasses at room temperature," *J. Soc. Glass Technology* **31**, 17–35 (1947).
13. "The Incredible Shrinking Man," [film] Universal-International, U.S. based on a novel by R. Matheson (1957).

Chapter 8
Figures of Merit

Figures of merit are values that denote the benefit of key material properties in terms of mechanical and thermal performance. Mechanical figures of merit relate to stiffness, strength, and mass. The stiffness of a material is proportional to its elastic modulus E, while the mass of a material is proportional to its density ρ.

The strength S of a material is defined as its yield point, fracture limit, ultimate failure point, or other such measure. Mechanical figures of merit (FOMs) are thus related to some function of these ratios as

$$\mathrm{FOM_k} = f\left(\frac{E}{\rho}\right) \tag{8.1}$$

for *stiffness* performance and

$$\mathrm{FOM_S} = f\left(\frac{S}{\rho}\right) \tag{8.2}$$

for *strength* performance. In either case, the higher the value the higher the merit.

Thermal FOMs generally relate to thermal conductivity K and coefficient of thermal expansion (CTE) α. Thermal FOMs are important for flux-induced gradient error and are therefore related to some function of this ratio as

$$\mathrm{FOM_t} = f\left(\frac{K}{\alpha}\right). \tag{8.3}$$

As is the case for mechanical FOMs, the higher the value the higher the merit.

8.1 Mechanical Figures of Merit

Mechanical FOMs are important for gravity sag, fundamental frequency, mount error, strength, and weight. Theoretical equations can be used to compare weights of various materials for equal stiffness or strength performance based on the use of appropriate mechanical FOMs. Performance and strength for equal weight can also be computed. Typically, these FOMs are reported as a linear relationship:

$$\mathrm{FOM_k} = \frac{E}{\rho} \tag{8.4}$$

for stiffness performance, called the *specific stiffness*, and

$$\mathrm{FOM_S} = \frac{S}{\rho} \tag{8.5}$$

for strength performance, called the *specific strength.*

However, in most cases, analyses show that these linear ratios of stiffness-to-weight and strength-to-weight are not always valid. Sometimes FOMs lie. Table 8.1 shows the linear relationship of specific stiffness for typical materials used in optical systems, both for structure and for optics. (Table 2.2 can be used to compute these values for other materials). Figure 8.1 depicts Table 8.1 graphically. All values are normalized to beryllium; again, the higher the value the better the performance. Note that the metals steel, titanium, aluminum, and magnesium all have virtually the same stiffness-to-weight ratio. Note also on the chart that spruce wood, which fares poorly (and is certainly not an optical material), has been included for comparison.

Table 8.2 shows the linear relationship of specific strength for typical materials used in optical systems, both for structure and for optics. Figure 8.2 depicts Table 8.2 graphically. All values are normalized to graphite cyanate

Table 8.1 Standard mechanical performance FOM based on the linear modulus-to-density ratio (Msi is megapounds/square inch; pci is pounds/cubic inch).

Material	Elastic Modulus (Msi)	Density (pci)	Stiffness/ Weight Ratio (normalized)
Beryllium	44	0.067	1.00
Silicon carbide	44.5	0.105	0.65
Aluminum beryllium	28	0.076	0.56
Graphite cyanate	15	0.063	0.36
Silicon	19	0.084	0.34
Aluminum silicon carbide	19	0.105	0.28
Aluminum silicon	14	0.094	0.23
Zerodur	13.1	0.091	0.22
ULE	9.8	0.079	0.19
Stainless steel	29	0.29	0.15
Titanium	16	0.16	0.15
Aluminum	10	0.1	0.15
Magnesium	6.5	0.065	0.15
Spruce	1.2	0.014	0.13
Invar	20.5	0.29	0.11

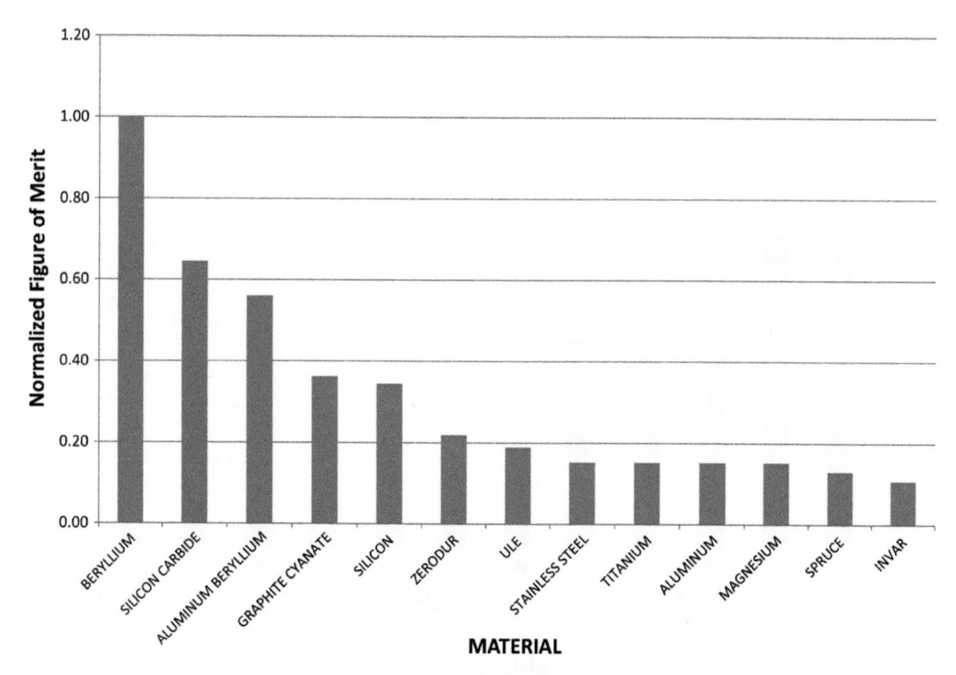

Figure 8.1 Standard mechanical performance FOM based on the linear modulus-to-density ratio.

Table 8.2 Standard mechanical strength FOM based on the linear modulus-to-density ratio.

Material	Strength (psi)	Density (pci)	Strength/ Weight Ratio (normalized)
Graphite cyanate	36000	0.063	1.00
Titanium	80000	0.16	0.88
Aluminum silicon carbide	45000	0.105	0.75
Aluminum beryllium	28000	0.076	0.64
Beryllium	22000	0.067	0.57
Aluminum	28000	0.1	0.49
Aluminum silicon	16000	0.094	0.30
Magnesium	9000	0.065	0.24
Invar	32000	0.29	0.19
Stainless steel	28000	0.29	0.17
Silicon carbide	8400	0.105	0.14
Silicon	6300	0.084	0.13
Spruce	1000	0.014	0.13
ULE	1200	0.079	0.03
Zerodur	1200	0.091	0.02

ester. Again, the higher the value the better the performance. Here, we choose strength based on yield for ductile materials, or fracture strength for brittle materials, including the usual applied safety factors as noted. Properties are

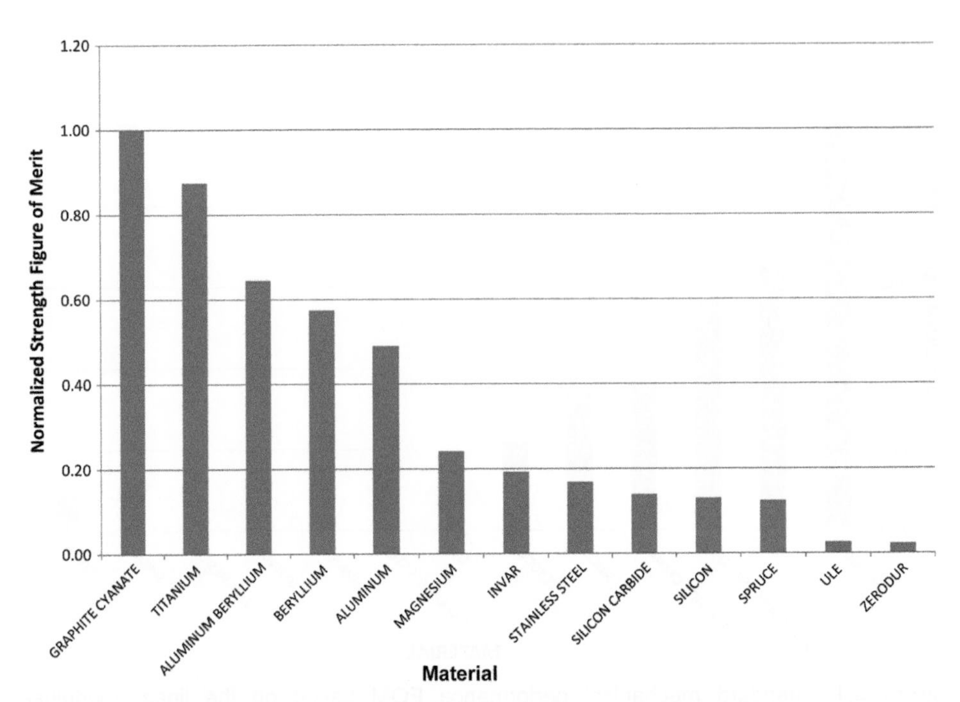

Figure 8.2 Standard mechanical strength FOM based on the linear modulus-to-density ratio.

general for purposes of this study, and it is recognized that variations on reported strength values may change as a function of alloy content or treatment, for example. Again, we include spruce wood, for comparison, which again fares poorly.

8.2 Thermal Figure of Merit

Thermal gradients can be very detrimental to optical performance, although if a material exhibits a high thermal conductivity, such gradients can be minimized. However, if a material exhibits a low thermal conductivity, performance may be acceptable even in the presence of gradients if its CTE is low. Using the linear relationship for thermal FOM, we have

$$\mathrm{FOM_t} = \frac{K}{\alpha}. \tag{8.6}$$

While other thermal FOMs can be produced to include specific heat and density in Eq. (8.6), or effects of radiation and convection, we concentrate here on thermal conductivity.

The linear thermal FOM is shown in Table 8.3 for typical materials used in optical systems, both for structure and for optics. Figure 8.3 depicts Table 8.3 graphically. All values are normalized to graphite cyanate ester,

Table 8.3 Standard thermal FOM based on the conductivity-to-CTE ratio.

Material	Thermal Conductivity (W/mK)	Coefficient of Thermal Expansion (ppm/K)	Conductivity/ CTE Ratio (normalized)
Graphite cyanate	31.5	0.2	1.00
Zerodur	1.46	0.02	0.46
ULE	1.31	0.02	0.42
Silicon	163	2.6	0.40
Silicon carbide	150	2.43	0.39
Beryllium	210	11.5	0.12
Aluminum beryllium	210	14	0.10
Aluminum silicon	130	15	0.06
Aluminum silicon carbide	130	16	0.05
Invar	10.4	1.3	0.05
Aluminum	151	22.5	0.04
Magnesium	96	26	0.02
Titanium	7.27	8.9	0.01
Stainless steel	16.3	17.3	0.01
Spruce	0.12	4.5	0.00

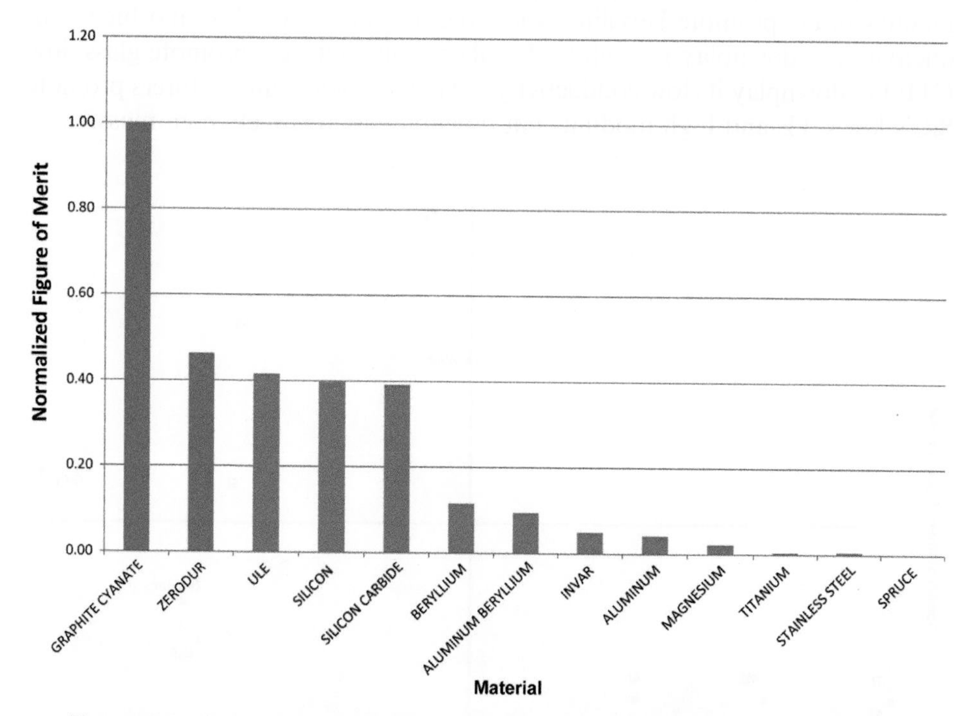

Figure 8.3 Standard thermal FOM (K/α) based on the conductivity-to-CTE ratio.

which leads the pack because of its near-zero CTE for pseudo-isotropic layup and its relatively good conductivity. Note that beryllium, which had topped the chart for mechanical stiffness, is relegated to a mediocre thermal FOM due to its high CTE. On the other hand, the near-zero CTE glasses, in spite of their poor thermal conductivity, fare quite well. Again, for comparison, we show spruce wood, which fails miserably.

8.3 Combined Figures of Merit

Using Tables 8.1 and 8.3 for mechanical stiffness and thermal FOMs, Fig. 8.4 plots mechanical-versus-thermal FOMs for the various materials we have considered. In this chart, since the higher the value the better, one focuses on materials that lie in the upper right quadrant. However, those in the lower right (or top left) quadrant may be sufficient, depending on design drivers. Materials in the lower left quadrant fare poorly on all accounts.

8.4 True Mechanical Figures of Merit

Although material manufacturers often like to promote the benefits of their material, they will, of course, downplay its drawbacks. For example, beryllium manufacturers promote beryllium's low density, high modulus, and high conductivity but downplay its high CTE. Glass manufacturers promote glass' low CTE but downplay its low conductivity. Silicon carbide manufacturers promote SiC's low CTE and high modulus but downplay its relatively high mass.

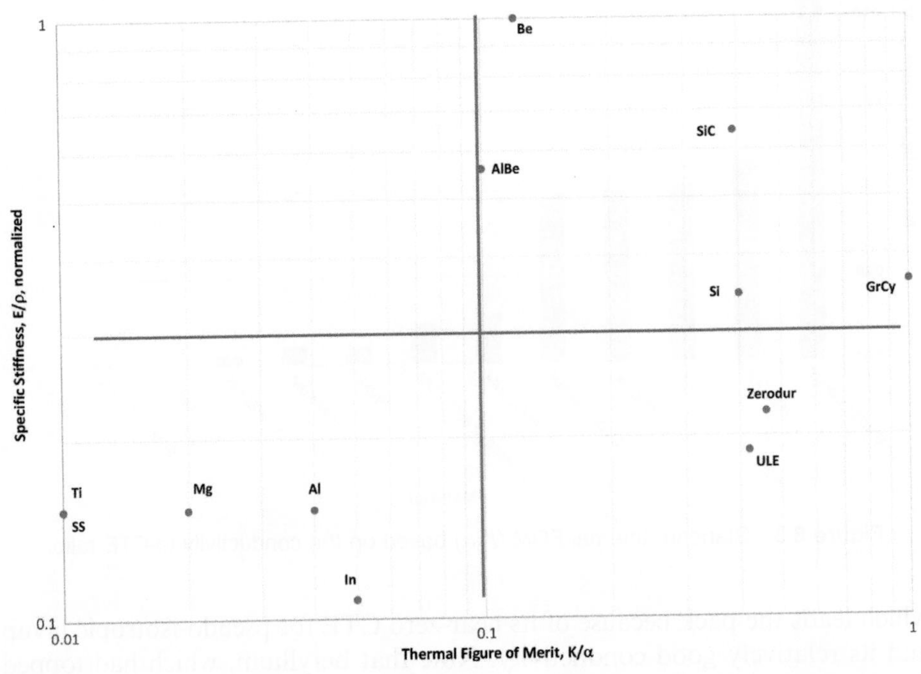

Figure 8.4 Typically reported standard linear mechanical and thermal FOMs.

Table 8.1 shows a comparison of mechanical stiffness merit properties of interest for select materials. Note, for example, that both beryllium and silicon carbide have a significantly higher stiffness-to-weight ratio than ULE™ glass, namely, 5:1 versus 3.7:1. This does not necessarily mean that beryllium and silicon carbide optics can be made lighter by that amount. In fact, the linear stiffness-to-weight ratio is often blatantly misleading, as we shall see. As a prelude, consider the following queries:

1. A ZERODUR® glass ceramic has a better stiffness-to-weight ratio than ULE glass. Which one can be made lighter for equal frequency? Which one deflects less under gravity for equal mass?
2. Beryllium exhibits a 50% lower thermal gradient in the presence of solar flux than silicon carbide for equal mass. Which one performs better in the presence of solar flux?
3. Silicon carbide has five times the stiffness of glass and three times the stiffness of graphite composites. Which will weigh less: a glass optics/graphite structure or an all–silicon carbide telescope?
4. Aluminum, steel, titanium, and magnesium all have the same stiffness-to-weight ratio. Which can be made lightest for equal stiffness? Which can be made stiffer for equal mass?
5. Silicon carbide has the same elastic modulus as beryllium but is 50% heavier. Does silicon carbide weigh 50% more than beryllium for equal frequency or gravity sag?
6. Silicon carbide has nearly five times the stiffness of glass. How much lighter can a silicon carbide optic be made compared to glass?

The answers are provided in Section 8.8. Solid optics and structure are reviewed in the following subsection through Section 8.5; a note on lightweight optics is given in Section 8.7.

8.4.1 Weight and performance figures of merit

Using theoretical equations, we first review weight for equal performance. Consider a structure under externally applied *tension* load P for a material of thickness t. Using the equations of Chapter 1, the structure will deflect by an amount

$$y = \frac{P}{Et}, \tag{8.7}$$

and its weight for a given density will be

$$W \sim \rho t. \tag{8.8}$$

For equal mass, comparing two materials (subscripts 1 and 2),

$$\rho_1 t_1 = \rho_2 t_2, \tag{8.9}$$

$$y_1 = \frac{1}{E_1 t_1},\ y_2 = \frac{1}{E_2 t_2}, \tag{8.10}$$

$$t_1 = \frac{\rho_2 t_2}{\rho_1}, \tag{8.11}$$

$$\frac{y_2}{y_1} \sim \frac{E_1 \rho_2 t_2}{E_2 t_2 \rho_1} \sim \frac{E}{\rho}. \tag{8.12}$$

Thus, the *tension* FOM is E/ρ (linear), as typically reported in the literature. However, most of our concern is not about tension, but rather about *bending*. Optics bend under gravity or mount load; similarly, a structure is often loaded in bending under applied or gravity acceleration loads. Bending, then, will most often dominate optical system performance.

8.4.1.1 Gravity weight: equal performance

Consider, for example, the effects of gravity on an optic. For a given gravitational error requirement, in bending with gravity and with the optical axis aligned, it has been shown in Chapter 4 [Eq. (4.7)] that sag is related as

$$y \sim \frac{\rho t}{E t^3}, \tag{8.13}$$

while its weight for a given density ρ is given by Eq. (8.8). In Eq. (8.13), ρ is the weight density (in pounds per cubic inch), E is the elastic modulus (in pounds per square inch), and t is the depth of the solid optic.

For equal performance, comparing mass for two materials,

$$\begin{aligned} W_1 &= \rho_1 t_1, \\ W_2 &= \rho_2 t_2, \\ y_1 &= \frac{\rho_1 t_1}{E_1 {t_1}^3}, \\ y_2 &= \frac{\rho_2 t_2}{E_2 {t_2}^3}, \\ t_1 &= \sqrt{\frac{\rho_1}{\rho_2}} \sqrt{\frac{E_2}{E_1}} t_2, \\ \frac{W_2}{W_1} &\sim \frac{\sqrt{E}}{\sqrt{\rho^3}}. \end{aligned} \tag{8.14}$$

The final line in Eq. (8.14) is the appropriate gravity weight FOM for equal performance. Using this relationship, a beryllium optic can be made 2.7 times lighter than glass for equal gravitational/frequency performance, while a silicon carbide optic can be made up to 1.5 times lighter.

The appropriate FOM for this beryllium requirement, along with FOMs for other select materials, is shown in Table 8.4 and Fig. 8.5. Note how the order of merit has changed from the linear FOM in Table 8.1! Interestingly, note how well spruce wood now fares! As seen in Eq. (8.14), the true mechanical FOM for equal performance is much more strongly driven by density than by modulus.

Returning to optical system candidates, however, it is worthy to note the comparisons of the metals aluminum, steel, titanium, and magnesium, all of which have the same specific stiffness, as evidenced in Table 8.1 for this linear FOM. Based on the nonlinear FOM for bending, however, [from Eq. (8.14) and Table 8.4] a magnesium structure gives the least weight for a given deflection requirement under gravity. For example, if a vibration fixture needs a certain stiffness to meet frequency requirements in order to avoid coupling to a unit under test, a magnesium structure can be made lighter, thereby avoiding taxing maximum load input constraints for a vibration shaker table. Magnesium fixtures, although costly, are ideal for such tests. Since fundamental frequency is directly related to gravity, the same FOM will apply as well.

8.4.1.2 Gravity performance: equal weight

Using Eqs. (8.8) and (8.13), we can review FOM for weight when weight is set equal. If weight for different materials is set equal, we calculate the weight-to-thickness ratio as

$$\frac{t_1}{t_2} = \frac{\rho_1}{\rho_2}, \tag{8.15}$$

and substitution shows the performance ratio of materials as

$$\text{FOM} \sim \frac{E}{\rho^3}, \tag{8.16}$$

which is the appropriate FOM for gravitational error requirements for equal-weight optics. Using this relationship, a beryllium optic can be made to perform seven times better than glass for equal weight, while a silicon carbide optic can be made to perform two times better. The appropriate FOM for this requirement, along with FOMs of other select materials, is shown in Table 8.4.

Table 8.4(a) Weight comparisons for equal performance error and performance comparisons for equal weight using true FOMs.

Performance Error (deflection) Figures of Merit for Various Materials										
Comparison	Weight					Deflection				
Constant	Thickness		Deflection			Thickness		Weight		
Direction	Tension		Bending			Tension		Bending		
Load	Gravity	External	Gravity	External	Coating	Gravity	External	Gravity	External	Coating
Figure of Merit	(E/ρ)	(E/ρ)	($E^{1/2}/\rho^{3/2}$)	($E^{1/3}/\rho$)	($E^{1/2}/\rho$)	(E/ρ)	(E/ρ)	$(E/\rho)^3$	(E/ρ^2)	$(E/\rho)^2$
Material										
Beryllium	1.00	1.00	1.00	1.00	1.00	1.00	1.00	1.00	1.00	1.00
SiC	0.73	0.73	0.54	0.66	0.68	0.73	0.73	0.30	0.47	0.47
ULE	0.20	0.20	0.37	0.52	0.41	0.20	0.20	0.14	0.17	0.17
Invar	0.11	0.11	0.08	0.18	0.16	0.11	0.11	0.01	0.03	0.03
Stainless	0.16	0.16	0.09	0.20	0.19	0.16	0.16	0.01	0.04	0.04
Titanium	0.16	0.16	0.17	0.30	0.26	0.16	0.16	0.03	0.07	0.07
Aluminum	0.16	0.16	0.27	0.41	0.33	0.16	0.16	0.07	0.11	0.11
Magnesium	0.16	0.16	0.41	0.55	0.41	0.16	0.16	0.17	0.16	0.16
Silicon	0.36	0.36	0.48	0.61	0.54	0.36	0.36	0.23	0.29	0.29
GraphiteCy	0.38	0.38	0.65	0.75	0.64	0.38	0.38	0.43	0.40	0.40
Aluminum beryllium	0.56	0.56	0.66	0.76	0.70	0.56	0.56	0.44	0.49	0.49
Spruce	0.14	0.14	1.77	1.46	0.81	0.14	0.14	3.13	0.65	0.65

Table 8.4(b) Material ranking comparisons for equal performance error and performance comparisons for equal weight using true FOMs.

Performance Error (deflection) Ranking for Various Materials										
	Weight					Deflection				
Comparison	Thickness		Deflection			Thickness		Weight		
Constant Direction	Tension		Bending			Tension		Bending		
Load	Gravity	External	Gravity	External	Coat	Gravity	External	Gravity	External	Coat
Figure of Merit	(E/ρ)	(E/ρ)	$(E^{1/2}/\rho^{3/2})$	$(E^{1/3}/\rho)$	$(E^{1/2}/\rho)$	(E/ρ)	(E/ρ)	(E/ρ^3)	$(E/\rho)^2$	$(E/\rho)^2$
Material	Beryllium	Beryllium	Spruce	Spruce	Beryllium	Beryllium	Beryllium	Spruce	Beryllium	Beryllium
	SiC	SiC	Beryllium	Beryllium	Spruce	SiC	SiC	Beryllium	Spruce	Spruce
	GraphiteCy	GraphiteCy	GraphiteCy	GraphiteCy	SiC	GraphiteCy	GraphiteCy	GraphiteCy	SiC	SiC
	Silicon	Silicon	SiC	SiC	GraphiteCy	Silicon	Silicon	SiC	GraphiteCy	GraphiteCy
	ULE	ULE	Silicon	Silicon	Silicon	ULE	ULE	Silicon	Silicon	Silicon
	Magnesium	Magnesium	Magnesium	Magnesium	ULE	Magnesium	Magnesium	Magnesium	ULE	ULE
	Aluminum	Aluminum	ULE	ULE	Magnesium	Aluminum	Aluminum	ULE	Magnesium	Magnesium
	Titanium	Titanium	Aluminum	Aluminum	Aluminum	Titanium	Titanium	Aluminum	Aluminum	Aluminum
	Stainless	Stainless	Titanium	Titanium	Titanium	Stainless	Stainless	Titanium	Titanium	Titanium
	Spruce	Spruce	Stainless	Stainless	Stainless	Spruce	Spruce	Stainless	Stainless	Stainless
	Invar	Invar	Invar	Invar	Invar	Invar	Invar	Invar	Invar	Invar

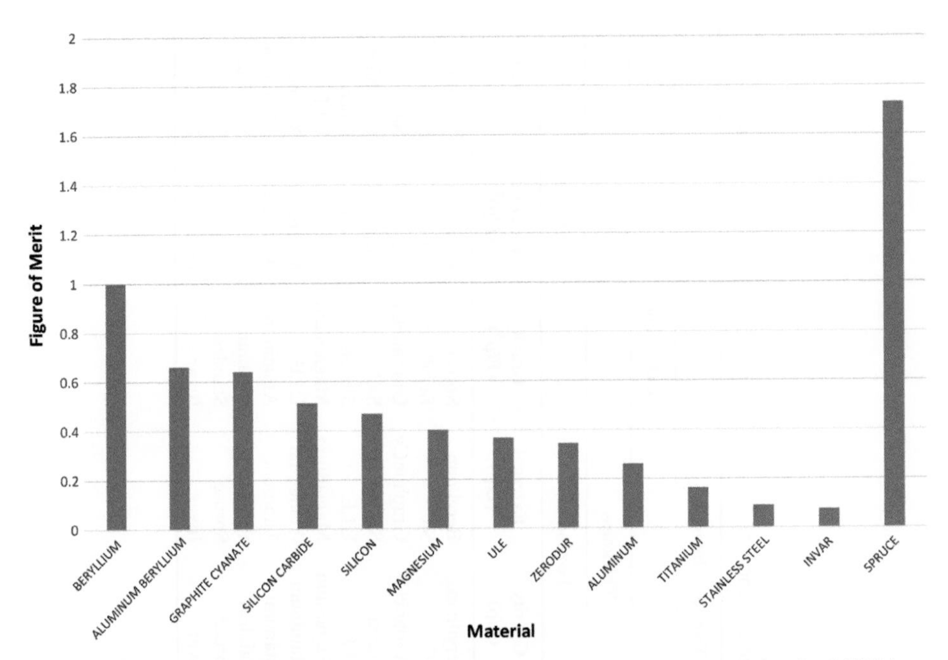

Figure 8.5 Weight for equal gravitational performance—a true mechanical FOM.

8.4.1.3 Mount weight: equal performance

Consider next the effects of mount or metrology error on the optic. For a given error requirement, it has been shown that sag is related as

$$y \sim \frac{1}{Et^3}. \tag{8.17}$$

If performance for different materials is set equal, we calculate the thickness ratio of the materials' proportionality as

$$\frac{t_2}{t_1} = \sqrt[3]{\frac{E_1}{E_2}}. \tag{8.18}$$

Since weight is proportional to the product of density and thickness, we find the weight ratio of the materials' proportionality as

$$\frac{W_2}{W_1} = \frac{\sqrt[3]{E}}{\rho}, \tag{8.19}$$

which is the appropriate FOM for mount-induced error requirements. Since metrology involving forces and loads is directly related to mount error, the same FOM applies as well. This FOM applies to any externally applied load.

The appropriate FOM for this requirement, along with FOMs of other select materials, is again shown in Table 8.4.

Note again how well wood behaves; while poor for optical use for other obvious reasons, it is quite efficient for houses (and sometimes for airplanes[1]). Using this relationship, a beryllium optic can be made 1.9 times lighter than glass for equal mount-error performance, while a silicon carbide optic can be made up to 1.25 times lighter.

8.4.1.4 Mount performance: equal weight

Using Eqs. (8.8), (8.15), and (8.17), we can review the FOM for performance when weight is set equal. We find the performance ratio of the materials' proportionality as

$$\frac{y_2}{y_1} = \frac{E}{\rho^3}, \tag{8.20}$$

which is the appropriate performance FOM for mount-error requirements for equal-weight optics.

Using this relationship, a beryllium optic can be made to perform six times better than glass for equal mount-error performance, while a silicon carbide optic can be made to perform up to three times better. The appropriate FOM for this requirement, along with FOMSs of other select materials, is shown in Table 8.4.

8.4.1.5 Coating and cladding FOMs

It has been shown in Section 4.6 and Eq. (4.36) that deflection induced by coating or cladding stresses under thermal soak condition (either residual stress or thermally induced) is related as

$$y \sim \frac{1}{Et^2} \tag{8.21}$$

for a given stress. Using similar arguments as before, we show the appropriate weight ratio FOM as

$$\text{FOM} \sim \frac{\sqrt{E}}{\rho} \tag{8.22}$$

for equal performance, and the appropriate performance ratio FOM as

$$\text{FOM} \sim \frac{E}{\rho^2} \tag{8.23}$$

for equal weight. Again, Table 8.4 depicts the FOMs for coating stress.

8.5 Strength-to-Weight Ratio

In a similar fashion, using theoretical equations, a review can be made to compare weight for equal-strength-margin optics and structures of various materials. A similar comparison can also be conducted to review the strength margin for equal-weight optics of various materials. Properties of all materials in this study are shown in Table 8.2.

As is the case for deflection performance, tensile (or compressive) loading—as achieved by a truss structure, axial bar load, or the like—results in the usual, linear, specific strength relation. For example, under gravity, in the axial direction, deflection due to self-weight is related as

$$y \sim \frac{\rho}{E}, \tag{8.24}$$

and the strength margin under self-weight is related as

$$S_{\mathrm{m}} \sim \frac{S}{\rho}. \tag{8.25}$$

In this case, the lightest design for equal strength performance or strength margin is achieved by minimizing thickness, and the appropriate FOM is given in Eqs. (8.24) and (8.25).

Similarly, for external tensile load, we find deflection related as

$$y \sim \frac{1}{Et}, \tag{8.26}$$

and the strength margin under external load related as

$$S_{\mathrm{m}} \sim St. \tag{8.27}$$

In these cases, for equal performance or strength,

$$\frac{t_2}{t_1} = \frac{E_1}{E_2}, \tag{8.28}$$

or

$$\frac{t_2}{t_1} = \frac{S_1}{S_2}. \tag{8.29}$$

The weight ratio is

$$\frac{W_2}{W_1} = \frac{\rho_2 t_2}{\rho_1 t_1}, \tag{8.30}$$

and upon substituting, the appropriate FOM is again given in Eqs. (8.4) and (8.5) for deflection and strength, respectively. These same relations also apply for equal weight. However, again, in bending, this linear relation is not adequate, as explained in the next subsection.

8.5.1 Gravitational acceleration: bending

Consider, for example, the effects of bending strength on the optic under inertial acceleration loading normal to its surface, and primarily producing bending. For a given strength and safety factor, it is readily shown[2] that stress is related as

$$\sigma \sim \frac{\rho}{t}, \tag{8.31}$$

and hence, for a given safety factor, the strength margin is related as

$$S_{\mathrm{m}} \sim \frac{St}{\rho}, \tag{8.32}$$

where S is the strength of the optic in pounds per square inch. If the strength margins for different materials are set equal, we calculate the thickness ratio as

$$\frac{t_2}{t_1} = \frac{S_1 \rho_2}{S_2 \rho_1}. \tag{8.33}$$

Using Eq. (8.8), we find the weight ratio of materials as

$$\mathrm{FOM} \sim \frac{S}{\rho^2}, \tag{8.34}$$

which is the appropriate FOM for gravitational strength in bending requirements.

The appropriate FOMs for structures bound by these requirements for several select materials are shown in Table 8.5 and Fig. 8.6. The graphite composite leads the pack by a substantial margin. Note that strong titanium is relegated to a lesser seat. Note also how effective spruce wood is!

In the case of the optics, we take strength as the minimum fracture or micro-yield allowable, whichever is lowest. In the case of the supporting structure, we generally use the 0.2% offset yield point or fracture allowable with an appropriate safety factor.

Table 8.5 (a) Material weight and strength comparisons and (b) material ranking using true FOMs.

(a)

Strength Figures of Merit for Various Materials - Structure

Comparison	**Weight**				**Strength**			
Constant	**Thickness**		**Strength**		**Thickness**		**Weight**	
Direction	**Tension**		**Bending**		**Tension**		**Bending**	
Load	**Gravity**	**External**	**Gravity**	**External**	**Gravity**	**External**	**Gravity**	**External**
Figure of Merit	(S/ρ)	(S/ρ)	$(S/\rho)^2$	$(S^{1/2}/\rho)$	(S/ρ)	(S/ρ)	(S/ρ^2)	(S/ρ^2)
Material								
Beryllium	1.00	1.00	1.00	1.00	1.00	1.00	1.00	1.00
SiC	0.23	0.23	0.15	0.38	0.23	0.23	0.15	0.15
ULE	0.04	0.04	0.03	0.18	0.04	0.04	0.03	0.03
Invar	0.38	0.38	0.09	0.30	0.38	0.38	0.09	0.09
Stainless	0.28	0.28	0.07	0.26	0.28	0.28	0.07	0.07
Titanium	1.52	1.52	0.64	0.80	1.52	1.52	0.64	0.64
Aluminum	0.97	0.97	0.65	0.81	0.97	0.97	0.65	0.65
Magnesium	0.42	0.42	0.43	0.66	0.42	0.42	0.43	0.43
Silicon	0.23	0.23	0.18	0.43	0.23	0.23	0.18	0.18
GraphiteCy	1.74	1.74	1.85	1.36	1.74	1.74	1.85	1.85
Aluminum beryllium	1.12	1.12	0.99	0.99	1.12	1.12	0.99	0.99
Spruce	0.22	0.22	1.04	1.02	0.22	0.22	1.04	1.04

(b)

Strength Figure-of-Merit Ranking for Various Materials - Structure

Comparison	**Weight**				**Strength**			
Constant	**Thickness**		**Strength**		**Thickness**		**Weight**	
Direction	**Tension**		**Bending**		**Tension**		**Bending**	
Load	**Gravity**	**External**	**Gravity**	**External**	**Gravity**	**External**	**Gravity**	**External**
Figure of Merit	S/ρ	S/ρ	S/ρ^2	$S^{1/2}/\rho$	S/ρ	S/ρ	S/ρ^2	S/ρ^2
Material	GraphiteCy	GraphiteCy	GraphiteCy	GraphiteCy	GraphiteCy	GraphiteCy	GraphiteCy	GraphiteCy
	Titanium	Titanium	Spruce	Spruce	Spruce	Spruce	Spruce	Spruce
	Al-Be	Al-Be	Beryllium	Beryllium	Al-Be	Al-Be	Beryllium	Beryllium
	Beryllium	Beryllium	Al-Be	Al-Be	Beryllium	Beryllium	Al-Be	Al-Be
	Aluminum	Aluminum	Aluminum	Aluminum	Aluminum	Aluminum	Aluminum	Aluminum
	Magnesium	Magnesium	Titanium	Titanium	Magnesium	Magnesium	Titanium	Titanium
	Invar	Invar	Silicon	Magnesium	Invar	Invar	Magnesium	Magnesium
	Stainless	Stainless	SiC	Silicon	Stainless	Stainless	Silicon	Silicon
	SiC	SiC	Beryllium	SiC	SiC	SiC	SiC	SiC
	Silicon	Silicon	Stainless	Invar	Silicon	Silicon	Invar	Invar
	Spruce	Spruce	Invar	Stainless	Spruce	Spruce	Stainless	Stainless
	ULE	ULE	ULE	ULE	ULE	ULE	ULE	ULE

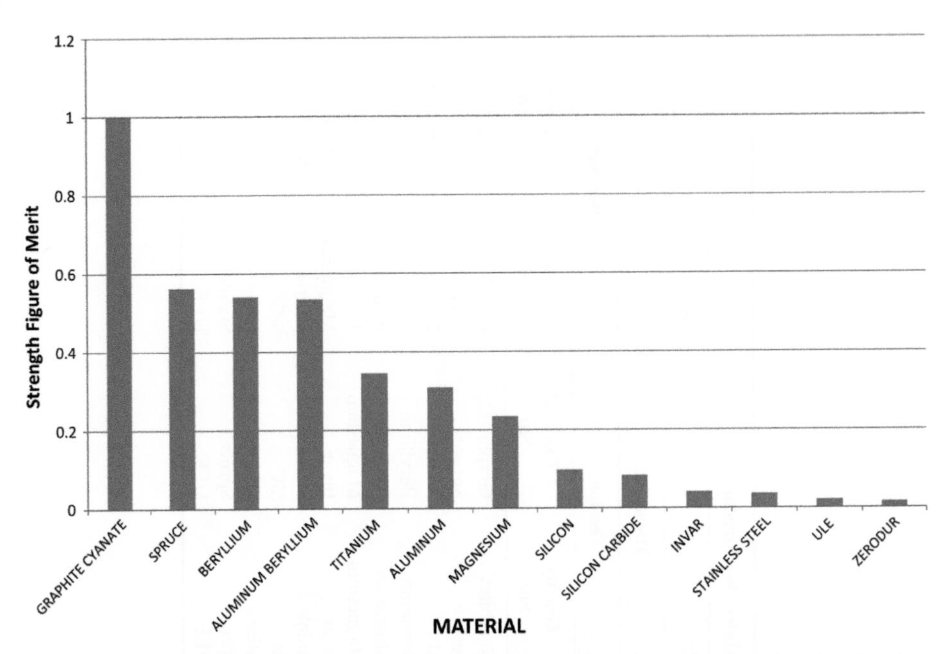

Figure 8.6 True mechanical FOM for self-weight strength under acceleration load.

8.5.2 External bending load and gravity acceleration

Next, consider the effects of bending strength on the optic under external load normal to its surface, and primarily producing bending. For a given strength and safety factor, it is shown that the strength margin is related as

$$\text{FOM} \sim \frac{\sqrt{S}}{\rho} \tag{8.35}$$

for external load strength in bending requirements. This FOM is shown in Table 8.5, along with FOMs of other select materials. Similarly, for equal weight, under gravity acceleration load, we find the stress margin related as in Eq. (8.34), as shown in Table 8.5.

8.6 Graphical Summary

It is of interest to now return to the combined thermal and mechanical chart of Fig. 8.4. Figure 8.7 shows the true, nonlinear mechanical FOMs, using self-weight equal performance. These FOMs can be compared to Fig. 8.4. While beryllium remains the best performer from a mechanical FOM standpoint, materials other than beryllium have moved up or down, or even changed positions! Steel and Invar have fallen off the chart. Note that all design drivers and criteria need identification and consideration for proper selection.

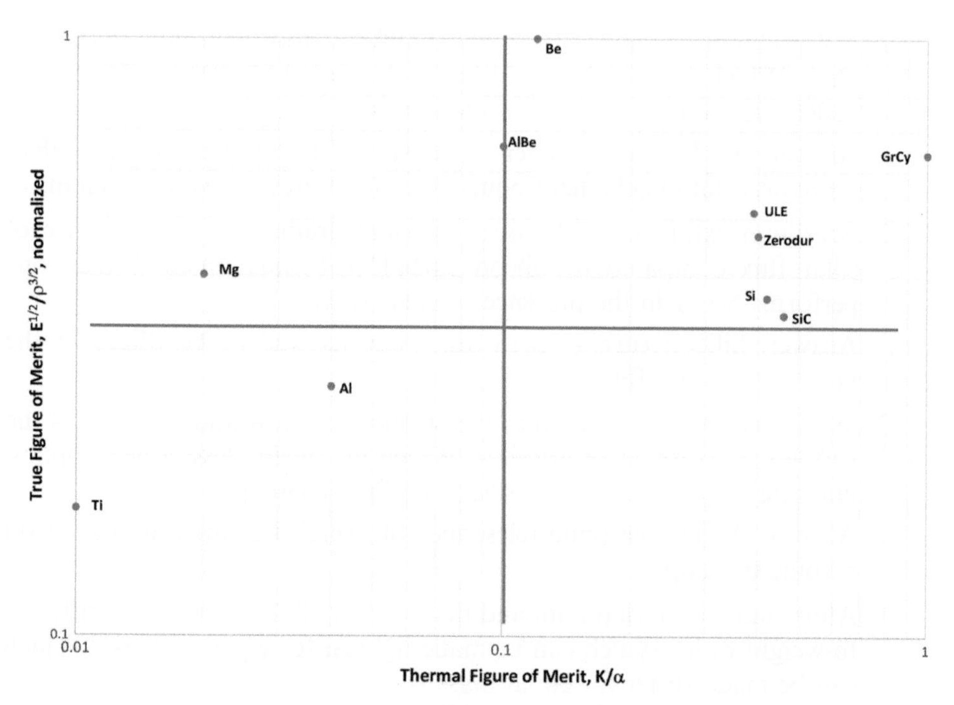

Figure 8.7 True mechanical and thermal FOMs.

8.7 Lightweight Optics

Using the theoretical relations shown in Chapters 5 and 6, a study similar to that made for solid optics is made to compare weight and performance for lightweight optics of various materials. The appropriate FOMs are essentially the same as those for solid optics, although this may not be intuitively obvious. However, since stiffness for lightweight optics is a function of equivalent thickness, the similarity becomes apparent. This relationship therefore applies to both open- and closed-back optics. Of course, as always, for any FOM, it is assumed that envelope or manufacturability are not constraints. For example, it is assumed that the design is not limited by core fillets, rib width, or facesheet thickness.

The equations of Chapter 5 for lightweight mirror optimization can be used to illustrate the similarity among the FOMs derived for solids. This will be seen in the examples that follow.

8.8 Examples

In Section 8.4, several questions were posed. These are repeated and answered here. Further examples for consideration follow to quantify the conclusions.

1. A Zerodur glass ceramic has a better stiffness-to-weight ratio than ULE glass. Which one can be made lighter for equal frequency? Which one deflects less under gravity for equal mass?

   Answer: ULE has a better effective stiffness-to-weight ratio than Zerodur, exhibiting higher frequency and less gravity sag for equal mass.

2. Beryllium exhibits a 50% lower thermal gradient in the presence of solar flux compared to silicon carbide for equal mass. Which one performs better in the presence of solar flux?

   Answer: Silicon carbide performs twice as well as beryllium in the presence of solar flux.

3. Silicon carbide has five times the stiffness of glass and three times the stiffness of graphite composites. Which will weigh less: a glass optics/graphite structure or an all–silicon carbide telescope?

   Answer: A glass graphite telescope will weigh less than an all–silicon carbide telescope.

4. Aluminum, steel, titanium, and magnesium all have the same stiffness-to-weight ratio. Which can be made lightest for equal stiffness? Which can be made stiffer for equal mass?

   Answers: For equal stiffness, magnesium can be made the lightest. For equal mass, magnesium can be made the stiffest.

5. Silicon carbide has the same elastic modulus as beryllium but is 50% heavier. Does silicon carbide weigh 50% more than beryllium for equal frequency or gravity sag?

   Answer: For frequency and gravity performance equal to beryllium, silicon carbide weighs 100% more (twice as much!).

6. Silicon carbide has nearly five times the stiffness of glass. How much lighter can a silicon carbide optic be made compared to glass?

   Answer: Silicon carbide can be made only 1.39 times lighter than glass.

8.8.1 Examples for consideration

8.8.1.1 Solid optics

1. An aluminum fixture is designed for a random vibration test to exhibit a fundamental mode resonance of 500 Hz in order to limit cross-axis responses during test. The fixture weighs 600 lbs, but the shaker capability after the unit under test is considered requires a limit of 500 lbs. Can a magnesium fixture solve the problem?

   While magnesium and aluminum gave the same specific stiffness, Eq. (8.14) and Table 8.4 are used to show the benefits of the lighter metal. For equal frequency, the weight FOM is 1.54; the inverse of the

Table 8.6 Weight benefit of beryllium and silicon carbide with respect to ULE glass for lightweighted optics.

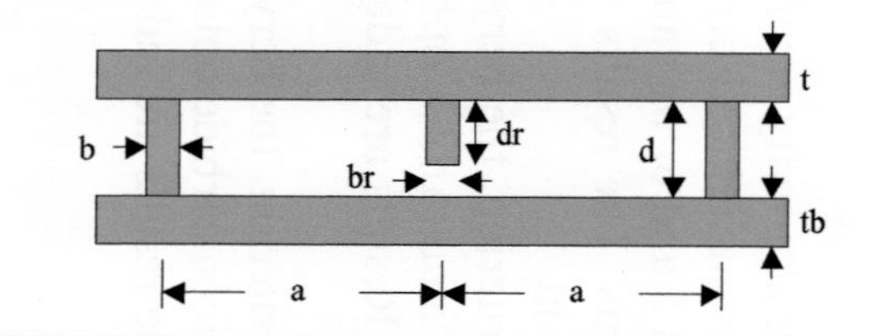

Material	Configuration	*d* (in.)	*dr* (in.)	*b* (in.)	*br* (in.)	*t* (in.)	*tb* (in.)	*a* (in.)	Gravity Error (μin.)	Mount Error (μin.)	Weight (pounds)	Inverse Weight Ratio (w.r.t. ULE)
ULE	Open	6	6	0.06	0.06	0.29		1.88	65.7	4.6	67.1	1.0
ULE	Closed	4.1	4.1	0.06	0.06	0.14		1.27	71.0	5.0	67.2	1.0
Beryllium	Open	2.5	2.5	0.06	0.06	0.14		1.59	64.2	N/A	27.9	2.4
Beryllium	Closed	1.75	1.75	0.06	0.06	0.07		1.10	67.5	N/A	28.3	2.4
Beryllium	Open	3.75	3.75	0.06	0.06	0.18		1.89	N/A	4.5	35.2	1.9
Beryllium	Closed	2.75	2.75	0.06	0.06	0.09		1.33	N/A	4.2	36.7	1.8
Silicon Carbide	Open	3	3	0.06	0.06	0.16		1.75	63.0	N/A	47.6	1.4
Silicon Carbide	Closed	2.2	2.2	0.06	0.06	0.08		1.23	60.1	N/A	49.6	1.4
Silicon Carbide	Open	3.5	3.5	0.06	0.06	0.17		1.89	N/A	4.4	53.0	1.3
Silicon Carbide	Closed	2.5	2.5	0.06	0.06	0.09		1.33	N/A	4.4	54.9	1.2

FOM is 0.65, so the magnesium fixture can be made to weigh (0.65) (600) = 390 lbs, which is less than the 500-lb requirement and is therefore okay.

2. A beryllium structure is designed to weigh 10 lbs but is found to be too costly. A significantly less costly aluminum structure designed for equal gravitational and frequency performance is found [from Eq. (8.14) and Table 8.4] to weigh 38 lbs, which is too heavy; additionally, (from Table 8.3) its thermal FOM is degraded threefold and is unacceptable. How does a "medium"-cost aluminum-beryllium structure fare?

   From Table 8.3, the thermal performance is degraded by 20%, and the structure is found to weigh 15 lbs. The engineer now must work the trade space to determine feasibility along with other design implications for optics, mounts, etc.

8.8.1.2 Lightweight optics

A 1-m (40-in.) optic is considered in which a normal gravity error of 65 μin. (±10%) peak surface error is required with a mount-induced error limit of 4.6 μin. (±10%). Using the equations of Chapters 5 and 6, we can size open- and closed-back optics to meet these requirements. The results are seen in Table 8.6. In all cases, rib thickness is set at 0.06 in.

When gravity sag performance is set as the criterion, the beryllium optic can be made 2.4 times lighter than glass, and the silicon carbide optic 1.4 times lighter, very close to the solid optic predictions. Results are independent of closed- or open-back configurations.

When mount-error performance is set as the criterion, the beryllium optic can be made 1.9 times lighter than glass, and the silicon carbide optic 1.25 times lighter, the same as the solid-optic predictions. The results are independent of closed- or open-back configurations.

References

1. "Flying Boat," Hughes H-4 Hercules, Evergreen Aviation and Space Museum, McMinnville, Oregon.
2. R. J. Roark and W. C. Young, *Formulas for Stress and Strain*, Third Edition, Table X, p. 216, McGraw-Hill, New York (1954).

Chapter 9
Adhesives

An adhesive is defined as any substance capable of holding objects together by surface attachment. This class of materials includes cement, glue, paste, epoxy, silicone, urethane, and any material causing one body to adhere to another. Adhesion is the state in which interfacial forces hold two surfaces together. These forces may consist of valence forces, interlocking actions, or both.

A list of adhesives commonly used in the optics industry is given in Table 2.3. Shown are those adhesives that are generally used for aerospace applications. The list is but a small portion of the over 1000 potential candidates, but it serves as a useful guide for those most commonly used. Descriptions of these adhesives are noted in Table 9.1. The sections in this chapter are intended to assist in proper selection of the adhesive candidate of choice for a particular application.

9.1 Mechanical Properties

Many mechanical and physical properties are required to determine the proper choice of adhesive. These include thermal expansion characteristics, tensile and shear strength, outgassing characteristics, modulus of elasticity, cure temperature and schedule, service temperature range, failure strain, glass transition temperature, viscosity, creep, and hardness.

We focus here on these properties, although other properties to consider are thermal conductivity, peel strength, shrinkage, solvent resistance, electrical conductivity, working life, shelf life, density, Poisson's ratio, bulk modulus, and compressive strength. Reference to these properties will be made as required.

9.1.1 Elastic modulus

The adhesive elastic modulus can be important in determining the stiffness of a bonded component assembly. The effect of the modulus is usually small for epoxies due to the thin bond lines used. However, the effect can be quite pronounced for silicones and urethanes because in this case the thickness may

Table 9.1 Description of typical adhesives for optical use.

Manufacturer	Adhesive	Classification	Typical Use	Color			Lap Shear Strength
				Resin	Hardener	Mixed	
3M	Scotchweld 2216	Structural epoxy	Room temperature to cryogenic	Gray	White	Gray	2500
Henkel	Hysol EA9394	Structural epoxy	Room temperature to 120 °C	Gray	Black	Gray	4200
Henkel	Hysol EA9361	Structural epoxy	Cryogenic	White	Black	Gray	3500
Huntsman	Epibond 1210	Structural epoxy	Room temperature to cryogenic	Tan	Blue	Blue	2500
Epo-Tek	301-2	Wicking epoxy	Thin bond	Clear	Clear	Clear	2000
Emerson & Cuming	Stycast 2850/24LV	Thermal epoxy	Thermal conductivity	Black	Clear	Black	4200
Emerson & Cuming	Eccobond CT5047	Electrical epoxy	Electrical conductivity	Silver	Amber	Silver	1000
Momentive	RTV 566	Silicone potting compound	Room temperature to cryogenic	Red	Yellow-brown	Red	450

be significant and the elastic moduli are reduced by up to three orders of magnitude compared to epoxy.

Figure 9.1 is a stress–strain plot for a typical polymeric epoxy adhesive. Note that the stress is proportional to the strain in a linear fashion (Hooke's law) over a small range and departs significantly thereafter as it enters the plastic zone. This means that the adhesive stresses will be redistributed as strain increases, allowing for plastic flow, at least at room temperature. The effective modulus of elasticity, known as the secant modulus, will determine stiffness under load. The secant modulus is simply the slope of the straight line connecting any two points of interest on the stress–strain diagram. With reference to the figure, the slope in the Hookean (linear) region is significantly higher than the secant modulus at its ultimate strength. The instantaneous modulus, better known as the tangent modulus, is the slope at any point of the stress–strain diagram. In the example of Fig. 9.1, the tangent modulus near-zero strain is 400,000 psi, while the secant modulus is only 300,000 psi at a strain of 0.016 (1.6%), which would indicate stiffness degradation if exercised

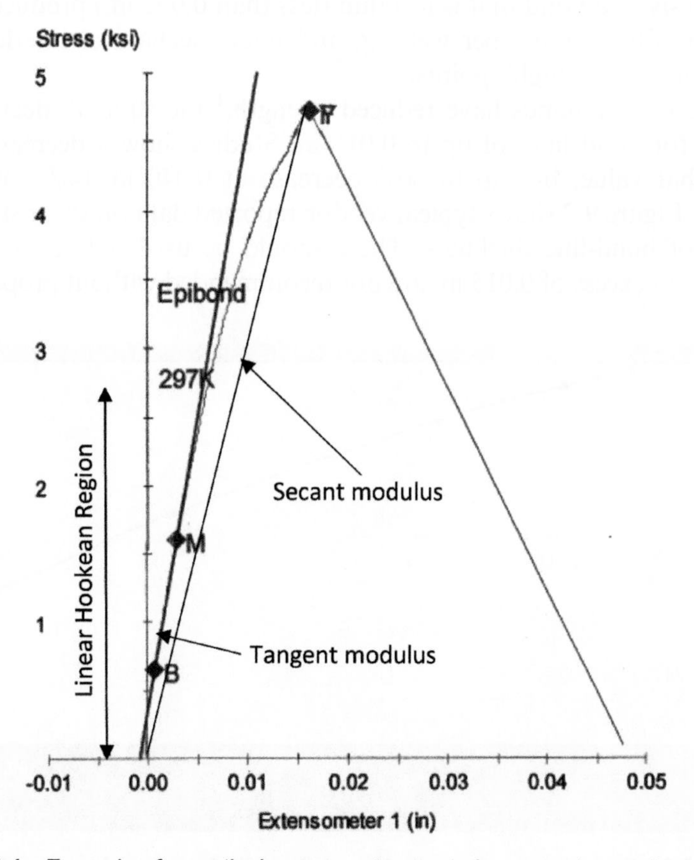

Figure 9.1 Example of a particular epoxy stress–strain curve at room temperature.

over that range. The proper use of the secant modulus will be discussed in more detail in Chapter 15.

The plastic nature of epoxy adhesives (at room temperature) helps to redistribute stress under conditions where stress is nonuniform, increasing the bonded-joint load-carrying capacity. We turn to this topic in Section 9.2. Modulus effects with temperature are given in Section 9.3, while Chapter 15 will expand the analysis with the proper use of secant properties.

9.1.2 Static strength

In general, increased thickness of an epoxy bond results in decreased strength. This is due in part to the increased potential for cavity air entrapment and in part to increased peel stress in shear strength testing due to load eccentricity. Most vendors of commonly used epoxy adhesives (rubbery silicones will be discussed separately) recommend a bond line of 0.003- to 0.005-in. thickness as optimum, although the more viscous epoxies (such as EA9394) are often specified nearer to 0.010 in. due to application limits, and the less viscous epoxies (such as EpoTek 301) will demand significantly smaller thickness. For most adhesives, a bond that is too thin (less than 0.002 in.) produces a risk of incomplete filling, improper wetting, and poor mechanical interlock due to surface roughness "high" points.

While thicker bonds have reduced strength,[1] the strength decrease is not dramatic for bond lines of up to 0.015 in. Studies show a decrease of about 15% at that value, but up to 60% decrease at 0.040 in. (40% of optimum strength). Figure 9.2 shows typical vendor-reported data on shear strength as a function of bond-line thickness. These should be used only as a guide, and bond lines in excess of 0.015 in. are not recommended without proper analysis.

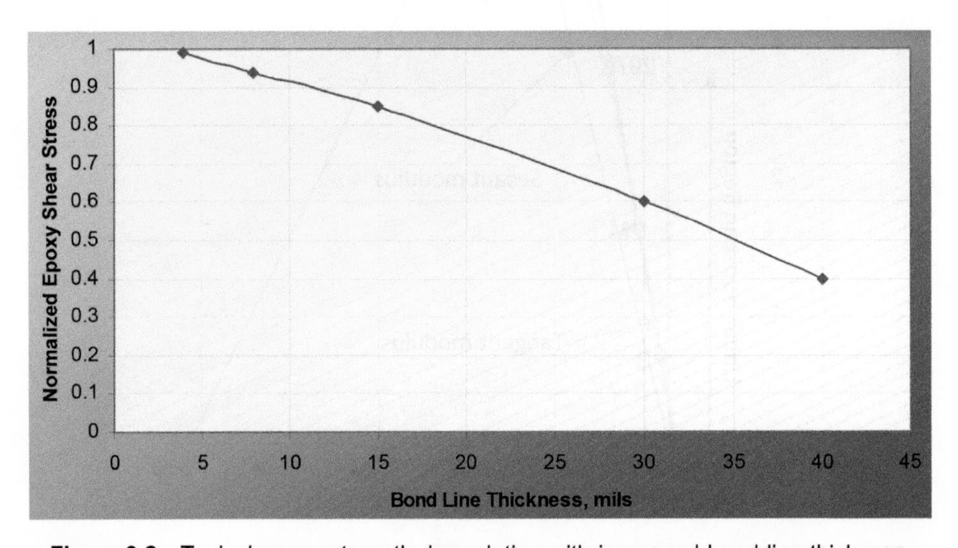

Figure 9.2 Typical epoxy strength degradation with increased bond-line thickness.

Table 9.1 gives some common adhesive shear strength properties for manufacturer-recommended thicknesses, which lie near 0.005 in. for most epoxies (except for wicking epoxies, which are much thinner). Thicknesses of elastomers and silicones can be significantly higher.

9.1.3 Peel strength

Peel strength is the ability of an adhesive to resist failure by stripping in a peeling mode. It is usually reported in terms of pounds per unit width of bond line and is thus the force required to separate two bonded adherends. Typical values are measured by standard ASTM T-Peel tests and range from 5 to 30 lb/in. of width.

For peel to occur, a peeling mechanism must exist in the form of a moment, which tends to separate the adherends. If such a mechanism is precluded, test values of failure may greatly exceed reported values. Furthermore, since peel strength is reported in pounds per inch, it is not often feasible to use stress results to determine joint adequacy. Cleavage, which is related to peel, is a more common cause of failure that typically reduces the allowable stresses to one-quarter of the tensile strength. The best way to design against failure is to develop joints that are peel resistant. Many examples of such designs are available in the literature. These designs, however, may not always be practicable. In the case where joints are loaded under peeling moments that cannot be avoided, it may be desirable to taper one of the adherends to the edges and increase the bond thickness slightly at the edges (but not beyond the allowable thickness discussed above). In this fashion, the moment is reduced at the edge (due to the increased flexibility of the adherend and adhesive), where peel stress is highest.

9.2 Load Stress Distribution

When loading in direct shear, as in a lap shear test, average stresses in the bond are calculated simply as the shear load divided by the shear area. However, finite element models show that stresses near the edges could approach 1.5 times this value or more. Notwithstanding, epoxy properties are such that, due to plastic flow, local yielding may serve to redistribute these stresses so that the use of "average" values is acceptable, as alluded to in Section 9.1. The average stress value can be used for comparison to the design allowable strength.

When loading in torsional shear, i.e., loading a bond under a torque moment that produces shear stress, the effect of averaging and yielding is more pronounced compared to when loading in direct shear. In this case, maximum stresses occur at the outside edges of the bond and are minimum at the bond center. Similarly, when loading a bond under combined torque moment and direct shear, as in a flexure bonded to an optic, stress maximums are pronounced at the side edge closest to the reactive load at the optic bezel mount end. Finite element modeling shows that these stresses are more than twice the average stress in the bond. Consequently, due to local yielding, the load-carrying capacity is increased twofold. For example, for an epoxy rated at 2500 psi in lap shear, stress calculations of 6000 psi at the edge are shown to be quite acceptable.

Consider, then, a flexure rigidly bonded to a substrate lens in which the bond area is subjected to shear and moment as produced by a cantilevered flexure with load along the optical axis, as in Fig. 9.3. The bond is subjected to shear in the direction of the optical axis and shear due to the reactive torsional moment (about a radial line) caused by the cantilever effect. As evidenced by the equations of equilibrium and Eq. (1.29), stresses are localized near the shear edge. Stresses produced in shear can be quite high. For a reactive load of 20 lbs over a 3-in. length and a 3/8-in. square bond, we calculate maximum stress using linear analysis as approximately

$$\tau \approx \frac{V}{A} + \frac{4.8M}{bt^2} \tag{9.1}$$

and find that $\tau = 5600$ psi.

This stress is far above the failure shear strength of the epoxy, which lies near 2500 psi, yet load failure tests under this loading conditions show the

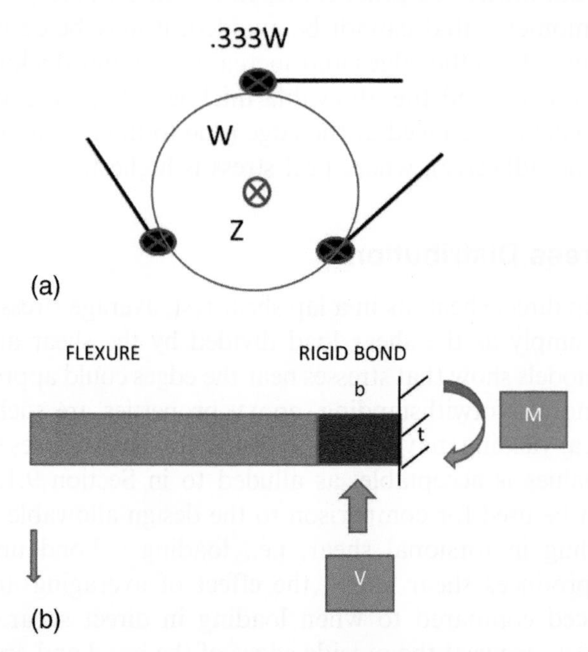

Figure 9.3 Bond shear stress under torsion M and shear load V: (a) force reaction to bond with optical axis vertical (*Z* is the optical axis coordinate system, W is the weight of the optic, and 0.333*W* (*W*/3) is the reactive force at each of the three flexures); (b) flexure force balance.

bond to be intact. Using linear theory, finite element analysis confirms the hand analysis. Thus, it follows that the plastic nature of the bond allows for stress redistribution, which the linear theory cannot predict. Using the finite element modeling results, the "average" stress over the entire cross-section lies under 2500 psi. Since we prefer not to use nonlinear analyses, such load redistribution approaches will suffice for first-order analyses in conjunction with linear finite element modeling techniques using average values. Thus, when stress distribution is localized in nature, the load-carrying capability can greatly exceed typically reported strength values. Caution is therefore required in determining the validity based on model analysis when, due to nonlinear effects, such stresses appear to exceed the usual design allowables.

9.3 Glass–Liquid Transition

Glass–liquid transition (or glass transition for short) occurs in polymers and is one phenomenon that make polymers unique. Each polymer has a unique temperature called its *glass transition temperature*, or T_g for short. When a polymer is cooled below its glass transition temperature, it becomes hard and brittle, like glass. Some polymers are used above their glass transition temperature, and some are used below it.

For example, since the glass transition temperatures of silicones are relatively low, it is desirable to use these materials *above* their glass transition temperature. This maintains flexibility (low elastic modulus) over a wide range. On the other hand, it may be desirable to use epoxies *below* their glass transition temperatures. This maintains high strength and relatively low CTE with a sacrifice of flexibility.

In many cases with epoxies, the glass transition temperature occurs not far from room temperature, so properties must be considered both below and above the glass transition temperature. The change in properties below and above the glass transition temperature are sometimes subtle and sometimes not. For example, a review of Huntsman Epibond 1210 data[2] shows a glass transition temperature of 328 K (55 °C). Values of thermal strain above and below this temperature are shown in Fig. 9.4. Note the slope change, which is noticeable but not extreme. On the other hand, Fig. 9.5 shows the CTE of RTV566 silicone as a function of thermal strain.[3] RTV (room-temperature vulcanized) silicone exhibits a glass transition temperature of 123 K (–150 °C), the lowest glass transition temperature of all adhesives. Notice in Fig. 9.5 that the CTE differences (the slope of the curve) above and below glass transition are marked—very high (more than 200 ppm/°C) above glass transition in its free (unconstrained) state. (Special consideration is required for the constrained state in which shape factor must be duly considered.) Below glass transition, its expansion characteristics are relatively low, near 60 ppm/°C, where it becomes glassy and typical of an epoxy. This is also evident in Fig. 9.6, which plots approximate elastic modulus versus temperature. The modulus is quite low

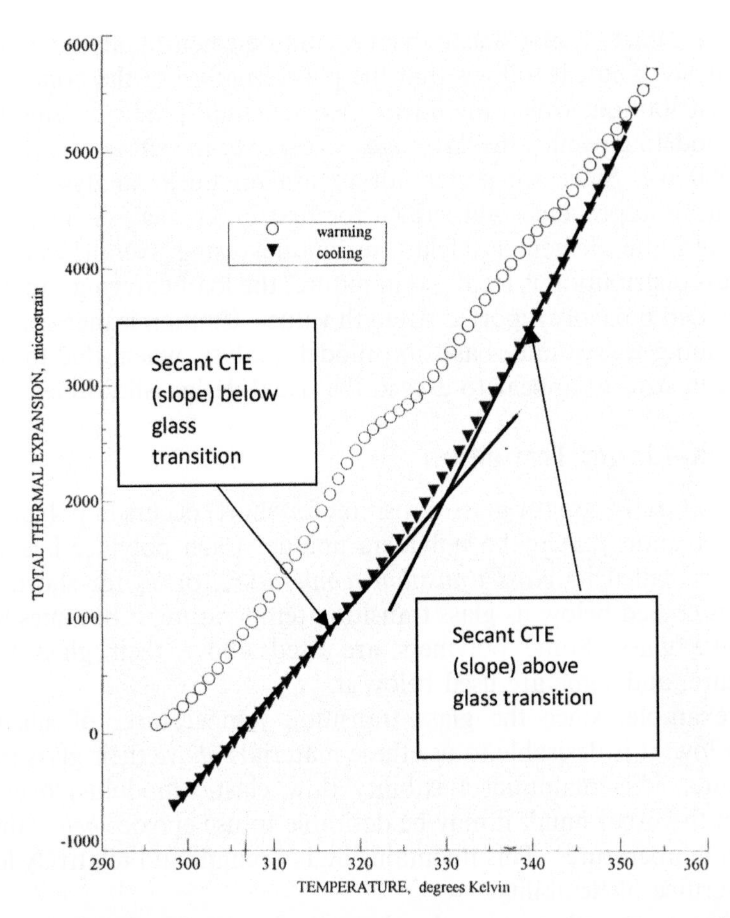

Figure 9.4 Glass transition of a select epoxy near 320–330 K evident during thermal expansion characterization. Note the phase change on warming and the higher secant CTE above glass transition temperature on cooling.

above the glass transition, on the order of 1000 psi, and quite high below the glass transition, near 1,000,000 psi, an increase of three orders of magnitude. Chapter 15 discusses analytical techniques for taking into account such variations in properties.

Most epoxy vendors will specify typical CTE values over a specified temperature range both above and below glass transition. Table 9.2 shows the expansion characteristics of several epoxies above and below these temperatures.

Table 9.3 provides elastic modulus changes and strength changes above and below glass transition for RTV566 silicone. Shown also are the typical strength and modulus changes for Epibond 1210 epoxy adhesive, again both above and below the glass transition temperature. Note again that both strength and modulus are decreased above glass transition. In general, a decrease in strength above glass transition does not affect strength below; i.e., there are no irreversible effects when passing through glass transition.

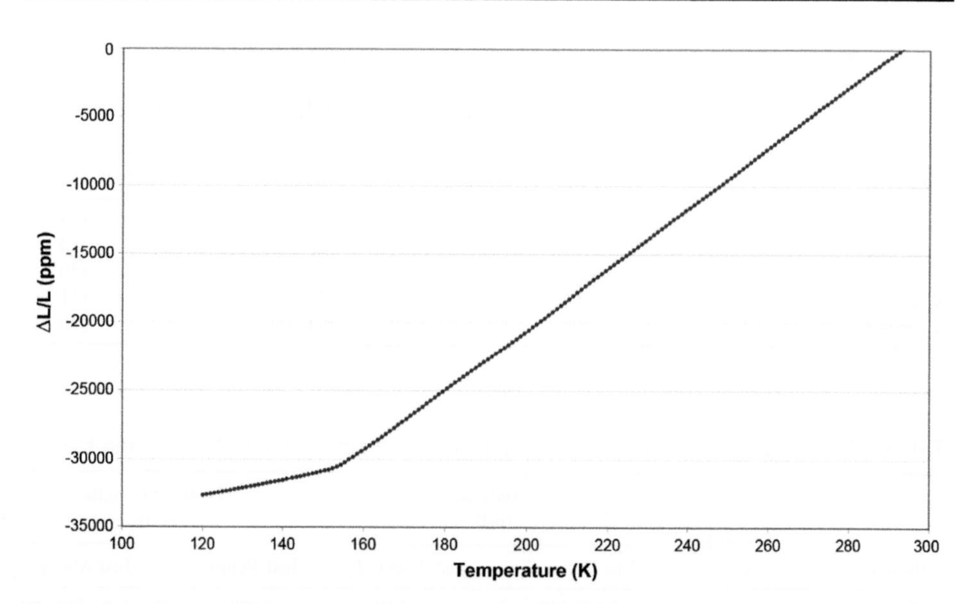

Figure 9.5 Pronounced glass transition temperature of RTV silicone undergoing thermal expansion.

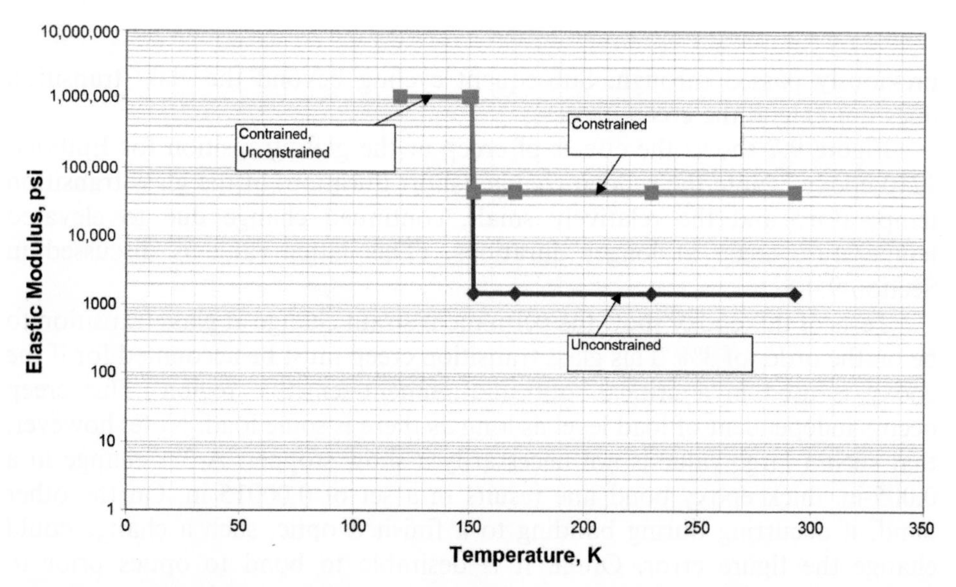

Figure 9.6 Pronounced change in the elastic modulus of RTV silicone at the glass transition temperature.

9.3.1 Glass transition temperature creep

Tests of epoxies at the glass transition temperature show a creep affect that is due to phase change rather than elevated temperature creep. This creep effect results in a permanent change in dimension; that is, a permanent set is developed in strain that is not reversible if the glass transition has not been

Table 9.2 CTEs and glass transition temperatures for select adhesives.

Adhesive	Glass Transition (°C)	Coefficient of Thermal Expansion (ppm/°C)	
		Just Below T_g	Just Above T_g
Scotchweld 2216	20	102	134
Hysol EA9394	78	55	80
Epibond 1210	55	100	140
Stycast 2850/24LV	68	39	111
RTV 566	−120	60	220

Table 9.3 Glass transition effect on modulus and strength for a select epoxy and silicone.

Adhesive	Glass Transition (°C)	Modulus (psi)		Tensile Strength (psi)	
		Just Below T_g	Just Above T_g	Just Below T_g	Just Above T_g
RTV 566	−120	1,000,000	750	6700	800
Epibond 1210	55	396,000	180,000	4700	2500

previously passed through. Subsequent cycling beyond the glass transition does not extend the creep process.

Figure 9.7 shows the effects of creep at the glass transition for Epibond 1210 epoxy.[4] Note the sudden change (short duration) at the glass transition temperature and the relatively small, continued change due to elevated temperature creep at longer durations. (This latter value is discussed in Section 9.4.)

Tests of this epoxy show the permanent strain change at glass transition to be on the order of 3%. This glass transition creep must be accounted for if the epoxy is used in a critical, optical-alignment-sensitive region. This creep occurs independent of load level as long as there is some load. Note, however, that such a large value is not necessarily a show stopper: A 3% change in a 0.005-in.-thick epoxy bond line results in a set of 0.00015 in. On the other hand, if occurring during bonding to a finished optic, such a change could change the figure error. Often, it is desirable to bond to optics prior to finishing to avoid this costly concern.

9.4 Temperature Creep

All materials exhibit creep to some degree if the temperature is high enough. Creep in adhesives can occur at room temperature but is more pronounced at elevated temperatures. To evaluate the effects of epoxy creep under load, we

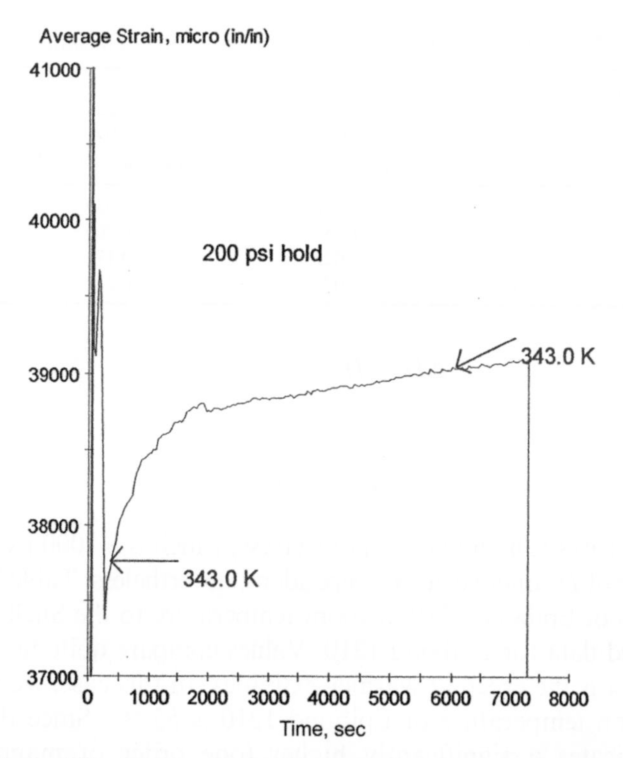

Figure 9.7 A large 3% creep strain (settling over several hundred seconds due to noise and drift) due to glass transition continues to creep due to temperature but in a significantly slower and asymptotic manner.

make use of the Boltzmann superposition integral for linear, viscoelastic, polymeric matrices such as epoxies. This gives rise to the equation

$$\varepsilon(t) = D(t)\sigma, \tag{9.2}$$

in which

$$D = D_0 + D_1 t^n, \tag{9.3}$$

where $\varepsilon(t)$ is strain as a function of time; D_0 is the initial elastic compliance [psi^{-1}] (inverse of Young's modulus E); D_1 is the creep coefficient [$\text{psi}^{-1}/\text{min}^n$]; t is time under load [min], and σ is applied stress [psi].

Tests[5] conducted for an epoxy resin (Shell 58-68R) used in fiberglass composites determine the values of D_0, D_1, and n as a function of temperature between 266 K and 343 K. The value of n is independent of stress and temperature, and is found to be 0.19. Table 9.4 gives the compliance constants as a function of temperature.

Since the initial compliance is included in the initial loading, it is only the net creep compliance that is of concern; thus,

Table 9.4 Temperature compliance creep coefficients.

Temperature			Compliance		
F	°C	K	D_0 psi^{-1} × 10^{-6}	D_1 psi^{-1}/min^n × 10^{-6}	Exponent n
20	−7	266	1.726	0.025	0.19
75	24	297	1.883	0.069	0.19
130	54	327	2.022	0.171	0.19
160	71	344	2.07	0.247	0.19

$$\Delta D = D - D_0 = D_1 t^n \tag{9.4}$$

and

$$\Delta\varepsilon = D_1 t^n \sigma, \tag{9.5}$$

where $\Delta\varepsilon$ is creep strain, and σ is epoxy stress, which is ~1000 psi from Ref. 2.

The value of D_1 may be epoxy dependent; nevertheless, Table 9.5 compares the properties of Epibond 1210 at room temperature to the Shell epoxy, along with measured data for Epibond 1210. Values compare quite favorably.

Of interest is the creep at warmer extremes. In this case, we note that the glass transition temperature of Epibond 1210 is 55 °C. Since the data from Fig. 9.7 indicates a significantly higher (one order of magnitude) creep coefficient than that reported in Table 9.4, this difference may be attributed to the marked drop in modulus beyond glass transition and associated creep during transition, as indicated in Table 9.6.

9.5 Lap shear strength

Lap shear strength is usually determined by single-lap shear testing. In this test, a tensile load is applied to each end of aluminum strips that are bonded

Table 9.5 Comparative epoxy creep constants at room temperature.

Epoxy	Modulus (psi)	Creep Coefficient D_1 × 10^{-6}
Shell 58-68R	530,000	0.069
Epibond 1210	395,000	0.088

Table 9.6 Modulus and creep constants at 60° C. Differences may be due to glass transition effects.

Epoxy	Modulus (psi)	Creep Coefficient D_1 × 10^{-6}
Shell 58-68R	495,000	0.191
Epibond 1210	180,000	2.2

together over a specified lap area per ASTM specifications. Because load is applied in tension but epoxy is measured in shear strength, stress results are often referred to as "tensile lap shear strength."

A caution with this method is that the load is offset by the metallic coupon and epoxy thickness; thus, an eccentric moment is applied that could cause peel failure. Further, due to bending of the aluminum coupons, load eccentricity is somewhat magnified. For this reason, stiffer adherends, such as steel, may yield higher strength results. For consistency, lap shear strength tests should be standardized and the stringent ASTM specifications adhered to.

To find a more indicative value of true shear strength, a double-lap shear test may be conducted. This test, for which the two sides of a coupon are bonded, eliminates the eccentric moment and peel tendency.

When single-lap shear tests are used as witness samples to verify epoxy adequacy on flight hardware, allowable stress should be set low enough to preclude rejection (e.g., due to adhesion failure) of otherwise good bonds and high enough to preclude inclusion of bad bonds. For an epoxy rated at 2000 psi, 1500 psi is a good cutoff point. Test results should be used for workmanship and proper cure validation rather than as an indication of exact allowable stress (allowable stress for flight hardware should be based on appropriate safety factors and/or fatigue considerations). Lap shear tests require proper surface preparation to avoid undue effects of contamination or premature adhesion failure. Use of hardness validation to ensure epoxy cure is a good backup to lap shear witness tests and should not be used in conjunction with lap shear testing to assure a good bond; as discussed below, in cases where the interface surface is poorly prepared for testing coupons, hardness is a better indicator of a proper cure.

9.5.1 Surface preparation

In any case, the most important parameter in obtaining good shear strength data lies in surface preparation. Ideally, surfaces to be bonded should be abraded, solvent cleaned, and primed. The abrasion serves a twofold purpose of cleaning difficult-to-remove contaminants and providing a mechanical interlock surface for proper bonding. Priming promotes proper surface wetting to obtain a good chemical bond.

9.5.1.1 Definitions

Epoxy joints can fail by surface adhesion, by bond cohesion, or in the adherends (substrate) themselves.

- An *adhesion* failure is one in which the bond separates completely from one of the adherends (see Section 9.5.1.2).
- A *cohesion* failure is one in which the bond fails within the adhesive (leaving epoxy on both adherends).

- A *substrate* failure is one in which the adherend fails because the epoxy bond is stronger than the parent substrate material.

9.5.1.2 Adhesion failure

If surfaces are properly prepared, epoxy failure should be cohesive; i.e., failure is within the epoxy itself and not at one of the surfaces, which would indicate an adhesion failure or a peel failure. Adhesion failures are caused by poor surface preparation; for properly prepared surfaces, the adhesion strength should be greater than the cohesion strength. However, experience dictates otherwise; most field failures of bonded joints are due to adhesion failure caused by poor surface preparation, wrong adhesive choice, or a nonreceptive surface. Examples of adhesion and cohesion failure are shown in Fig. 9.8.

Typically, after a part for bonding has been machined, it will be cleaned by use of an agitation ultrasonic bath containing deionized water and

Figure 9.8 (a) Cohesive epoxy failure of primed surfaces: (left) aluminum to Invar and (right) and Invar to Invar. Note the epoxy on both surfaces. (b) Adhesion failure of unprimed surfaces. Note the epoxy on one side only.

detergent. This will be followed by solvent cleaning with chemicals such as acetone and isopropyl alcohol. However, these chemicals by themselves may be insufficient to remove surface contaminants.

Solvent cleaning may need to be followed by abrading. *Abrading* serves multiple purposes: it removes contaminants, provides a mechanical interlock for a proper bond, increases surface area, and forms a more reactive surface. The surface, however, should be not too smooth and not too rough. An overly smooth (as in polished) surface does not provide the key interlock, while an overly rough (as in 80 grit) surface weakens the bond since it does not provide for good wetting and filling.

Priming promotes surface wetting for proper bond (chemical) adhesion. Additionally, it protects the surface from corrosion. Chemical *etching* is another step that assists in good adhesion. Etching removes material and unstable oxides, and is ideal for smoothing fracture *crack* tips and increasing adherend strength. It is used for materials that form oxides readily in air, such as aluminum and beryllium, and for fracture-critical brittle surfaces (like glass) in order to remove residual stress. *Anodizing* (aluminum), *tiodizing* (titanium), or passivating (steel) are other methods to improve bonding while protecting from corrosion.

Chemical film (aluminum) or use of Alodine® or Iridite® for beryllium serves to protect from corrosion. However, it is not the best finish to improve bond adhesion, so caution is advised here for critical applications under high stress. Similarly, bonding over electroless nickel, a phosphor-based corrosion inhibitor, results in a weak bond.

Air oxidation is an enemy to many metallic materials, including those often used for optical systems, which includes Invar (Chapter 2). The oxides immediately attach to the surface and are unstable. Being loosely bound, they provide poor surface adhesion. They are readily removed by etching or abrading.

Primers are excellent materials to promote a strong interface bond that will not fail in adhesion. Primers promote wetting, thereby increasing adhesion strength, and also serve as a corrosion inhibitor. The standard metal primer is Cytec BR®-127, while many other primers exist for glass and ceramics. Silicone adhesives, which adhere to nothing but themselves without primer, demand their own special primers.

Application of pressure after bonding is generally not required, but light pressure promotes wetting. As a caution, one should not change or apply pressure while curing, as this could affect proper crosslinking.

Tests have shown that for surfaces that are abraded, solvent cleaned, etched, and primed, the joints are minimally 25% stronger than those that are solvent cleaned only, the former consistently failing in cohesion and the latter in adhesion. For poorly cleaned surfaces, failure strengths can be significantly lower.

In cases where strength tests are performed on brittle, nonmetallic materials, such as glass or engineering ceramics, failure may occur in the

Table 9.7 Surface preparation and strength for a select epoxy bonding beryllium to beryllium, for which priming is paramount.

Surface	Epoxy Tensile Strength (psi)
Etch, primed within 4 hours	7500
No etch, primer only	6650
Etch, delay prime 7 days	5150
Etch only, no primer	3400
Bare, solvent clean only	2600

substrate prior to occurring in the epoxy. This is an acceptable failure, although it does not prove out the epoxy strength capability.

As an example of the importance of proper surface preparation, Table 9.7 shows strength improvement for a beryllium test sample compared to a surface with no preparation except solvent cleaning. Note that priming alone can double the adhesion strength, and etching followed by immediate priming can improve strength by 250%! Note also that delaying the priming even after etching, or delaying the bonding after etching, does little to improve strength. This shows the power of oxides! Surface preparation and timing are paramount.

9.6 Thermal Stress

The CTE of the adhesive, or, more precisely, the thermal strain over the temperature range of interest, is important for several reasons. A mismatch between the CTE of the adhesive and adherends can cause motion and distortion of the optical elements, bond failure, or fracture within the adherend.

Therefore, it is best to match the adhesive expansion characteristics to the adherends being bonded. This reduces stress to the adherend and reduces the potential for fracture over large thermal extremes. It also minimizes optical element critical motions and stress-induced optical birefringence. The stress in the adhesive itself is reduced if the CTE is of low value, which may prevent cracking. Matching the adherend CTE also relieves the effects of bond-line thickness variations.

Unfortunately, for most epoxies, room-temperature CTEs lie in the range of 50 to 100 ppm/°C—well above those of most common engineering materials—and are particularly higher than most optical components that demand low thermal expansion. In this case, selection of adhesive properties and bond-line thickness becomes critical in the design.

The elastic modulus in adhesives can vary significantly over temperature. This can cause one to underestimate stresses if the modulus under ambient conditions alone is considered, and can cause one to overestimate stresses if only the final modulus at temperature is considered (Chapter 15). Force and thermal stress acting across the bond line are increased in direct proportion to the adhesive modulus, CTE difference, and temperature variation. A low modulus over the temperature range reduces stress across the bond.

This combination of factors is called the modulus–thermal strain product. As we saw in Chapter 4, for bond lines that are thin relative to the substrate, the maximum adhesive stress is given as [Eq. (4.44)]:

$$\sigma = \frac{E\Delta\alpha\Delta T}{(1-\nu)}.$$

This formula is based on purely linear analysis and applies away from the edges of the bond. Since the stress is driven by thermal strain and not load, a forgiving material that strains well into the plastic zone can greatly mitigate stress. However, in reality, there is very little plasticity as temperature decreases, and very little strength as temperature increases. Both the modulus and CTE will change markedly with temperature, increasing and decreasing, respectively, as temperature is lowered. In this case, we can make use of the secant material properties, which are discussed more fully in Chapter 15. For the moment, we rewrite Eq. (4.44) as

$$\sigma = \frac{E_{sec}\Delta\alpha_{sec}\Delta T}{(1-\nu)}, \tag{9.6}$$

where E_{sec} is the secant (effective) modulus over the temperature range, and $\Delta\alpha_{sec}$ is the secant (effective) CTE over the temperature range, the latter of which is discussed in Section 4.4.

For example, consider two rigid, zero-expansion adherend substrates bonded with an adhesive and undergoing a temperature change from room temperature (293 K) to 173 K, and exhibiting the following properties: $E_{sec} = 600{,}000$ psi, $\alpha_{sec} = 6 \times 10^{-5}$/°C, $\Delta T = 120$ °C, and $\nu = 0.35$. The tensile strength of this particular epoxy is given as 4700 psi at room temperature and 9000 psi at the cold extreme of 120 K. From Eq. (9.6), $\sigma = 6{,}650$ psi. This is significantly higher than the ultimate breaking strength in tension at room temperature but well below the cryogenic strength. Thus, the adhesive does not crack and accommodates the temperature excursion.

9.6.1 Thermal stress at boundaries

Funny things happen near the edges of adhesives. As Timoshenko[6] so aptly notes, near the edges, additional shear and normal stresses exist that add to the stress locally. To quote, he says, "The distribution of shearing stress along the bearing surface cannot be determined in an elementary way, and it can be stated only that they are of "local" type and concentrated near the ends of the strip along a distance, the magnitude of which is on the order of the thickness of the strip. The magnitude of the stresses may be on the same order as the normal stress; there will also be normal stresses of local character at the surface boundary." [6]

A complement to this statement is given by Saint-Venant's principle, which states that stresses away from edge boundaries and concentrated forces will approach their nominal values a few characteristic distances away from the edge (or, as a corollary, nothing can be known at the edges without exotic analysis). We turn, briefly, to these exotic analyses.

First, consider the usual formulation of shear stress for a substrate, such as a metal pad at a flexure mount end, bonded to a relatively rigid optic. Here, we find, from Hookes's law for shear stress [Eq. (1.8)], that $\tau = G\gamma$, and, therefore,

$$\tau = G(\alpha 2 - \alpha 1)\Delta T L/t. \tag{9.7a}$$

This formulation, however, does not account for the edge boundary conditions and is quite conservative for long bonds of small thickness. We need to apply the free-edge considerations for equilibrium. These stresses are determined by applying equations of equilibrium at the free edge, as has been done by various authors.[7–9] These equations are not intended for the faint of heart.

Here, for the 1D case, at the bonded interface, the interlaminar shear stress to which Timoshenko refers is given as[7]

$$\tau = \frac{(\alpha_1 - \alpha_2)\Delta T G \sinh(\beta x)}{\beta t \cosh(\beta L)}, \text{ and at the edge,}$$

$$\tau = \frac{(\alpha_1 - \alpha_2)\Delta T G \tanh(\beta L)}{\beta t}; \text{ here,} \tag{9.7b}$$

$$\beta^2 = \frac{G}{t\left(\frac{1}{E_1 t_1} + \frac{1}{E_2 t_2}\right)}; \text{ at the edge, } \tau = \frac{(\alpha_1 - \alpha_2)\Delta T G}{\beta t}$$

if tan$h(\beta L)$ approaches infinity, as it typically does. In the above equation, x is the distance from center, t is the epoxy thickness, $t_{1,2}$ is the adherend thickness, $E_{1,2}$ is the adherend modulus, G is the epoxy shear modulus, and L is the epoxy half-length.

These stresses increase significantly at the edges. Note that the CTE of the epoxy is not accounted for in this equation; if the epoxy is less rigid than the substrates, this may be a good approximation as long as the adherends are significantly different in their CTE values. When the adherend CTE differences approach zero, the epoxy CTE dominates and the equations do not apply.

Note that if the value of βL is small, i.e., $\beta L \ll 1$, then $[\tanh(\beta L)]/\beta L)$ approaches unity in Eq. (9.7b) and reduces to Eq. (9.7a). However, for thin bond lines, typically in the 0.005- to 0.010-in. range, βL is not small, and Eq. (9.7a) will not apply.

If the bond length is small, however, Eq. (9.7b) does approach Eq. (9.7a), again if $\beta L \ll 1$. Thus, bonding of optics using "dots," or small areas, as opposed to a continuous length, may be beneficial. However, the reduced bond area must be duly considered for strength under externally applied load.

The axial stress is given as[8]

$$\begin{aligned}
\sigma_1(x) &= E_1 E_2 \Delta T t_2 \left(\frac{\alpha_2 - \alpha_1}{t_1 E_1 + t_2 E_2}\right)\left[1 - \frac{\cosh(cx)}{\cosh\frac{cL}{2}}\right],\\
\sigma_2(x) &= E_1 E_2 \Delta T t_1 \left(\frac{\alpha_2 - \alpha_1}{t_1 E_1 + t_2 E_2}\right)\left[\frac{\cosh(cx)}{\cosh\frac{cL}{2}} - 1\right],\\
\tau_3(x) &= \frac{G_3}{c t_3}\sinh(cx)\frac{\Delta T}{\cosh\frac{cL}{2}}\frac{\cosh(cx)}{\cosh\frac{cL}{2}}(\alpha_2 - \alpha_1).
\end{aligned} \tag{9.8}$$

This stress is normal to the bond-line cross-section and tends toward zero at the boundaries. However, normal stresses perpendicular to the bond line will exist at the bonded edge interface and are the interlaminar normal stresses to which Timoshenko refers. These normal stresses cannot be computed by either Eq. (9.7a) or (9.7b) but can be computed in terms of a series of Bessel functions, which at best are unwieldy.[7] These stresses act in a direction perpendicular to the axial stress distribution, i.e., in a peel (flatwise tensile) direction. Finite element analyses bear this out.

An ingenious but more complex solution[9] that avoids Bessel functions shows such normal stresses to be given as

$$\begin{aligned}
\sigma_{yi} = -\sum_{n=1}^{\infty}\Bigg\{\left(\frac{n\pi}{l}\right)^2 &[A_{ni}\cosh n\lambda_i y + B_{ni}\sinh n\lambda_i y + C_{ni}\cosh n\mu_i y\\
&+ D_{ni}\sinh n\mu_i y] + E_{yi}\alpha_{yi}t_{ni}\Bigg\}\sin\frac{n\pi x}{l}.
\end{aligned} \tag{9.9}$$

Here, the constants A_{ni}, B_{ni}, etc., and λ_i are functions of the modulus and CTEs of all three layers, *including* the CTE of the epoxy. Solution to the equation, however, makes evident the benefit of understanding Bessel functions, and is even more unwieldy. In any case, the equation solutions show that interlaminar and normal stresses at the edge are dramatically increased compared to those near the center, and, in fact, can be reversed in sign. For a thin bond line sandwiched between a thick optic and a thin adherend, edge stresses can be an order of magnitude (!) increased from the center, and opposite in sign. For example, consider an optic of near-zero CTE bonded to a metal bipod of high CTE and subjected to a cold soak. The bipod at the bonded interface shrinks relative to the glass and is nominally in tension, while the optic is nominally in compression. However, at the edge, the optic stress rises dramatically and is in tension; for brittle materials, tension loads are problematic, as will be discussed in Chapter 12. Examples of the dramatic edge increase are shown in Fig. 9.9.

In order to compute stresses in an easier fashion than the nearly intractable equations suggest, finite element analysis may be used. However, this needs to be done with extreme caution, as described in Section 9.7.

In lieu of these analyses, to the first order, an approximate formulation (as confirmed by both the equations and finite element analyses) can be made for edge stresses in an optic that is thick relative to the adhesive and adherend to which it is bonded. This edge stress is given as

$$\sigma = E\Delta\alpha\Delta T, \tag{9.10}$$

where E is now the modulus of the *optic*. How elegantly non-exotic and simple! For a thin bond line of the epoxy, we can approximate the stress using Eq. (1.50) as $\sigma = E\alpha\Delta T$ in one dimension. In two dimensions, we can approximate the stress using Eq. (1.7a) to give

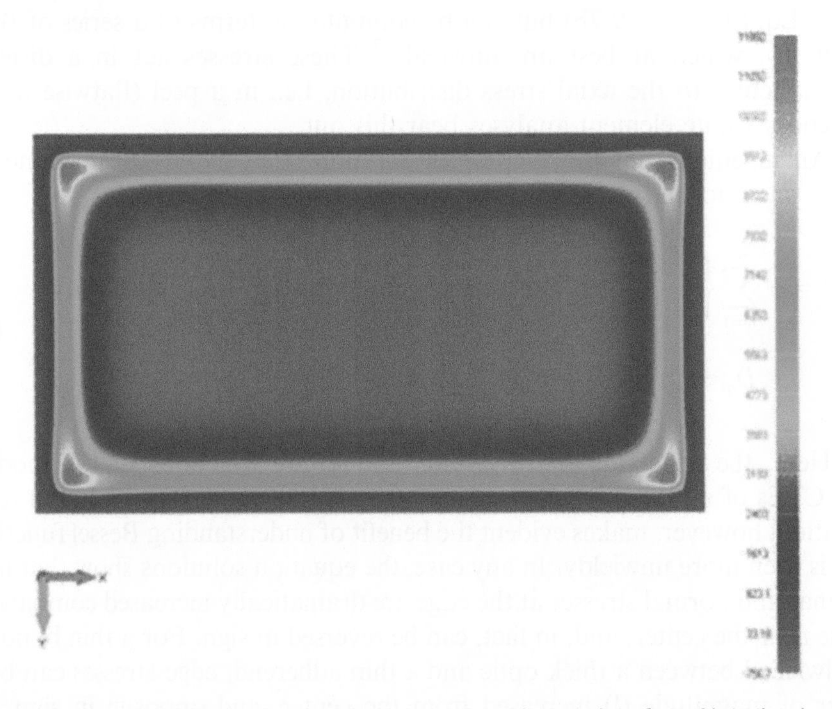

Figure 9.9 Thermal stresses in a bonded joint at the substrate interface. Note the dramatic increase in stress contour from low compressive (1000 psi) in the central zone to high tensile (12,000) at the edge.

$$\sigma = E\alpha\Delta T/(1-\nu), \tag{9.10a}$$

where E and α are now the modulus and CTE of the *epoxy*, respectively.

9.6.1.1 Cemented doublet

While Eq. (9.7b) applies to bonded pads, when two circular substrates are bonded as in a lens doublet, 2D analysis is required. Here, Bessel functions can be used to determine shear stresses.[10] In this formulation, as in the 1D case, edge stresses increase dramatically at the edge and are a function of the bonded substrates' diameter. In general, larger substrates exhibit higher edge stresses. The equations are given[10] as

$$\begin{aligned}
\sigma_S &= \frac{(2)(\alpha_1-\alpha_2)(\Delta T)(\sigma_e)[I_1(x)]}{(t_e\beta)(C_1+C_2)},\\
\beta &= \left\{\left(\frac{\sigma_e}{t_e}\right)\left[\frac{(1-\nu_1^2)}{E_1t_1}+\frac{(1-\nu_2^2)}{E_2t_2}\right]\right\}^{1/2},\\
x &= \beta R,\\
C_1 &= -\left[\frac{2}{(1+\nu_1)}\right]\left\{\left[\frac{(1-\nu_1)I_1(x)}{x}\right]-I_0(x)\right\},\\
C_2 &= -\left[\frac{2}{(1+\nu_2)}\right]\left\{\left[\frac{(1-\nu_2)I_1(x)}{x}\right]-I_0(x)\right\}.
\end{aligned} \tag{9.11}$$

Here the functions I_0 and I_1 are modified Bessel functions of the first kind and require detailed computation.

Note that if the substrates are *radially* restrained, as in a lens cell undergoing differential expansion, a further stress is developed in accordance with Eq. (1.60a). Here the lens stress is *reduced* as diameter increases but is generally compressive. The engineer must act accordingly to combine the stresses.

Once again, to preclude complex analysis, the use of Eqs. (9.10) and (9.10a) will give a quick and close approximation of epoxy and substrate stresses.

Again, how elegantly simple!

9.6.1.2 Example for consideration

Consider a thick glass optic with modulus of 1.0×10^7 psi and secant CTE of 2.0×10^{-7}/°C bonded to a titanium flexure with modulus of 1.6×10^7 psi and secant CTE of 8.0×10^{-6} °C over an excursion from room temperature (293 K) to 100 K (–173 °C) for an excursion of 193 °C. The epoxy is 0.005-in. thick with a secant modulus of 600,000 psi and a secant CTE of 5.0×10^{-5} °C. It exhibits a cryogenic tensile strength of 12,000 psi and a Poisson ratio of 0.4. The glass allowable tensile strength for its finish is given as 3000 psi. Approximate the tensile stress in the glass and adhesive.

From Eq. (9.6), the stress in the epoxy is $\sigma = E_{sec}\Delta\alpha_{sec}\Delta T/(1 - \nu) =$ 9600 psi, which is less than the allowable stress and is thus acceptable. From Eq. (9.10), the tensile stress in the glass is $\sigma = E\Delta\alpha\Delta T = (10)(7.8)(193) =$ 15,000 psi, which will break the glass! Away from the edges, using Eq. (9.8), we find the stress to be on the order of 1000 psi in *compression*. Note the dramatic and reversed stress increase at the edge. If the titanium flexure is replaced with Invar 36, exhibiting an expansion coefficient of 1.3 ppm/°C, we find that $\sigma = E\Delta\alpha\Delta T = (10)(1.2)(193) = 2700$ psi, and the glass survives.

The above formulations apply to the rigid epoxies of Table 9.1. If more-elastic adhesives such as urethanes or silicones are used, stress will be reduced provided that glass transition temperatures are below the temperature range of interest.

9.7 Modeling Techniques

As we have noted, stresses in adhesives and adherends in the vicinity of the bond-line boundaries are difficult to calculate. Saint-Venant's principle suggests that these high, localized stresses will decay to uniform values within several characteristic lengths. Unfortunately, failure begins in the area of the edges or near concentrated loads where simple formulas give inaccurate predictions of stress. We can turn to finite element models, but such modeling needs to be understood, as element stresses will depend on the element size chosen. Since high stresses occur near the edge, too big an element will yield too low a stress, missing the solution entirely and rocking it out of the cradle; too small an element may result in stresses approaching the infinite that are meaningless. With an appropriate element size, the model can then be used to determine margins of safety. Thus, while this text is not about finite element analysis, it is worthwhile to point out some of its benefits and downfalls in modeling adhesives.

9.7.1 Element size

To obtain accurate results when modeling adhesives, solid elements must be used and need to be at least "two deep." Many optimum bond lines lie near 0.005-in. thickness, so element size in this direction will be only 0.0025 in. This poses a dilemma in choosing element sizes in the other dimensions. Small element sizes are required. Yet small elements make large models and may poorly estimate stress and/or tax memory.

For example, a 1-in.-square bond line may be modeled with solid elements of 0.1 in. square and 0.0025 in. thick. In this case, there are 600 degrees of freedom, and the element aspect ratio reaches 40:1. Due to the high aspect ratio, stress values may be inaccurate and understated. Alternatively, if an aspect ratio of 1:1 is chosen, element size is reduced to 0.0025 in. square. This results in over 5,000,000 degrees of freedom, which may tax memory constraints. Also, as element sizes become smaller and smaller, stresses tend to go higher and higher, approaching the infinite in some cases. A dilemma occurs in that too few elements understates stress, while too many elements taxes system limitations while overstating stress.

To define stress margins for a bonded joint such as in a lens mount, it is first necessary to make several admissions:

1. Due to a numerical singularity, finite element analysis results at the edge of a bond line are not realistic, no matter how many elements are used.
2. In a real structure, there is a peaking effect at the edge of a bond line that can be a driver in determining the failure load.
3. In a real structure, the stress profile and the associated peaking are dependent on the relative stiffness of the constitutive materials, their inherent plasticity, and the materials' fracture properties.
4. Plastic effects typically reduce stress peaking, and edge failures are more appropriately modeled using a fracture mechanics approach. Fracture mechanics properties, however, are rarely measured. Thus, failure is typically characterized by a stress exceedance at a critical distance from the edge of the bond line.
5. Analytical models can predict the relative shape of the stress profile, which depends on the relative stiffness of the constitutive materials. (This allows their use in sensitivity studies.) Due to the singularity, however, they cannot predict the actual stress amplitude at the edge of the bond. This value will change with mesh density.
6. If the predicted stress profile from an analytical model is correlated with test data to define the critical point in the stress profile, it can be used to predict failure in similar configurations.

Once the above admissions are understood, it is possible to develop an analytical technique (based on test data) that produces stress margins.

9.7.2 Thermal stress

Thermal stresses in the adhesive and adherends are induced under thermal soak due to CTE mismatch, as we have noted, and the interlaminar stress is a maximum at the adhesive/adherend(s) interface edges. Equations (9.7) and (9.8) are insufficient to determine the normal (flatwise) tensile stresses in the adherends, particularly if the substrate is brittle, as failure will occur in tension (Chapter 12). While Eq. (9.9) can accurately predict these stresses, its derivation is well beyond the scope of this (and most other) texts. Equation (9.10), however, can be used despite the fact that it is approximate, maximum, and first order. The following case study compares this use to test experiments and detailed finite element analyses.

9.7.2.1 A case study

An infrared transmittal lens is bonded to a metallic flexure using a cryogenically approved, 0.005-in.-thick, epoxy adhesive for operational use at 120 K. Experimental data for the material properties were obtained for the lens [made of hot isostatically pressed (HIP) zinc sulfide], the adhesive, and two types of metal flexures. The first type of flexure was a titanium alloy not well matched in CTE to the lens, while the second type was a better-matched, low-expansion iron-nickel alloy. The ceramic strength is highly dependent on flaw size (Chapter 12) and therefore is also highly dependent on the finish. Thus, a controlled grinding procedure was specified to obtain the desired surface. Table 9.8 lists the effective properties of the lens, adhesive, and flexures. The zinc sulfide exhibits an average strength near 5000 psi at its finish; tests show that failure is expected at this level at least half of the time. At a stress level of 2500 psi, failure is expected less than 1% of the time.

From Eq. (9.10), and as a first-order approach, we find the maximum substrate stress to be

$$\sigma = E\Delta\alpha\Delta T = (10.8)(3.1)(173) = 5800\,\text{psi at the titanium mount, and}$$
$$\sigma = E\Delta\alpha\Delta T = (10.8)(0.6)(173) = 1120\,\text{psi at the iron-nickel mount.}$$

Here, recall that the elastic modulus property E is that of the lens. For the adhesive, from Eq. (9.6),

$$\sigma = E\Delta\alpha\Delta T/(1-\nu) = (0.6)(55)(173) = 5700\,\text{psi}.$$

Here, recall that the elastic modulus property of the modulus is that of the epoxy.

We can see that one can generally expect failure at the titanium interface in the lens, but no failure at the iron–nickel interface in the lens. The adhesive stress is below its allowable level. Actual tests on multiple samples bear this out, i.e., no failure in the adhesives in any case, but failure in the lens substrate when titanium is attached, and no failure when the iron-nickel alloy is attached. Stresses in the metallic flexures themselves are not an issue and are well below allowable levels.

Table 9.8 Effective properties of the bonded lens assembly at 120 K.

Item	Material	Secant Modulus (Msi)	Secant CTE (ppm/K)	Strength (psi)
Lens	Zinc sulfide	10.8	4.9	5000
Adhesive	Epoxy	0.6	55	7000
Flexure	Titanium	16	8	80,000
Flexure	Alloy 42	21	5.4	40,000

A finite element model was then constructed to determine thermal stress levels for comparison to the first-order equations and the experimental results. Several different models were constructed for which element sizes were varied from 0.001 to 0.020 in. The element thickness in all cases for the adhesive was 0.0025 in.

In all cases, the average stress at the element centroid was used for comparison. Figure 9.10 shows the stress level as a function of element size. Note that at the smallest element size, stresses are greatly overstated; failure would occur in all cases. Note further that even at element sizes of 0.010 in., failure would still be shown for the lens if bonded to the iron-nickel alloy, even though such failure did not occur. At an element size of 0.020 in., we see more accurate predictions, showing failure at the lens if bonded to titanium and none at the lens if bonded to the iron-nickel alloy.

A review of the models with the smaller element sizes shows that stresses approach that of the largest chosen element size if they are reviewed two to four bond-line thicknesses removed from the boundary. Thus, it would appear that use of element sizes near 0.020 in. (with aspect ratio on the order of relative thickness) are not only sufficient, but are more in line with expectations. Note further that these stresses are in the approximate vicinity of the Eq. (9.10) predictions.

This analysis implies that modeling to small element levels is neither required nor desired. Further analytical studies have indicated that a reasonable approximation to stress can be made with element sizes on the order of 0.025 in. using element centroidal stress (not corner stress). It is not desirable to use greater sizes as aspect ratio and inaccuracies can potentially understate stress.

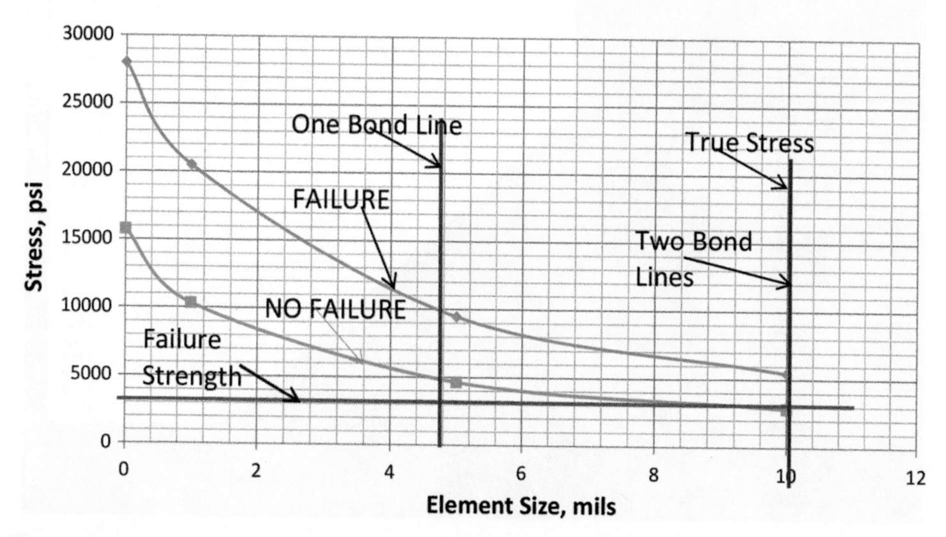

Figure 9.10 Average stress as a function of element size: too many small-size (fine) elements at the edge will overstate the stress at the boundaries.

Although the stress results from the model are dependent on element size, and stresses are difficult to estimate at a bond line, models and analysis confirm that the use of adherends with mismatched thermal expansion along with high-expansion/low-strain adhesives may cause failure in brittle adherends. Matching expansion properties reduces both adhesive and adherend stress. Use of a low-strain, flexible adhesive further decreases the stress.

9.8 Fillets

The static strength of an adhesive joint is improved with bond fillets by an effect similar to that achieved by welding of metals with filler material. However, thermal stress levels developed in epoxies and adherends increase significantly due to excessive bond-line fillets. This can produce failure in the adhesive or substrate under extreme soak conditions. If the adherend is weaker than the adhesive and stresses rise above the design-allowable level, the substrate material could fracture and pull out. If the adhesive is weaker than the adherend and stresses rise above the design-allowable level, the adhesive could fracture and crack. The stress increase due to fillets is generally up to 20% in both adhesives and adherends (Fig. 9.11). This could be significant when stress levels are high to begin with. On the other hand, negative fillets may be desirable, i.e., less adhesive at the interface with the adherend edges. Similar studies show that stress decreases by approximately the same amount, up to 10–20%, which adds to the design margin.

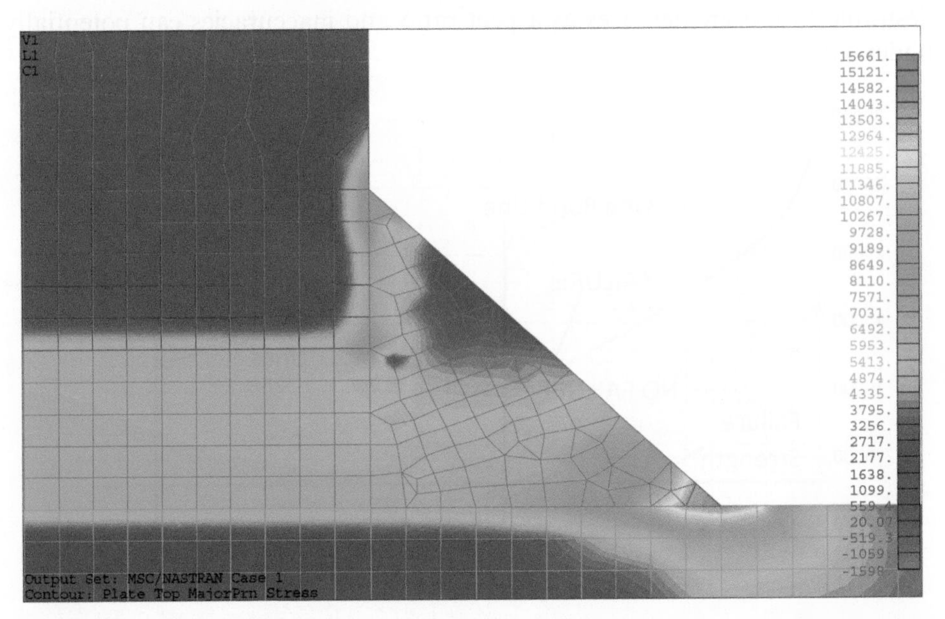

Figure 9.11 Fillets increase stress in both the epoxy and the substrate when temperatures decrease and should be minimized at cryogenic temperatures.

The above discussion on reducing fillets in a thermal environment applies in general to low-expansion adherends and high-expansion epoxies when operating at cold extremes below 200 K. If the adherends are ductile metals, local yielding may preclude fracture. Nonetheless, it is good practice to avoid excess fillets at cryogenic extremes of operation. In general, an excess fillet is one in which the fillet side dimensions exceed three bond-line thicknesses.

9.9 Soft Elastomers

We turn our attention to soft elastomers, i.e., materials that have low elastic moduli and large strain-to-failure rates. Silicones and some urethanes fall into this category. A common elastomer is RTV silicone—a silicon-based rubber polymer adhesive. This material is nearly incompressible; i.e., it exhibits a Poisson ratio of almost 0.5. As such, its elastic modulus is highly dependent on its loaded area and thickness. For extremely thin layers, the elastic modulus will approach the bulk modulus of the material. The bulk modulus B and elastic modulus E are related as

$$B = \frac{E}{3(1 - 2\nu)}. \tag{9.12}$$

It is evident that as Poisson's ratio approaches 0.5, the bulk modulus approaches infinity, resisting compression entirely. RTV silicone exhibits a Poisson ratio of 0.4997 and an elastic modulus of about 500 psi; substituting into Eq. (9.12) yields $B \sim 275{,}000$ psi. While experimental values show a bulk modulus closer to 200,000 psi, nonetheless, this greatly exceeds the elastic modulus. By way of comparison, metal properties are not appreciably affected by the bulk modulus. For example, setting the bulk modulus equal to the elastic modulus, we have

$$B = \frac{E}{3(1 - 2\nu)} = E, \tag{9.13}$$

and $\nu = 0.33$, which is Poisson's ratio of aluminum. The Poisson ratios of many metals lie in the range of 0.25 to 0.33. Epoxy adhesives exhibit a Poisson ratio near 0.4, but even this value will not significantly increase the bulk modulus above the compression modulus.

Because of the high-bulk-modulus effect in silicones and rubbers, the effective elastic modulus can be found[10] as

$$E_c = E(1 + 2S^2), \tag{9.14}$$

where S is the shape factor, which is given as the loaded area divided by the load-free area. Equation (9.14) is based on the nearly incompressible volume constraints and the theory of elasticity for infinite strips. Thus, the equation is applicable only to very high-Poisson-ratio materials, as noted above.

For a circular loaded section of diameter D, the shape factor is

$$S = \frac{\pi D^2}{4\pi D h} = \frac{D}{4h}, \tag{9.15a}$$

and for a square loaded section of side L, the shape factor is

$$S = L^2/(4Lh) = L/4h. \tag{9.15b}$$

For a rectangular loaded section of length L and width b, we have

$$S = \frac{Lb}{2h(L+b)}. \tag{9.15c}$$

For a long strip, where $L \gg b$, we have

$$S = \frac{b}{2h}. \tag{9.15d}$$

In practice and based on experiments, the effective modulus equation [Eq. (9.14)] needs modification when the elastomer durometer, or hardness, increases due to additives and other effects.[10] The modified effective elastic modulus equation is

$$E_c = E(1 + 2kS^2), \tag{9.16}$$

where the correction factor k varies from 0.5 to 1.0, depending on the hardness. A table of hardness, elastic modulus, shear modulus, and correction factor[11] for typical rubbers is given in Table 9.9. Note that Eq. (9.16) applies to the effective modulus in both compression and tension but does not apply to shear. Figure 9.12 plots the shape factor as a function of elastomer hardness.[11] This plot should be used in place of Eq. (9.16) when the shape factor exceeds 10.

Table 9.9 Correction factor k applied to shape factor to account for elastic modulus increase due to nonrubber constituents (reprinted from Ref. 11 with permission).

Durometer Shore A	Elastic Modulus (psi)	Shear Modulus (psi)	k
30	130	43	0.93
40	213	64	0.85
50	310	90	0.73
60	630	150	0.57
70	1040	245	0.53

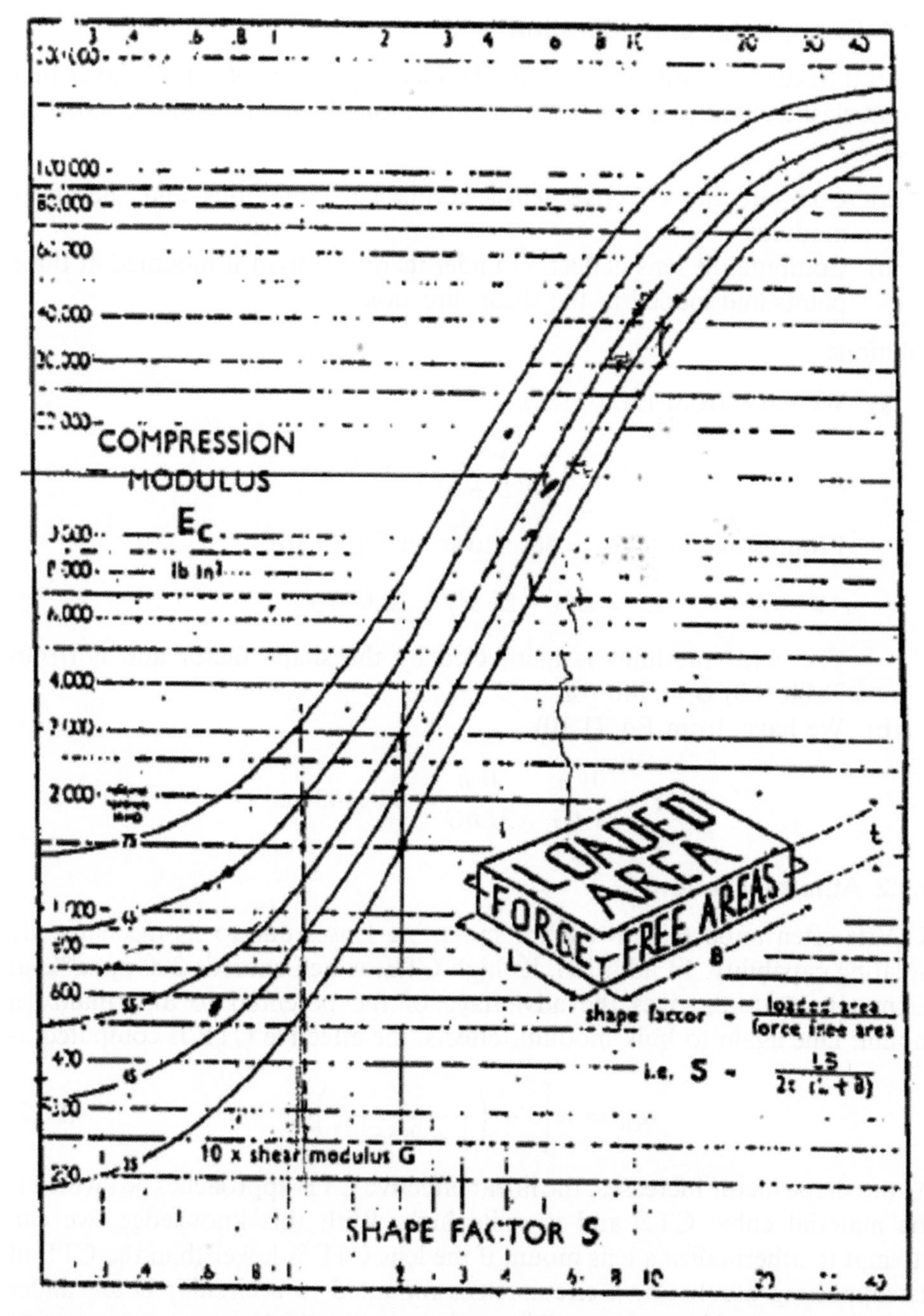

Figure 9.12 Shape factor S as a function of elastomer hardness E_c. Elastomeric materials with extremely high Poisson's ratios approaching 0.5 exhibit a marked increase in stiffness when approaching a constrained state (thin layer) (reprinted from Ref. 11 with permission).

9.9.1 Example for consideration

A metal flexure is bonded to a 10-lb glass lens by means of a RTV layer that has an elastic modulus of 600 psi and is 1 in. long by 0.5 in. wide by 0.030 in. thick.

a) Compute the effective tensile/compressive modulus and the shear modulus.
b) Compute the lens deflection under its own weight if mounted at three points and loaded in the shear direction.

Solutions:

a) We have, from Eq. (9.15c),

$$S = \frac{Lb}{2h(L+b)} = 5.55.$$

From Table 9.9, $k = 0.58$, and from Eq. (9.16),

$$E_c = E(1 + 2kS^2) = 22{,}000\,\text{psi}.$$

The shear modulus is unaffected by the shape factor and is (from Table 9.9) $G = 140$ psi.

b) We have, from Eq. (1.10),

$$Y = \frac{Wh}{3AG} = \frac{Wh}{3LbG} = 0.00143\,\text{in}.$$

9.9.2 Athermalization

RTV is often important to use for mounting lenses due to its mechanically isolating capability. In addition, its high CTE, which exceeds 200 ppm/°C at room temperature, gives the advantage of the potential to athermalize a mount. Due again to bulk modulus effects, the effective CTE is computed as

$$\alpha_{\text{eff}} = \alpha\left[3 - \left(\frac{2}{1 + 2.5S^{1.75}}\right)\right]. \tag{9.17}$$

As the shape factor increases, the linear effective CTE approaches 3α, which is the material cubic CTE and is quite high. With this knowledge, we can attempt to athermalize a lens mount if the lens CTE is lower than the CTE of its supporting bezel. For a circular lens mounted to a circular bezel, under temperature soak, from compatibility relations, the thickness t_p of the potting adhesive (RTV) is

$$t_p = R\frac{(\alpha_b - \alpha_l)}{(\alpha_{\text{eff}} - \alpha_l)}, \tag{9.18}$$

where R is the lens radius, α_b is the CTE of the bezel, and α_l is the CTE of the lens.

9.9.2.1 Example for consideration

A circular lens having of radius 4 in. and CTE of 4 ppm/°C is supported in an aluminum bezel having a CTE of 22.5 ppm/°C by RTV adhesive at three points with bond dimensions of 1 in. by 0.5 in. Neglecting the radius difference between the lens and the bezel, athermalize the mount. The RTV expansion coefficient is 222 ppm/°C. From Eq. (9.15c),

$$S = \frac{Lb}{2(L+b)t_p}.$$

Note that the solution is iterative as we need to assume an adhesive thickness. We try $t_p = 0.10$ in. Then, $S = 1.66$, and from Eq. (9.17),

$$\alpha_{eff} = \alpha\left[3 - \left(\frac{2}{1+2.5S^{1.75}}\right)\right].$$

$$\alpha_{eff} = 2.72,\ \alpha = 603\ \text{ppm/°C},$$

and from Eq. (9.18),

$$t_p = R\frac{(\alpha_b - \alpha_l)}{(\alpha_{eff} - \alpha_l)} = 0.13\ \text{in.}$$

Note that our thickness assumption was close; we can iterate to home in on the true thickness. If this is thicker than one would care to spread, a pad of nearly that thickness can be cast and cured, and then bonded with thin RTV adhesive. Note that if our calculations are off a bit, we still have a nearly athermalized joint, as the RTV "softness" has an isolating effect. But if the shape factor is too high, we could rupture the bond, as it is not very strong, exhibiting a tensile strength of only 400 psi. Or, the load may be too high for the optic to bear in wavefront performance.

For example, if producibility issues allow for a bond thickness of only 0.065 in., what would be the load to the lens and stress in the bond for a 100 °C soak change?

From Eqs. (9.15c) and (9.17), we find that $S = 2.56$ and $\alpha_{eff} = 634$ ppm/°C. The net growth is

$$Y = R(\alpha_b - \alpha_l)\Delta T - (\alpha_{eff} - \alpha_b)t_p\Delta T = 3.279 \times 10^{-3}\ \text{in.}$$

From Eq. (1.5), the force to the lens is

$$P = \frac{YAE_c}{t_p}. \tag{9.19}$$

From Eq. (9.14), $E_c = 5160$ psi. Substituting into Eq. (9.18), we find that $P = 130$ lbs, which may be more force than the lens can withstand. The stress in the RTV adhesive is [from Eq. (1.2)]

$$\sigma = \frac{P}{A} = 260\,\text{psi} < 400\,\text{psi, so is okay.}$$

This example points out the need to check designs carefully.

References

1. "Investigation of thick bondline adhesive joints," DOT/FAA/AR-01/33, Office of Aviation Research, Washington, D.C. (2001).
2. T. Altshuler, "Thermal expansion of Epibond 1210 from 296K to 353K," AML-TR-2001-5 (2001).
3. T. Altshuler, "Thermal expansion of RTV 566 from 120K to 293K," AML-TR-2000-31 (2000).
4. T. Altshuler, "Creep and tensile tests of Epibond 1210 at 297K, 333K, and 343K," AML-TR 2001-02 (2001).
5. S. W. Beckwith, "Viscoelastic creep behavior of filament wound case materials," *J. Spacecraft* **21**(6), 546 (1984).
6. S. Timoshenko, "Analysis of bi-metal thermostats," *J. Optical Society of America* **11**(33), 233 (1925).
7. W. T. Chen and C. W. Nelson, "Thermal stress in bonded joints," *IBM J. Research Develop.* **23**(2), 179–188 (1979).
8. S. Ney, E. Swensen, and E. Ponslet, "Investigation of compliant layer in SuperGLAST CTE mismatch problem," HTN-102050-018, Hytec, Inc., Los Alamos, NM (2000).
9. S. Cheng and T. Gerhardt, "Laminated beams of isotropic or orthotropic materials subjected to temperature change," FPL-75, Forest Products Laboratory, U.S. Dept. of Agriculture (1980).
10. P. R. Yoder, Jr. and D. Vukobratovich, "Shear stresses in cemented and bonded optics due to temperature changes," *Proc. SPIE* **9573**, 95730J (2015) [doi: 10.1117/12.2188182].
11. P. B. Lindley, *Engineering Design with Natural Rubber*, Third Edition, NR Technical Bulletin #8, Malayan Rubber Fund Board, National Rubber Producers' Research Association, London (1974).

Chapter 10
Simple Dynamics

Opto-structural analysis generally involves statics (things that do not move) but is not complete without involving dynamics (things that move a little). Almost all optical systems will be subject to vibration during ground transport, handling, shock, aircraft operation, rocket launch, space flight, and in other scenarios.

For several reasons, it is important to note both frequency and acceleration response to optical systems. One needs to ensure that the system does not fail by strength limitations under amplified loading; in addition, displacements cannot be so high as to cause impact or interference with neighboring components or housings. The opto-structural engineer must ascertain that frequencies do not couple within the system to amplify acceleration. The system must exhibit a fundamental mode to decouple it from its platform or payload, and from its vibration source and frequencies, such as from a launch vehicle. Accordingly, we address the response of such systems in a simplified fashion in this chapter.

10.1 Basics

Since vibration is oscillatory, it is best represented in terms of sinusoidal variation with time (Fig. 10.1) and applies to all forms of vibration (shock, random, etc.). Recall from Chapter 1 [Eq. (1.1), which is Hooke's law], that force is proportional to deflection, or $F = kx$, where F is the applied force (in pounds), k is the stiffness constant (in pounds per inch), and x the deflection (in inches).

Consider Fig. 10.2, which shows a mass–spring system in equilibrium under a static force. From Newton's law, when acceleration a is applied, an additional force is applied:

$$F = ma, \tag{10.1}$$

where m is the mass of the object under acceleration. This force is resisted by the spring such that

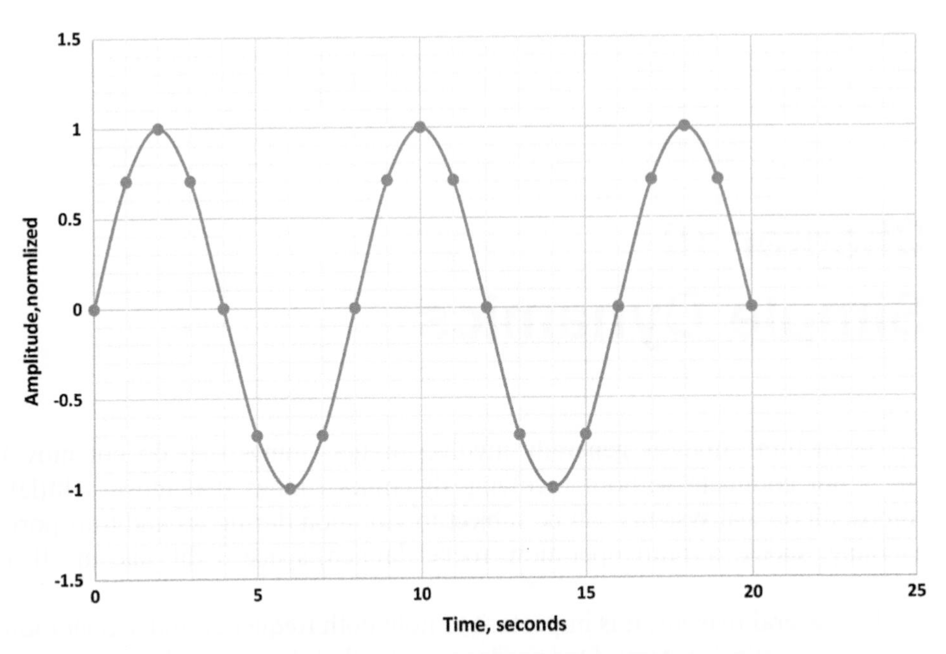

Figure 10.1 Sinusoidal vibration: full reversal of amplitude versus time.

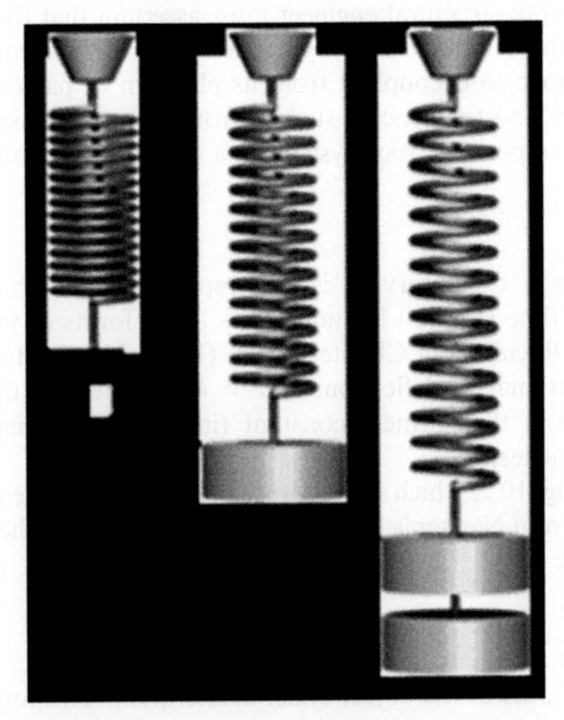

Figure 10.2 Mass–spring system: equilibrium condition (adapted from *Encyclopedia Britannica* (2012), topic: Hooke's law).

$$\Sigma F = ma + kx = 0, \tag{10.2}$$

or

$$m\frac{d^2x}{dt^2} + kx = 0. \tag{10.3}$$

Solution to Eq. (10.3) in the time domain gives rise to a natural frequency f:

$$f = \frac{1}{2\pi}\sqrt{\frac{k}{m}}. \tag{10.4}$$

The natural frequency is given in units of cycles per second, or hertz (Hz). It is sometimes referred to as the fundamental frequency, or resonant frequency, and applies to a single-degree-of-freedom (DOF) system. While almost all systems have many modes beyond the fundamental, the natural frequency will serve well in many applications using first-order principles.

Acceleration response to vibration sources will depend on the natural frequency and damping characteristics of the spring. Without damping, response at resonance to a given input is infinite. In general, damping will limit response amplification to a value between 3 (rubber shock mount) and 50 (metals and ceramics), and sometimes higher (up to 100) for low-input applications. This amplification is designated as Q, or transmissibility, where

$$Q = \frac{1}{2\zeta}, \tag{10.5}$$

and ζ is the viscous damping value of the spring. The viscous damping value generally ranges between 0.5% and 15% for common structural materials. Acceleration response a_r at resonance is therefore

$$a_r = Qa. \tag{10.6}$$

Response for a single-DOF system to a sinusoidal input source is shown in Fig. 10.3 for various damping levels. Note that at an octave below resonance (an octave represents a doubling of frequency), the response essentially equals the input. It then rises dramatically at resonance and drops to near the input level at $\sqrt{2}$ beyond resonance. One octave beyond resonance, response is greatly reduced. Any input that is much greater than resonance will not be seen in the system response, as it is significantly isolated.

It is also useful to define the time period T of the frequency as

$$T = \frac{1}{f} \tag{10.7}$$

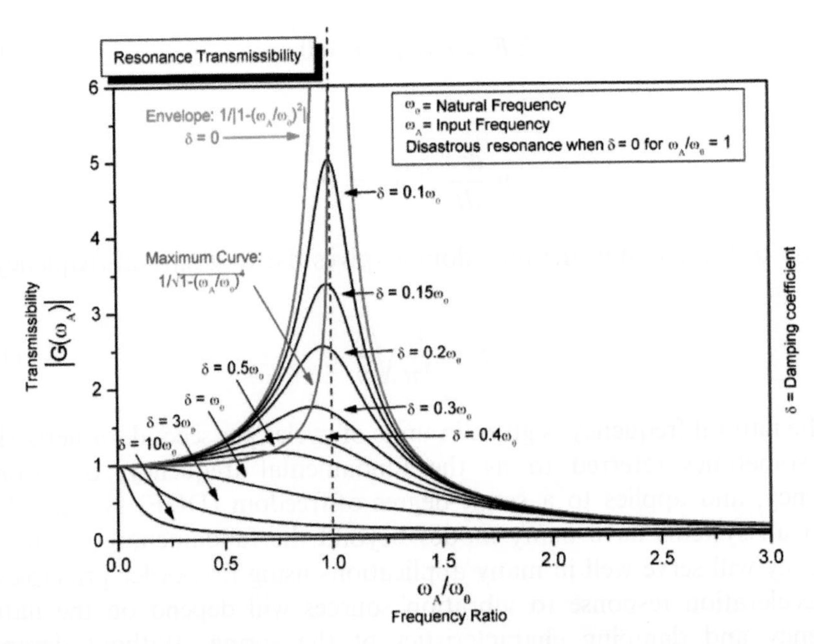

Figure 10.3 Transmissibility Q versus the fundamental frequency ratio as a function of damping (adapted from Ref. 1).

such that one complete reversal cycle occurs. The natural frequency is related to the rotational natural frequency ω as

$$\omega = 2\pi f. \tag{10.8}$$

The displacement x and velocity v, and the acceleration a under sinusoidal vibration are related (from differential calculus) according to

$$x = A\sin(\omega t), \tag{10.9a}$$

$$v = \frac{dx}{dt} = \omega A\cos(\omega t), \tag{10.9b}$$

$$a_r = \frac{dv}{dt} = -\omega^2 A\sin(\omega t), \tag{10.9c}$$

where A is the displacement peak magnitude.

10.2 A Useful Relationship

One of the most useful relations in dynamics is computation of natural frequency from static deflection of an optic or structure under its own weight. We note that mass m is given as

$$m = \frac{W}{g}, \tag{10.10}$$

where W is the weight of an object in a gravitational acceleration field g. From Hooke's law [Eq. (1.1)], we have

$$W = kx. \tag{10.11}$$

Substituting Eqs. (10.10) and (10.11) into Eq. (10.4), the natural frequency is

$$f = \frac{1}{2\pi}\sqrt{\frac{kg}{W}},$$
$$f = \frac{1}{2\pi}\sqrt{\frac{g}{x}}. \tag{10.12}$$

In earth's gravitational field, g = 386.4 (in./sec)/sec, in which case Eq. (10.12) becomes

$$f = \frac{3.13}{\sqrt{x}}. \tag{10.13}$$

Thus, knowing the static deflection of a system, and using Eq. (10.13), one can obtain a good approximation of the fundamental mode of the system. Note that the value 3.13 is close to π, but this is purely coincidental. However, it may help to remember this important relation as

$$f \approx \frac{\pi}{\sqrt{x}},$$

$$\omega = 2\pi f = \frac{2\pi^2 f}{\sqrt{x}}.$$

Note that Eq. (10.13) applies to a single-DOF system. Although in practice the optical system will consist of multiple modes, Eq. (10.13) is a good first-order approximation. How elegantly simple!

10.2.1 Rotational frequency

Similar to the natural frequency of a single-DOF system, defined in translation by Eq. (10.14), where

$$f = \frac{1}{2\pi}\sqrt{\frac{k}{m}},$$

we can derive the rotational natural frequency of a single-DOF system as

$$f_{\Theta} = \frac{1}{2\pi}\sqrt{\frac{k_{\Theta}}{I_{\mathrm{m}}}}, \tag{10.14}$$

where k_{Θ} is the rotational stiffness, and I_{m} is the mass moment of inertia. Here,

$$k_{\Theta} = \frac{M}{\Theta},$$

where M is the static moment equivalent of Hooke's law relation as applied to rotation rather than translation.

10.2.2 Example

A relatively rigid optic is supported on a three-point bipod mount. Calculations from Chapter 3 based on the optic weight and bipod stiffness show a self-weight deflection in the worst-case orthogonal axis of 0.001 in. Compute the natural frequency of the optic on its mount.

From Eq. (10.13), we find that $f = \frac{3.13}{\sqrt{0.001}} = 99\ \mathrm{Hz}$.

How elegantly simple!

10.3 Random Vibration

In many applications, vibration is not purely sinusoidal in nature but rather is random. Random vibration from ground transport, aircraft vibration, rocket launch, etc., is probabilistic in nature, with both amplitude and frequency varying throughout.

To obtain random vibration input, the vibration source is passed through a series of filters over a given bandwidth frequency, and over a spectrum of frequencies, e.g., over a range of 20–2000 Hz with a bandwidth of 10 Hz. Acceleration response [Eq. (10.6)] over each filter is noted, squared, and divided by the bandwidth frequency, producing a response-versus-frequency plot, known as the power spectral density (PSD), or, more appropriately, the acceleration spectral density (ASD), since power is converted to acceleration from accelerometer data. Units of ASD are in g^2/Hz, where g is the g-force. In this formulation, where probability of amplitude is generally Gaussian (bell-shaped), input and output responses are obtained by taking the square root of the area under the curve, resulting in one standard deviation (1σ) or a g-force RMS value. Typically, 3σ output responses are then obtained for analysis. Here, using the Gaussian distribution, 68.3% of the response occurs within the 1σ value, 95.5%

occurs within the 2σ value, and 99.7% occurs within the 3σ value. In other words, 1σ values occur 68.3% of the time, while 1σ to 2σ values occur 27.2% of the time, and 2σ to 3σ values occur 4.2% of the time, approximately.

While response to a given PSD input is mathematically complex, Miles[2] gives an approximation for a one-DOF system under a constant (white-noise) spectral density over an infinite range of frequencies. Although a system is rarely one degree of freedom and is rarely infinite in frequency spectrum, the single-DOF Miles response is quite useful in determining the first-order acceleration response based on the system fundamental mode, and is given as

$$a_{\mathrm{r}} = \sqrt{\frac{\pi G f Q}{2}}, \tag{10.15}$$

where G is the ASD input (in g^2/Hz) at fundamental frequency f, and a_{r} is the 1σ response value in g. 3σ values, which are used in the analysis to determine stress level, are simply three times the value of Eq. (10.15).

10.3.1 Example

A typical ASD curve used for rocket launch of spacecraft is known as the General Environmental Vibration Standard (GEVS)[3] and is shown in Fig. 10.4. The input is an envelope of rocket vehicle inputs and is worst case.

Consider, for example, an optical system weighing 50 kg (110 lbs), exhibiting a fundamental mode of 200 Hz and a transmissibility of 25. The acceleration response [from Eq. (10.15)] at the 3σ level is

$$a_{\mathrm{r}} = 3\sqrt{\frac{\pi G f Q}{2}} = 75\,\mathrm{g},$$

which would be applied statically for use in the stress analysis.

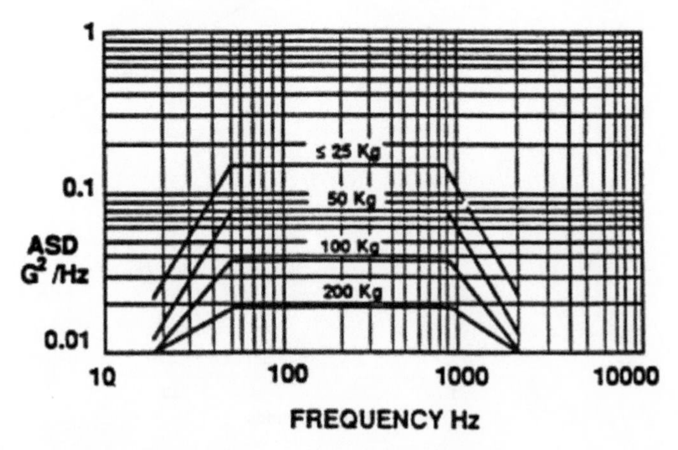

Figure 10.4 General Environmental Vibration Standard (GEVS) random vibration envelope as a function of system mass (reprinted from Ref. 3).

In practice, these loads are generally conservative and apply to primary structure components (more-complex analysis for stress and displacement, not discussed herein, is required to produce random stresses, which are calculated for individual modes and take phase into account). Secondary structure components and system centroid locations are typically analyzed to lower levels, as shown on the typical mass–acceleration curve (MAC)[4] of Fig. 10.5.

10.3.2 Decibels

Perhaps one of the more demanding tasks for opto-structural engineers of flight systems is not the operational and non-operational vibration environment itself, but rather the ability to get the system out of the facility in the first place. To ensure hardware integrity, laboratory vibration loads are significantly higher than flight loads. The ratios of the magnitude of random vibration loads can be compared by use of decibels (dB). For power spectral densities given in terms of magnitude squared per hertz, we have

$$\mathrm{dB} = 10\log\left(\frac{G_{12}}{G_{22}}\right), \tag{10.16}$$

where G_{12} and G_{22} are the magnitudes (displacement, velocity, or acceleration) of the PSD ordinate in g^2/Hz. For example, if the ratio of PSD is 4, the

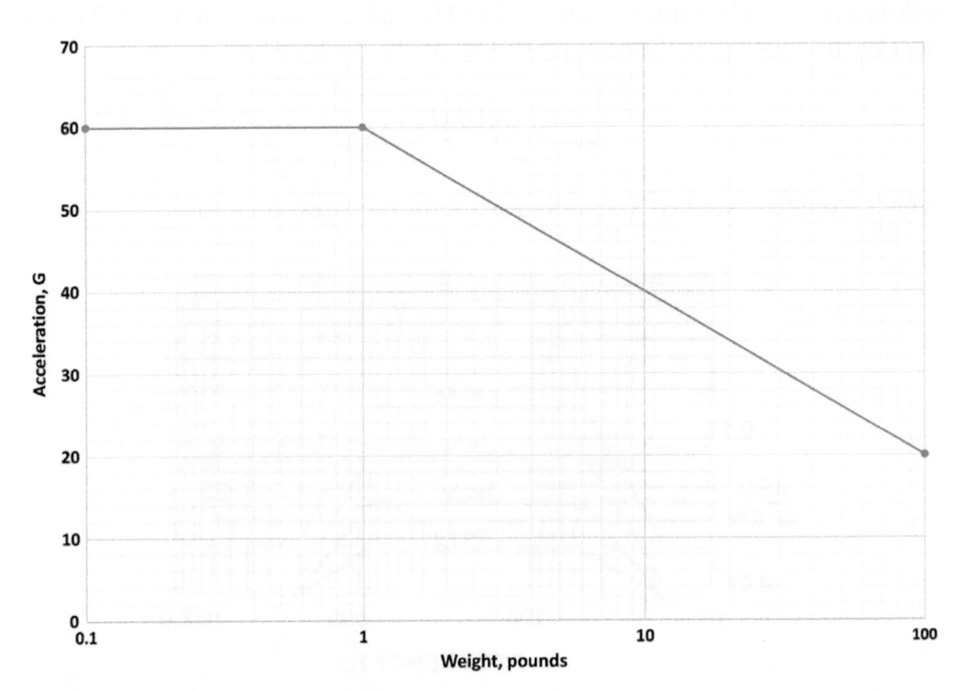

Figure 10.5 Typical mass–acceleration curve (MAC) as applied to secondary structure components (adapted from Ref. 4).

increase in decibels is dB = 10log(4) = 6.02 (approximately 6). For RMS values (G_1 or G_2), which are related to the square root of the PSD spectrum, we have

$$dB = 20\log\left(\frac{G_1}{G_2}\right). \tag{10.17}$$

Table 10.1 gives a summary of these various ratios.

Qualification loads intended for engineering development may be 3–4.5 dB higher (1.4–1.7 g RMS) than acceptance loads, while proto-qualification loads, which may be used on first article flights, are 1.5–3 dB times higher (1.2–1.4 g RMS).

Acceptance test loads themselves, intended to be more "flight-like," are generally higher than true flight loads, largely due to enveloping specifications (vehicle to payload to optical systems), which result in excessive system response.

If the primary structure loads are too high, then a different, less conservative, approach may need to be taken, such as response limiting, or force limiting, provided one can justify such limiting, as is reviewed next.

10.4 Force Limits

Random vibration input spectra to an optical system tend to be very conservative, as they envelope peak measured responses from the driving vehicle and do not take impedance into account. To illustrate this, consider a 1-g sine input acceleration from a space vehicle. If a payload of mass m and frequency f with transmissibly Q is excited, its peak response will be Q. If this response is then used as input to an optical system of similar mass, frequency, and damping, its response at resonance will be $Q \times Q$. Yet, analysis of a two-DOF vibration system would show response to be only $1.6Q$.

For example, for a payload mass and optical system mass exhibiting equal mass and frequency with a Q of 50, the envelope approach yields a response of 2500 g, while the true response is only 80 g—a difference factor of more than

Table 10.1 Decibel (dB) ratios and random vibration equivalencies.

dB	PSD Ratio	RMS Ratio
0	1	1
0.5	1.122	1.059
1	1.259	1.122
1.5	1.413	1.188
2	1.585	1.259
3	1.995	1.413
6	3.981	1.995
9	7.943	2.818
12	15.85	3.981
18	63.09	7.943

30! This is shown graphically in Fig. 10.6, where the envelope response and the more-realistic two-DOF system are compared.

This shortcoming is addressed by Scharton[5] in his classic approach of force limiting. Scharton realized that enveloping responses of a payload system and applying them to an optical system can severely overtest the system, as applied forces have no limits on a hard shaker table (see Fig. 10.7). Note that the interface acceleration (payload response) to a unit acceleration input at resonance results in a split mode, with the lower frequency being approximately 60% of the input frequency, and responding at 50 g. But typically, this response

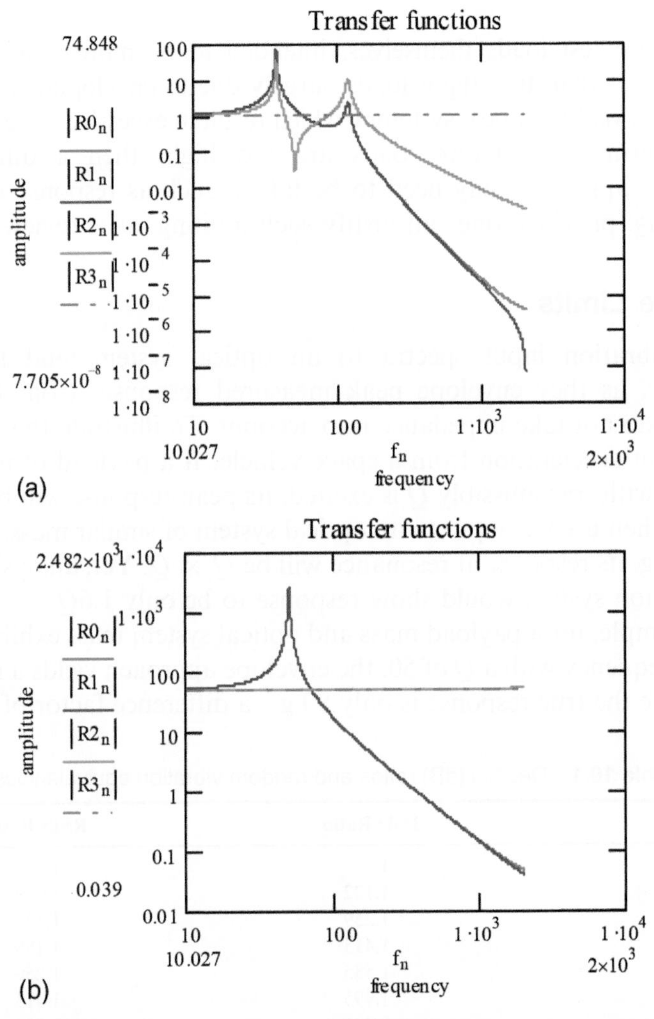

Figure 10.6 Enveloping response. (a) 1-g input of a 2-DOF system, each of equal mass and spring, results in 80-g response for Q = 50. (b) Response to a single-DOF system with enveloped 50-g input from payload is 2500 g for Q = 50, greatly exaggerating the response (adapted from Ref. 5).

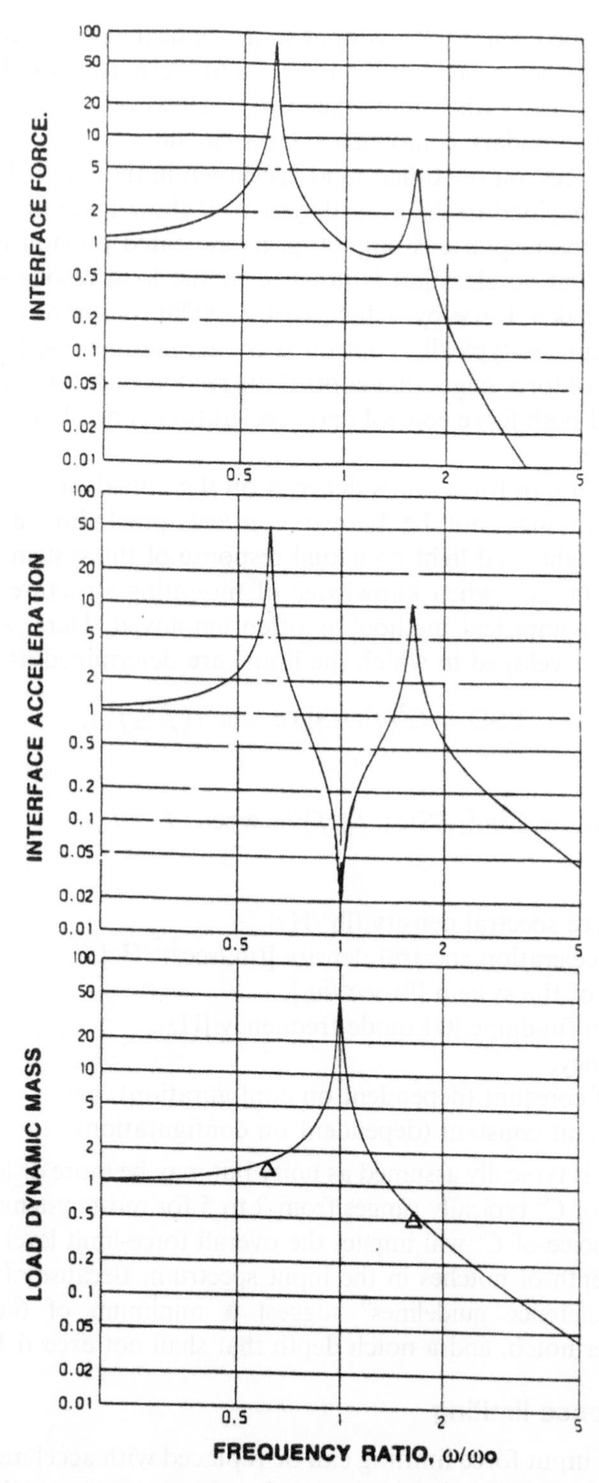

Figure 10.7 Force-limited response of a 2-DOF system with equal masses and frequencies: (bottom) base response (vehicle level) of 50 g from 1-g input results in (middle) split acceleration response at (top) payload-level-limiting forces (reprinted from Ref. 5 with permission from American Institute of Aeronautics and Astronautics).

would be enveloped, and the 50-g input to the optical system would result in a 2500-g response (shown previously in Fig. 10.6). Perhaps it would be better to envelope the valleys rather than the peaks! However, this could result in undertesting of secondary components and structure.

Interface forces, on the other hand, as shown in the top of Fig. 10.7, limit force from the payload to the optical system. If these force limits are applied to a test, then the response is only 80 g, as indicated previously. With force limiting, the input acceleration is notched at the load resonance frequency, reducing the shaker force by a factor of 2500/80, or more than 30. While acceleration input is typically controlled by accelerometers, force limits are applied by use of force gages that control the response. A force specification is calculated, and both force control and acceleration control are applied by the vibration shaker.

Determination of force limits depends on the impedance of the mounting platform, which may not be known *a priori*, precluding a coupled-load analysis that might shed light on actual response of the system under test. In lieu of real data, i.e., when knowledge of mounting structure impedance is lacking, a semi-empirical method[6] is often employed. Here, a force–power relationship is developed in which the limits are determined as

$$\text{FSD} = C^2 m^2 (\text{ASD}), \text{ when } (f \leq f_0), \tag{10.18a}$$

and

$$\text{FSD} = C^2 m^2 (\text{ASD}) (f_0/f)^{2n}, \text{ when } (f > f_0), \tag{10.18b}$$

where

FSD is the force spectral density [lb^2/Hz],
ASD is the acceleration spectral density [$(\text{in./sec}^2)^2$/Hz],
m is the mass of the system [lb-sec^2/in.],
f_0 is the system fundamental mode frequency [Hz],
f is the frequency,
n is the rolloff constant (dependent on configuration), and
C^2 is a force limit constant (dependent on configuration).

The value of n is typically assumed as unity but may be more or less than unity, and the value of C^2 typically ranges from 2 to 5 for most systems but could be higher. The choice of C^2 will impact the overall force-limit level and therefore impacts the depth of notches in the input spectrum. Because of uncertainties, the NASA reference guidelines[6] suggest a minimum of 6-dB resonance remaining after notch, and a notch depth that shall not exceed 14 dB.

10.4.1 Response limiting

Alternatively, input force limiting can be replaced with acceleration response limiting by enforcing notches in the acceleration spectrum at the split modes of interest. This approach is less reliable, as it depends more on analysis and modeling techniques compared to force limiting. It does, however, result in more realistic system response and is often employed, with customer approval.

10.5 Shipping Vibration

The next most-difficult task in getting a system out of the facility is designing for a shipping and transportation environment. Over-the-road acceleration, aircraft environment, handling shock events, and potential for dropping of shipped items can result in high g-loading, requiring isolation. Typical specifications for handling and shipping are given in Table 10.2. These include aircraft loads, handling shock, and over-the-road transport vibration.

If a low-frequency isolation system is provided, the over-the-road input specifications will be attenuated at high frequencies but will amplify at low frequencies. This will be examined in Example 10.5.1.2 in conjunction with the design of shipping fixtures (Example 10.5.1.1) and associated handling.

Table 10.2 Typical specifications for handling and shipping: (a) truck and aircraft shock, (b) transport shock, and (c) transport over-the-road vibration.

(a) Handling Shock

Gross Weight (pounds)	Drop Height (inches)
0–20	30
21–60	34
61–100	18
101–150	12
151–250	10
>250	8

(b) Transportation Shock

Gross Weight (pounds)	Truck (G)	Aircraft (G)
0–50	10	8
51–100	8	6
Duration (msec)	0.8–40	0.8–40

(c) Truck Road Sinusoidal Vibration

Acceleiation (peak g)	Frequency (hertz)
3	2–10
4	10–100
6	100–1000

10.5.1 Drop shock

In order to calculate the g-response from a drop event, we turn to the physics of the impact of an isolated system of given stiffness characteristics to a rigid ground, or, of a given stiffness system to a rigid ground. Consider, then, a single-DOF system in which a rigid mass m is accelerated in a gravitational field g over a drop height h to a spring of stiffness k. This deforms the spring by an amount x on impact. Before the drop, the system has potential energy (PE) of

$$\mathrm{PE} = mghG. \tag{10.19}$$

Note here that in an earth gravitational field, $g = 386.4$ (in./sec)/sec, while G is the acceleration g-level, equal to 1 in a standard drop. Upon impact, the potential energy is converted to kinetic energy (KE) of

$$\mathrm{KE} = \frac{mv^2}{2}, \tag{10.20}$$

where v is the impact velocity.

Setting Eq. (10.19) equal to Eq. (10.20), we have

$$v = \sqrt{2ghG}. \tag{10.21}$$

After impact, the spring deflects by an amount x, and energy is converted and preserved to

$$\mathrm{PE} = \frac{kx^2}{2} = mghG. \tag{10.22}$$

Also, momentum is conserved as

$$Ft = mv, \tag{10.23}$$

where F is the impact force, and t is the time to stop after initial impact. We have

$$F = \frac{mv}{t} = ma = mAg, \tag{10.24}$$

where A is the response acceleration in g. Also, and again from Hooke's law, $F = kx = mAg$, so

$$x = \frac{mAg}{k}. \tag{10.25}$$

Substituting Eq. (10.25) into Eq. (10.22) gives

$$mghG = \frac{k(\frac{mAg}{k})^2}{2}$$

such that, solving for acceleration A, we have

$$A = \frac{k}{mg}\sqrt{\frac{2mghG}{k}}. \tag{10.26}$$

From Eqs. (10.3) and (10.8), we have

$$\frac{k}{m} = \omega^2 = (2\pi f)^2$$

such that

$$A = 2\pi f\sqrt{\frac{2hG}{g}}. \tag{10.27}$$

Again, A is the response acceleration in g, and G is the initial g-level in g (=1 in a drop environment).

Figure 10.8 is a drop shock nomogram that determines acceleration, velocity, and displacement—it does the work for you. Simply enter the drop height on the ordinate axis and determine frequency at the output acceleration level, along with displacement and velocity.

10.5.1.1 Examples

Example 1. A 100-lb optical telescope assembly must survive a 14-in. drop with acceleration response limited to 15 g. Determine the natural frequency required for a shock mount to properly isolate the assembly.

From Eq. (10.27), we solve for f, where $A = 15$, $G = 1$, $h = 14$, and $g = 386.4$, and find that $f = 8.87$ Hz. One can readily determine the displacement of the spring from Eq. (10.9a) as $x = 1.87$ in. and design for envelope excursion and the clearance required for shipping.

Note that there may be conflicting requirements for transportation. For example, a low-frequency, over-the-road, sinusoidal vibration input will require a high-frequency, isolation resonant system in order to prevent amplification, while a drop shock requirement would require a low-frequency, isolation resonant system to reduce amplification. Thus, a balanced isolation system that is not too soft and not to stiff is required.

Example 2. Consider the design of Example 1, in which an appropriate isolation system for handling shock drop is determined. Determine the response to an over-the-road truck transport based on the Table 10.2 specifications.

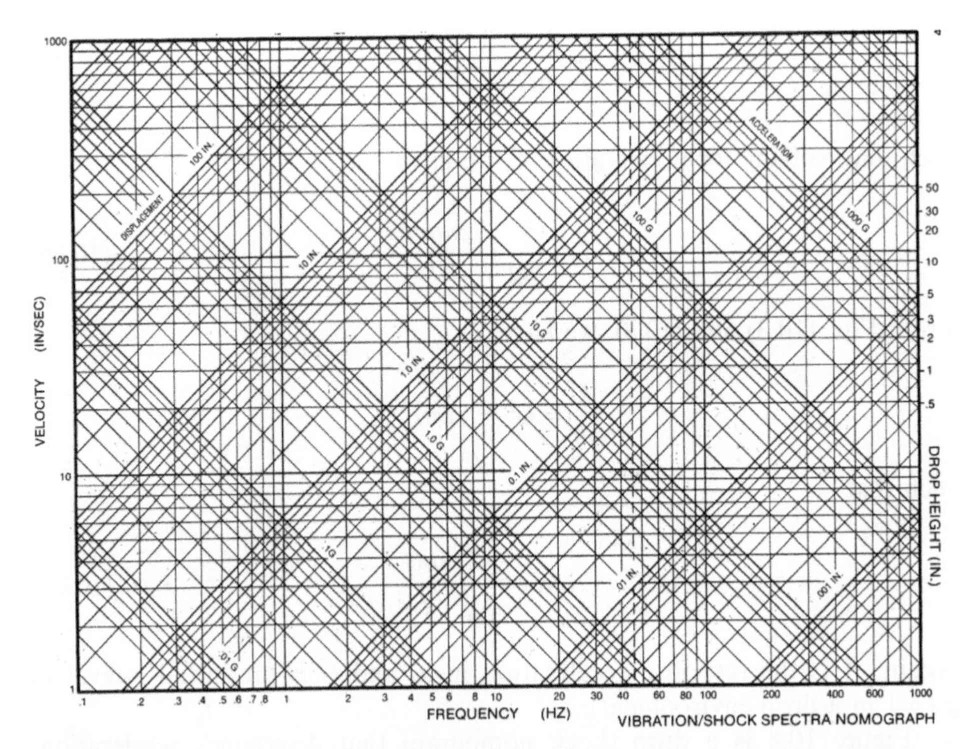

Figure 10.8 Drop shock nomograph. Acceleration, velocity, and displacement are determined for a given drop height and isolation frequency.

The response of the 8.87-Hz frequency at the 3-g road input of Table 10.2 is dependent on the shock mount transmissibility due to damping. To limit response to 15 g, shock mount isolators will need to exhibit a maximum transmissibility Q of 5, and from Eq. (10.5), we have $Q = \frac{1}{2\zeta} = 15/3 = 5$, or structural damping (2ζ) of 0.20 (20%), requiring a highly damped isolator, such as can be obtained from silicone rubber.

10.6 Acceleration Shock

During high-acceleration loading, tight envelope requirements may leave little clearance between components in an optical assembly, causing impact. These impact acceleration forces can be attenuated with relatively soft bumpers. For small excursions, where h approaches zero in Eq. (10.27), we have no acceleration response. In this case, we need to take into account the fact that the body force continues to perform more work on the mass after the instant the spring starts to deflect. Here, we have the potential energy as

$$PE = mgG(h + x). \tag{10.28}$$

With this energy, we can solve for A as before, with some manipulation, as

$$A = G + \sqrt{G^2 + (2\pi f)^2\left(\frac{2hG}{g}\right)}. \tag{10.29}$$

10.6.1 Example

Consider a 5-lb lens cell that accelerated at 20 g during launch and impacts a bumper of frequency 10 Hz after an excursion of 0.10 in. Determine the g-response from Eq. (10.29) and compare the response to Eq. (10.27).

We find from Eq. (10.29) that $A = 41$ g. Using the large-drop Eq. (10.27), we find that $A = 6.4$ g. We see that the small-drop equation results in a significantly higher response than what would be erroneously calculated if the work of the bumper were neglected. For larger drops or higher frequencies, the values of each equation will approach one another.

10.6.2 Variable acceleration

Often, a mass spring system is accelerated in a variable-gravity-acceleration field, as occurs in sinusoidal or random vibration environments. Again, the spring's motion is limited to the travel of the spring as [from Eq. (10.13)]

$$x = G\left(\frac{3.13}{f}\right)^2. \tag{10.30}$$

If this motion is less than the gap between the mass–spring system and the ground, then no impact will occur. If it is greater than this gap, impact will occur, but Eq. (10.29) needs to be modified because acceleration, and therefore velocity, is not constant.

In this case we note that it is impact velocity, *not* acceleration, that is responsible for the impact force and resulting response acceleration; i.e., using the impulse momentum balance,

$$Ft = mv,$$

$$F = \frac{mv}{t} = ma,$$

where a is the acceleration response. Note that for sinusoidal (and random) vibration, the peak displacement [from Eq. (10.9a)] is

$$x = Y\sin(\omega t),$$

where Y is the peak amplitude; and from Eq. (10.8),

$$\omega = 2\pi f.$$

The velocity is obtained from Eq. (10.9b) as

$$\frac{dx}{dt} = v = \omega Y \cos(\omega t),$$

and the acceleration a, from Eq. (10.9c) as

$$\frac{d^2x}{dt^2} = a = -\omega^2 Y \sin(\omega t).$$

Note that when amplitude is maximum, velocity is minimum while acceleration is maximum but opposite in sign, i.e., moving away. Thus, if the gap before impact is the same as the maximum motion, velocity is zero and there is no increase in response as the unit impacts. If the gap is less than the motion, impact will occur but at a reduced velocity compared to the impact in the constant-acceleration field.

For a gap h and peak amplitude motion Y ($h < Y$), we have

$$x = Y \sin(\omega t) = h \text{ and thus,}$$

$$(\omega t) = \sin^{-1}\left(\frac{h}{Y}\right). \tag{10.31}$$

The velocity is

$$v = \omega Y \cos\left(\sin^{-1}\frac{h}{Y}\right) = \cos\left(\sin^{-1}\frac{h}{Y}\right) V, \tag{10.32}$$

where V is the maximum velocity.

While Eq. (10.29) does not contain a velocity term, we note that if we set potential energy equal to kinetic energy, we have $mgh = \frac{mv^2}{2}$. We can now re-compute the energy at impact and solve for an effective gap h' as

$$\cos\left(\sin^{-1}\frac{h}{Y}\right) V = \cos\left(\sin^{-1}\frac{h}{Y}\right)\sqrt{2gh}, \tag{10.33}$$

and compute the effective gap h' as

$$\sqrt{2gh'} = \cos\left(\sin^{-1}\frac{h}{Y}\right)\sqrt{2gh}, \tag{10.33a}$$

$$h' = \left[\cos\left(\sin^{-1}\frac{h}{Y}\right)\right]^2 h. \tag{10.33b}$$

We now enter Eq. (10.29) with $h = h'$ to compute the impact acceleration as

$$A = G + \left\{ G^2 + (4\pi f)^2 \left[\cos\left(\sin^{-1} \frac{h}{Y} \right) \right]^2 \frac{hG}{g} \right\}^{\frac{1}{2}}. \quad (10.34)$$

We use Eq. (10.1) to compute impact force.

10.6.3 Lift equipment

Note from the small-drop Eq. (10.29) that when h is zero or very small, the response is 2 g. This is an important relation for handling fixtures using sling equipment. A sudden tug on the sling, causing it to deform, will impact the system at 2 g. In most codes, for safety, lift equipment must be designed to 3 g against yield strength and 5 g against ultimate strength. Most lift equipment devices have a specified design load criterion[7] defined as the maximum load for which the device or equipment is designed and built. This load is also referred to as the "rated load" or "working load." This load includes a high safety factor of 5.0 against ultimate strength failure. Proof tests (using a factor of 2 overload) are generally performed prior to first-time use, and subsequent periodic testing is performed at predetermined intervals with load greater than or equal to the rated load but less than the proof load.

10.6.4 Pyrotechnic shock

Shock is a sudden, severe transient acceleration that can produce high g-loading. Severe shock is caused in particular by pyrotechnic events such as explosive bolts, rocket stage separation, and the like. These events take on the form of a pulse, such as a half-sine, that lasts only milliseconds. Typically, the shock is converted to a shock response spectrum (SRS), in which the enveloped magnitude of acceleration is plotted versus frequency, each of which corresponds to a single-DOF system.

A SRS plot is shown in Fig. 10.9, where it can be seen that the acceleration increases with frequency up to 1200 g. Such values make the analyst shudder, and the immediate response is to pass this on to the next guy (the payload or vehicle integrator), even if specifications allow for interface or distance reduction factors. However, the buck *always* stops somewhere.

Modeling of shock response spectra is not a task for the faint of heart, as, generally speaking, the SRS needs conversion to a series of transient inputs, and then there is the problem of understanding modal stresses. Perhaps a more gentle approach makes use of techniques[8] developed that show that velocity, not acceleration, drives the design.

Work by Gaberson[9–11] concludes that stress due to a severe shock is related to the component modal velocity. This research has resulted in shock severity limits for aerospace and military equipment. The empirical threshold limit velocity is typically 100 in./sec. This limit is based on Gaberson's work on actual shock testing and analytical yield-stress–velocity relationships for mild

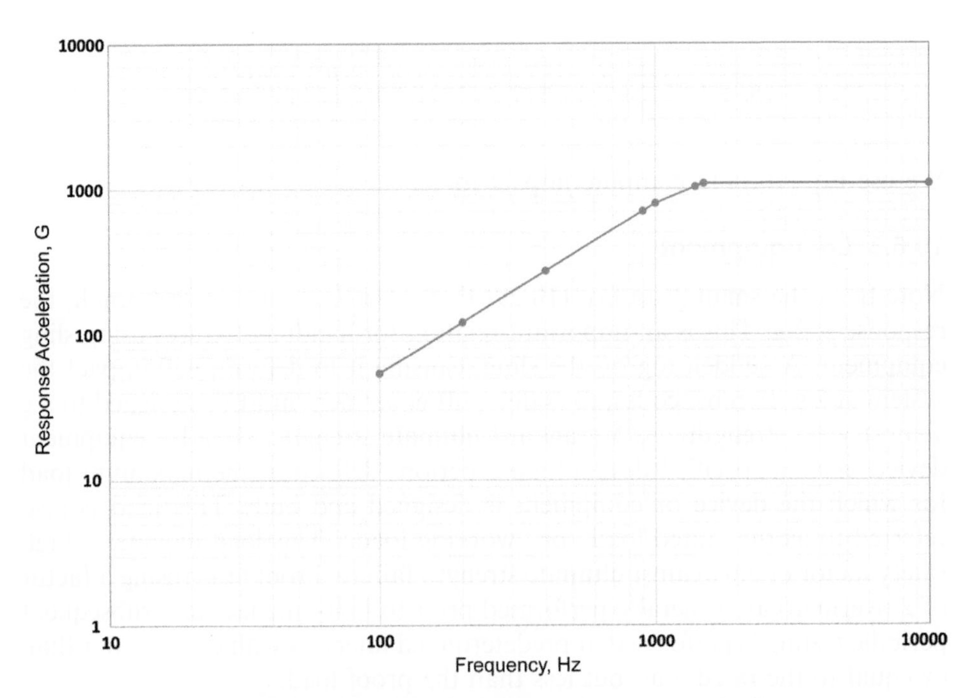

Figure 10.9 Typical pyrotechnic shock response spectrum.

steel. Other observations on military equipment confirm the threshold. For conservatism, a 6-dB reduction is applied to this level, reducing the threshold to 50 in./sec. Modal velocity relations for aluminum and other materials show a threshold of greater than 200 in./sec, resulting in even more conservatism.

Acceleration-versus-time shock spectra plots are readily converted to velocity. The 50 in./sec threshold is equivalent to about a 3-in. drop shock. This is evident from the nomograph of Fig. 10.8.

Note that at 50 in./sec over a 2-in. drop, [from Eqs. (10.7) through (10.9)] we have a velocity of

$$v = \frac{a}{\omega}. \tag{10.35}$$

Substituting,

$$\begin{aligned} v &= \frac{386G}{2\pi f} = 50;\ \text{therefore,} \\ \frac{G}{f} &= 0.8, \end{aligned} \tag{10.36}$$

where f is the natural frequency, and G is the peak acceleration.

Thus, if the response g-level is less than 0.8 times the frequency, a detailed analysis, or test, may be deemed unnecessary. This spectrum applies to almost all ductile materials and many other components.

Severity limits for electronics[12] are based on displacement limits and natural frequency from which velocity limits can then be derived. In this case it has been shown that, for printed circuit boards, the velocity limit is

$$v = \omega Z. \tag{10.37}$$

Here, $\omega = 2\pi f$, where f is the board natural frequency, and Z is given as (for a worst-case component and location)

$$Z = \frac{0.0001L}{h\sqrt{b}}, \tag{10.38}$$

where L is the board length, h is its thickness, and b is the component length.

The velocity limit calculated from Eq. (10.37) is then compared to the velocity shock spectrum to determine acceptability. If the calculated velocity exceeds the limit for circuit boards, or exceeds the 50-in./sec limit for structure, then more-detailed analysis and testing or redesign for shock isolation may be warranted.

References

1. S. Timoshenko and D. Young, *Vibration Problems in Engineering*, Third Edition, D. Van Nostrand Co., New York, p. 79 (1955).
2. J. W. Miles, "On structural fatigue under random loading," *J. Aeronautical Sciences* **21**(11), 753–762 (1954).
3. "General Environmental Verification Standard (GEVS) for GSFC Flight Programs and Projects," NASA GSFC-STD-7000A, Goddard Technical Standard (2013).
4. M. Trubert, "Mass acceleration curve for spacecraft structural design," JPL D-5882, NASA Jet Propulsion Laboratory (1989).
5. T. D. Scharton, "Vibration-test force limits derived from frequency-shift method," *J. Spacecraft and Rockets* **32**(2), 312–316 (1995).
6. *NASA Technical Handbook: Force Limited Vibration Testing*, NASA HDBK-7004B, Washington, D.C. (2003).
7. NASA Technical Lifting Standard, NASA-STD-8719.9A, Washington, D.C. (2015).
8. T. Irvine, "An introduction to the shock spectrum," Revision S, Vibrationdata website www.vibrationdata.com (2012).
9. H. A. Gaberson, "Pseudo velocity shock spectrum rules for analysis of mechanical shock," *Proc. Modal Analysis Conference 2007* (IMAC-XXV), pp. 633–668 (2007).

10. H. A. Gaberson, "Shock severity estimation," *Sound & Vibration Magazine*, January 2012, pp. 12–19 (2012).
11. H. A. Gaberson and R. H. Chalmers, "Modal velocity as a criterion for shock severity," *Shock and Vibration Bulletin* **40** part 2, U.S. Naval Research Lab, Washington, D.C., pp. 31–49 (1969).
12. T. Irvine, "Shock severity limits for electronic components," Revision B, Vibrationdata website www.vibrationdata.com (2014).

Chapter 11
Fatigue

Fatigue is a process in which material strength is degraded under fluctuating or continuous load with time. Cyclic fatigue is a phenomenon in which premature failure occurs with time in the presence of fluctuating load, or stress reversal. Stress reversal does not need to be full tension to compression; any fluctuating load about a nominal value can cause a degree of cyclic fatigue. Static fatigue, on the other hand, is a phenomenon in which premature failure occurs with time under a constant, nonfluctuating load in the presence of a chemically active environment. Static fatigue is associated with slow crack growth and is more properly known as stress corrosion. Corrosion fatigue is a fracture that occurs under a combination of both cyclic fatigue and stress corrosion.

The dominant concern for metals and ductile materials is cyclic fatigue. The prevailing phenomenon for crystals, ceramics, glass, and brittle materials is stress corrosion. As we shall see in Chapter 12, water is the chemically active environment that produces stress corrosion. Materials that are brittle at ambient temperatures do not exhibit the same dislocation phenomena common in cyclic fatigue failure. While cyclic fatigue is possible in brittle materials, we limit our discussion to stress corrosion for such materials. Similarly, while stress corrosion is possible in ductile materials, we limit our discussion to cyclic fatigue for such materials.

11.1 Cyclic Fatigue

Cyclic fatigue can be an important contributor in opto-structural analysis where vibration is involved, as in airborne or space flight applications. For example, consider a rocket flight where high stresses are applied over a period of a minute or so due to random vibration. Typically, before flight, the optical system will have undergone ground test for several minutes at various load levels to qualify the structure. After integration to the payload, the system is again subjected to testing, and perhaps once again after vehicle integration. If a total of 10 minutes of test and flight is accumulated and the system responds, say, at 400 Hz, we have experienced nearly one-quarter of a million cycles.

Using a scatter factor of 4 for analysis on time (a typical requirement), this results in nearly 1,000,000 cycles.

Cyclic fatigue can occur with fluctuating load after a number of cycles. Low-cycle fatigue is seen when plastic strains (i.e., stresses beyond the material elastic yield point) occur, generally below 1000 cycles, and is beyond the scope of this text. High-cycle fatigue is seen in the elastic region, generally well beyond 1000 cycles. High-cycle fatigue causes failure well below the material ultimate strength and usually below its yield strength. Again, this phenomenon is most pronounced in ductile materials due to slip dislocation at boundary imperfections. These imperfections may or may not be induced by a pre-existing crack; most fatigue data are obtained from polished specimens. Pre-existing cracks will exacerbate fatigue.

11.1.1 High-cycle fatigue

High-cycle fatigue occurs in three stages: crack initiation, crack propagation (stable crack growth), and catastrophic failure (unstable crack growth). Cracks propagate under tensile loading, not compression, unless nominal compressive stresses are reversed to tensile stresses during stress fluctuation.

Cracks will initiate at an imperfection if stress is high enough and will begin at the highest stress point near such imperfections. Holes, discontinuities, abrupt cross-sectional changes, and stress risers are prime initiation sites. At continued stress reversals, the crack will grow as stress intensity increases until a critical value is reached and failure occurs. The crack growth rate is most pronounced immediately before failure. Crack growth is explained by the theory of fracture mechanics and involves an inherent property of the material, known as the critical stress intensity factor K_{Ic}, or fracture toughness, which will be discussed in Section 11.4.1. A more common approach is a method based on experimental data that records strength S versus cycles N at various levels of reversed stress amplitude. This approach is commonly called the S-N method and uses an S-N diagram or S-N curve.

11.2 S-N Method

Consider, for example, the S-N curve shown in Fig. 11.1. The strength at 1,000,000 cycles is greatly reduced from that at 1000 cycles, or near its static strength. Many ductile materials will exhibit an endurance limit that is defined as that level of stress below which fatigue failure will not occur after an infinitely large number of cycles. In practice, a value of 10–20 million cycles or more is chosen as an endurance limit.

The curve shown in Fig. 11.1 is for an unnotched specimen undergoing fully reversed stress, i.e., a mean stress of zero with an equal and opposite peak tension and compression stress reversal. If stress risers exist, values need to be reduced by the stress concentration factor.

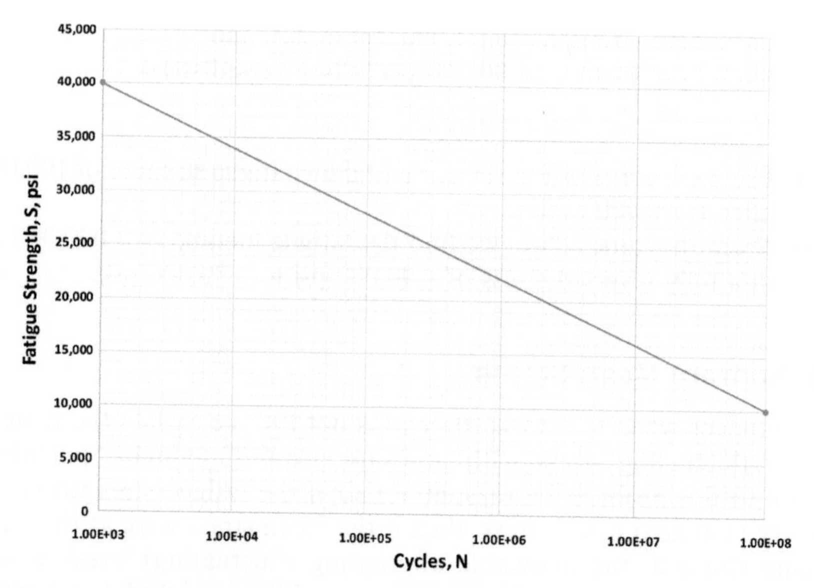

Figure 11.1 Simplified S-N curve for fully reversed fatigue of aluminum 6061-T6. Note the marked drop in fatigue strength with increased number of cycles.

11.2.1 Example for consideration

Consider the S-N diagram of Fig. 11.2 depicting a material under complete stress reversal about a nominal mean stress of zero. The material's ultimate strength is 160,000 psi, and its fatigue strength after 10,000,000 cycles is 80,000 psi. Determine

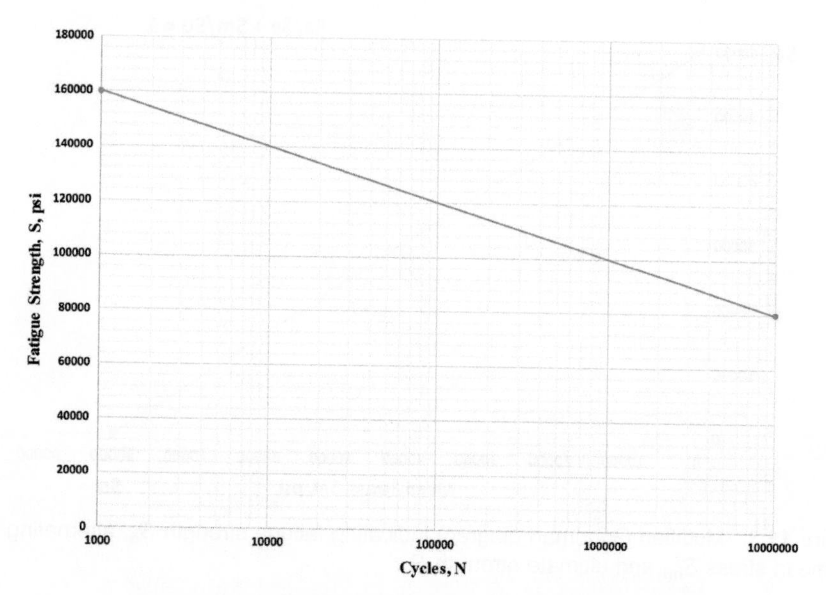

Figure 11.2 Fatigue S-N curve for high-strength steel.

a) the fatigue strength after 1,000,000 cycles, and
b) the fatigue lifetime at 90,000 psi fully reversed stress.

Solutions:

a) From the curve of Fig. 11.2, we find the fatigue strength of 100,000 psi after 1,000,000 cycles.
b) From the same curve, we find the fatigue lifetime of 3,000,000 cycles at a fully reversed stress of −90,000 psi to +90,000 psi.

11.3 Nonzero Mean Stress

What happens when the stress reversal is not full, i.e., when the material is under a fluctuating stress with nonzero mean? Goodman[1] postulated a solution to this problem by reasoning that if the fatigue strength is known for reversal at zero mean stress, then if the mean stress were at the material ultimate strength, the allowable alternating (fluctuating) stress would be zero. He simply connected these end points with a straight line (in standard *x-y* Cartesian coordinates), as shown on the modified Goodman chart of Fig. 11.3, estimating the value at nonzero mean from the straight line curve. Tests show his assumptions to be on the conservative side, with allowable fluctuations in stress lying above the line. Thus, the conservative Goodman approximation is a good indicator of allowable failure with some margin.

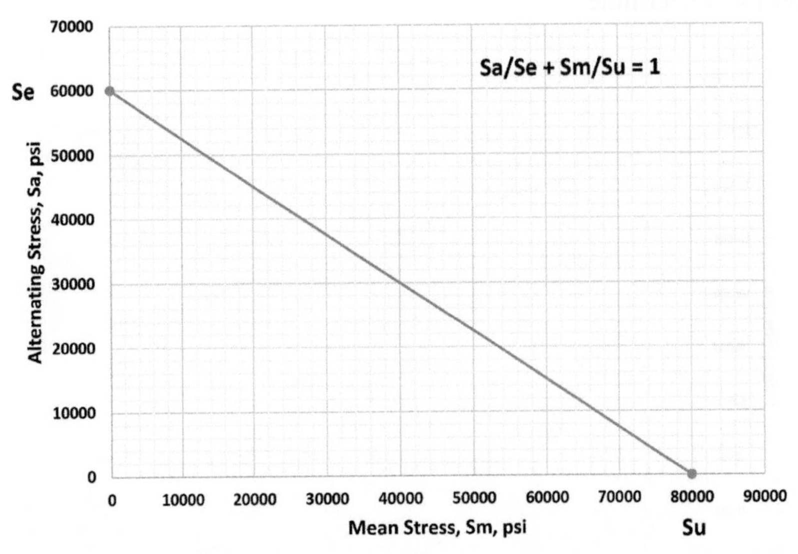

Figure 11.3 Modified Goodman diagram indicating fatigue strength S_e, alternating stress S_a, mean stress S_m, and ultimate strength S_u.

When the Goodman linear relation is put into standard form, we see that

$$S_a = \frac{-S_e S_m}{S_u} + S_e,$$

or

$$\frac{S_a}{S_e} + \frac{S_m}{S_u} = 1, \tag{11.1}$$

or

$$S_a = S_e\left(1 - \frac{S_m}{S_u}\right), \tag{11.2}$$

where S_a is the allowable alternating stress, S_e is the fully reversed fatigue strength at a given number of cycles, S_u is the ultimate strength in tension, and S_m is the mean stress. Mean and alternating stresses are depicted in Fig. 11.4.

11.3.1 Example

From the example provided in Section 11.2.1, we found a fatigue strength of 1,000,000 cycles at 90,000 psi. Determine the allowable stress fluctuation S_a at 1,000,000 cycles if the mean stress were 40,000 psi.

Here we solve for S_a when $S_e = 90{,}000$, $S_u = 160{,}000$, and $S_m = 40{,}000$. From Eq. (11.1) or (11.2), we find that $S_a = 67{,}500$ psi. The allowable peak stress is thus $67{,}500 + 40{,}000 = 117{,}500$ psi.

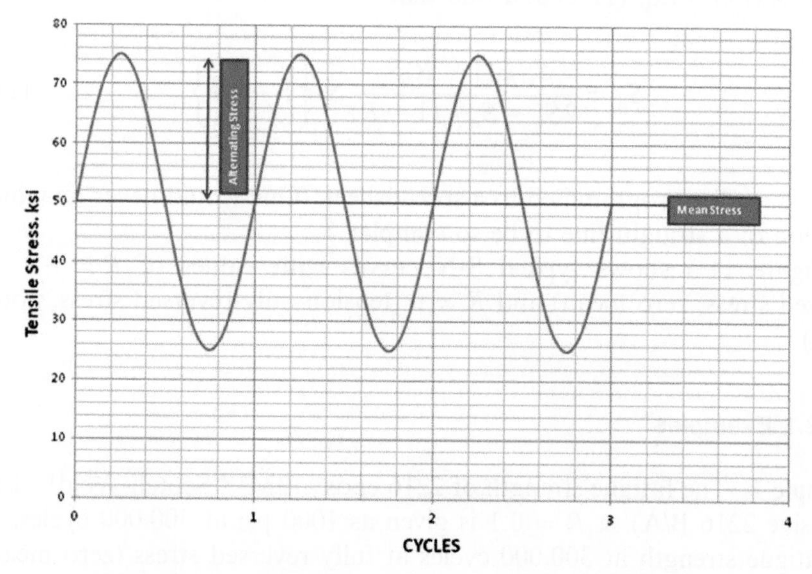

Figure 11.4 Alternating stress cycle about mean stress.

11.3.2 *R* ratio

We can generalize the Goodman equation even further by introducing a stress ratio factor, called the R ratio, where R is the ratio of minimum stress S_{min} to maximum stress S_{max} with sign duly considered:

$$R = \frac{S_{min}}{S_{max}}. \tag{11.3}$$

For fully reversed tension–compression loading, $R = -1$; for loading from 0 to maximum tension, $R = 0$. At zero mean stress and complete reversal of load, the maximum allowable stress is lowest because cracks open and close significantly with cycles. In this case ($R = -1$), the maximum stress equals the alternating stress. At a high tensile mean stress, the maximum allowable stress increases, but allowable alternating stress decreases ($R = 0$). In general, we have $-1 \leq R \leq +1$.

We can now modify the Goodman equation for full-reversal fatigue strength for any value of R by manipulating Eq. (11.1) (exercise left for student) to read

$$S_e = \frac{S_{e(R)}(1-R)S_u}{2S_u - S_{e(R)}(1+R)}, \tag{11.4a}$$

where $S_{e(R)}$ is the fatigue strength at value R. We can now substitute Eq. (11.4a) into Eq. (11.2) and find that

$$S_a = \frac{S_{e(R)}(1-R)S_u}{2S_u - S_{e(R)}(1+R)} S_e \left\{1 - \frac{S_m}{S_u}\right\}, \tag{11.4b}$$

which is readily programmed to a spreadsheet. (One would never imagine the equation of a straight line to be so complex.)

Figure 11.5 shows typical S-N curves with values of $R = -1$ (fully reversed stress, zero mean) and $R = 0$ (tension-only-reversed stress, nonzero mean).

11.3.2.1 Examples

Example 1. The fatigue strength of 2216 epoxy (3M™ Scotch-Weld™ Epoxy Adhesive 2216 B/A) at $R = 0.1$ is given as 1000 psi at 300,000 cycles. Find the fatigue strength at 300,000 cycles at fully reversed stress (zero mean) at $R = -1$. The static strength is 2500 psi.

From Eq. (11.1),

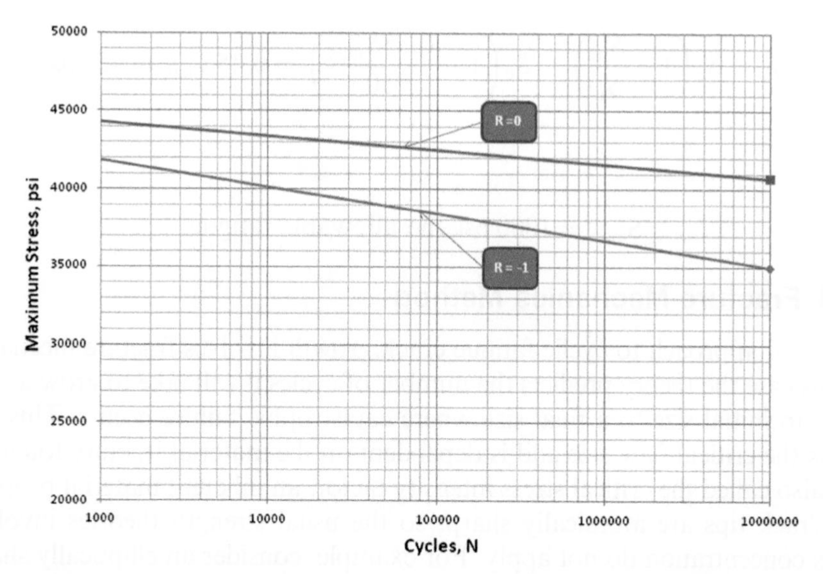

Figure 11.5 Example of a fatigue S-N plot comparing fully revered stress ($R = -1$) to tension-only-reversed stress ($R = 0$), the latter indicating longer lifetime for a given maximum stress.

$$S_e = \frac{S_a}{\left(1 - \frac{S_m}{S_u}\right)}$$

$$S_{max} = 1000\,\text{psi}$$

$$S_m = 1.1(S_{max}/2) = 550$$

$$S_u = 2500$$

$$S_a = 450\,\text{psi} = 1000 - 550$$

$$S_e = 577\,\text{psi}.$$

Example 2. The fatigue strength of 2216 epoxy at $R = -1$ is given as 577 psi at 300,000 cycles. Find the fatigue strength at 300,000 cycles at $R = 0.1$. The static strength is 2500 psi.

$$S_e = 577$$

$$S_a = S_{max} - S_m$$

$$S_m = \frac{(S_{max} + 0.1S_{max})}{2}$$

$$S_a = 0.45S_{max}$$

From Eq. (11.2),

$$S_a = S_e\left(1 - \frac{S_m}{S_u}\right) = S_e\left(1 - \frac{0.55S_{max}}{2500}\right) = 0.45S_{max} = 450\,\text{psi},$$

and

$$S_{max} = 1000\,\text{psi, the allowable fatigue.}$$

11.4 Fracture Mechanics Method

A second approach to cyclic fatigue crack growth involves fracture mechanics. In this case the theory predicts the number of cycles it will take to grow a crack from an initial size to a final size where catastrophic failure occurs. This final size is the critical flaw size and is dependent on the material fracture toughness K_{Ic}, also called the critical stress intensity factor, an inherent material property.

Crack tips are atomically sharp, so the usual strength theories involving stress concentration do not apply. For example, consider an elliptically shaped notch of depth $2c$ embedded in an infinite plate specimen, as shown in Fig. 11.6. The notch has a major diameter of $2c$ and a minor diameter of $2a$. If $c \gg a$, we have the case of a rounded (blunt) crack, as depicted in Fig. 11.7 for a

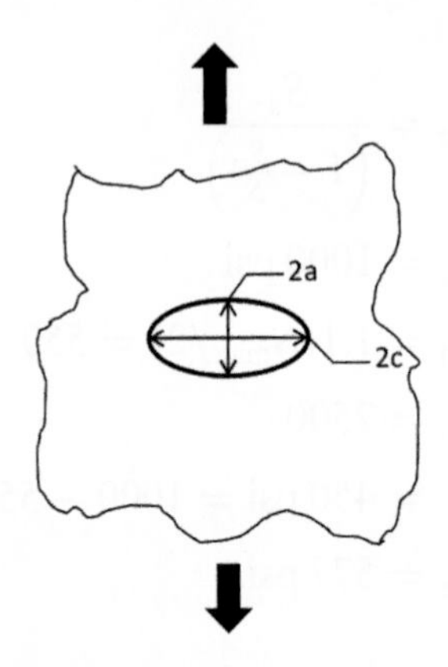

Figure 11.6 Elliptical cutout in an infinite body.

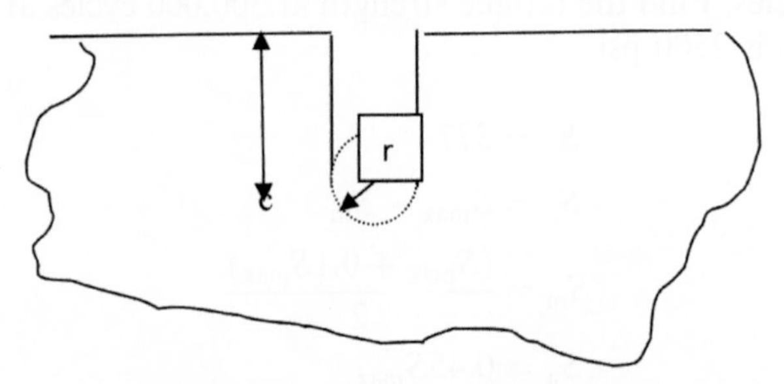

Figure 11.7 Notch tip. Stress concentration for notch depth *c* and radius *r* results in a dramatic rise in stress as the tip radius decreases toward zero.

surface crack, where the tip radius r is now shown. Elasticity theory applied around such notches shows that this condition gives rise to stress concentrations at the tip; the deeper the notch and the smaller the radius the higher the stress.

For the subject case, it can be shown[2] that

$$\sigma_k = \sigma\sqrt{\frac{c}{2r}}, \tag{11.5}$$

where σ is applied stress, and $\frac{\sigma_k}{\sigma}$ is the stress concentration factor. However, as we have noted, crack tips are not blunt; rather, they have an atomically small radius—at the nanometer level. The casual observer will quickly note that, as such a crack tip radius approaches zero, the actual stress amplification approaches the infinite. Infinite stresses do not fare well here. The usual strength of materials theory does not apply.

11.4.1 Stress intensity

Although the usual strength theory results in an infinite stress, mathematicians have no problem with singularities. If you do not like the answer, substitute. Consider then defining a term

$$K = \sigma_k\sqrt{2\pi r} \tag{11.6a}$$

and substitute into Eq. (11.5) to yield

$$K = \sigma\sqrt{\pi c}, \tag{11.6b}$$

where K is a stress field, or the so-called mode-I stress intensity factor (see Chapter 12 for more discussion).

When K reaches a critical value K_{Ic}, spontaneous failure occurs. Here, K_{Ic} is denoted as the mode-I critical stress intensity factor. This factor is an inherent property of the material and is a measured value for a given material, as is achieved in a Chevron notched specimen test.[3] K_{Ic} is also known as the fracture toughness of a material. [As an aside, the term K_{Ic} is properly referred to as "KAY-ONE-SEE," not "KAY-EYE-SEE," as it is the mode-I critical intensity factor, separating it from modes K_{IIc} and K_{IIIc}].

Denoting the material strength as S, and noting that failure occurs at the critical stress intensity, we have [from Eq. (11.6b)],

$$K_{Ic} = S\sqrt{\pi c}. \tag{11.7}$$

S is also known as the material modulus of rupture. Because it is flaw size dependent, a detailed description of the surface finish is required for S to have meaning when using vendor-provided values. Because of this, caution needs to be used in manufacture-provided data, as these data are quite dependent on shape and size of the defect, humidity, and other test conditions. Indeed, the strength of ceramics and glasses is not an intrinsic material property, as it can be greatly reduced to below theoretical strength when flaws are present.

In true fracture mechanics terms, S is properly called the inert strength, which is independent of moisture and designated as σ_i. Thus, if we know the fracture toughness of a material, we can calculate its strength for a given flaw depth and shape, or its maximum flaw depth for a given strength and shape.

Note in the stress intensity formulation the introduction of π. (Mathematicians seem to like this as much as they do singularities.) Although π appears to be arbitrary, the original formulation by Griffith[4] uses an energy approach to derive Eq. (11.6b) in which π does indeed occur. There is no need to fret over the supplied formulation from stress concentration, since in the measurement of K_{Ic}, π is taken into account. And since K is a fraction of K_{Ic}, there is a self-canceling effect. The Griffith law describes a through crack, which is the most severe of flaws and is used here. We will discuss other types of flaws in Chapter 12.

The value of fracture toughness K_{Ic} is available for many materials. Table 11.1 shows a fracture toughness comparison of metals and brittle materials. Ductile metals have significantly higher K_{Ic} values, indicating their tolerance for larger defects.

11.4.2 I love Paris

We now need to see how the flaw grows with stress intensity and cycles. Crack growth is given by a power relationship developed by P. C. Paris and is given as

$$\frac{dc}{dN} = A(\Delta K)^n, \tag{11.8}$$

where c is the flaw size (depth), N is the number of cycles, ΔK is the change in stress intensity, and A and n are material-dependent constants called Paris contants.

11.4.3 Case study

We now review a particular example of the Paris law. Consider a structure made of S200F beryllium exhibiting a fundamental mode resonance of

Table 11.1 Approximate values of the critical stress intensity for ductile and brittle materials.

Material	Critical Stress Intensity Factor ($MPa\text{-}m^{1/2}$)	Class
High-strength steel	60–100	Ductile
Aluminum	20–40	Ductile
Beryllium	10	Semi-brittle
Engineering ceramics	1–4	Brittle
Glasses	0.7–1	Brittle

350 Hz. Stress margins against failure in fatigue fracture can be calculated assuming an initial flaw size, and knowing the crack growth rate and anticipated cyclic loading during sinusoidal vibration.

Failure from fatigue in itself is not an issue, as beryllium S200F fatigue strength essentially equals its yield strength at 10,000,000 cycles of fully reversed loading.[5] However, fatigue fracture may be of concern if an initial flaw is present in the beryllium. In this case, crack growth rates are reported by Lemon,[6] and their effect on strength versus cyclic loading for various flaw sizes can be calculated. Note that data were obtained for beryllium S200E and are conservative for S200F, where reported rates[5] are noted. However, we use the more conservative data.

While the frame assembly is inspected using dye penetrant, we assume a minimum detectable flaw at 0.005 in. and often use 0.010 in. to be conservative in the assumption of an initial flaw. Based on flaw size and stress level, analysis is performed to compute lifetime, as below discussed.

Calculations are based on growth rate data,[6] which indicate a value of the Paris constants [Eq. (11.8)] as $A = 4.6 \times 10^{-13}$, and $n = 7.3$, which are determined for fully reversed loading ($R = -1 = S_{min}$-to-S_{max} stress ratio) and

$$\Delta K = \frac{\Delta\sigma}{\sqrt{\pi a}}, \tag{11.9}$$

where a is the flaw depth, $\Delta\sigma$ is the range of maximum stress to minimum stress in tension only, and ΔK is the change in stress intensity. We have assumed a worst-case Griffith flaw.

Using the maximum stress at resonance, we can now compute the number of fatigue cycles N as

$$N = 4ft, \tag{11.10}$$

where f is the fundamental mode (single DOF) = 350 Hz, t is time in seconds, and 4 is the scatter factor. We then compute the allowable number of cycles based on the fracture growth equation [Eq. (11.8)], which is readily integrated to determine allowable cycles N as

$$N = \frac{1}{\left[A\sigma Y^n\left(1-\frac{n}{2}\right)\right] * \left[\frac{1}{a_f^{\left(\frac{n}{2}-1\right)}} - \frac{1}{a_i^{\left(\frac{n}{2}-1\right)}}\right]}, \tag{11.11}$$

where σ is the maximum stress at resonance, a_f is the final flaw size at failure, a_i is the initial flaw size, and a_f is determined from the critical stress intensity factor K_{Ic} from Eq. (11.7) as

$$a_f = \frac{\left(\frac{K_{Ic}}{\sigma}\right)^2}{\pi}. \tag{11.12}$$

Here, $K_{Ic} = 10.4$ MPa-m$^{1/2}$ (9.5 ksi-in.$^{1/2}$ when flaw size is in inches, and stress is in ksi).

We now can plot allowable cycles versus initial crack size as a function of stress, which is shown in the four plots of Fig. 11.8. For an initial crack size of 10 mils, for example, Fig. 11.8(a) shows the allowable cycles at a stress level of 16,000 psi to be about 1,500,000 cycles, or time to failure from Eq. (11.10) to be about 32 minutes. For an initial crack size of 5 mils, lifetime improves to 9,000,000 cycles, or over 180 minutes.

It is of interest to make use of Eqs. (11.11) and (11.12) to show how the crack size propagates with increasing cycles until reaching critical flaw size. Figure 11.8(b) shows the crack propagation for an initial 0.010-in. flaw defect in beryllium at a stress level of 22,000 psi. Note that failure can occur in as few as (on the order of) 150,000 cycles. Figure 11.8(c) depicts the growth at a smaller initial defect of 0.005 in., again at 22,000 psi. Here, one can approach 1,000,000 cycles before failure. The value of 22,000 psi is often taken as the fracture limit for beryllium after 1,000,000 cycles, as the 0.005-in. flaw can be readily detected by dye penetrant methods. Finally, Fig. 11.8(d) depicts the growth of a finely ground beryllium surface with an initial defect of 0.0015 in., this time at a higher stress level of 35,000 psi. Here, more than 10,000,000 cycles can occur before failure. Note that the stress value approaches the yield point of beryllium, so often it is the yield point that drives the design of a

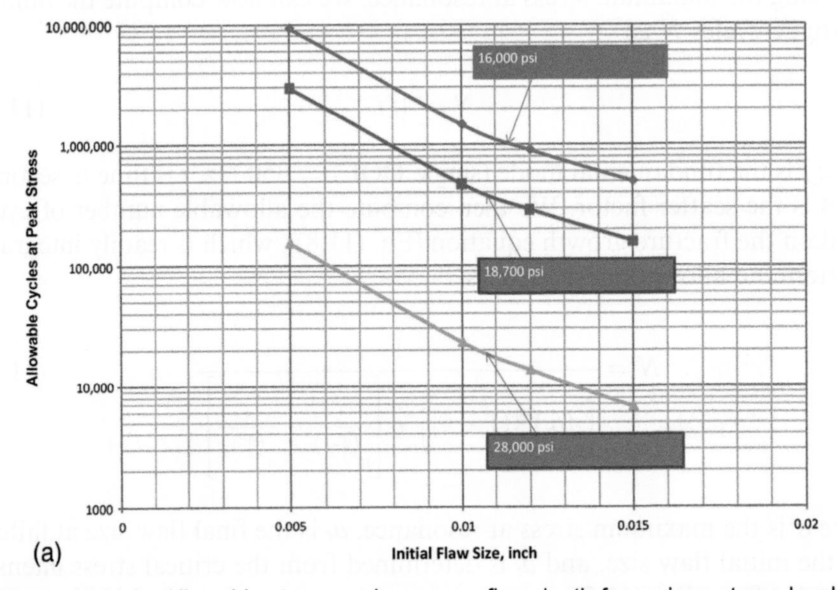

Figure 11.8(a) Allowable stress cycles versus flaw depth for various stress levels.

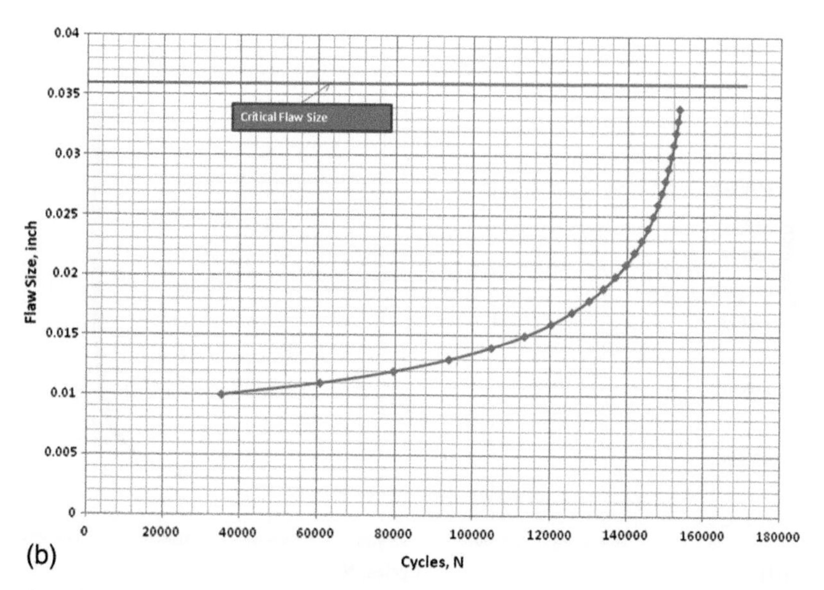

Figure 11.8(b) Flaw propagation after cycling for beryllium with an initial 0.010-in. flaw defect and a stress level of 22 ksi.

well-polished beryllium optic. Note that in all cases the major portion of the growth takes place in the latter stages of cycles just prior to failure.

11.5 Random Vibration Fatigue

While the previous discussion on fatigue relates to a single-DOF system under sinusoidal vibration, this analysis can be readily extended to a generalized

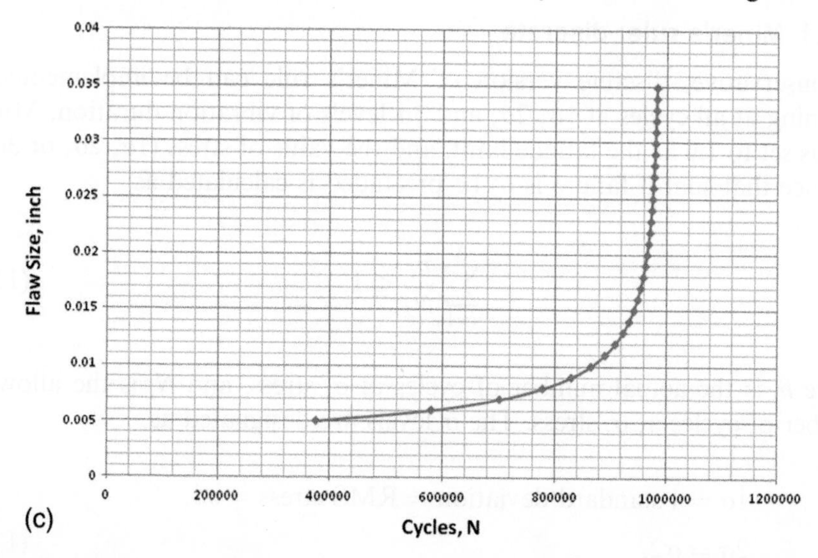

Figure 11.8(c) Flaw propagation after cycling for beryllium with an initial 0.005-in. flaw defect and a stress level of 22 ksi.

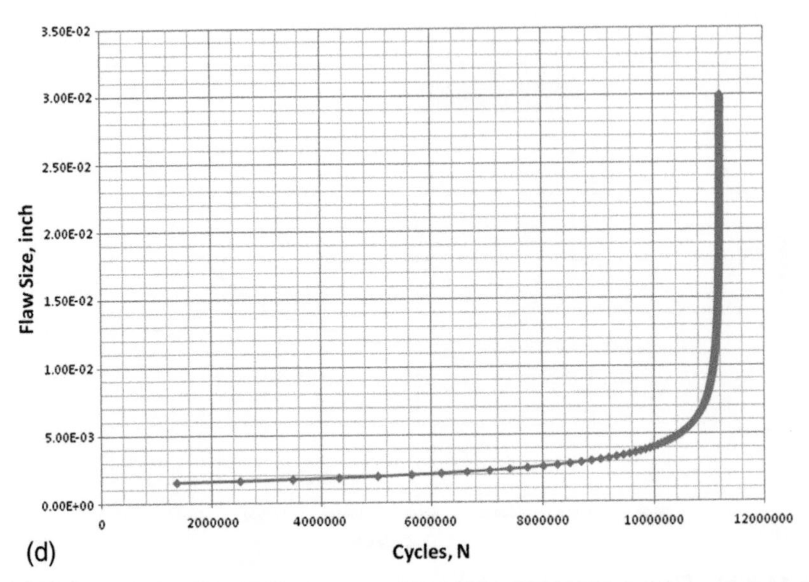

Figure 11.8(d) Flaw propagation after cycling for beryllium with an initial 0.0015-in. flaw defect and a stress level of 35 ksi.

method to compute fatigue lifetime for both random vibration loading and multiple-DOF systems. The method is based on application of the Palmgren–Miner[7,8] rule using both discrete and exact formulations. In random vibration, stress magnitude and frequency vary throughout the duration of loading. First we review random vibration for a single-DOF system.

11.5.1 Miner's rule: discrete

A conservative, discrete version of Miner's rule can be implemented by summing at all cycles at 1σ, 2σ, and 3σ levels of vibration duration. Miner's sum is set to 1.0 as the fatigue limit, and the value of stress (1σ, 2σ, or 3σ) to produce that sum if failure is to be precluded is calculated as

$$\sum_{i=1}^{3} \frac{n_i}{N_i} < 1, \tag{11.13}$$

where n_i is the actual number of cycles at σ_i stress, and N_i is the allowable number of cycles at σ_i stress. The σ levels are designated as

$$\begin{aligned} 1\sigma &= 1 \text{ standard deviation} = \text{RMS stress} = \sigma 1, \\ 2\sigma &= \sigma 2, \\ 3\sigma &= \text{maximum stress} = S_{\max} = \sigma 3. \end{aligned} \tag{11.14}$$

This method is called the discrete Palmgren–Miner rule and assumes that stresses occur 68.3% of the time at 1σ, 27.1% of the time at 2σ, and 4.3% of the time at 3σ. This assumption is a bit conservative since these stresses actually occur at 0σ to 1σ, 1σ to 2σ, and 2σ to 3σ, respectively, but are applied at the full σ level. (A more exact formulation is discussed in Section 11.5.2.)

Consider, then, a fatigue curve plot of strength versus cycles (S-N) plotted semi-logarithmically, an example of which is shown in Fig. 11.9. The example plot is for 2216 epoxy undergoing fully reversed stress at zero mean stress ($R = -1$), but the method applies to any fatigue curve.

The curve equation can be obtained by taking any two convenient points of stress (S_1 and S_2) and cycles (N_1 and N_2) and determining slope b on the semi-log plot, and then taking any two convenient points, S_1 (stress) and N_1 (cycles) to compute

$$\sigma = \sigma_1 \left(\frac{N_1}{N}\right)^{\frac{1}{b}}, \tag{11.15a}$$

$$\text{where } b = -\frac{\log\left(\frac{N_1}{N_2}\right)}{\log\left(\frac{\sigma_1}{\sigma_2}\right)}. \tag{11.15b}$$

For ease of manipulation, value of b should be rounded to the nearest integer.

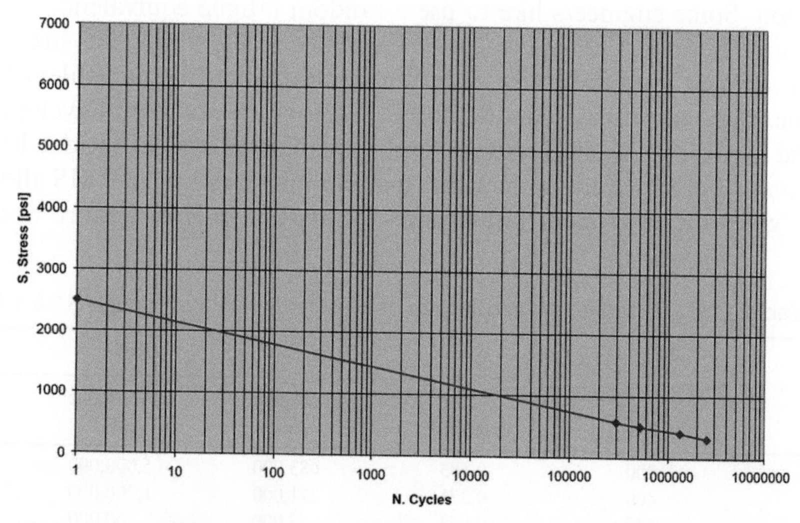

Figure 11.9 Sample fatigue plot from which Miner's rule can be used to compute lifetime to failure in random vibration. Curve is adjusted for $R = -1$ (fully reversed stress at zero mean).

Using Eqs. (11.13) though (11.15), and letting $K = \sigma_1 (N_1)^{\frac{1}{b}}$,

$$1 = \frac{0.683N}{\left(\frac{3K}{\sigma_3}\right)^b} + \frac{0.271N}{\left(\frac{1.5K}{\sigma_3}\right)^b} + \frac{0.043N}{\left(\frac{K}{\sigma_3}\right)^b}. \tag{11.16}$$

After some manipulation we can now solve for σ3, the peak allowable (3σ) stress for a given number of cycles. This can get a bit messy, so it is often easier to select values from the fatigue curve and iterate on a solution.

11.5.1.1 Example for consideration

Consider the fatigue curve of Fig. 11.9 for an epoxy adhesive in shear. The peak (3σ) shear stress in a particular joint is 600 psi, and the system undergoes 1,000,000 cycles with a scatter factor included. Using Miner's rule, determine the fatigue adequacy of the bonded joint.

Using Eqs. (11.13) and (11.14) and the noted probabilistic percentages, Table 11.2 shows the actual cycles at each standard deviation stress level along with the allowable cycles from the plot of Fig. 11.9. We find Miner's sum as

$$\sum_{i=1}^{3} \frac{n_i}{N_i} = 0.50, \text{which is less than 1, so is okay.}$$

11.5.1.2 Random fatigue equivalency

The Palmgren–Miner rule can be used to compute equivalent static loads, as well as allowable stresses and cycles to determine failure during random vibration. Some engineers like to use a random fatigue equivalency factor to determine lifetime. The *equivalent stress* method determines the stress value that, if applied over the entire duration (cycles), gives the equivalent fatigue lifetime. The *equivalent cycles* method determines the number of cycles that, if applied at the maximum 3σ stress level, gives the equivalent fatigue lifetime. The *lambda factor* λ defines the value that, if multiplied by the RMS allowable stress, gives the equivalent failure stress if applied over the entire duration,

Table 11.2 Random vibration fatigue using Miner's rule for Example 11.5.1.1.

			Cycles		
Level	Stress (psi)	Probability	Actual (n_i)	Allowable (N_i)	Ratio (n_i/N_i)
1σ	200	0.683	683,000	5,000,000	0.143
2σ	400	0.271	271,000	1,200,000	0.226
3σ	600	0.043	43,000	300,000	0.137
Miner's sum					0.506

and is employed in conjunction with the equivalent stress method for comparative purposes. All methods are equivalent to each other and to Miner's rule [Eq. (11.13)]. Thus, since more computation is required, there is little added value, except for reporting purposes; however, a sample problem is shown next.

11.5.1.3 Sample random fatigue equivalency

Consider the approximate fully reversed cycle S-N fatigue curve for a heat-treated aluminum component (Fig. 11.1). The component undergoes 1,000,000 cycles in random vibration with a maximum (3σ) stress of 30,000 psi. Determine the equivalent number of cycles, equivalent stress, and λ factor.

The 1σ RMS stress is 10,000 psi, and the 2σ stress 20,000 psi. From the probabilistic distribution and the S-N curve, and using Miner's sum of Eq. (11.13), we find that

$$\sum_{i=1}^{3} \frac{n_i}{N_i} = 43{,}000/50{,}000 + 271{,}000/2{,}100{,}000 + 683{,}000/100{,}000{,}000 = 1.$$

Therefore, in this case, we are at the fatigue lifetime.

The equivalent number of cycles from the S-N curve, if applied at the 3σ level, is 50,000. The equivalent stress from the S-N curve, if applied over the full number of cycles (i.e., 1,000,000), is 20,000 psi. The λ factor in this case is the equivalent stress divided by the RMS stress, or $\lambda = 20{,}000/10{,}000 = 2.0$.

11.5.2 Miner's rule: continuous

While the discrete method is simplistic, it is a bit conservative, and its use is a bit unwieldy if determining an allowable stress. Alternatively, the continuous method is complex but more exact in its computation, and the equations are more tractable.

Miner's continuous rule sums the cumulative damage to 1, as does the discrete method [Eq. (11.13)] but integrates over the entire duration as

$$1 = \int \frac{n(s)}{N(s)} ds. \tag{11.17}$$

When considering the random environment, the number of cycles at stress s is obtained by considering the Rayleigh probability density function (bell curve) distribution as

$$p(s) = \left(\frac{s}{\sigma^2}\right) e^{-\frac{s^2}{2\sigma^2}}, \tag{11.18}$$

where s is the maximum stress, and $\sigma = 1\sigma$ stress. Therefore, the number of cycles at stress s can be expressed as

$$n(s) = f_n t p(s), \tag{11.19}$$

where f_n is natural frequency, and t is time in seconds.

We again use the S-N curve to determine slope b as the nearest *even* integer (for integration) on a semi-log plot from two points s_1 and N_1 on curve, and s and N_s, where

$$N_s = \left(\frac{s_1}{s}\right)^b N_1. \tag{11.20}$$

Substituting into Eq. (11.17) and integrating yields, approximately,

$$1 = \left(\frac{f_n t}{N_1}\right)\left(\frac{\sigma}{s_1}\right)^b \left(2^{\frac{b}{2}}\right)\left(\frac{b}{2}\right)! \tag{11.21}$$

Time to failure is

$$t = \frac{N_1}{2^{b/2} f_n \left(\frac{b}{2}\right)! \left(\frac{\sigma}{s_1}\right)^b}. \tag{11.22}$$

We then solve for the RMS stress σ as

$$\sigma = \left[\frac{N_1 s_1^b}{60 f_n t \left(\frac{b}{2}\right)! 2^{b/2}}\right]^{1/b}, \tag{11.23}$$

where we divided by 60 using the time in minutes. This is readily programmed to a spreadsheet.

11.5.3 Multiple degrees of freedom

While results from the previous subsections apply to random vibration fatigue of a single-DOF system, more advanced analyses are required for multiple-DOF systems. In the latter case, an approximation can be made for equivalent cycles using a positive crossing technique from finite element analysis. Exercising a system dynamic model, crossings are interrogated at a particular component under investigation. Crossings per second are statistically recovered and equal the "equivalent" response frequency.

Lacking finite element data, however, we turn to other tools. One such tool is the complex Dirlik[9] spectral moment method; however, without the necessary software, this method is not so useful. An approximation of the Dirlik method is based on the Rice[10] method, as simplified by Steinberg.[11] This useful tool called the Rice approximation has been developed to compute an equivalent frequency over which to apply the multiple-DOF response. This is particularly useful when response data is received from a test condition.

The Rice approximation takes the form

$$N_0 = \frac{1}{2\pi}\sqrt{\left(\frac{\frac{\left(\frac{\pi}{2}\right)G_1 f_1 Q_1}{\omega_1^2} + \frac{\left(\frac{\pi}{2}\right)G_2 f_2 Q_2}{\omega_2^2} + \ldots}{\frac{\left(\frac{\pi}{2}\right)G_1 f_1 Q_1}{\omega_1^4} + \frac{\left(\frac{\pi}{2}\right)G_2 f_2 Q_2}{\omega_2^4} + \ldots}\right)}, \tag{11.24}$$

where N_0 is the number of zero positive crossings and equals the effective frequency [Hz]; G_i is the ASD (acceleration spectral density) [g^2/Hz]; Q_i is transmissibility; $\omega_i = 2\pi f_i$ [rad/sec]; and f_i is the modal response frequency [Hz]. The following example demonstrates the use of Eq. (11.24).

11.5.3.1 Example for consideration

Consider the response spectrum to the input random vibration spectrum shown in Fig. 11.10. Input and output ASDs at the four key resonant frequencies are shown in Table 11.3. From the data we can approximate

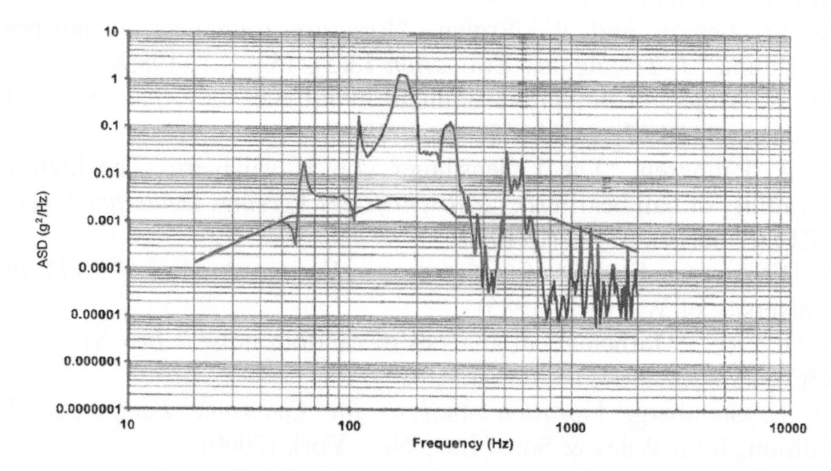

Figure 11.10 Multiple-DOF random spectral density response from which the Rice approximation can be used to determine effective frequency for fatigue calculations.

Table 11.3 Effective frequency (zero positive crossings) for a multiple-DOF system under the random vibration profile of Example 11.5.3.1.

Frequency	ASD Input	ASD Output	Transmissibility
110	0.0015	0.1	8.0
180	0.003	1.2	20.0
286	0.002	0.11	7.5
600	0.0015	0.02	4.0
Effective frequency (Hz)			160.0

transmissibility as the square root of the ratio of output ASD to input ASD, as discussed in Section 10.3. Substitution into Eq. (11.24) gives an equivalent single-DOF frequency of approximately 160 Hz.

References

1. J. Goodman, *Mechanics Applied to Engineering*, Longman, Green & Co., London (1899).
2. B. Gross, J. E. Srawley, and W. F. Brown, Jr., "Stress-intensity factors for a single-edge-notch tension specimen by boundary collocation of a stress function," NASA Technical Note D-2395 (1964).
3. ASTM, "Standard test methods for determination of fracture toughness of advanced ceramics at ambient temperature," ASTM C1421-10, American Society for Testing an Materials (ASTM) International, West Conshohocken, Pennsylvania (2000).
4. A. A. Griffith, "The phenomenon of rupture and flow in solids," *Philosophical Transactions of the Royal Society of London Series A* **221**, pp. 163–198 (1921).
5. "Physical and Mechanical Properties of Beryllium," Brush Wellman Inc., Mayfield Heights, Ohio (2009).
6. D. D. Lemon and W. Brown, "Fracture toughness of hot-pressed beryllium," *J. Testing and Evaluation* **13**(2), 152–161 (1985).
7. M. A. Miner, "Cumulative damage in fatigue," *J. Appl. Mech.* **12**(3) [*Trans. ASME* **67**], pp. A159–A164 (1945).
8. A. Z. Palmgren, "Die Lebensdauer von Kugellagern (The lifetime or durability of ball bearings)," *Zeitschrift des Vereines Deutscher Ingenieure (ZVDI)* **68**(14), 339–341 (1924).
9. T. Dirlik, "Application of Computers in Fatigue Analysis," Ph.D. thesis, University of Warwick (1985).
10. S. O. Rice, "Mathematical analysis of random noise," *Bell Sys. Tech. J.* **23**(3), 282–332 (1945).
11. D. E. Steinberg, *Vibration Analysis for Electronic Equipment*, Third Edition, John Wiley & Sons, Inc., New York (2000).

Chapter 12
Brittle Materials

A material is considered to be brittle if it exhibits low strain at failure when subjected to stress. In other words, unlike ductile materials, it has very little plastic capability and hence no specific yield point. Many optics—both refractive lenses and reflective mirrors—consist of brittle materials. All glasses (fused silica, ULE™, ZERODUR®, BK-7, borosilicate, etc.) as well as most engineering ceramics (zinc sulfide, zinc selenide, germanium, silicon, silicon carbide, etc.) fall into the brittle category.

Amorphous brittle materials are quite strong; however, all brittle materials generally exhibit low fracture toughness, also known as the critical stress intensity factor, as Table 11.1 indicates. Thus, they are subject to severe strength degradation in the presence of flaws. Furthermore, should residual stress exist in these flaws (as they generally do), strength is further reduced. Also, almost all glasses and ceramics are subjected to further strength reduction in the presence of moisture, making this issue a "triple whammy." We explore these topics in the subsequent sections of this chapter.

12.1 Theoretical Strength

The theoretical strength of amorphous materials is in excess of 7,000 MPa ($\sim$1,000,000 psi) due to strong covalent bonds. For example, silica glasses (SiO_2) have strong tetrahedral bonds between the silicon and oxygen atoms. Indeed, strengths of 3,500 to 14,000 MPa (500,000 to 2,000,000 psi) have been attained[1] by chemical polishing with hydrofluoric acid solutions and by flame polishing small silica rods. One might consider the only "defect" in such materials to be the interstitial molecular spacing, which is on the subnanometer level. Recent advances in nanotechnology further bear this out. However, manufacturing processes, such as those involving generating, grinding, or lapping, are such that defects of much larger proportion (on the micron level) are introduced, which greatly reduces strength. A. A. Griffith,[2] in noting that failure of glass occurs orders of magnitude below its theoretical atomic strength, was the first to postulate that microscopic cracks exist in every material.

Griffith believed these cracks to be greater than the interatomic distance and hypothesized that they lower the overall strength of the material. He presented experimental results on glass to prove this by introducing defects of various sizes and showing that it was these defects that reduce the strength of the glass. His work is the basis of modern fracture mechanics, which describes the failure of glass, ceramics, and other materials.

12.2 Failure Modes

There are three potential modes of flaw failures, denoted as mode I, mode II, and mode III, as depicted in Fig. 12.1. Mode I is an opening mode, while modes II and III are shear modes, indicating sliding and tearing, respectively. A general state of stress will excite all three modes; however, while there are special cases of failures in modes II and III, these are rare and beyond the scope of this text. Mode I is the most important and dominant, so we concentrate only on this opening-mode failure. Note that mode I is an opening mode, not a closing mode. Flaws will fail at a critical depth under external tension load, not in compression. To be sure, cases of compression failures have been recorded, but, again, these are rare and beyond the scope of this text. In general, the compressive strength of glasses and ceramics far exceeds flaw-reduced tension strength. However, compressive stress can be accompanied by shear stresses, producing tension along the principal plane. Compressive stresses can also be accompanied by tensile stresses beneath the surface, as in Hertzian contact.

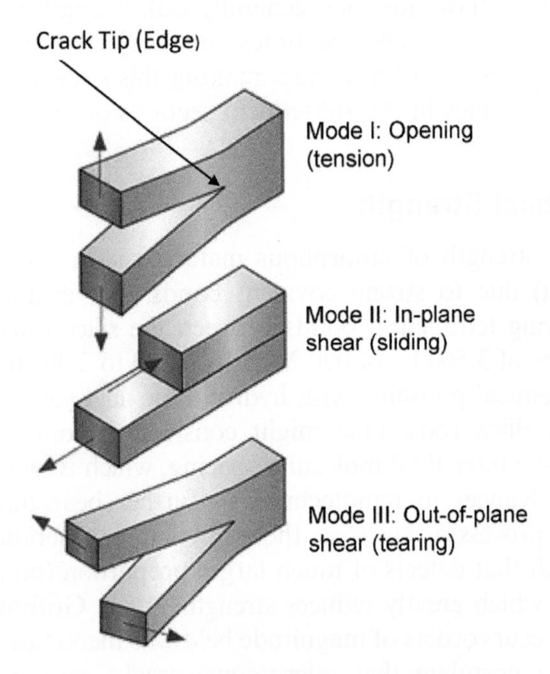

Figure 12.1 Flaw failure modes (reprinted from Ref. 3).

Suffice it to say that cracks grow in tension. Thus, we concentrate on mode I tensile-load-induced failures only.

12.2.1 Mode I failure description

Consider mode I loading as shown in Fig. 12.2. The flaw has a depth $2c$ and a width $2b$, where b is appropriately called the half-width. When the flaw width extends throughout the piece, as shown in the figure, the flaw is referred to as a through crack, or Griffith crack, defined as a long, shallow flaw. Essentially, since $b \gg c$, the ratio b/c is infinite, so a Griffith crack is the most severe type of flaw.

When the flaw width extends only partially through the piece, the flaw is called a partial crack. Figure 2.3 compares the Griffith through crack to a partial crack. The special case where $b/c = 1$, where c is the maximum, describes what is known as a penny crack, as it is shaped like a penny. When such a crack is at a surface, as shown in the Fig. 12.4, it is actually a half-penny flaw. In the case of surface cracks, the flaw depth is c, while the flaw width remains as $2b$. This type of flaw, along with other flaw types with differing b/c ratios, is less severe than the Griffith through crack and is the most common to occur. Note that, while c varies with the penny radius, in the discussions to follow, it is the maximum value that controls flaw growth, with

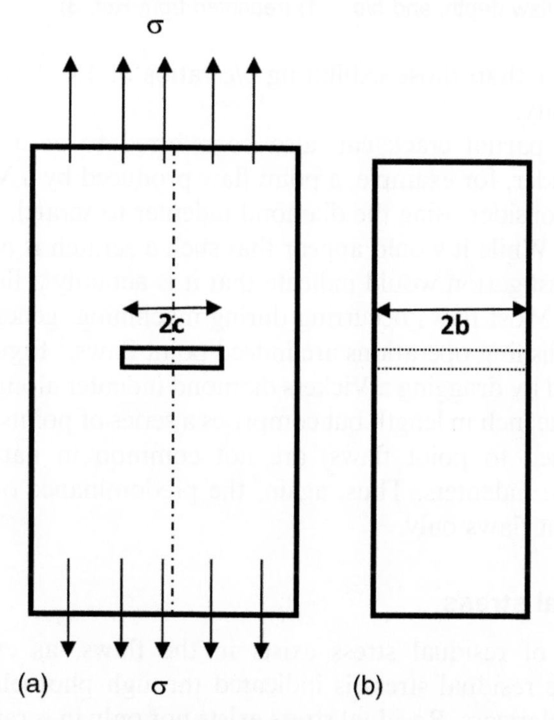

Figure 12.2 Mode I embedded Griffith crack (depth of $2c$) extends through the entire thickness $2b$: (a) top view and (b) side view (reprinted from Ref. 3).

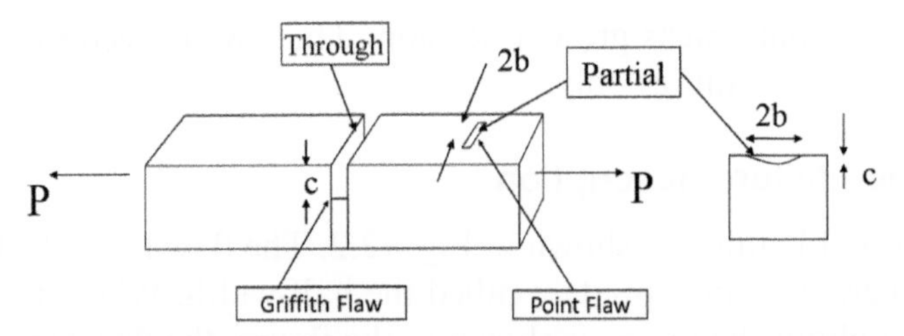

Figure 12.3 Through crack and partial surface crack. Note that while the through crack is shown as a square cornered notch for clarity, it will actually form an atomically small radius at its tip (adapted from Ref. 3).

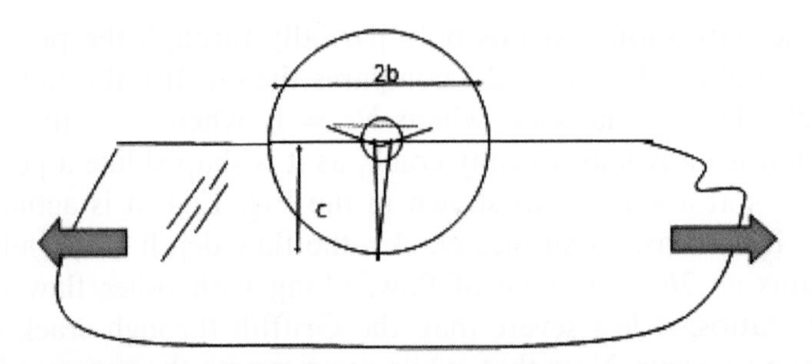

Figure 12.4 Vickers indenter penny crack at a surface. Note the small crack tip at the maximum depth. The stress field is actually normal to the plane of the figure (*b* is the flaw half-width, *c* is the flaw depth, and $b/c = 1$) (reprinted from Ref. 3).

flaw shapes other than those exhibiting b/c ratios of 1.0 developing more or less stress intensity.

Finally, the partial crack can also be referred to as a point flaw (see Fig. 12.3). Consider, for example, a point flaw produced by a Vickers diamond indenter. Now consider using the diamond indenter to scratch a surface over a specified length. While it would appear that such a scratch is now a line crack, microscopic investigation would indicate that it is actually a line made up of a series of points. Most flaws occurring during machining, generating, grinding, figuring, and polishing operations are indeed point flaws.[3] Figure 12.5 shows a scratch produced by dragging a Vickers diamond indenter along a glass surface. The scratch is one inch in length but comprises a series of points, as shown. Line flaws (as opposed to point flaws) are not common in nature but can be produced by line indenters. Thus, again, the predominance of this discussion will involve point flaws only.

12.2.2 Residual stress

A high degree of residual stress exists in the flaws, as evident again in Fig. 12.5, where residual stress is indicated through photoelastic techniques using crossed polarizers. Residual stress exists not only in scratches but also in the manufacturing effects of grinding. It is now common practice to assume that all flaws contain a degree of residual stress, including those produced by

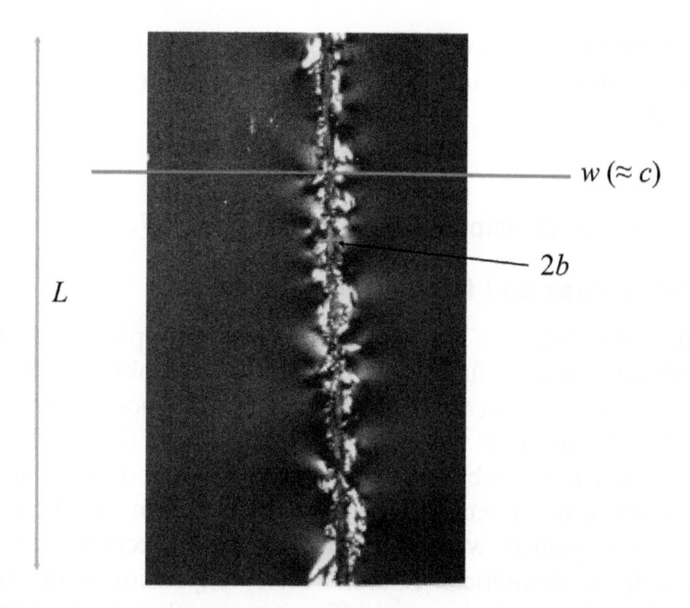

Figure 12.5 Diamond-scratched point flaws (reprinted from Ref. 3).

machining, generating, grinding, figuring, and polishing. This is important to note, as we will see in Section 12.4.

12.3 Strength Theory

Chapter 11 discussed the strength of materials based on a fracture mechanics approach. The formulation therein assumes no residual stress in the flaws that initiate failure. However, for glass and ceramics, which exhibit relatively low fracture toughness, this is a rather poor assumption.

We saw from Eqs. (11.6b) and (11.7) (repeated here) that without residual stress we have

$$K = \sigma\sqrt{\pi c}$$

and

$$K_{\mathrm{Ic}} = S\sqrt{\pi c}.$$

These definitions will require some modification in the presence of residual stress, which occurs in almost all manufacturing processes. We review first the case of no residual stress, then turn to the general case in which residual stress is included.

12.3.1 General strength equation: residual stress free

In the determination of K_{Ic}, a through crack (Griffith crack) is utilized. Recall that the strength of glass and ceramics is a function of flaw depth c and a flaw shape factor, which we here designate as Y. The value of $\sqrt{\pi}$ is the flaw shape factor of a Griffith crack, i.e., where the ratio of flaw width to flaw depth approaches the infinite:

$$Y = \sqrt{\pi}. \tag{12.1}$$

This is the most severe shape of flaws, where $Y = 1.77$.

12.3.2 Finite bodies and free-surface correction

The Griffith formulation of the shape factor above applies to through flaws that are imbedded in an infinite body. Since we are looking at flaws that are small relative to the size of components, this is a very good assumption. Many handbooks[4] and texts show modifications to the shape factor when the body in which the crack is embedded is not infinite, but this is not required here, since most of our interest is in surface flaws, as in Fig. 12.4. For small flaw sizes at the surface, we find a free-surface correction factor that can be obtained from theoretical Laurent series expansion formulation with appropriate boundary techniques.[5] Suffice it to say that due to an increase in strain energy at the boundary, the value of the free-surface correction factor, which is to be multiplied by Y, is given as 1.122. Thus, in general, for a free surface, we find the Griffith shape factor to be

$$Y = 1.122\sqrt{\pi} = 1.98. \tag{12.2}$$

12.3.3 General point flaws

Most flaws are not Griffith flaws; i.e., they are not through flaws but are point flaws. An example is the penny crack, so called because its ratio of half-width to depth is 1:1 (actually, a half-penny at the surface), where $b/c = 1$. We can solve for the value of Y by using advanced fracture mechanics techniques. The flaw shape factor for the penny flaw is given[6] as an elliptical integral of the second kind with free-surface correction:

$$Y = 1.12\sqrt{\pi}/\varnothing, \tag{12.3}$$

where

$$\varnothing = \int_0^{\frac{\pi}{2}} \sqrt{\cos^2\theta + \left(\frac{c^2}{b}\right)\sin^2\theta\, d\theta}, \text{ when } c \le b,$$

and

$$\varnothing = \frac{c}{b}\int_0^{\frac{\pi}{2}} \sqrt{\cos^2\theta + \left(\frac{c^2}{b}\right)\sin^2\theta\, d\theta}, \text{ when } c > b. \tag{12.3a}$$

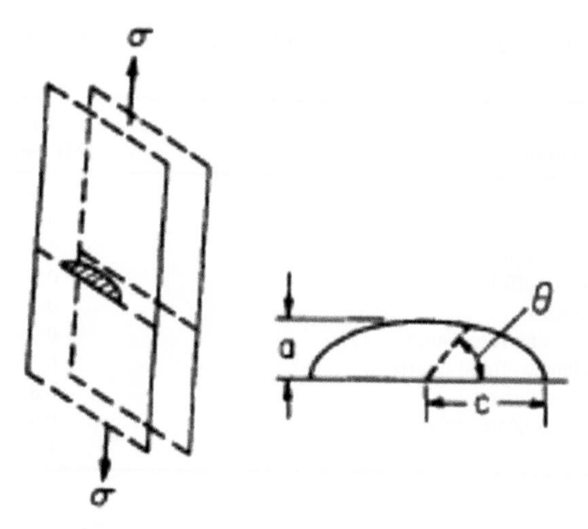

Figure 12.6 The flaw shape factor is a function of the integral over the flaw half-angle.

The integral is carried over the half-angle subtended by the shape of the crack (see Fig. 12.6). The integral is readily solved for Griffith and half-penny flaws. For the Griffith surface flaw, we have seen from Eq. (12.2) that $Y = 1.122\sqrt{\pi} = 1.98$. For the half-penny surface crack,

$$Y = 1.12(1.13) = 1.26. \qquad (12.4)$$

In most instances, the value of Y ranges from 1.0 to 1.98, with the half-penny crack solution being the usual assumption.

12.3.4 The basic fracture mechanics equation

We now rewrite Eq. (11.6b) in a more general form,

$$K = \sigma Y \sqrt{c}, \qquad (12.5)$$

and we rewrite Eq. (11.7) as

$$K_{\mathrm{Ic}} = SY\sqrt{c}. \qquad (12.6)$$

The value of the fracture toughness K_{Ic} is readily available for many glasses and ceramics. A small list of the K_{Ic} values for some common engineering materials is given in Table 12.1. Note that the values lie in a rather small range. By comparison, Table 11.1 compares the K_{Ic} values of glasses and ceramics to those of some common metals. The metals have significantly higher values, indicative of their more ductile nature and tolerance of larger defects. One might note that if the fracture toughness values of most ceramics lie in a similar range, they will have similar strengths for similar size flaws. That is a true observation; however, some ceramics may be more susceptible

Table 12.1 Critical stress intensity factors K_{Ic} for select glasses and ceramics.

Material	K_{Ic} (MPa-m$^{1/2}$)
Schott F2	0.6
Vycor	0.7
Soda lime	0.7
Zinc sulfide	0.7
Silicon	0.7
Corning fused silica	0.75
Hereaus fused quartz	0.75
Borosilicate	0.8
Borosilicate crown	0.9
Schott Zerodur	1.1

to scratches and flaws than others due to hardness, while others may be so tough that grain size limits the strength.

12.3.5 Example for consideration

Consider a fused-silica optical component in which a penny-shaped cracked with a half-width of 50 μm occurs in the surface of the optic. Calculate the strength of this crack.

From Table 12.1 we have $K_{Ic} = 0.75$ MPa-m$^{1/2}$, and we know that $b = 50$ μm. For a penny shape, $b/c = 1$; therefore, $c = 50$ μm. With a surface correction factor of 1.12, we find that $Y = 1.26$. From Eq. (12.6), we compute

$$S = \frac{K_{Ic}}{Y\sqrt{c}} = 84.2\,\text{MPa}\,(12{,}200\,\text{psi}).$$

This, of course, assumes no residual stress.

12.4 Strength with Residual Stress

When a flaw is created by an indentation force, the flaw has an associated residual stress at the crack tip. This is true for almost all processes, whether the flaw is induced artificially, accidentally, or in manufacturing processes. Without any applied stress, a residual stress intensity field exists, similar to that produced by the applied stress. Again, removing the singularity caused by the infinitesimal value of the crack radius, and for unit consistency, we introduce residual stress intensity as

$$K_r = \frac{XP}{\sqrt{c^3}} \tag{12.7}$$

for point flaws, where P is the indentation force N, and X is a dimensionless constant, where $0 \leq X \leq 1$, depending on the degree of residual stress, and

$$K_r = \frac{XP}{\sqrt{c}} \tag{12.8}$$

for line flaws, where the value of P is in newtons per meter (N/m). Note that in both Eqs. (12.7) and (12.8) the units are consistent with those of the applied stress intensity discussed earlier.

12.4.1 Combined residual stress and applied stress

When an external stress is applied, we have

$$K = K_r + K_a, \tag{12.9}$$

where K_a is the redefined applied stress intensity K of Eq. (12.5). Using a value r devised by Fuller,[7] we can rewrite Eq. (12.9) as

$$K = \frac{XP}{\sqrt{c^r}} + \sigma Y \sqrt{c}. \tag{12.10}$$

Here, the value r will be 1 for line flaws, 3 for point flaws, and ∞ for the residual-stress-free case, which recovers the latter-case solution. Again, for dimensional consistency, these are the only acceptable values for r; however, Fuller recognized that r can be treated as a continuous variable for ease of use in application.

12.4.2 Crack stability

The astute reader will note that, if a mechanism can be found to increase flaw depth, the residual stress intensity will *decrease* (stress relief) while the applied stress intensity will *increase*. Note that since residual stress decreases with flaw growth, the flaw, even though it starts its life at fracture, is stable. Thus, we can define some terms: A flaw is considered to be both *stable* and *subcritical* when $K < K_{Ic}$ and/or $\frac{dK}{dc} < 0$. Conversely, a flaw is considered to be both *unstable* and *critical* when $K > K_{Ic}$ and $\frac{dK}{dc} \geq 0$ (the stress intensity must be growing). Both conditions must apply for catastrophic failure to occur.

12.4.3 Strength with residual stress and applied stress

With the stability criterion above noted, we see that at fracture[8] [from Eq. (12.10)],

$$K_{Ic} = \frac{XP}{\sqrt{c^r}} + SY\sqrt{c}. \tag{12.11}$$

Since $\frac{dK}{dc} = 0$ when $K = K_{\mathrm{Ic}}$, differentiating Eq. (12.10) yields

$$\begin{aligned} \frac{dK}{dc} &= 0.5rXPc^{-0.5r-1} + \frac{0.5SY}{\sqrt{c}} = 0, \\ 2c\frac{dK}{dc} &= \frac{-rXP}{c^{0.5r}} + \frac{0.5SY}{\sqrt{c}} = 0. \end{aligned} \tag{12.12}$$

Solving for strength by adding Eqs. (12.11) and (12.12), we have

$$S = \frac{rK_{\mathrm{Ic}}}{(r+1)Y\sqrt{c}}. \tag{12.13}$$

For point flaws ($r = 3$), Eq. (12.13) reduces to

$$S = \frac{0.75K_{\mathrm{Ic}}}{Y\sqrt{c}}, \tag{12.14}$$

which is 0.75 times the residual-stress-free strength of Eq. (12.6).

Without any applied stress, we have

$$K = \frac{XP}{c_{\mathrm{i}}^{r/2}}, \tag{12.15}$$

where c_{i} is the initial crack depth prior to extension under applied stress (the flaw will grow stably under applied stress). From Eqs. (12.11) and (12.13), we can compute

$$\frac{c}{c_{\mathrm{i}}} = (r+1)^{2/r} \tag{12.16}$$

and

$$S = \frac{rK_{\mathrm{Ic}}}{Y\sqrt{c_{\mathrm{i}}}(r+1)^{\frac{(r+1)}{r}}}, \tag{12.17}$$

which, for point flaws ($r = 3$), is

$$S = \frac{0.47K_{\mathrm{Ic}}}{Y\sqrt{c_{\mathrm{i}}}}. \tag{12.18}$$

For the less-common line flaw, where $r = 1$,

$$S = \frac{0.25K_{\mathrm{Ic}}}{Y\sqrt{c_{\mathrm{i}}}}. \tag{12.19}$$

From Eq. (12.13) for the line flaw, this is also equivalent to

$$S = \frac{0.50 K_{\mathrm{Ic}}}{Y\sqrt{c}}, \tag{12.20}$$

as the crack will grow under the applied load.

For the residual-stress-free case, where $r = \infty$, we recover the Griffith solution:

$$S = \frac{K_{\mathrm{Ic}}}{Y\sqrt{c_{\mathrm{i}}}}. \tag{12.21}$$

From Eq. (12.16), the flaw is seen to grow by a factor of 2.52 for the most common point flaws when external stress to failure is applied. The strength is reduced by approximately one-half compared to the initial flaw size.

12.4.4 Example for consideration

Recalling Example 12.3.5, consider a fused-silica optical component in which a penny-shaped crack with a half-width of 50 μm is formed by a diamond indenter to the free surface of an optic. When the optic is externally loaded to failure, calculate its failure strength in the presence of the initially induced residual stress.

From Table 12.1 we have $K_{\mathrm{Ic}} = 0.75$ MPa-m$^{1/2}$, and we know that $b = 50$ μm. For a penny shape, $b/c = 1$; therefore, $c = 50$ μm. With surface correction factor of 1.12, we find that $Y = 1.26$. From Eq. (12.18) we compute $S = \frac{0.47 K_{\mathrm{Ic}}}{Y\sqrt{c_{\mathrm{i}}}} = 39.6\,\mathrm{MPa}\,(5740\,\mathrm{psi})$. We can also use Eq. (12.14) in conjunction with Eq. (12.16) to produce the same result.

12.5 Stress Corrosion

So far, we have learned that flaws reduce the strength of glass and ceramics to below their theoretical strength values, and that the residual stress present in those flaws further reduces these values. Now we shall see that time and moisture are enemies of these brittle materials because they further reduce strength due to stress corrosion.

12.5.1 Definitions

- *Stress corrosion* is a phenomenon in which a chemical reaction causes premature failure with time under constant load in the presence of a chemically active environment. Stress corrosion cracking is subcritical crack growth, which, in the presence of a chemically active environment, is also known as slow crack growth, since growth is not sudden but time dependent. Stress corrosion is also known as static fatigue (since stress is constant with time) to distinguish it from cyclic fatigue, which varies with time.

- *Cyclic fatigue* is a phenomenon in which premature failure occurs with time in the presence of cyclic loading for both fully reversed stress (tension–compression) and any oscillatory stress change.
- *Corrosion fatigue* is a combination of stress corrosion and cyclic fatigue.

The prevailing phenomenon for crystals, ceramics, glass, and brittle materials is *stress corrosion*. Materials that are *brittle* at ambient temperature do not exhibit the same dislocation phenomena common in cyclic fatigue failures. Conversely, the prevailing phenomenon for metals (e.g., steel, aluminum) that are *ductile* at ambient temperature is *cyclic fatigue*. To be sure, metals are subject to corrosion (salts, etc.), and glasses and ceramics can be subject to cyclic fatigue,[9] but in the latter case stress corrosion is dominant. Thus, we concentrate on this type of failure only.

12.5.2 Chemically active environment

The chemically active environment that causes stress corrosion (slow crack growth) in glass and ceramics is *water*, not only in liquid form but in high-humidity air as well. The process involves a chemical reaction between water and the material composition. Fused silica (SiO_2), for example, reacts with water (H_2O) to produce a silanol (SiOH) byproduct[10] in conjunction with electron and proton donations:

$$H_2O + \text{Si-O-Si} \rightarrow 2(\text{Si-OH}).$$

In order to produce this reaction, high activation energy is required; this energy is provided in the form of tensile (crack opening) stress at the crack tip. Thus, for stress corrosion to occur, three requirements on the material are mandated: the material must be (1) susceptible to corrosion, (2) in a corrosive environment, and (3) in the presence of stress.

Note that all three conditions are required for stress corrosion to occur. Figure 12.7 is a schematic of the corrosion process. Of course, different exchanges occur for materials other than fused silica, but the process is the same. While other chemicals such as ammonia and hydrochloric acid[11] are indeed corrosive to glass and cause slow crack growth, water is the most common environment to which glass and ceramics are exposed and has primarily been studied. Typical velocity curves are given in Fig. 12.8 for select glasses as studied by Wiederhorn.[12]

12.5.3 Reaction rates

The rate of reaction with water will increase with temperature increase and, conversely, will decrease as temperature drops below room temperature. Thus, we expect higher crack velocity and lower lifetimes at warm temperatures.

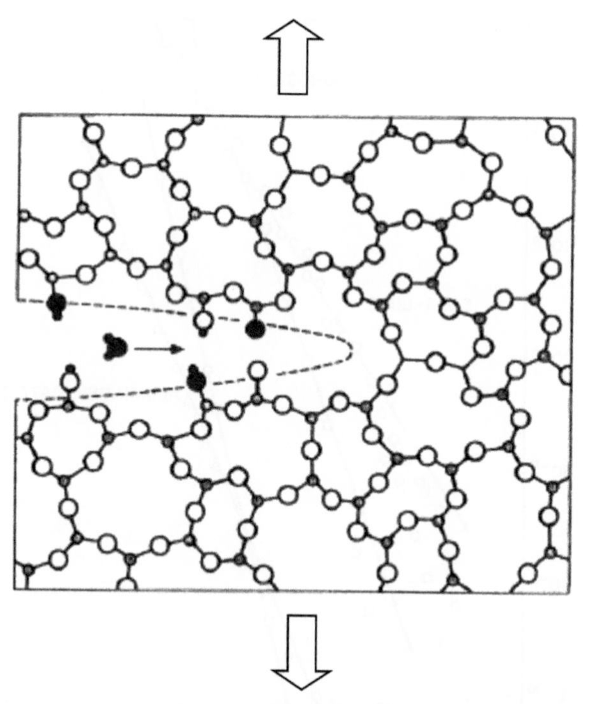

Figure 12.7 Stress corrosion process for fused silica. The shaded circles are silicon, while the open circles are oxygen; filled circles are water. The crack is represented by the dashed curve. Arrows represent the stress field (adapted from Ref. 10 with permission).

While it had been known since the early 1900s that water affects the strength of glass with time, crack growth rates were first quantified by Wiederhorn[13] in the late 1960s. Wiederhorn used direct observation techniques by applying a diamond-induced crack to a glass specimen, immersing the specimen in water, subjecting it to tensile stress, and inspecting crack growth with time using a microscope. He found a good fit to the data by using the exponential law,[13] in which velocity is given by

$$\frac{dc}{dt} = Ve^{nK}, \tag{12.22}$$

where V and n are material constants, and K is the stress intensity.

12.5.4 I love Paris

Alternatively, a power law similar in form to the Paris law[14] for cyclic fatigue (Chapter 11), where the governing variable is now growth/time not growth/cycles, can be used:

$$\frac{dc}{dt} = AK^N, \tag{12.23}$$

where A and N are material-dependent properties, and

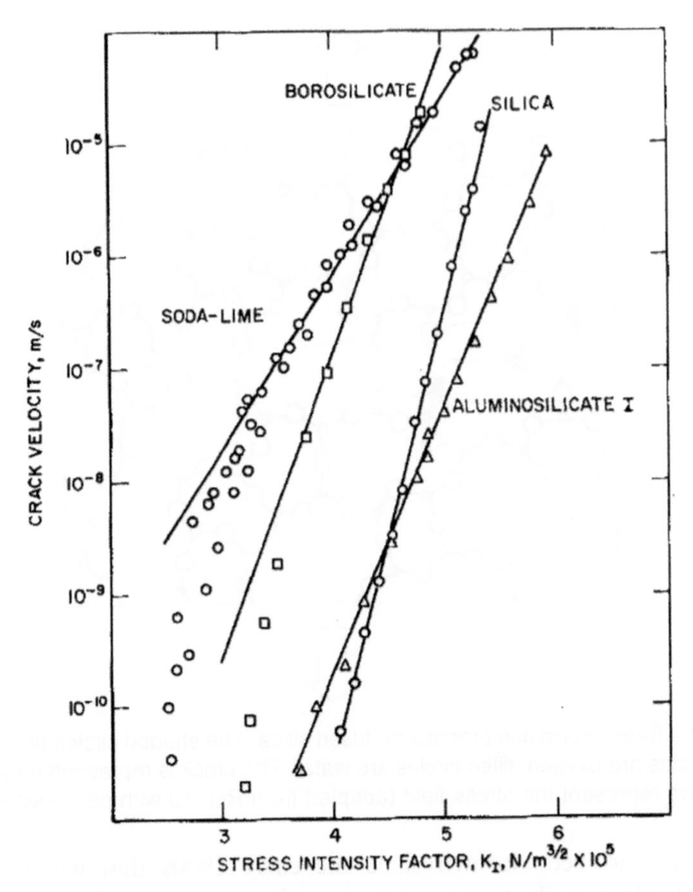

Figure 12.8 Crack velocity versus stress intensity factor (V-K curve) showing stress corrosion in various glasses. In the presence of water, crack velocity increases with stress intensity (adapted from Ref. 12 with permission).

$$A = \frac{v_o}{K_{Ic}^N}, \tag{12.24}$$

where v_o is a material-dependent velocity property. N is referred to by a number of terms; we choose to call it the flaw growth exponent, or flaw growth susceptibility factor.

Note from the relation of Eq. (12.23) that A is the intercept value of velocity when $K = 1$ unit of its dimensional term, as determined from the V-K curve (e.g., Fig. 12.8). Table 12.2 gives a few typical values of the crack velocity constant A. These are not easy to come by, and the A value is quite dependent on material composition. Alternative methods that avoid its use will be presented in Section 12.8.

Studies have shown that both Eqs. (12.22) and (12.23) are well matched to test results; however, the latter power law equation is preferred, since it is

Table 12.2 Crack velocity constant *A* for select glass materials.

Material	A (m/sec) × (Mpa-m$^{1/2}$)$^{-N}$
Fused silica	1.42E+05
Borosilicate crown	~10
Soda lime	~50

an easier expression to deal with mathematically when computing time to failure during integration.

12.5.5 Crack growth regions

Use of the power law equation [Eq. 12.23)] applies to the region where stress corrosion is dominated by the moisture-induced reaction in which velocity increases with stress intensity. To visualize this region of crack velocity, refer to Fig. 12.9, where the life of a crack under stress and moisture is depicted. In region 0, there is no crack velocity and therefore no crack growth until a threshold stress intensity is reached. Once this occurs, the Paris law applies, and velocity increases with stress intensity (region I). Here, the velocity increases even without an increase in stress, as the crack is growing. In region II, the crack velocity slows (idealized here as constant) because it is diffusion limited;

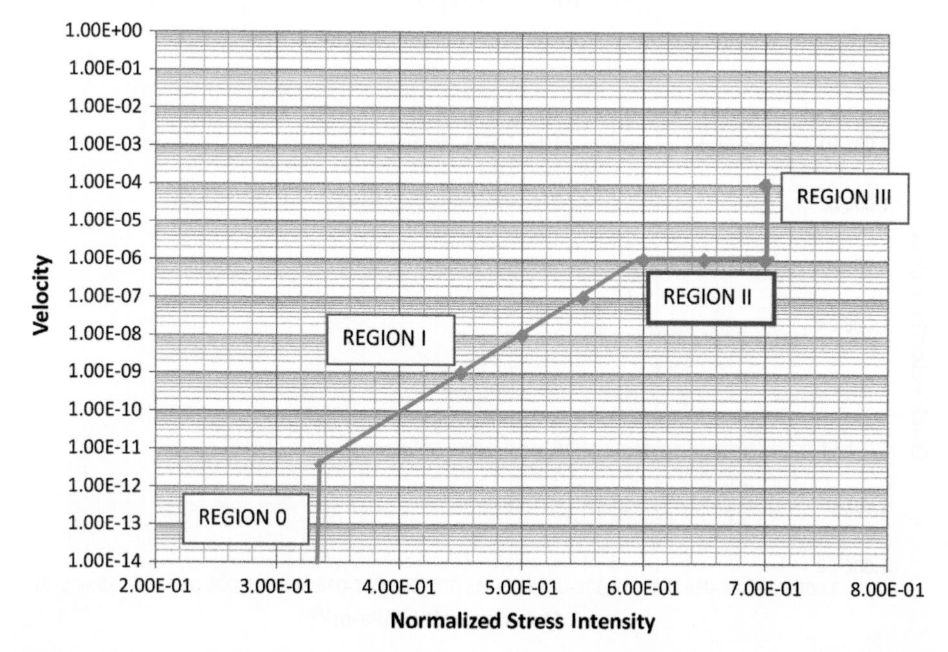

Figure 12.9 Idealized crack velocity versus stress intensity for stress-corrosion-susceptible materials.

that is, the crack velocity cannot keep up with the water diffusion rate. Finally, in region III, the crack grows at a very fast rate independent of water. The growth rate is idealized in Fig. 12.9 as infinite in velocity; in practice, it is still quite high, approaching the speed of sound before catastrophic failure occurs at the critical, unstable stress intensity.

While threshold (region 0) values have been measured for select materials (there is evidence that, unlike fused silica, glasses with alkali oxides such as sodium oxide, lithium oxide, and potassium oxide exhibit a fatigue threshold limit), it is common practice to assume no threshold. Indeed, crack velocity has been measured for silica glasses at low stress intensities in the vicinity of less than 1 pm/sec. A more typical plot of the stress intensity regions (for which region I is typically made linear) is shown schematically in Fig. 12.10.

12.5.6 Region I relation

Note that in the power law relation, time to failure is proportional to the Nth power, where again N is material dependent. Upon first looking at Eq. (12.23), it would appear that the higher the N the higher the velocity and therefore the lower the lifetime. However, when Eq. (12.24) is substituted into Eq. (12.23), we have

$$\frac{dc}{dt} = v_o \left(\frac{K}{K_{\text{Ic}}} \right)^N . \tag{12.25}$$

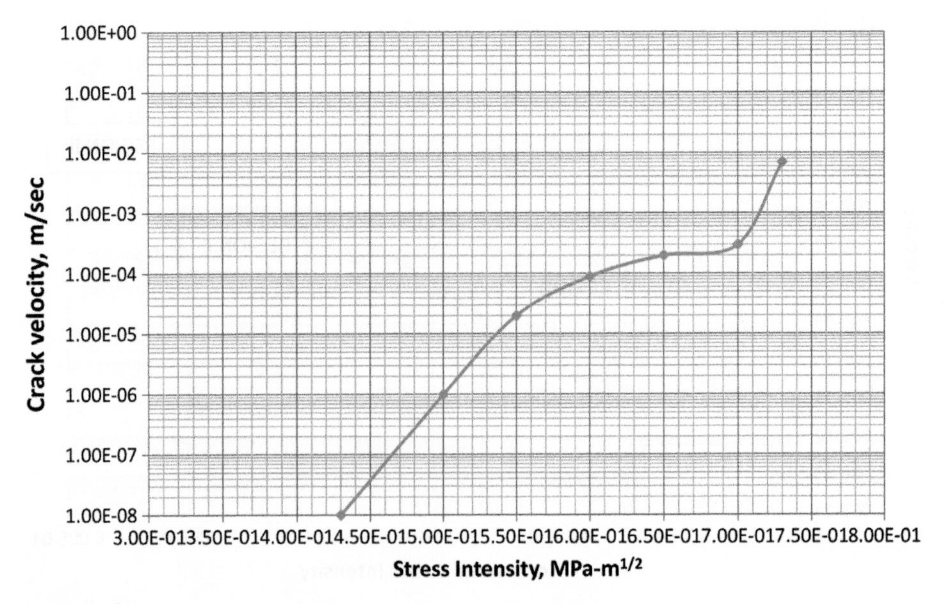

Figure 12.10 Typical crack velocity versus stress intensity for fused-silica crack growth in the presence of water.

Since the fraction on the right side of Eq. (12.5) is always less than unity, we see that the higher the value of N the longer the lifetime.

12.5.7 Example for consideration

Consider a flaw in a fused-silica optical component that is subjected to moisture and stress. The stress level is 30 MPa, the initial flaw size is 100 μm, and the flaw is penny shaped. Given a value of A from Table 12.2 and a value of N of 36, compute the initial crack velocity and the crack velocity when the flaw has grown by factor of 2 under constant stress.

From Eq. (12.5), we initially have

$$K = \sigma Y \sqrt{c},$$

$$K = (30)(1.26)(0.0001)^{1/2} = 0.378\ \text{MPa-m}^{1/2}.$$

From Eq. (12.23), we find the initial velocity to be

$$\frac{dc}{dt} = AK^N = (1.4 \times 10^5)(0.378)^{36} = 8.75 \times 10^{-11}\ \text{m/sec}.$$

When the crack has doubled in size, we find that

$$K = (30)(1.26)(0.0002)^{1/2} = 0.535\ \text{MPa-m}^{1/2},$$

$$V = (1.4 \times 10^5)(0.535)^{36} = 2.36 \times 10^{-5}\text{m/sec}.$$

Thus, the crack velocity has increased by five orders of magnitude.

12.6 Stress Corrosion Free of Residual Stress

We now turn to the derivation of lifetime assuming region I behavior. We first determine lifetime in the residual-stress-free case. While almost all flaws have residual stress (Chapter 16), we need to understand the basic formulation before undergoing the more torturous calculations demanded in the residual stress case.

We can use the calculus technique of separation of variables to solve for the time to failure of Eq. (12.25):

$$V = \frac{dc}{dt} = \frac{dc}{dK}\frac{dK}{dT}. \tag{12.26}$$

From Eq. (12.5),

$$c = \frac{K^2}{Y^2\sigma^2}. \tag{12.27}$$

Differentiating Eq. (12.27),

$$\frac{dc}{dK} = \frac{2K}{Y^2\sigma^2}, \tag{12.28}$$

and substituting into Eq. (12.26),

$$V = \frac{2K}{Y^2\sigma^2}\frac{dK}{dt} \tag{12.29}$$

so that

$$dt = \frac{2K}{Y^2\sigma^2 V}dK. \tag{12.30}$$

Integrating over the limits from initial to final (fail) stress intensity yields

$$t = \frac{2}{Y^2\sigma^2}\int \frac{K}{V}dK. \tag{12.31}$$

Substituting Eq. (12.23) for V in Eq. (12.31) after integration gives

$$t = \frac{2(K^{2-N} - K_{Ic}{}^{2-N})}{(N-2)AY^2\sigma^2}, \tag{12.32}$$

which is the time to failure.

While this calculation may seem formidable, it is really just a "plug and chug" operation to determine the failure time, provided one knows the following parameters: the intrinsic properties of the material (K_{Ic},N); the stress intensity, and therefore the flaw depth c and applied stress σ; the flaw shape factor Y; and the material's crack velocity constant A.

Equation (12.32) can be simplified somewhat by noting that, unless K approaches K_{Ic}, the K_{Ic} power term is very small compared to the K power term and can be neglected. Further noting that when Eqs. (12.5) and (12.6) are substituted into Eq. (12.32), we arrive at

$$t = B\frac{\sigma_i{}^{N-2}}{\sigma^N}, \tag{12.33}$$

where

$$B = \frac{2K_{Ic}{}^{2-N}}{(N-2)\,Y^2 A}. \tag{12.34}$$

Although we need to know the inert strength of the material, we do not need to know the flaw depth, which does not enter the equation, since flaw depth is already included in the strength determination. Remember that for Eq. (12.33) to be valid, the flaw distribution, shape, and depth must be representative of the process used in determining the inert strength.

In any case, while one may be able to beg, borrow, or steal a value of N from the literature for a given material and determine inert strength, finding a value of A or B is more problematic, as is indicated in Table 12.2. This rather meager list is slightly expanded in Table 12.3 to include the B value, which, as indicated in Eq. (12.34), is a function of the flaw growth factor, the flaw shape, and the critical stress intensity. (An expanded table of N values is presented in Section 12.7.2.) For uncharacterized materials, a method to find the values of B, N, and inert strength using a dynamic fatigue experiment is discussed in Section 12.8.

12.6.1 Examples for consideration

Example 1. Consider a fused-silica optical component that has been ground and polished, and exhibits an inert strength of 60 MPa (8700 psi). Assuming no residual stress, compute its life expectancy under an applied stress of 15 MPa (2175 psi).

Using the flaw growth exponent N and alternative crack constant B from Table 12.3 along with Eq. (12.33), we find that

$$t = B\frac{\sigma_i^{N-2}}{\sigma^N} = (0.005)(60)^{34}/15^{36} = 6.56 \times 10^{15} \text{ sec.}$$

That equates to more than two-hundred million years—much too long a time to worry about! In the next example, we review the failure time in the presence of a more severe flaw.

Example 2. Consider a fused-silica optical component that has been scratched with a 150-μm diamond. Tests show an inert strength of 35 MPa (5000 psi). Assuming no residual stress, compute the optics' life expectancy under an applied stress of 15 MPa (2175 psi).

Using the alternative crack constant B from Table 12.3 and Eq. (12.33), we find that

Table 12.3 Select material constants for penny-shaped cracks.

Material	A (m/sec) × (Mpa-m$^{1/2}$)$^{-N}$	K_{Ic} (MPa-m$^{1/2}$)	N	B (MPa^{2}sec)
Fused silica	1.42E+05	0.75	36	0.005
Borosilicate crown	~10	0.9	20	0.047
Soda lime	~50	0.73	21	0.524

$$t = B\frac{\sigma_i^{N-2}}{\sigma^N} = (0.005)(35)^{34}/15^{36} = 7.2 \times 10^7 \text{ sec}.$$

That equates to 2.3 years. We shall see in the next section that residual stress will greatly reduce this value.

12.7 Stress Corrosion with Residual Stress

Starting with the same formulation as presented earlier in the chapter for crack velocity [i.e., Eq. (12.23)], repeated here as

$$V = AK^N, \tag{12.35}$$

we can now include residual stress in which the value of K is given by both the residual and applied stress intensities from Eq. (12.10), repeated here as

$$K = \frac{XP}{c^{0.5r}} + \sigma Y\sqrt{c}. \tag{12.36}$$

12.7.1 A complex integration

Substituting into Eq. (12.35) and performing the integration to compute time to failure is now anything but straightforward. If you thought the residual-stress-free integral of Eq. (12.31) was formidable, it pales in comparison to the residual stress integral. In fact, for many years, numerical integration was the best way to solve it until ingenious work by Fuller et al. in 1983[7] led to a solution. Namely, they found that

$$t = \frac{2\int_0^1 K_e^{N'-1}(1-K_e)^{N-N'-1}dK_e}{(r+1)A(Y\sigma)^{N'}[K_{Ic}(c)^{r/2}]^{N-N'}}, \tag{12.37}$$

where the integral is the β function,

$$K_e = \frac{K_a}{K_a + K_r}, \tag{12.38}$$

and

$$N' = \frac{rN+2}{r+1}. \tag{12.39}$$

The mathematical β function is related to its cousin, the Γ function. Fortunately, Eq. (12.37) can be manipulated similarly to the manipulation of the equation without residual stress to yield the familiar form of Eq. (12.33):

$$t = \frac{B' \sigma_i^{N'-2}}{\sigma^{N'}}, \tag{12.40}$$

where

$$B' = \left[\frac{(N-2)4^{N-3}}{3^{N'-2}} \frac{\Gamma(N')\Gamma(N-N')}{\Gamma(N)} \right] B, \tag{12.41}$$

and $\Gamma(N)$ is the Γ function of N. In case the Γ function sounds formidable, for integers of N and N', it can be given as

$$\Gamma(N) = (N-1)! \text{ (factorial)}. \tag{12.42}$$

12.7.2 Computation of constants and resulting time to failure

We can now compute B'/B for various values of N and N'. A list of typical values of N for select materials is given in Table 12.4, along with the value of N' for the point flaws (again, where $r = 3$). Figure 12.11 plots the values of the ratio of B'/B to N. (When N and N' are not both integers, one has to take care in computing the factorial expression, but the figure properly accounts for this.) Note that the value of B' is not far removed from B for most common values; i.e., it is within an order of magnitude when computing failure times. However, from Eq. (12.40), the time to failure is reduced in proportion to the stress to the power of N', resulting in several orders of magnitude lower failure times compared to the residual-stress-free case. This is rather astounding!

Table 12.4 Approximate flaw growth exponents for select materials in the presence of water (HIP is hot isostatic pressing).

	N	N'
Magnesium fluoride	10	8
Fluorohafnate	11	8.8
Schott BK-7 glass	20	15.5
Soda-lime glass	21	16.3
Corning ULE 7971	27	20.8
Borosilicate	29	22.3
Schott ZERODUR®	31	23.8
Corning fused silica 7940	35	26.7
Corning fused silica 7957	36	27.5
Heraeus Infrasil® 302	36	27.5
Zinc selenide	40	30.5
HIP zinc sulfide (Cleartran™)	46	35
Polycrystal alumina	47	35.8
Calcium fluoride	50	38
Single-crystal alumina	67	50.8
Zinc sulfide	76	57.5
Silicon	>100	>100

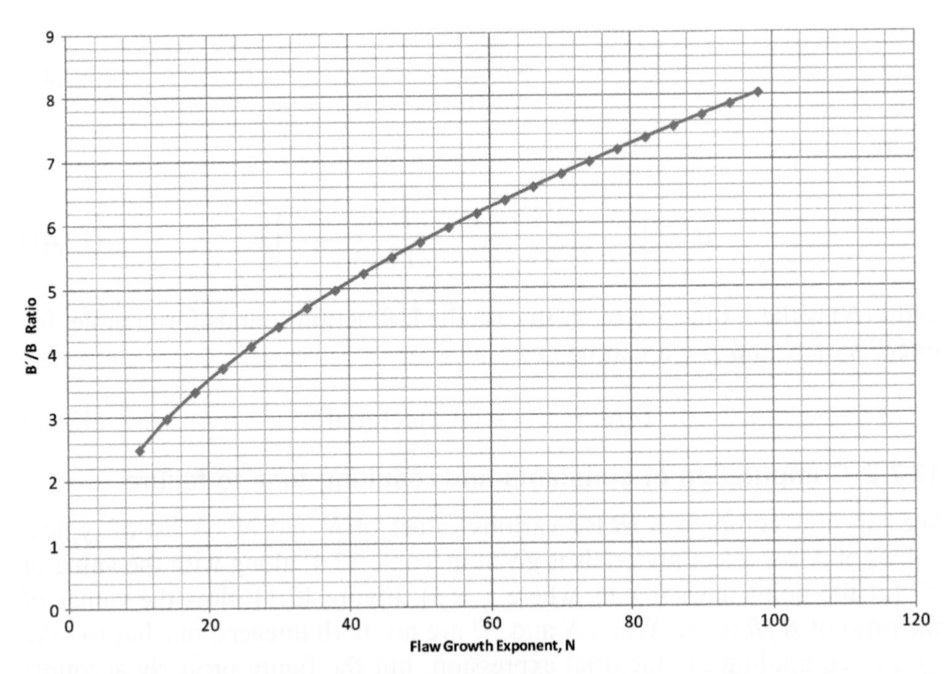

Figure 12.11 Residual/free-stress alternative crack growth constant versus flaw growth exponent.

12.7.3 Examples for consideration

Example 1. Consider a fused-silica optical component that has been ground and polished, and exhibits an inert strength of 60 MPa (8700 psi). With residual stress included, compute its life expectancy under an applied stress of 15 MPa (2175 psi). Compare the results to the residual-stress-free results of Example 2 in Section 12.6.1.

Using the flaw growth exponent N and alternative crack constant B from Table 12.3, we find from Eq. (12.39) for point flaws that $N' = \frac{3N+2}{4} = 28$, and from Eq. (12.41), or Fig. 12.11, that $B' = 4.8B = 4.8(0.005) = 0.024\,\text{MPa-sec}^2$.

Then, from Eq. (12.40), we compute time to failure as

$$t = \frac{B'\sigma_{\mathrm{i}}^{N'-2}}{\sigma^{N'}} = (0.024)(60)^{26}/15^{28} = 4.8 \times 10^{11}\,\text{sec}.$$

That equates to over 15,000 years. When compared to Example 2 in Section 12.6.1 with no residual stress, where lifetime was 200 million years, we see that the residual stress has a lifetime that is decreased by four orders of magnitude! However, this failure time is still too long to be of concern. In the next example, we review the failure time in the presence of a more severe flaw.

Example 2. Consider a fused-silica optical component that has been scratched with a 150-μm diamond. Tests show an inert strength of 35 MPa (5000 psi). With residual stress included, compute the component's life expectancy under an applied stress of 15 MPa (2175 psi). Compare the results to the residual-stress-free results of Example 2 in Section 12.6.1.

Here we find [from Eq. (12.40)] that

$$t = \frac{B'\sigma_i^{N'-2}}{\sigma^{N'}} = (0.024)(35)^{26}/15^{28} = 3.9 \times 10^5 \text{ sec}.$$

That equates to about 110 hours, or 4.5 days. When compared to Example 2 in Section 12.6.1 with no residual stress, where lifetime was 2.3 years, we see that residual stress has a lifetime that is decreased by two orders of magnitude. Now this failure time is certainly of concern and points out the detrimental effect of residual stress.

For any applied stress, it is useful to plot this comparison, as given in Fig. 12.12. Note the large difference in failure times at all stress levels.

12.7.4 Obtaining constants and failure time

To compute time to failure, only the values of B', N', inert strength, and applied stress need to be known, as long as the inert strength represents the flaw distribution in question. Certainly, the applied stress is readily calculated, and strength testing can bear out the inert strength of a material. You may be

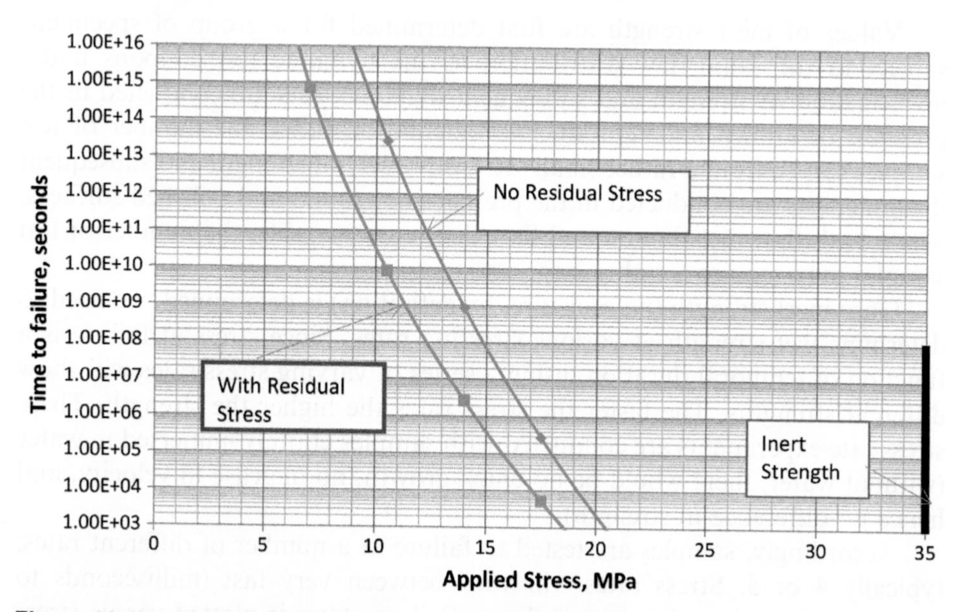

Figure 12.12 Effect of residual stress on failure time: comparison of residual-stress failure time to no-residual-stress failure time in the presence of moisture and applied stress for 150-μm scratched fused silica (inert strength is 35 MPa).

able to beg, borrow, or steal a value of N from Table 12.3 or from the literature, and hence easily calculate N' from Eq. (12.39). Getting a value of B', however, may be more difficult. (How many tables of B' values have you ever seen?) You can, however, obtain B' values from a table of A or B values (as in Tables 12.1 and 12.3, or as otherwise provided). B' is computed by converting A or B to B' by Eq. (12.41). Short of that, you may need to obtain this value by test, as we will see in the next section.

12.8 Dynamic Fatigue

It is apparent that obtaining values of N' and B' (or A and their counterparts N, B, and A) is not easily accomplished. The usual experiments for determining A and N, for example, would involve microscopic inspection of crack growth with time and water, and could prove rather costly; and these values would then need conversion to their residual stress values.

A better method that computes the values of B' and N' directly utilizes a technique called dynamic fatigue. Dynamic fatigue is based on the integration of Eq. (12.25) under a constant stress rate. Since we are concerned about failure under constant stress, not constant stress rate, a slight modification[10] to the determination of B is required, in which time to failure under constant stress rate is increased by $(N+1)$; i.e.,

$$t_{\text{dynamic}} = (N+1)t_{\text{static}}. \tag{12.43}$$

Values of inert strength are first determined for a group of specimens subject to the same flaw distribution as the dynamic test coupons under investigation. If strength and subsequent dynamic tests are conducted in the presence of known and reproducible indentation flaws, the number of test samples can be from a rather small set. Conversely, if strength and subsequent dynamic tests are conducted in the presence of ground and polished surfaces, where both flaw distribution and flaw depth are variable, the number of test samples may require a rather large set.

Once inert strength (no moisture, by definition) is determined, we need to determine wet strength at various stressing rates. Since stress at failure is a function of temporal duration, using a series of varying stress rates will show different strengths. The faster the stress rates the higher the strength. These stress rate experiments are conducted with samples entirely immersed in water (without water, there would be no crack growth, no increase in velocity, and hence no difference in strength).

Accordingly, samples are tested to failure at a number of different rates, typically 4 or 5. Stress rates will vary between very fast (milliseconds to failure) to very slow (hours to failure). Failure stress is plotted versus stress rate on a log-log chart from which the constants N and B can be solved. Again, the preference is to produce consistent cracks made by diamond indenters to minimize samples, on the order of three or four samples at each stress rate. If polished specimens are used, samples could number in the tens at each rate due to the statistics of flaw depth variability.

12.8.1 Example for consideration

An example of such a dynamic fatigue diagram[15] is shown in Fig. 12.13. The value of N' is determined as

$$N' = -\frac{\log\left(\frac{t_i}{t_2}\right)}{\log\left(\frac{\sigma_i}{\sigma_2}\right)}, \tag{12.44}$$

or

$$m = \frac{1}{N' + 1}, \tag{12.45}$$

where σ_2 is the reference strength at time to failure t_2, σ_i is the strength at time to failure t_i, and m is the slope of the log-fail-stress/log-stress-rate diagram.

The value of B' is determined as

$$B' = \frac{\sigma_f^{N'+1}}{\sigma_{dot}(N' + 1)\sigma_i^{N'-2}}, \tag{12.46}$$

where σ_f is the moist failure stress at constant stress rate σ_{dot}, and σ_{dot} is the stress rate (stress/second).

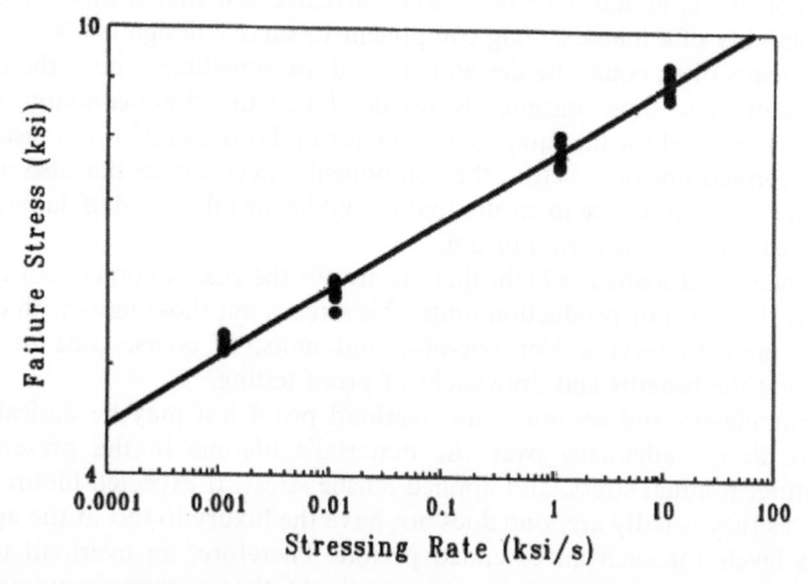

Figure 12.13 Dynamic fatigue data for BK-7 glass using Vickers-indented specimens. The value of N' determined from this curve is 15.39 (adapted from Ref. 15).

With the dynamic fatigue test, we see that we have all of the constants we need to determine time to failure in the presence of residual stress. These constants are readily substituted into Eq. (12.40). Constants determined in this way require no knowledge of the stress intensity factor, flaw depth, or flaw shape factor, since we have computed inert strength, which is the controlling parameter. This is rather nice!

12.9 An Approximation Technique

Lacking a dynamic fatigue test, most analysts will not have access to the value of B for any material, let alone a new one. However, the analyst may have access to the value of N, and hence N', from the literature. Of course, inert strength data may be readily attainable by test or vendor supplier information. To this end, an approximation technique developed by Pepi[16] may be used. It is found that

$$t = \left(\frac{\sigma_a}{\sigma_i}\right)^{-(N'-2)}(0.0001). \tag{12.47}$$

This approximation is based on a minimal set of data, so caution in its use is advised. However, if we apply a safety factor of 1.2 to the applied stress, we can achieve a good first-order, albeit conservative, approximation.

12.10 Overload Proof Test

A proof test is intended to be a nondestructive test that demonstrates the acceptability of a load-carrying component to survive design loads. As such, proof loads often equal the design load and are sometimes above the design load in order to show margin. Overloads of 1.25 to 1.5 are common; where safety is required for lift equipment, a factor of 2.0 is used.[17] Proof tests may be performed not only before the component enters service but also after a period of time in service to ensure that no hidden or other kind of damage has occurred during that period of use.

Since proof loads are high, there is always the risk of component failure during the test. For production units, this screens out those units with design issues prior to service. For one-of-a-kind units, of course, one needs to ascertain the benefits and drawbacks of proof testing.

For glasses and ceramics, an overload proof test may be desirable to ensure design adequacy over the material's lifetime in the presence of moisture, residual stress, and applied tensile stress. If expected lifetimes are long, as they usually are, one does not have the luxury to test at the applied stress levels for such an extended period. Therefore, an overload test of short duration can be applied, ensuring the lifetime over the required long duration.

In this case, we cannot choose an arbitrary overload factor of 1.25, 1.5, or any other number, as we can for static loading. Stress corrosion of glass and ceramics is anything but static, with gradual (slow) crack growth occurring over the component life. However, we already have the tools to compute the required overload factor.

12.10.1 Application to ceramics

Mathematical determination of the required proof factor for glasses and ceramics was first proposed by Wiederhorn.[18] In this light, we have seen the equation of lifetime in the presence of moisture in Eq. (12.40):

$$t = \frac{B'\sigma_{\mathrm{i}}^{N'-2}}{\sigma^{N'}}.$$

Here, of course, t is the expected lifetime. For required lifetime t_{r}, we simply substitute into Eq. (12.40) and determine, rather than inert strength, the required proof strength σ_{p}, to yield

$$t_{\mathrm{r}} = \frac{B'\sigma_{\mathrm{p}}^{N'-2}}{\sigma^{N'}}, \tag{12.48}$$

or, defining the overload proof factor (PF), as

$$\mathrm{PF} = \frac{\sigma_{\mathrm{p}}}{\sigma_{\mathrm{a}}}. \tag{12.49}$$

We find that

$$t_{\mathrm{r}} = \frac{B'(\mathrm{PF})^{N'-2}}{\sigma_{\mathrm{a}}^{2}},$$

or

$$\mathrm{PF} = \left(\frac{\sigma_{\mathrm{a}}^{2}t_{\mathrm{r}}}{B'}\right)^{1/(N'-2)}. \tag{12.50}$$

Thus, if we know the required lifetime and the applied stress, we can readily calculate the overload proof factor required to ensure that the lifetime is met. As in the determination of inert strength, this proof strength is statically applied.

12.10.2 Examples for consideration

Example 1. Using the fused-silica data for the alternative crack velocity constant B and flaw growth exponent N from Table 12.3, compute the required overload proof factor to ensure a lifetime of 3 years under a continuously applied stress of 6.9 MPa (1000 psi).

Using Fig. 12.11, we find that $B' = 0.018$, and from Eq. (12.39), $N' = (rN+2)/(r+1) = 28$. From Eq. (12.50), we find (converting time to seconds using the units provided; 3 years = 9.5×10^7 sec) that PF = 2.75 and σ_p = 19 MPa (2750 psi).

Figure 12.14 is a plot of the desired lifetime versus overload proof factor as a function of the applied stress level of 6.9 MPa (1000 psi) for the fused-silica optic in question. Note that since the proof factor is a strong function of the flaw growth exponent N', even higher values are required for glasses and ceramics with lower values of the flaw growth exponent.

Example 2. During dynamic fatigue testing, a soda-lime glass is found to exhibit values of $B' = 1.3$ MPa-sec^2 and $N' = 15$. Compute the required overload proof factor to ensure a lifetime of 3 years under a continuously applied stress of 6.9 MPa (1000 psi).

From Eq. (12.50) we find that PF = 5.42 and σ_p = 37.4 MPa (5420 psi). Such proof factors make one shudder, but that is the requirement to ensure lifetime adequacy.

As may be expected, there are drawbacks to proof testing. In the first place, in order to ensure that no subcritical crack growth occurs prior to failure, the test must be conducted in an inert environment, in the same fashion as inert strength tests are conducted. At such high stress levels, any

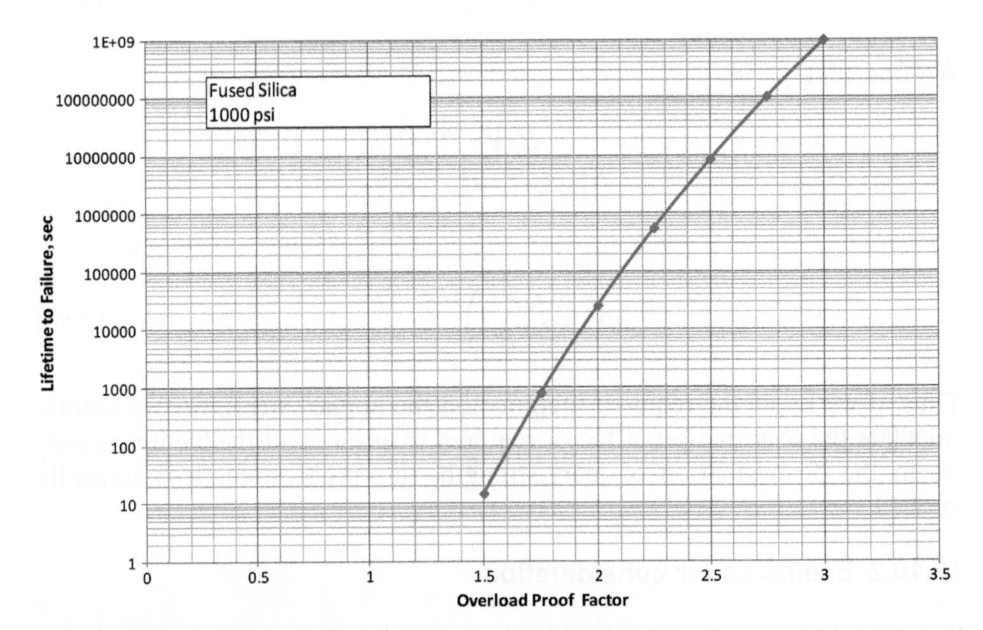

Figure 12.14 Guaranteed lifetime versus overload proof factor for a constant stress of 6.9 MPa (1000 psi) in fused-silica glass.

moisture will rapidly degrade strength and render the test useless, doing more harm than good. Secondly, such inert strengths (1) must be conducted at rapid load rates, (2) must not be held at load, and (3) must be rapidly unloaded since no environment is completely inert.

Furthermore, the component must be mounted the same way as it is mounted during operation; i.e., the boundary conditions must be the same to ensure that stresses are developed properly throughout the part and, in particular, at the edges. Additionally, any subsequent damage after proof test will negate the test, requiring another proof test if subsequent damage is suspected.

Thus, for multiple units, such testing may be effective in truncating the strength distribution and even allowing for less heavy components by pushing the design envelopes. For one-of-a-kind types, the benefits of the test may be negated by the drawbacks if the material is subject to relatively low stress levels. For example, consider the following case of the same soda-lime glass used in Example 2 but with a lower applied stress level.

Example 3. During dynamic fatigue testing, a soda-lime glass is found to exhibit values of $B' = 1.3$ MPa-sec^2 and $N' = 15$. Under a stress of 4.83 MPa (700 psi), compute the overload test requirement for a severely scratched component under tension for a lifetime of 30 years. Inert strength tests of the scratched component show a Weibull A value of 34.5 MPa (5000 psi) and a polished Weibull A strength of 82.7 MPa (12,000 psi).

From Eq. (12.50) we find that, where $t = 9.5 \times 10^8$ sec, PF $= 6.12$ and $\sigma_P = 29.65$ MPa (4300 psi). Note that the proof stress is still below the safe allowable stress level, even if the glass is severely scratched. With stress levels so low, a proof test does not seem warranted, as there may be more risk involved than in the analysis; after all, 30 years in water is a long time.

References

1. S. M. Wiederhorn, "Environmental stress corrosion cracking of glass," National Bureau of Standards Report 10865, Washington, D.C., pp. 2–3 (1971).
2. A. A. Griffith, "The phenomenon of rupture and flow in solids," *Philosophical Transactions of the Royal Society A* **221**, 163–198 (1921).
3. J. W. Pepi, *Strength Properties of Glass and Ceramics*, SPIE Press, Bellingham, Washington (2014) [doi: 10.1117/3.1002530].
4. H. Tada, P. C. Paris, and G. R. Irwin, *The Stress Analysis of Cracks Handbook*, Third Edition, ASME Press, New York (2000).
5. B. Gross and J. E. Srawley, "Stress-intensity factors for a single edge notch tension specimen by boundary collocation of a stress function," NASA TN D-2395 (1964).

6. P. C. Paris and G. C. Sih, "Stress Analysis of Cracks," ASTM STP381, pp. 51–52, West Conshohocken, Pennsylvania (1965).
7. E. R. Fuller, B. R. Lawn, and R. F. Cook, "Theory of fatigue for brittle flaws originating from residual stress concentrations," *J. American Ceramic Society* **66**(5), 314–321 (1983).
8. E. R. Fuller, private notes, October, 2008.
9. S. Bhowmick, J. J. Melendez-Martinez, and B. R. Lawn, "Contact fatigue of silicon," *J. Materials Research* **23**(4), 1175–1184 (2008).
10. B. R. Lawn, *Fracture of Brittle Solids*, Second Edition, *Cambridge Solid State Science Series*, Cambridge University Press, Cambridge, p. 172 (1993).
11. S. M. Wiederhorn and H. Johnson, "Influence of sodium hydrogen ion exchange on crack propagation in soda-lime silicate glass," *J. American Ceramic Society* **56**(2), 108–109 (1973).
12. S. M. Wiederhorn and L. H. Bolz, "Stress corrosion and static fatigue of glass," *J. American Ceramic Society* **53**(10), 543–548 (1970).
13. S. M. Wiederhorn, "Influence of water vapor on crack propagation in soda lime glass," *J. American Ceramic Society* **50**(8), 407–414 (1967).
14. S. W. Freiman, "Stress-Corrosion Cracking of Glasses and Ceramics," Chapter 14 in *Stress-Corrosion Cracking: Materials Performance and Evaluation*, R. H. Jones, Ed., ASM International, Materials Park, Ohio, pp. 337–344 (1992).
15. E. R. Fuller, Jr., S. W. Freiman, J. B. Quinn, G. D. Quinn, and W. C. Carter, "Fracture mechanics approach to the design of glass aircraft windows: A case study," *Proc. SPIE* **2286**, 419–430 (1994) [doi: 10.1117/12.187363].
16. J. W. Pepi, "A method to determine strength of glass, crystals, and ceramics under sustained stress as a function of time and moisture," *Proc. SPIE* **5868**, 58680R (2005) [doi: 10.1117/12.612013].
17. Occupational Health and Safety Administration (OSHA), Fall Protection Code 1910.66, Appendix C (1974).
18. A. Evans and S. Wiederhorn, "Proof testing of ceramic materials—an analytical basis for failure prediction," *International J. Fracture* **10**(3), 379–392 (1974).

Chapter 13
Performance Analysis of Optical Structures

All of the optics described in the previous chapters require a support system. Although it is not the intent of this text to describe mechanical design details (Yoder[1] provides an excellent and comprehensive source on this topic), here we briefly touch on the opto-structural analyses of such structures.

13.1 Supporting Optics

Lenses are made of glass, ceramic, or another material that is transparent to the electromagnetic spectrum of light, dependent on the wavelength. As such, when one or both faces of a lens are curved, the light is refracted and transmitted through the lens depth and is used for magnification, focus, or high-resolution imaging. Often, multiple lenses are required. These are supported in a lens cell. The purpose of a lens cell is to support and maintain the position of the lens or lenses housed within it in its operational and non-operational environments. To maintain centration and to avoid despace, lenses are often preloaded within the lens cell. If this is the case, care must be taken to preclude high contact, or Hertzian stress, to brittle optics (contact stress is discussed in Chapter 16). Under a thermal environment, lenses can be attached by flexure edge mounts (Chapter 3) or supported by soft adhesives, such as silicones, as is described more fully in Chapter 9. The lens cell itself must be strong and stiff, and standard structural design principles apply.

Reflective optics are supported by an edge structure, commonly called a bezel, or by a back structure, commonly referred to as a bulkhead. Again, such structures must be strong and stiff, and standard structural design principles apply.

Structures that separate and support one or more optics are called metering structures. We briefly examine some of these structures as they pertain to opto-structural analysis.

13.2 Metering Despace

Metering of reflective optics is of extreme importance for optical systems since minute displacements between one optic and another can wreak havoc with an optical performance error budget. Despace (defocus) of such optics can be caused by temperature soak and gradient conditions, and is dominated by the difference in CTE between the structure and the optic. Or, in the case of a hygroscopic structure in the presence of a moisture change, despace occurs due to the structure's coefficient of moisture expansion (CME), discussed further in Section 13.4.

In either case, despace between two optics under temperature change is computed simply as

$$y = \Delta\alpha\Delta TL, \tag{13.1}$$

where $\Delta\alpha$ is the difference in effective CTE over the range of temperature ΔT, and L is the distance between optics.

When a linearly varying axial gradient is also present, its contribution to despace is

$$y = \frac{\alpha L[\Delta(\Delta T)]}{2}, \tag{13.2}$$

where $\Delta(\Delta T)$ is the difference in temperature of the structure at each optic, and α is the structure CTE.

Under moisture change, we have, similarly,

$$y = C_M \Delta ML, \tag{13.3}$$

where C_M is the structure CME in units per percent moisture change, and ΔM is the moisture change in percent. Despace motions are then converted to WFE based on optical sensitivity analysis for the particular system.

13.2.1 Example for consideration

A primary fused-silica optic is metered to a secondary fused-silica optic over a distance of 20 in. by a hygroscopic metering structure. After alignment at room temperature (293 K) and humidity (50% relative), compute the WFE under a thermal soak to 193 K in a vacuum environment. The optical motion sensitivity analysis shows that a 0.001-in. despace yields an error of 0.50 visible waves RMS. The following properties are given:

$$\alpha_{\text{eff optic}} = 0.32\ \text{ppm}/^\circ\text{C},$$

$$\alpha_{\text{eff structure}} = 0.2\ \text{ppm}/^\circ\text{C},$$

$$C_M = 100\ \text{ppm}/\Delta M, \text{ and}$$

$$\Delta M = 0.2\%\ \ (50\%\ \text{humidity to vacuum}).$$

From Eq. (13.1), we have $y = \Delta\alpha\Delta TL = (0.12)(10^{-6})(100)(20) = 0.00024$ in. (contraction), and from Eq. (13.3), we have $y = C_M\Delta ML = (100)(10^{-6})(0.2)(20) = 0.0004$ in. (contraction). The total contraction is thus 0.00064 in., or 0.32 waves RMS without focus capability.

13.3 Decentration and Tip

Decentering and tip of one optic metered to another can occur when a diametrical gradient exists across the structure. Using the relationships of Chapter 4 [Eq. (4.28)], for a cantilevered system we find that

$$y = \frac{\alpha\Delta TL^2}{2D}, \tag{13.4}$$

and

$$\theta = \frac{\alpha\Delta TL}{D}. \tag{13.5}$$

Here, D is the system aperture, and ΔT is the diametrical gradient, while y is the relative decenter and the relative rotation of the optics. Decenter and tip space motions are then converted to WFE based on optical sensitivity analysis for the particular system.

13.3.1 Example for consideration

A primary fused-silica optic of 15-in. diameter is metered to a secondary fused-silica optic over a temperature range. Compute the WFE if the object is subjected to an operational diametrical gradient of 10 °C. The optical motion sensitivity analysis shows that a 0.001-in. decenter yields an error of 0.20 visible waves RMS, while a 0.001-rad tip yields an error of 0.20 visible waves RMS.

From Eq. (13.4), we have

$$y = \frac{\alpha\Delta TL^2}{2D} = \frac{(0.2)(10^{-6})\,(10)\,(20)^2}{2(15)} = 0.000027\,\text{in.} = 0.005\,\text{waves RMS},$$

which is the contribution from the decenter. From Eq. (13.5), we have

$$\theta = \frac{\alpha\Delta TL}{D} = \frac{(0.2)(10^{-6})(10)(20)}{15} = 00000027\,\text{rad} = 0.0005\,\text{waves RMS},$$

which is the contribution from the tip.

13.3.2 Gravity and frequency

In addition to the thermal effects described above, gravitational sag of a metering structure needs to be kept low; this is particularly important if fundamental frequency requirements demand a stiff structure. Furthermore, in a variable-gravity environment, deflection (in both despace and decentration) must be kept low. For example, when aligning an orbital space system on the ground, error will occur in zero gravity. Similarly, for ground-based systems such as for astronomical telescopes, gravity error will change relative to the system orientation. Once gravitational sag is calculated, optical sensitivities will determine degradation in wavefront quality. From such sag, the simple techniques of Chapter 10 can be used to get a quick, first-order glance at fundamental frequency. Various structure forms are next discussed.

13.4 Structure Forms

Structures designed to meter optics can take on several forms: truss, shell, or frame. The design choice depends largely on packaging constraints. A three-point truss design, or its cousin, the six-point hexapod structure, can be expediently stiff and strong due to minimization of bending. On the other hand, a shell structure, which has a high diameter-to-length ratio, is also quite stiff and strong since most of its deformation and load is in shear, i.e., produces ovalization. Shell structures can also serve as optical baffles.

For off-axis designs, when a symmetrical shell or truss design is precluded, often a frame structure is required. Frames are less efficient and may be bending-mode dominated but can be made to be stiff and strong enough with appropriate analyses. Since near-kinematic mount schemes are important, frames must be closed out to provide proper support. For example, consider the 2D frame structures shown in Fig. 13.1, illustrated here for comparative purposes. Values of deflection and stress under side load are normalized. Case (a) depicts a simply supported frame. Its normalized deflection and stress are unity. Case (b) depicts a fixed-support frame in which normalized displacement and stress values are 24% and 57%, respectively, of the simply supported case, but the fixed-support frame is not kinematic in nature. Case (c) on the other hand, depicts a simply supported frame but adds a member connecting the simply supported mounts with rigid attachment; this frame still replicates a near-kinematic arrangement, as desired. In this case, normalized displacement and stress values are one-third and one-half, respectively, of the case (a) values, resulting in a significant increase in stiffness and strength. The case (c) arrangement is referred to as a Vierendeel[2] truss, named after its developer. It is not a true truss, as it is dominated by bending forces, but its advantages over the other frame types are obvious.

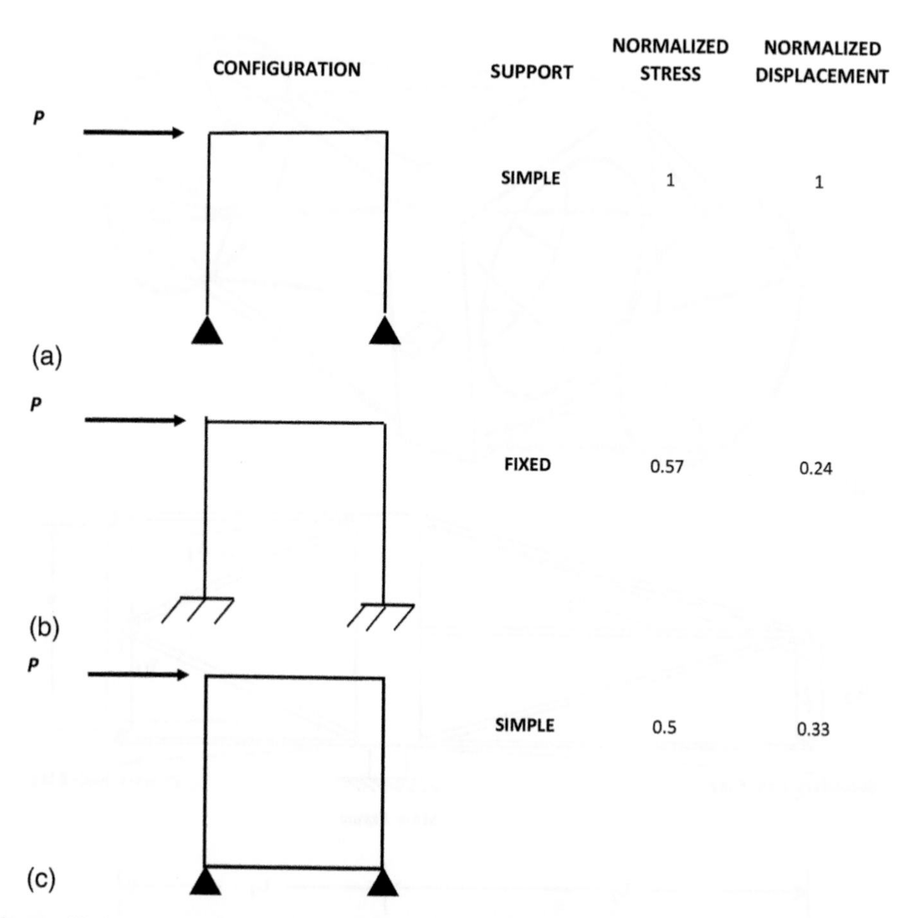

CONFIGURATION
SUPPORT
NORMALIZED STRESS
NORMALIZED DISPLACEMENT
P
SIMPLE
1
1
(a)
P
FIXED
0.57
0.24
(b)
P
SIMPLE
0.5
0.33
(c)

Figure 13.1 (a) Simply supported and (b) fixed-support rigid frame structures showing the benefits of (c) a simply supported frame with an added member, which maintains kinematic support while minimizing stress and deflection.

13.5 Metering Truss Design

A metering truss consists of struts and rings that carry load dominantly in tension and compression. With bending minimized, the truss design is quite rigid. This is important for dynamic stability, frequency requirements, and gravity sag.

13.5.1 Serrurier truss

Many astronomical telescopes,[3] which operate in earth's gravity field, make use of a Serrurier[4] metering truss design named after its inventor. A Serrurier truss is basically a truss that is self-centering under gravity load, as schematically depicted by Fig. 13.2(a). The primary and secondary optics are balanced about a central pivoting structure such that the gravity sag of both are equal and remain

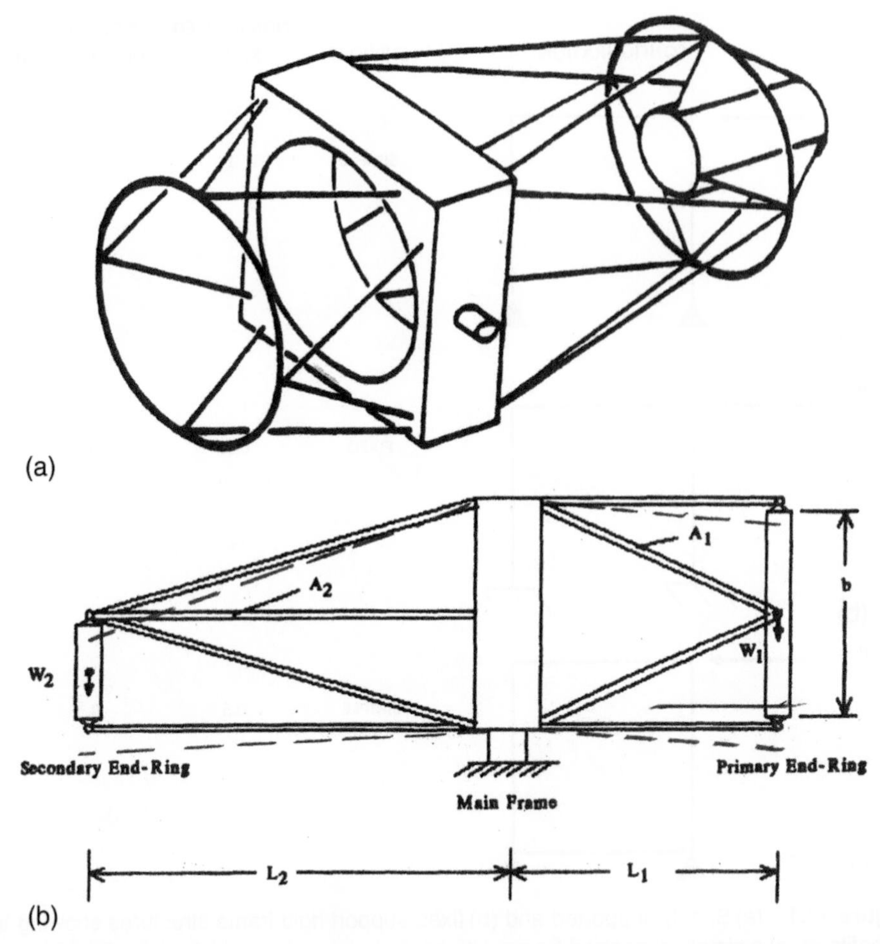

Figure 13.2 Serrurier truss: (a) Typical balanced design showing optical elements and main bulkhead. (b) Schematic of the deflected shape of a Serrurier truss under gravity loading. Note that primary and secondary mirrors have neither relative decenter nor relative rotation to one another (adapted from Ref. 4).

parallel. This can be simply demonstrated by the equations of beam deflection. For example, we see in Table 1.1 that a beam supported simply at it ends under its own weight will deflect its maximum at the center where its rotation is zero, while there is no displacement at its end points where rotation is maximum. One can envision that if the support points are moved inboard toward the center, the center displacement is reduced and the end displacements increased, along with a corresponding decrease in rotation. Thus, at some distance of supports, end rotations can be made zero and the displacements becomes identical, yielding parallel motion. If unequal weights are added at each end of the beam, the support overhang distances will not be symmetrical, but pivot points may be found to yield identical end displacement and zero rotation.

In this fashion, the optical axis remains in line, without tip, tilt, or decenter, as seen in Fig. 13.2(b). In this case the beam is actually a truss structure, so tensile and compressive forces balance the end displacements and minimize end rotations. Since the end weights of the supported optics are likely to be quite different, the design is nonsymmetrical. The "pivot" point is generally a square frame for optimization of the truss motions, which remain parallel and equal regardless of the gravity orientation. These designs are not optimized for the stiffness-to-weight ratio and can result in relatively low frequencies. Thus, such a design, although not uncommon for large, ground-based telescopes, is less common for telescopes designed for space use. In this latter case the optical system is generally tested in a particular orientation to minimize the zero-gravity orbital release condition.

13.5.2 Thermal expansion

In addition to stiffness and strength requirements, metering trusses must maintain a proper distance between optics during temperature change. If the truss expansion characteristics match the reflective optics expansion characteristics, the design is self-metering. This athermalized design works because of the self-compensating effect of the radius-of-curvature change for the optics. Thus, aluminum optics use aluminum metering structures, beryllium optics use beryllium metering structures, etc. These structures are not without drawbacks, however. For example, an all-beryllium telescope may be costly, and, while it has a superior stiffness-to-weight ratio, its thermal figure of merit (Chapter 8) is not exceedingly attractive—given its relatively high CTE despite its high thermal conductivity—and is subject to gradient error. An all-aluminum telescope may be less costly but has relatively poor mechanical and thermal figures of merit, with its very high CTE subjecting it to gradient error. An all–silicon carbide telescope is a happy medium between aluminum and beryllium telescopes, being "between" in terms of cost and mechanical figure of merit but superior in thermal figure of merit due to its low CTE. It does have size limitations, however, in terms of metering structure fabrication. Glass (and glass ceramic) optics desirably require glass metering structures—but because of strength and stiffness requirements, these do not fare very well (a unique use of this is discussed in Section 13.5). Alternatively, structures closely (but not exactly) matched in CTE to low-expansion glass optics, such as Invar, are sometimes used if mass budgets can support the choice; graphite (carbon fiber)- reinforced composites (Section 13.5.4), on the other hand, are attractive in this regard, although moisture expansion is sometimes of greater concern. Finally, an all-composite telescope is attractive in terms of mechanical and thermal figures of merit, but fabrication of precision optics in this regard is still in its fledgling stage.

Table 13.1 gives a qualitative summary of the benefits and drawbacks of several optic and metering structure combinations. Table 13.2 quantifies the properties of these materials. When these properties are used in conjunction with optical sensitivities and appropriate figures of merit (Chapter 8), mass budgets, loading requirements, optical tolerance criteria, cost, and schedule, an appropriate choice can be made.

Table 13.1 Optics and metering structure combinations with associated drawbacks.

Optic	Metering Structure	Comment
Ule	Graphite composite	Moisture backout
Zerodur	Graphite composite	Moisture backout
Fused silica	Graphite composite	Moisture backout
Fused silica	Invar	Heavy; temperature limited
Fused silica	Fused silica	Risky
Aluminum	Aluminum	Inexpensive; control gradients
Beryllium	Beryllium	Expensive; ultralite
Silicon	Silicon carbide	Size limited
Silicon carbide	Silicon carbide	Structure joints; structure size
Silicon carbide	Graphite composite	Focus may be required; lightest design (ex. all Beryllium)
Silicon carbide	Invar	Focus may be required; heavy design

Table 13.2 Quantified physical properties of optics and metering candidates: (a) thermal expansion, (b) modulus, (c) density, (d) thermal conductivity, and (e) strength.

(a)

Optic	Cte (ppm/K)	Metering Structure	Cte (ppm/K)
Ule	0.03	Graphite composite	−0.2
Zerodur	0.03	Graphite composite	−0.2
Fused silica	0.52	Invar	1.3
Aluminum	22.5	Aluminum	22.5
Beryllium	11	Beryllium	11
Silicon	2.6	Silicon carbide	2.4
Silicon carbide	2.4	Silicon carbide	2.4
Silicon carbide	2.4	Graphite composite	−0.2

(b)

Optic	Modulus (Msi)	Metering Structure	Modulus (Msi)
ULE	9.8	Graphite composite	15
Zerodur	13.1	Graphite composite	15
Fused silica	10.6	Invar	20.5
Aluminum	10	Aluminum	10
Beryllium	44	Beryllium	44
Silicon	20	Silicon carbide	44.5
Silicon carbide	44.5	Silicon carbide	44.5
Silicon carbide	44.5	Graphite composite	15

(continued)

Table 13.2 Continued

(c)

Optic	Density (lbs/in^3)	Metering Structure	Density (lbs/in^3)
Ule	0.08	Graphite composite	0.06
Zerodur	0.091	Graphite composite	
Fused silica	0.08	Invar	0.3
Aluminum	0.1	Aluminum	0.1
Beryllium	0.07	Beryllium	0.07
Silicon	0.08	Silicon carbide	0.105
Silicon carbide	0.105	Silicon carbide	0.105
Silicon carbide	0.105	Graphite composite	0.06

(d)

Optic	Conductivity (W/m-K)	Metering Structure	Conductivity (W/m-K)
Ule	1.31	Graphite composite	32
Zerodur	1.64	Graphite composite	32
Fused silica	1.38	Invar	10
Aluminum	150	Aluminum	150
Beryllium	200	Beryllium	200
Silicon	125	Silicon carbide	150
Silicon carbide	150	Silicon carbide	150
Silicon carbide	150	Graphite composite	32

(e)

Optic	Strength (psi)	Metering Structure	Strength (psi)
Ule	1500	Graphite composite	40,000
Zerodur	1500	Graphite composite	40,000
Fused silica	1500	Invar	71,000
Aluminum	42,000	Aluminum	45,000
Beryllium	35,000	Beryllium	35,000
Silicon	6500	Silicon carbide	12,000
Silicon carbide	12,000	Silicon carbide	12,000
Silicon carbide	12,000	Graphite composite	40,000

13.5.3 Athermalized truss: a design before its time

In order to properly meter low-CTE-glass optics, an idea was developed years ago to do so with a metallic athermalized metering truss design for use with Cassegrain systems. The truss consists of a series of rings, struts, and bays, as shown schematically in Fig. 13.3. The truss members comprising the ring are made of one material, while the strut members joining the ring members are

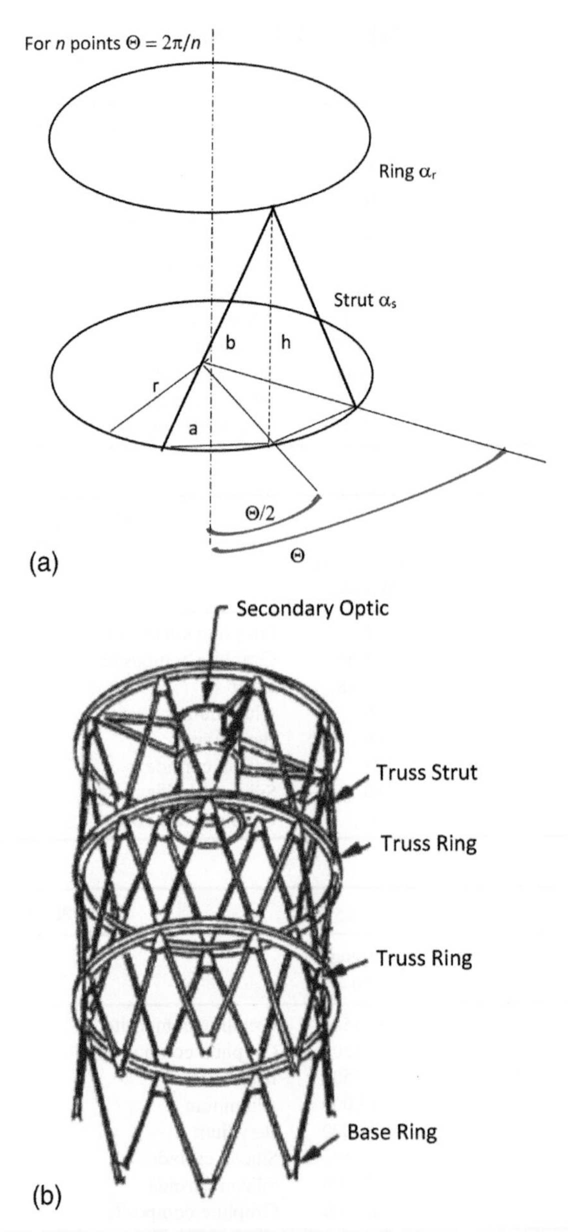

Figure 13.3 Athermalized truss bay concept. (a) Ring and strut CTE can be chosen to result in expansion characteristics comparable to the optics. (b) Example of a four-bay athermalized truss concept.

made of another. The ring members expand under temperature change, causing contraction of the structure, while the strut members expand, causing elongation of the structure. With the right balance of strut expansion α_s and ring expansion α_r, the truss becomes self-metering, as explained next.

From Fig. 13.3(a), let a' denote a new member length due to a temperature soak ΔT. It follows that

$$
\begin{aligned}
r' &= r(1 + \alpha_r \Delta T), \\
a' &= 2r' \sin\left(\frac{\pi}{2n}\right) = a(1 + \alpha_r \Delta T), \\
b' &= b(1 + \alpha_s \Delta T), \\
h' &= h + \Delta h, \text{ and} \\
(h + \Delta h)^2 &= b^2(1 + \alpha_s \Delta T)^2 - a^2(1 + \alpha_r \Delta T)^2.
\end{aligned}
$$

Expanding, and eliminating higher-order terms in Δh and ΔT,

$$\Delta h = \frac{(b^2 \alpha_s - a^2 \alpha_r)\Delta T}{h},$$

or

$$\Delta h = \frac{\left[b^2\left(\frac{\alpha_s}{\alpha_r}\right) - a^2\right]\alpha_r \Delta T}{h}. \tag{13.6}$$

For zero-expansion glasses, we set $\Delta h = 0$; therefore,

$$\frac{\alpha_s}{\alpha_r} = \frac{a^2}{b^2}.$$

We now define N as the number of bays and the overall truss length as $\therefore h = \frac{L}{N}$. Also,

$$b^2 = h^2 + a^2 = \frac{\alpha_r}{\alpha_s} a^2,$$

or

$$h = a\sqrt{\frac{\alpha_r}{\alpha_s} - 1} = \frac{L}{N}, \tag{13.7}$$

where

$$
\begin{aligned}
a &= 2r \sin\frac{\pi}{2n}, \\
\frac{L}{r} &= 2N\sqrt{\frac{\alpha_r}{\alpha_s} - 1}\,\sin\frac{\pi}{2n}.
\end{aligned} \tag{13.8}
$$

We can now plot various values of L/r as a function of number of bays N and ring/strut intersection points n, computing the required strut-to-ring CTE to give net zero structure expansion. Figure 13.4 gives an example of a

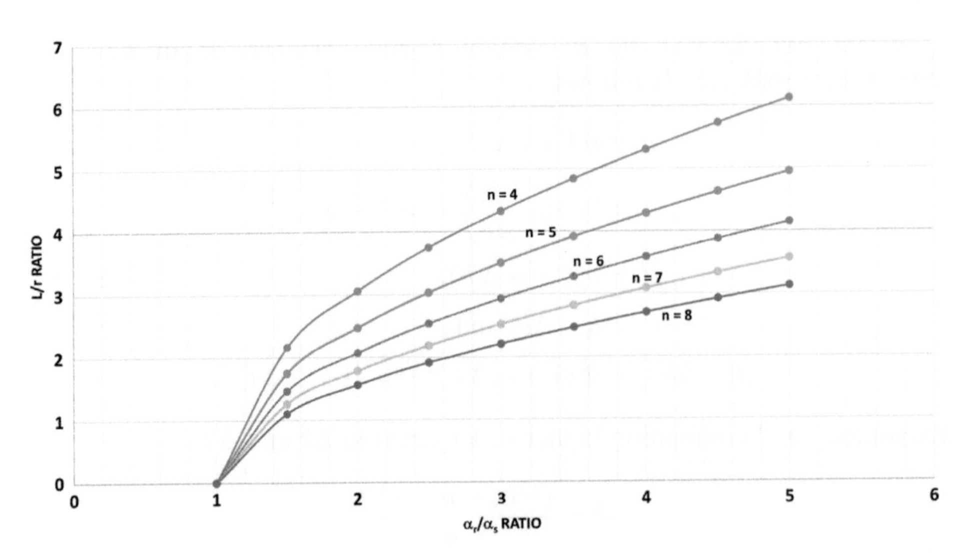

Figure 13.4 Four-bay athermalized truss design plot. For a given overall truss-length-to-truss-radius ratio, the number of ring intersection points n can be chosen for a given ring-CTE-to-strut-CTE ratio.

four-bay athermalized truss design plot. Notice that a wide range of values is possible provided that materials with proper CTE ratios are found. For example, consider a truss having an L/r ratio of 2.5 with aluminum rings and titanium struts. Near room temperature, we have $\alpha_r = 2.25 \times 10^{-5}$ ppm/K and $\alpha_s = 9.0 \times 10^{-6}$ ppm/K. Therefore, $\frac{\alpha_r}{\alpha_s} = 2.5$. From Fig. 13.4, we have an athermalized design for $n = 6$.

Other L/r ratios will yield similar results when both n and N are varied. Note that for a Cassegrain system, the L/r ratio is a function of the system F-number $f/\#$. The relation is $r_s = 4r(f/\#)$, where r_s is the mirror radius of curvature. The effective focal length (EFL) of the system is EFL $= r_s/2 = 2r(f/\#) = L$, where L is the distance from the primary optic to the secondary optic. Hence,

$$L/r = 2(f/\#). \tag{13.9}$$

For very fast systems (e.g., $f/\# = 0.5$), $L/r = 1$, while for fast systems (e.g., $f/\# = 1.0$), $L/r = 2$, whereas for slower systems (e.g., $f/\# = 3.0$), $L/r = 6$.

The theory of athermalized truss design relies on pin truss joints; that is, each member is a two-force member carrying tension or compression only. In reality, secondary moments exist if joints are bolted or welded together. This is common for bridge designs, for example, where such moments are inconsequential; however, when millionths-of-an-inch precision is required, these moments can become meaningful. When these secondary moments are accounted for, along with manufacturing length tolerances, CTE uncertainty, etc., obtaining truly zero expansion is not possible. However, analyses show that ± 0.2 ppm/K uncertainty is feasible if one can live with such a value, depending on system sensitivity and temperatures soak range.

With the advent of low-expansion composites, for which CTEs can be tailored to better than 0.2 ppm/K, an athermalized design is not generally required. The Hubble telescope, which uses low-expansion glass and a composite truss, makes use of a tailored-CTE, athermalized design. As noted, however, often the moisture-expansion characteristics of composites can negate the benefit of low-CTE composites, so the metalized athermal concept is useful but not often practiced.

Of course, many choices of reflective mirror materials are available to the opto-structural engineer. Silicon carbide, for example, with its high stiffness and relatively low thermal expansion coupled with its high thermal conductivity, is an ideal candidate in many instances. Here, for self-metering, a silicon carbide structure is demanded. This has been readily accomplished; however, for large optical systems, such structures may be more than manufacturers wish to take on. Alternatively, an athermalized metal truss may be considered. In this case, the value of Δh [(Eq. (13.6)] is not zero but rather

$$N\Delta h = \alpha_{\mathrm{m}}\Delta TL,$$
$$\Delta h = \frac{\alpha_{\mathrm{m}}\Delta TL}{N}, \tag{13.10}$$

where α_{m} is the mirror CTE.

We now have *three* variables of expansion coefficients, so the ratio $\frac{\alpha_{\mathrm{r}}}{\alpha_{\mathrm{s}}}$ is no longer a constant. We can, however, choose a value of the mirror CTE and the ring member CTE, and solve for the required strut CTE. Substituting Eq. (13.10) into Eq. (13.8), after some manipulation, gives

$$\alpha_{\mathrm{s}} = \frac{\left[\alpha_{\mathrm{m}}\left(\frac{L}{r}\right)^2 + 4N^2[G(n)]^2\alpha_{\mathrm{r}}\right]}{\left[\left(\frac{L}{r}\right)^2 + 4N^2[G(n)]^2\right]}, \tag{13.11}$$

where $G(n) = 2\sin\left(\frac{\pi}{2n}\right)$.

13.5.3.1 Example for consideration

Consider a silicon carbide optic with a temperature change over a small range near room temperature and $\alpha_{\mathrm{m}} = 2.43$ ppm/K, with aluminum ring members $\alpha_{\mathrm{r}} = 22.5$ ppm/K. For an L/r ratio of 3, we vary the bay number N and strut points n to see whether titanium struts are feasible. Titanium's CTE is 9.0 ppm/K.

With reference to Table 13.3, we see that for $N = 3$ and $n = 6$, the required strut CTE is 9.84 ppm/K, close to titanium, while for $N = 4$ and

Table 13.3 Strut CTE requirements for net zero expansion of a silicon carbide optic using aluminum bay rings. Values computed from athermal truss design equations.

Bays (N)	Points (n)	Length/Radius (L/r)	Ring CTE (α_r)	Strut CTE (α_s)
3	4	3	22.5	9.84
3	6	3	22.5	6.67
3	8	3	22.5	5.08
2	4	3	22.5	6.58
2	6	3	22.5	4.57
2	8	3	22.5	3.70
4	4	3	22.5	12.67
4	6	3	22.5	8.90
4	8	3	22.5	6.70

$n = 6$, the required strut CTE is 8.9 ppm/K, even closer. How close is close? We can solve for the effective CTE of the truss from Eq. (13.11) as

$$\alpha_{\mathrm{eff}} = \frac{N^2}{\left(\frac{L}{r}\right)^2}\left\{\left[\left(\frac{L}{rN}\right)^2\frac{\alpha_s}{\alpha_r} + 4[G(n)]^2\frac{\alpha_s}{\alpha_r} - 4[G(n)]^2\right]\alpha_r\right\}. \tag{13.12}$$

We can now substitute the titanium CTE value (9.0 ppm/K) into Eq. (13.12) to compute the effective CTE, as we do in Table 13.4. It is seen that for $N = 3$ and $n = 6$, we have an effective CTE of 1.09 ppm/K (a difference of 1.34 ppm/K from the silicon carbide optic), while for $N = 4$ and $n = 6$, we have an effective CTE of 2.57 ppm/K (a difference of 0.14 ppm/K from the silicon carbide optic).

An optical sensitivity study will tell us whether this closeness is sufficient. In either case, the value differences are better than those achieved by low-expansion composites and result in much lower weight than heavy Invar 36, which has a CTE of 1.6 ppm/°C.

With the advent of newer exotic materials for optics—materials that may be less amenable for use as structures—athermalized truss design may have some merit. Perhaps, someday, we will look back fondly on these things.

13.5.4 Composite metering structure

Ideally, to maintain an athermal design, a metering structure is made of the same material as the optics it houses. This is not always possible or desirable.

Table 13.4 Good CTE matches for athermalization of a silicon carbide optics using aluminum rings and titanium struts.

Bays (N)	Points (n)	Length/Radius (L/r)	Effective CTE (α_e) (ppm/°C)	Mirror CTE (α_m) (ppm/°C)	Delta CTE ($\Delta\alpha$) (ppm/°C)
3	6	3	1.09	2.43	1.34
4	6	3	2.57	2.43	−0.14

Glass optics, for example, are rarely housed in glass structures. As noted in the previous subsections, a silicon carbide structure may not be the first choice for manufacturers looking to house silicon carbide optics. In these cases, closely matched structure thermal expansion characteristics are mandated. While an athermalized truss design may be an option in limited-centered-on-axis telescope configurations (Section 13.5.3), materials such as Invar 36 exhibit a fairly good CTE match to silicon carbide and glasses, or glass ceramics, over a wide range of temperatures, as seen in Table 13.2. Such a structure, however, becomes heavy and costly.

To maintain metering and light weight, a low-CTE carbon-fiber-reinforced composite is an attractive alternative. In this case, the carbon fiber is a high-modulus graphite that is carbonized at a high temperature—a relatively pure carbon fiber.

Such composites use graphite fibers in a matrix, the latter of which can comprise a resin or metal system. Resin systems, by their nature, are hygroscopic; i.e., they can take on and release moisture, causing both swelling and outgassing, both of which are detrimental to optical telescope designs. Metal matrices, on the other hand, are not sensitive to moisture changes, but their manufacturing processes are demanding and costly. Over the years, development of better resin systems has led to graphite resin systems dominating the matrix choice, and metallic matrices have fallen out of favor. Here we concentrate on graphite resin systems.

Graphite fibers are bundled in tows with many filaments, each filament being on the order of a few microns in diameter. These filaments are incorporated into a resin matrix with a pre-determined fiber-to-resin ratio, depending on the desired properties. The resulting fiber–resin unidirectional layup is called a "ply," or a lamina, and is generally a few mils in thickness; a series of stacked plies at various orientations forms a laminate.

While the mechanical and thermal properties of a ply are readily determined by the rule of mixtures (Chapter 4), analysis of the laminate properties, where ply angles vary, requires the use of anisotropic elastic analysis, using the constituency relationships of Hooke's law in three dimensions. By varying angle and material constituent properties, a variety of mechanical properties can be achieved to optimize strength, stiffness, conductivity, and thermal expansion properties.

Graphite composites are quite attractive for use as a lightweight material to meet spacecraft and aircraft payload limitations. These composites are often used as a bulkhead or bezel with high-strength and high-stiffness supporting optics, as they can be fabricated to almost any shape one can achieve with conventional design. They exhibit low thermal expansion over a wide range of temperatures, through the cryogenic region, and in particular can be made to closely match the thermal strain of select glasses over cryogenic soak. They also have fairly good thermal conductivity in the range of many metals, and select fibers (K type) can be utilized to greatly increase thermal conductivity.

Importantly, composites can be utilized as optical metering structures when the optics exhibit low-thermal-expansion characteristics. Since the topic at hand is metering, we now touch briefly on expansion properties.

Each ply is defined by (1) modulus-of-elasticity constants in the two planar directions, (2) an in-plane shear modulus, and (3) Poisson's ratio. Effective properties are computed by the rule of mixtures. Since graphite fibers exhibit a rather unique, negative CTE and a high modulus, and the resin exhibits a low modulus but a high, positive CTE, laminates can be tailored to exhibit near-zero CTE by using the constituency relationships. The analysis for this is outside the scope of this text but is readily calculated using widely available software.

While unidirectional composites exhibit very high strength, a very high modulus, and a low CTE, in the orthogonal plane, these composites have low strength, a low modulus, and a high CTE. Thus, it is often desirable to develop laminates with quasi-isotropic properties, resulting in equal, or nearly equal, properties in the planar directions. This can be achieved by arranging symmetrical layups of plies oriented at angles of 0, +45, −45, and 90 deg, as well as mirror images of these four plies for a total of eight plies. These plies can be stacked to achieve the desired thickness. Alternatively, one can use symmetrical layups of plies oriented at angles of 0, +60, and −60 deg, plus mirror images of these three plies for a total of six plies.

In either case, to obtain near-zero CTE, a review of the constituency equations shows that, for a quasi-isotropic design, the graphite fiber needs to exhibit a very high modulus—in excess of 70,000,000 psi. Quasi-isotropic layups composed of fibers with a modulus below that value will not have zero expansion.

Early designs of quasi-isotropic graphite composites developed in the late 1970s to early 1980s consisted of graphite fibers in an epoxy-based resin system. An example of such a composite is GY70/934 graphite epoxy with fibers manufactured by Celanese Corp. (the 70 denotes modulus in millions of psi) and resin manufactured by Fiberite. The plies are nominally 0.005 in. thick, so an 8-layer quasi-isotropic laminate is 0.040 in. thick. The high-modulus fibers and low-CTE laminate, however, were subjected to a phenomenon called translaminar stress relief (TSR), which is a euphemism for micro-cracking. TSR occurs when temperatures decrease because the stress that is built up due to the fiber–epoxy CTE mismatch causes fiber separation. The result is that, after thermal cycling, the fiber CTE (which is negative) dominates.[5] This effect is shown in Fig. 13.5, where the near-zero ideal expansion changes to a negative value with temperature decrease. Continued cycling may produce additional micro-cracks, which then tend to be asymptotic (having saturated crack density) after repeated cycling, and the negative CTE value becomes rather stable. While micro-yield properties change significantly, strength properties on a macro-scale are affected somewhat, but minimally. Tests have shown a degradation in tension and compression strength on the order of 10%, and degradation of

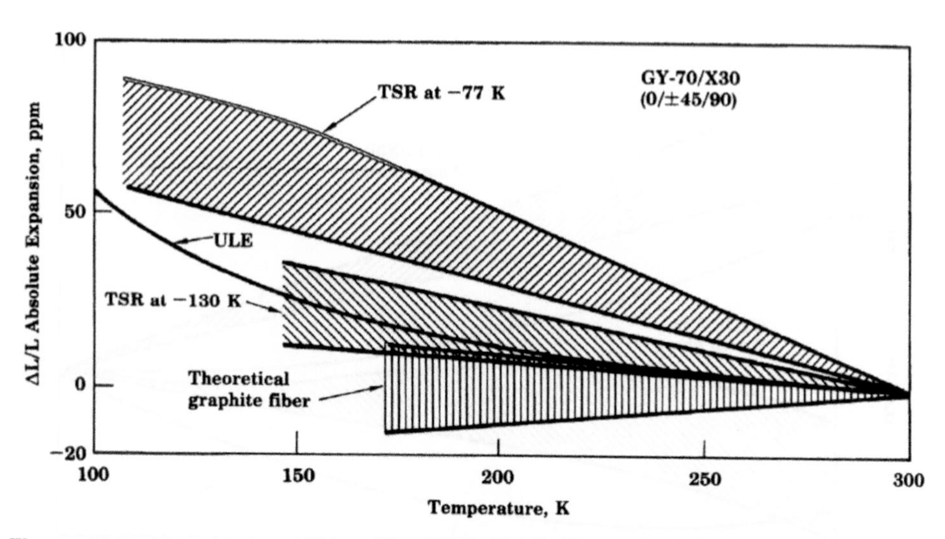

Figure 13.5 Translaminar stress relief (TSR) of a thick-ply composite causing a negative shift in expansion characteristics over cold soak as fibers dominate (reprinted from Ref. 5).

interlaminar shear strength on the order of less than 20%. Nonetheless, degradation in CTE may not be desirable.

Methods to reduce micro-cracking effects were developed in the early 1980s. It was found that epoxies having more-uniform ply thickness, a higher modulus with improved strain capability, and pre-impregnated softer fabric (called pre-preg) could minimize such effects. For example, an eight-harness weave approach for laminates reduced micro-cracking, as did development of thinner plies on the order of 0.0025 in.—one-half of the usual nominal value. This approach results in increased and more-uniform plies for a given laminate thickness. A comparison of woven fibers and thin-ply fibers is shown in Fig. 13.6.

Development of better resin systems has now considerably reduced the effects of micro-cracking. These new resins take the form not of epoxy but rather of cyanate ester or cyanate siloxane. In these cases, even the nominal, thicker 5-mil plies have been found to be acceptable for the cryogenic extremes of the thermal soak. Typical mechanical properties for unidirectional and quasi-isotropic layups are given in Table 13.5.

13.5.4.1 Moisture

The hygroscopic nature of resin systems leads to both outgassing and dimensional change when graphite composites undergo a change from an ambient air environment to other moister or vacuum conditions. Outgassing requirements depend on the system requirement, which is dependent on the volume of material. As a starting point, most specifications will dictate total loss requirements in terms of allowable mass loss of condensable materials, as well as moisture loss. Typically, total mass loss (TML) is set at less than

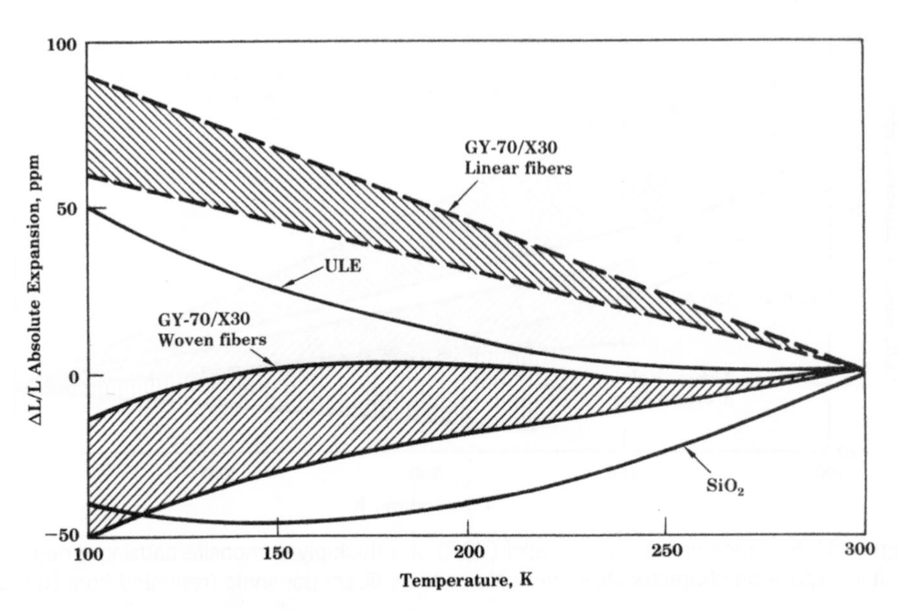

Figure 13.6 Woven fibers and thin-ply fibers mitigate the micro-crack phenomenon in epoxy resin graphite composites (reprinted from Ref. 5).

1.00%, and the collected volatile condensable material (CVCM) values are set at less than 0.10%.

Along with moisture loss or gain comes dimensional change. When moisture levels drop, the composite shrinks; as moisture levels rise, the composite expands according to Eq. (13.3). While a total dimensional change is indicated, the time required to achieve full dimensional change depends on the matrix. For example, Fig. 13.7 shows dimensional change versus time for a quasi-isotropic epoxy–based matrix. Note that to achieve 90% saturation takes several months, and full dimensional change can continue for many months beyond this time. Additionally, percent-moisture loss for these composites is quite high, making it difficult to "set and forget" for vacuum conditions, or to make predictions for diurnal moisture variations.

On the other hand, Fig. 13.8 shows dimensional change versus time for a quasi-isotropic cyanate ester–based matrix. Note that it requires on the order of only one week to achieve 90% saturation, and the time for full dimensional change is considerably shortened. Along with significantly lower percent-moisture loss when compared to epoxies, pre-set adjustments are more readily made in this matrix. Table 13.6 gives a comparison of moisture expansion characteristics for select matrix systems.

13.6 Case Study: Teal Ruby Telescope

Composite structures are quite attractive for stiffness and light weight, and for their expansion characteristics. However, they are not a panacea. As is

Table 13.5 Approximate properties of graphite composite laminates (cyanate ester).

		Tension				Compression						
		Strength (ksi)		Modulus (Msi)		Strength (ksi)		Modulus (Msi)		Thermal expansion (ppm/K)		Conductivity (W/m-K)
	Angle, deg →	0	90	0	90	0	90	0	90	0	90	0
Fiber	Layup											
K 13C2U	Unidirectional	200	3	75	0.7	50	16	75	0.7	−1.3	33	270
K 13C2U	Quasi-isotropic	80	80	25	25	25	25	25	25	−0.9	−0.9	160
M 55J	Unidirectional	290	5	45	0.8	130	25	40	0.8	−1	35	55
M 55J	Quasi-isotropic	100	65	15	15	45	60	15	15	−0.2	−0.2	31

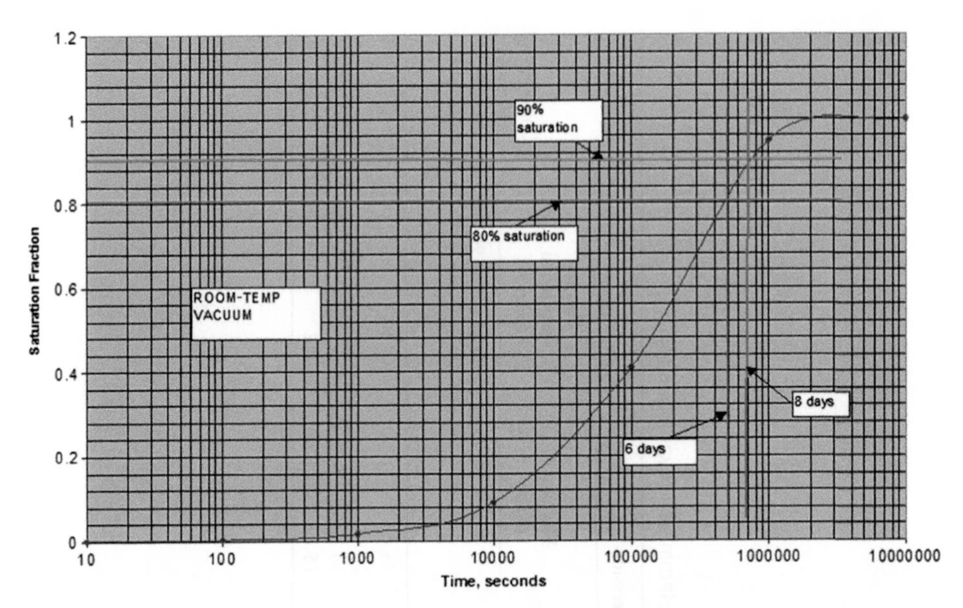

Figure 13.7 Moisture dry out (dimensional change versus time) for a graphite epoxy composite.

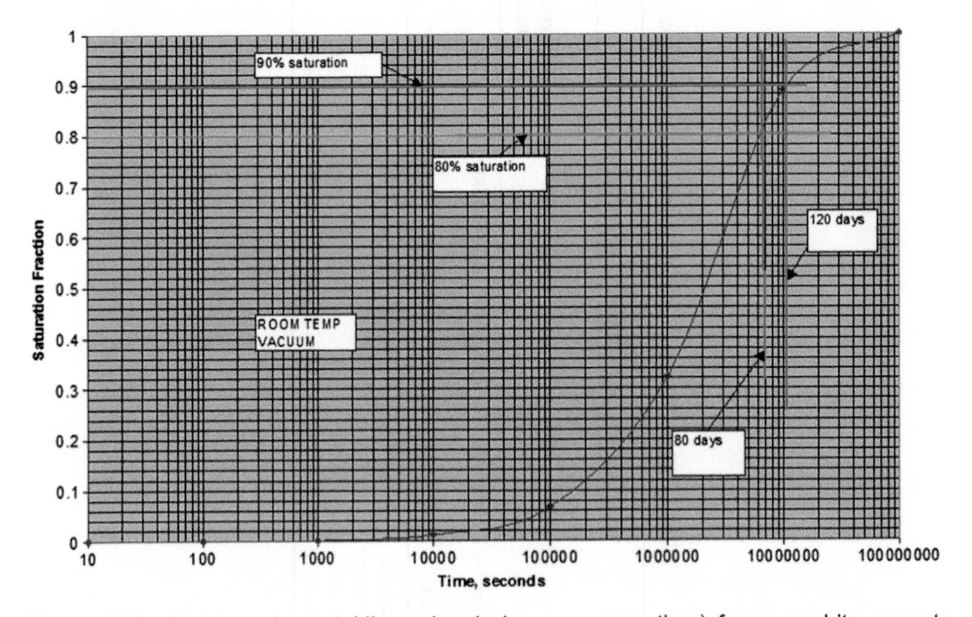

Figure 13.8 Moisture dry out (dimensional change versus time) for a graphite cyanate ester composite.

evident, when the support structure is not of the same material as the optics, metering can be affected for optical systems demanding performance quality to a fraction of wavelength of light—even for near-infrared systems, let alone those operating in the visible or ultraviolet regime. This will, of course, be

Table 13.6 Moisture expansion and time to 90% saturation for select graphite matrices.

Matrix	Cme (ppm/%M)	Moisture (%)	Strain (ppm)	Time (days)
Epoxy	162	0.42	68	120
Cyanate ester	105	0.14	14	8
Cyanate siloxane	78	0.09	7	7

dependent on the particular optical design sensitivities; indeed, many systems have been made to operate with a mismatch in material properties, in particular, those with active or passive (set-and-forget) focus capability.

To illustrate this point, consider a telescope design developed in the 1980s for space use called the Teal Ruby experiment.[5] The Teal Ruby was an infrared telescope designed to passively operate in a cryogenic and orbital environment. As such, it had to be shown to be capable of maintaining integrity under a severe set of design criteria. Spacecraft payload capabilities required minimum optical element and structure weight, while sufficient support strength was required to resist stresses developed under severe launch loading. Good stiffness characteristics were necessary to preclude excessive dynamic excursions as well as to minimize optical element motions caused by gravity release. Finally, low-thermal-expansion characteristics were a must if subassembly relative motion and cryogenic mirror distortions were to be held to the required optical tolerances. The telescope made use of a woven graphite epoxy composite structure housing lightweight fused-silica mirrors. Its structural design and analysis are discussed below.

Figures 13.9(a) and (b) and Fig. 13.10 show the infrared telescope unit (ITU), which is a four-mirror, curved-field centered design. Light enters from the outside world into the forward assembly of the telescope, which consists of the primary mirror optics and the secondary mirror housed within composite supportive rings and a light-tight shell enclosure. The secondary mirror assembly and shell are integrated by means of a thin, three-legged spider arrangement (Section 13.5.1) in order to achieve minimum obscuration. (Metering between the primary and secondary optics is discussed in Section 13.2.) Light is then reflected through a hole in the primary into the aft assembly (relay optics) of the telescope (consisting of the tertiary and quaternary mirrors), where it is then reflected to the focal plane. Again, all elements are housed within composite rings connected by a shell enclosure that also provides optical metering. The aft assembly in turn is connected to the primary ring of the forward assembly through another composite structure by a network of high-thermal-resistance tripods, since the focal plane, consisting of a series of sensors, is cryogenically cooled to well below 70 K. These subassemblies are shown in Figs. 13.11 through 13.14.

In order to meet performance in terms of optical resolution, the overall error of the design must be held to one-tenth of one wavelength or less (RMS)

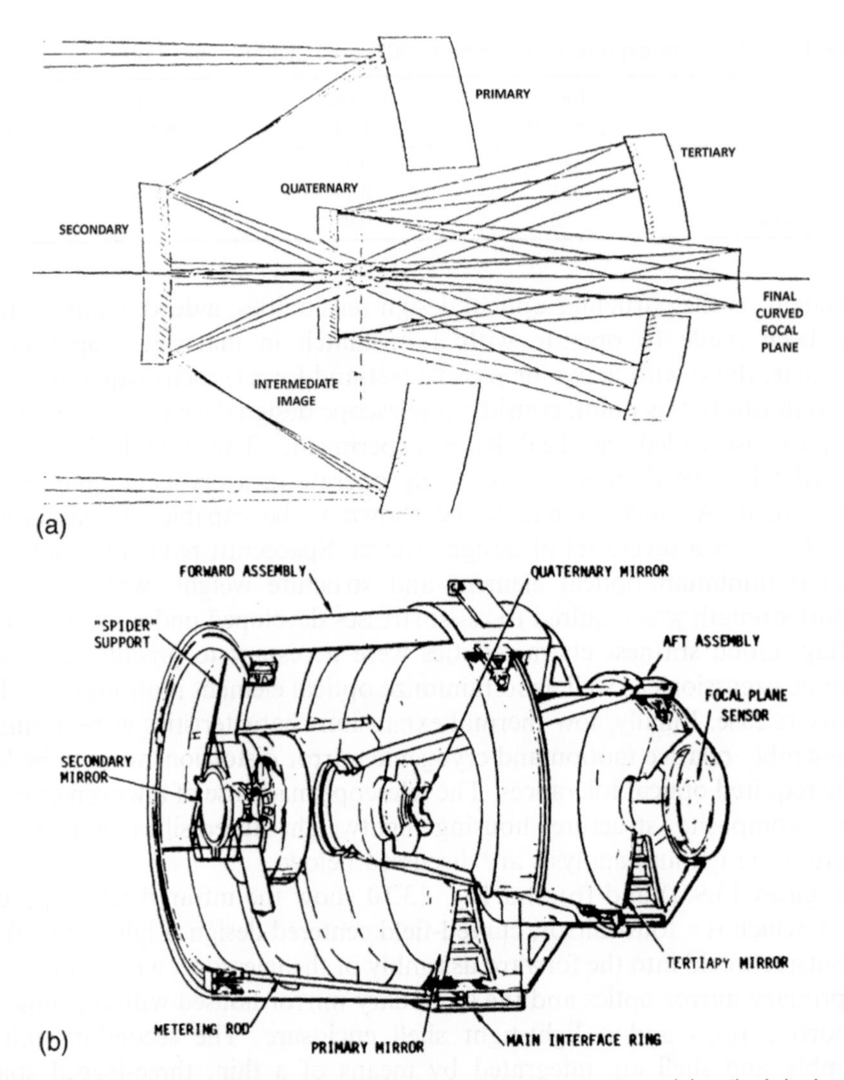

Figure 13.9 Teal Ruby optical telescope using composite structure: (a) optical design and (b) structural assembly (reprinted from Ref. 5).

of near-infrared light. A significant driver for the design is the thermal environment under which the telescope must operate. The forward structure must operate passively (without active focus) at 140 K. This assembly would experience orbital temperature swings between 130 and 185 K. The aft relay optics are cooled to 70 K with orbital swings of ±10 K. Shims are used to preset focus at these operating temperatures after test.

At the set temperature, the expansion characteristics of the optic and structure are matched to the order of 0.2 ppm/K. The motion sensitivity of the relay optics to one another and to the forward assembly itself allows this match

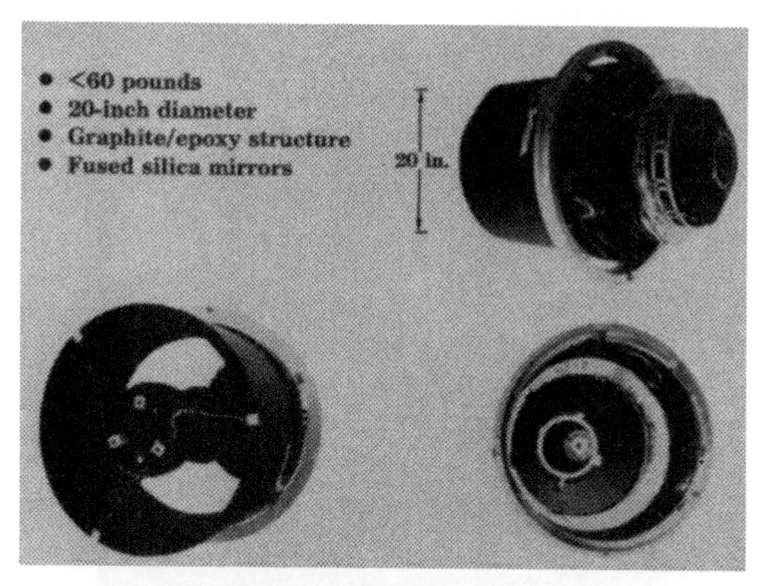

Figure 13.10 Teal Ruby structural assembly (reprinted from Ref. 5).

Figure 13.11 Teal Ruby light enclosure and lateral support structure (reprinted from Ref. 5).

to be readily made. However, the optical sensitivity of the secondary to primary motion limits the maximum excursion to 40 μin. over a 15-in. metering distance. Note from Eq. (1.47) that, over a 45 K nominal swing, $y = \Delta\alpha\Delta TL = (0.2)(45)(15) = 135$ μin., which greatly exceeds the requirement.

Figure 13.12 Teal Ruby aft composite support structure (reprinted from Ref. 5).

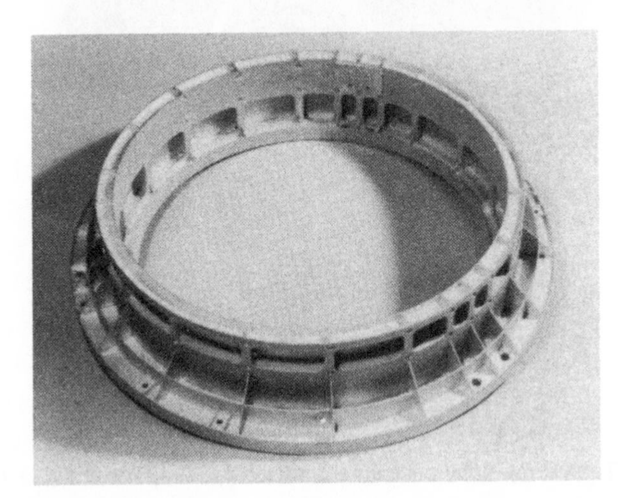

Figure 13.13 Teal Ruby main composite interface ring (reprinted from Ref. 5).

Note further that moisture strain (from Table 13.1) is quite high—on the order of 60 ppm from humidity to vacuum, resulting in a despace of [from Eq. (1.48)] $y = (60)(15) = 900\ \mu\text{in.}$, which easily exceeds the requirement, even for the aft optics.

Figure 13.14 Teal Ruby forward assembly composite spider support structure (reprinted from Ref. 5).

A proprietary, moisture-barrier, nickel-based seal coat for the composite was developed by Composite Optics of San Diego, California. This coating reduced moisture outgassing and dimensional change tenfold, allowing for acceptable performance of the aft and aft-to-forward structures. However, this reduction was still unacceptable for the forward structure; both thermal excursion error and reduced moisture error far exceeded the requirement.

An alternative approach using a metal athermalized truss (Section 13.5.3) would solve the moisture problem but, with a nominal CTE match to the optics of 0.2 ppm/K, would not solve the thermal problem. And since a light-tight shell enclosure was required, the added weight of the athermal truss would be a driving factor.

The enigma was solved by metering the fused-silica optics with a structure also comprising fused silica. As discussed in Chapter 12, using glass as a supportive structure for launch has problems of its own. Thus, three fused-silica glass rods that are identically CTE matched to the optic (being manufactured from the same boule) carried the axial load directly with low direct tensile (P/A) and compressive stresses. Such a structure would not be able to carry later loads in bending, but the composite shell would readily handle these.

The composite shell was divorced from the metering by a series of three axial blade flexures (Chapter 3) that carried the necessary lateral loads but divorced moisture and thermal motions from the metering path by providing axial flexibility. Tests under simulated launch and thermal conditions proved out the design. Thus, a blend of matched-expansion structure and closely matched composite design resulted in meeting the critical design specifications.

13.7 Support Structure

The preceding sections on structure support have concentrated on metering of the optics. However, both the optics and the metering structure need to be supported as well. For example, the secondary mirror is attached to the metering structure through a spider assembly (Fig. 13.14), and the metering structure is attached to a bulkhead, or to the interface structure of Fig. 13.13. The interface structure is then supported via flexures to attach to the payload bench.

Support structures such as bulkheads must be designed for strength and stiffness, and according to required detailed analyses using the usual strength-of-materials formulations that have been refined by finite element analyses. Spider support frames also must be designed for strength and stiffness. In this case, a three- (or four-) vane spider [if it is designed to point toward the supportive optic center, as shown schematically in Fig. 13.15(a)] will be stiff in the optical and orthogonal axes. Along the optical axis, stiffness is achieved in bending using a set of guided cantilever beams. In the orthogonal axes, stiffness is dominated in tension and compression. In both cases, the spider support can be sufficiently stiff. However, in a torsional mode about the optical axis, the vane stiffness is quite low due to bending about the weak axis. The vanes are necessarily thin to avoid obscuration issues. The low torsional frequency can result in large displacements. Although the system is generally tolerant of these rotations, misalignment of the optical axis can occur. Consequently, a tangential spider assembly [Figs. 3.14 and 13.15(b)] rather than radial spider vane arrangement can greatly improve this mode. The vanes now act as bicycle wheel spokes. Rather than bending, the torque is resisted in tension and compression by the vanes, greatly increasing torsional stiffness with little effect on the orthogonal axes' stiffness. The tangential angle relative to the radius does not need to be large (on the order of 15 deg) to increase torsional stiffness by an order of magnitude.

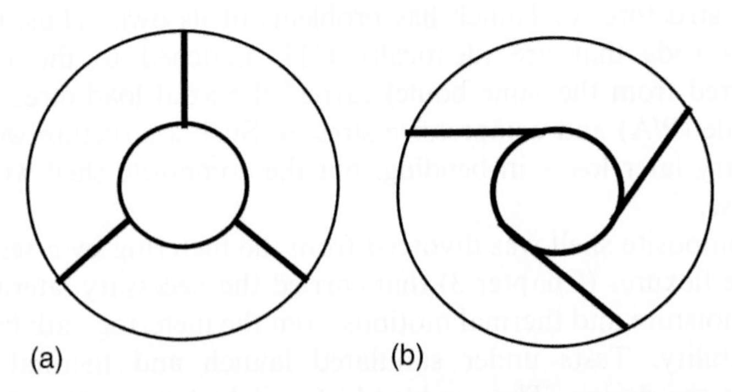

Figure 13.15 Spider support schematics: (a) Traditional spider assembly pointing to the optical center has minimal torsional stiffness. (b) Canted-angle spider assembly not pointing to optical center maximizes torsional stiffness.

References

1. P. R. Yoder, Jr., *Mounting Optics in Optical Instruments*, Second Edition, SPIE Press, Bellingham, Washington (2004) [doi: 10.1117/3.785236].
2. Arthur Vierendeel (1852–1940) was a civil engineer who served as Director of the Ministry of Public Works, West Flanders, Belgium and was a professor at the Catholic University of Leuven, Belgium.
3. P. R. Yoder, Jr. and D. Vukobratovich, *Opto-Mechanical Systems Design*, Fourth Edition, Vol. **2**, CRC Press, Boca Raton, Florida, Chapter 10, p. 415 (2015).
4. M. Serrurier, "Structural features of the 200-inch telescope for Mt. Palomar Observatory," *Civil Engineering* **8**(8), S24 (1938).
5. J. W. Pepi, M. A. Kahan, W. H. Barnes, and R. J. Zielinski, "Teal Ruby: design, manufacture, and test," *Proc. SPIE* **0216**, pp. 160–173 (1980) [doi: 10.1117/12.958459].

Chapter 14
Nuts and Bolts

When it comes down to nuts and bolts, these devices are not as simple as the colloquial term may imply. The opto-structural analyst is concerned about maintaining proper preload to prevent loosening, gapping, overtightening, fatigue, strength failure, stripping, seizing, and the like during both assembly and severe operational and non-operational environments. Proper analysis is crucial to optical performance. If properly designed, a good bolt that did not fail during initial tightening should never fail. When failure does occur, a bolt will generally fail in the threads under tension or shear if overloaded.

14.1 Terminology

Technically, a threaded fastener is called a bolt when it is used with a nut and is called a screw when it is used with a tapped hole or insert.[1] However, most engineers use the terms bolt and screw interchangeably. At any rate, some definitions are in order.

The maximum diameter of the screw thread or nut thread is called its major diameter D_{maj}, while its minimum diameter D_{min} is called the minor diameter. The average of the two is the mean diameter D_{m}. The lead l is the distance between threads, also known as the pitch p; N is the inverse of pitch, or the number of threads per inch (see Figs. 14.1 and 14.2). The thread series comprises both the bolt size (nominal diameter) and N count. The lead angle λ is the lead l divided by the developed length of the mean diameter (i.e., the mean circumference) or, more precisely,

$$\tan\lambda = \frac{l}{\pi D_{\mathrm{m}}} \sim \lambda. \tag{14.1}$$

The lead angle is typically on the order of 3 deg, while the thread angle α, defined as the angle made between the thread and the screw axis, is 60 deg for Unified National (UN—the U.S industry standard) threads. Square threads, on the other hand, have zero thread angle.

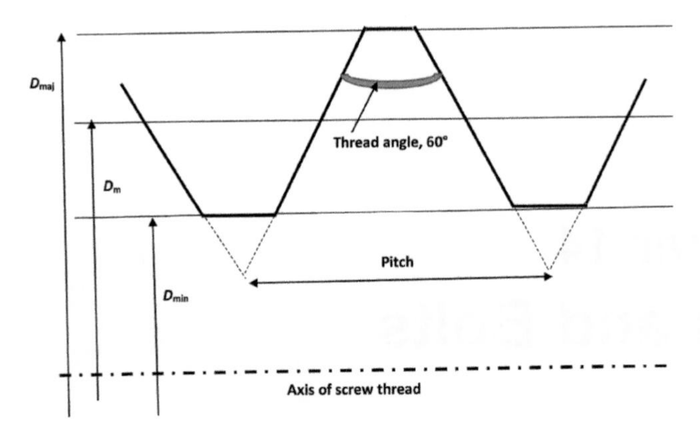

Figure 14.1 Unified National screw thread diagram showing pitch, diameters (major D_{maj}, minor D_{min}, and mean D_m), and thread angle.

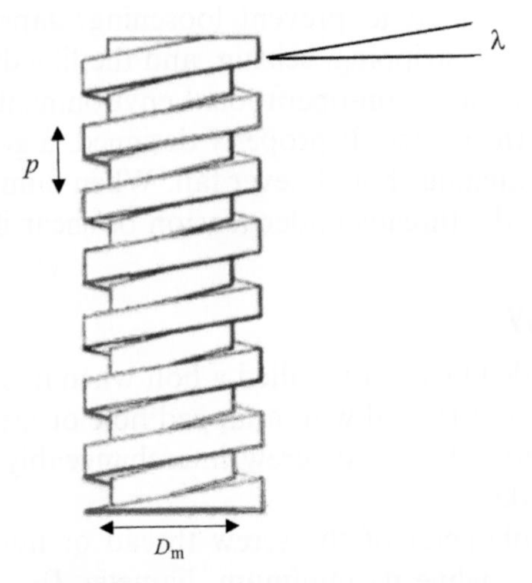

Figure 14.2 Screw thread showing the lead angle λ, which is determined by dividing the pitch by the running circumference.

Bolts (screws) are classified by their diameter at ¼ inch and higher, or by their bolt number N at diameters less than ¼ inch. In the numbering system (old mechanical engineer's secret!), the bolt diameter D is defined as

$$D = 0.013N + 0.060\,\text{in.} \quad (0.060 < D < 0.25). \tag{14.2}$$

For diameters less than 0.060 in., $D = 0.060 - [0.013 \times (\text{number of zeros after the first zero})]$. Table 14.1 shows this relation for various sizes.

Additionally, *thread class* is an alphanumeric designation to indicate thread tolerance, where 1A is the lowest precision, 2A is standard precision, and 3A is the highest precision. The letter A applies to external thread *screw*, while the letter B applies to internal thread *nut* or *insert*. Mix or match of internal and external thread classes is fully allowed. Class 1 is generally used

Table 14.1 Bolt thread sizes N and corresponding bolt diameters D for UN series bolts that are less than ¼ inch in diameter.

Bolt Number N $D = 0.013N + 0.060$	Bolt Diameter D (inch)
12	0.216
10	0.19
8	0.164
6	0.138
4	0.112
2	0.086
0	0.06
00	0.047
000	0.034

in noncritical environments where cleanliness is secondary, while class 2 is the standard callout. Class 3 is used where tight tolerance is demanded.

Further, *thread form* is designated as UN for unified screws, which have a flat root at the thread, UNR for screws having a radiused root at the thread, or UNJ for screws having a highly radiused root at the thread. The forms UNR and UNJ improve fatigue resistance.

Finally, *thread series* are designated as follows: C (coarse) is the standard thread pitch, F (fine) is a fine thread pitch, which increases strength and adjustment precision, and EF (extra fine) is an extra-fine thread pitch, which greatly increases strength and precision, e.g., for use in precision actuation, as with a stepper-motor-driven screw device. Figure 14.3 is an example that labels all of the parts included in bolt terminology.

14.2 Bolt Material

Bolts can be made from a wide selection of materials ranging from plastics to metals. The most commonly used material, however, is steel in a variety of

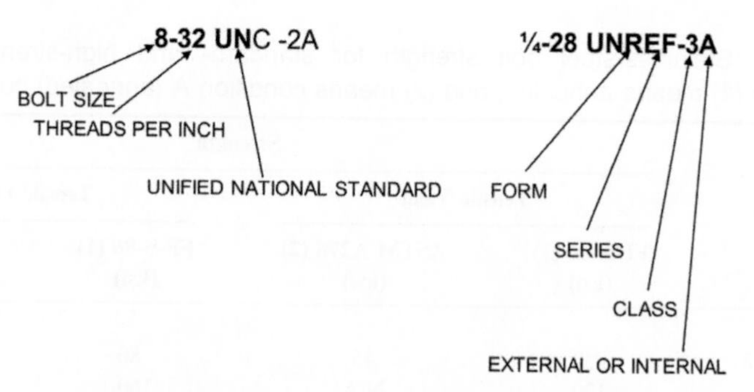

Figure 14.3 Bolt nomenclature parts defined.

forms (carbon, stainless, alloy, etc.). Choice of a particular steel depends on strength and corrosion requirements.

The selection of material is designated in procurement specifications and drawing notes. As an example, a national code [National Aerospace Standards (NAS)] for socket head cap screws may designate specification NAS1352, denoting alloy steel, or 1352C, designating corrosion-resistant (stainless) steel of standard strength, or 1352N, a high-strength, heat-treated stainless steel.

For space and flight applications of optical systems, use of the corrosion-resistant steels is most common. Standard-strength steel bolts are usually 18-8 stainless steels, while high-strength bolts are generally of the A286 steel class. High-strength bolts are nearly double in strength compared to standard-strength bolts. While most specifications designate a low yield strength for standard screws and bolts, these are generally cold drawn and thread rolled, greatly increasing minimum yield strength (by more than a factor of 2) and increasing ultimate strength as well. Table 14.2 gives a comparison of the bolt strengths for various stainless-steel varieties.

14.3 Bolt Stress

In tension, bolt stress σ is given as bolt load P divided by bolt stress area A_t:

$$\sigma = \frac{P}{A_t}. \tag{14.3a}$$

Tests[1] have shown that the stress area A_t can be computed from the average of the screw minor and mean diameters as

$$A_t = \frac{\pi}{4}\left(\frac{D_{min} + D_m}{2}\right)^2. \tag{14.3b}$$

However, in order to avoid this calculation, tables for stress area are readily available, a sample of which is given in Table 14.3.

Table 14.2 Stainless-steel bolt strength for standard- and high-strength fasteners. [Designation (1) means annealed, and (2) means condition A (annealed) cold finished.]

	Strength			
	Tensile Yield		Tensile Ultimate	
Specification	FF-S-86 (1) (ksi)	ASTM A276 (2) (ksi)	FF-S-86 (1) (ksi)	ASTM A276 (2) (ksi)
Stainless steel				
300 series (18-8)	30	45	80	90
A286	120	N/A	160	N/A

Table 14.3 Bolt stress areas for various bolt sizes used in tensile stress calculations.

Bolt Size	Diameter (in.)	N	Series	Stress Area (in.^{2})
2	0.086	56	C	0.0037
4	0.112	40	C	0.0060
6	0.138	32	C	0.0090
8	0.164	32	C	0.0139
10	0.19	32	F	0.0199
1/4	0.25	20	C	0.0318
1/4	0.25	28	F	0.0364
5/16	0.3125	18	C	0.0524
3/8	0.375	16	C	0.0775
7/16	0.4375	14	C	0.1063
1/2	0.5	13	C	0.1419

14.3.1 Shear

In shear, i.e., when the load is perpendicular to the bolt axis, stress is given as

$$\tau = \frac{V}{A}, \tag{14.4}$$

where V is the shear load, and A is the bolt shank area, which is computed from the nominal bolt diameter provided that the bolt is body bound. This is not usually the case, as clearance holes required for threaded fasteners will likely exist, causing shear load to be carried by friction (Section 14.8). Carrying shear load in the fastener is not recommended; ordinarily, pins are used for that purpose. Nonetheless, in cases where the bolt thread lies against its bearing surface, we use the thread minor diameter to calculate stress as

$$\tau = \frac{4V}{\pi {D_{\text{min}}}^2}. \tag{14.5}$$

The bolt shear strength F_{su} is derived from the ultimate tensile strength F_{tu} according to the maximum shear distortion (von Mises) theory (Chapter 1) as

$$F_{su} = \frac{F_{tu}}{\sqrt{3}}. \tag{14.6}$$

14.3.2 Thread shear

When the fastener is loaded in tension, the thread at the insert hole or nut is loaded in shear and will tend to strip if overloaded. In this case, we calculate the thread shear stress as

$$\tau = \frac{P}{A_s}, \tag{14.7}$$

where A_s is the shear area of the thread engagement. Conservatively, as developed for square threads, where only one-half of the engagement is in contact, the shear area is

$$A_s = \frac{\pi D_{min} L}{2} \tag{14.8a}$$

for external threads, and

$$A_s = \frac{\pi D_{maj} L}{2} \tag{14.8b}$$

for internal threads, where L is the thread engagement depth.

UN series threads have a larger contact area at the minor diameter cylinder and more precise external/internal contact control. Here,

$$A_s = \pi N D_{min_n} L \left[\frac{N}{2} + \frac{(D_p - D_{min_n})}{\sqrt{3}} \right] \tag{14.9a}$$

for external threads, and

$$A_s = \pi N D_{maj} L \left[\frac{N}{2} + \frac{(D_{maj} - D_{p_n})}{\sqrt{3}} \right] \tag{14.9b}$$

for internal threads, where n denotes the nut or insert, and D_p is the pitch diameter, which can be assumed to equal the mean diameter D_m for ease of calculation. Here, N denotes the number of threads per inch of the screw.

For a variety of screw sizes, using Eqs. (14.9a) and (14.9b), the thread engagement length [numerator of Eq. (14.8a)] can be multiplied by about 0.55–0.63 for external threads and about 0.63–0.68 for internal threads. Calculations for the UN thread shear area are conservative since minimum tolerances are used. For ease of use, a factor of 2/3 (0.67) can be used as

$$A_s = \frac{2\pi D_{min} L}{3} \tag{14.10a}$$

for external threads, and

$$A_s = \frac{2\pi D_{maj} L}{3} \tag{14.10b}$$

for internal threads.

With regard to engagement length L, and due to nut or tapped-hole rigidity, many codes only allow for a maximum of three engaged threads to be utilized. Three-thread engagement develops the full strength of the screw. In reality, nuts are manufactured to be softer than the screw to allow for more thread distribution due to shear yield. However, using the three-thread rule, Eqs. (14.10a) and (14.10b) become

$$A_s = \frac{2\pi D_{min}}{N} \tag{14.11a}$$

for external threads, and

$$A_s = \frac{2\pi D_{maj}}{N} \tag{14.11b}$$

for internal threads, where N is the number of threads per inch of the screw.

When the tapped-hole material is not as strong as the bolt material, helical wire inserts can be used. These are very soft springs that allow more engagement—up to three bolt diameters—due to the even load distribution, which rigid nuts and inserts do not allow. Helical wire insets are required in brittle materials such as beryllium and are required for many low-strength materials such as aluminum. An engagement factor of 0.85 can be used to account for the coil engagement lead.

14.4 Stress Examples

Example 1. Consider a #8-32 high-strength, corrosion-resistant, heat-treated A286 steel bolt-and-nut combination that has been preloaded to 1120 lbs. Compute the tensile stress in the bolt and the thread shear stress at the nut.

For tensile stress, from Eq. (14.3a) and Table 14.3, we have $\sigma = P/A_t$ and $\sigma = 80{,}000$ psi, which is well below the ultimate strength from Table 14.2.

For thread shear stress, from Eqs. (14.7) and (14.11a), we have $\tau = P/A_s$ and $\tau = PN/2\pi d_{min}$. The minor diameter[2] for a #8-32 screw is 0.126 in. We find that $\tau = 45{,}300$ psi, which is well below the shear strength (ultimate tensile strength/$\sqrt{3}$) of the screw.

Example 2. Consider a #8-32 high-strength, corrosion-resistant, heat-treated A286 steel screw threaded to a tapped aluminum 6061-T6 part preloaded to 1120 lbs. Compute the thread shear stress in the tapped hole.

For thread shear stress, from Eqs. (14.7) and (14.11b), we have $\tau = P/A_s$, $\tau = PN/2\pi D_{maj}$. From Table 14.3, $d = 0.164$ in. and $\tau = 34{,}800$ psi. This is well above (see Table 1.2) the shear strength of 24,000 psi (ultimate tensile strength/$\sqrt{3}$) of the aluminum thread, so the thread will strip. Thus, a helical wire insert is required to distribute the load.

If we use a diameter thread engagement that is 1.5 times the diameter of the screw or bolt along with a steel helical wire insert, where $\tau = P/A_s$, we find that $A_s = 0.85\pi(0.164)(1.5)(0.164)$, and $\tau = 10{,}400$ psi, which is less than 24,000 psi, so is okay.

14.5 Bolt Load

Bolt load consists of the *preload* developed during assembly of the joined parts during torque, *externally applied* static or dynamic acceleration forces, and *thermal load* (Section 14.6) due to joint/bolt thermal expansion differences.

14.5.1 Preload

When bolts are torqued and tightened to compress the members being attached, a tension load is developed in the bolt. Typically, a bolt is tightened to achieve stress of about 75% of its yield point or 50% of its ultimate strength. Friction plays an important role in achieving the desired preload. We see this in Fig. 14.4, from which, through considerations of equilibrium, we find the torque–preload relation as

$$T = \frac{FD_\mathrm{m}}{2}\left(\frac{l + \pi\mu D_\mathrm{m}\sec\alpha}{\pi D_\mathrm{m} - \mu l \sec\alpha}\right) + \frac{F\mu_\mathrm{c}D_\mathrm{c}}{2} \tag{14.12}$$

when tightening, and

$$T = \frac{FD_\mathrm{m}}{2}\left(\frac{-l + \pi\mu D_\mathrm{m}\sec\alpha}{\pi D_\mathrm{m} + \mu l \sec\alpha}\right) + \frac{F\mu_\mathrm{c}D_\mathrm{c}}{2} \tag{14.13}$$

when loosening. Here, μ and μ_c are the coefficient of friction of the thread and head, respectively, and D_c is the head diameter. Note that Fig. 14.4 considers a lead angle for a square thread in which the thread angle is zero. Equations (14.12) and (14.13) modify this to account for a nonzero thread angle, as in the UN thread series.

Typically, from Eqs. (14.12) and (14.13), the loosening torque will be about 60 to 70% of the tightening torque. When checking whether torque was relieved after static, vibratory, or thermal testing, this percentage difference would need to be accounted for if loosening, as the breakaway (loosening) torque is lower than the torque required to tighten the screw; a better way to

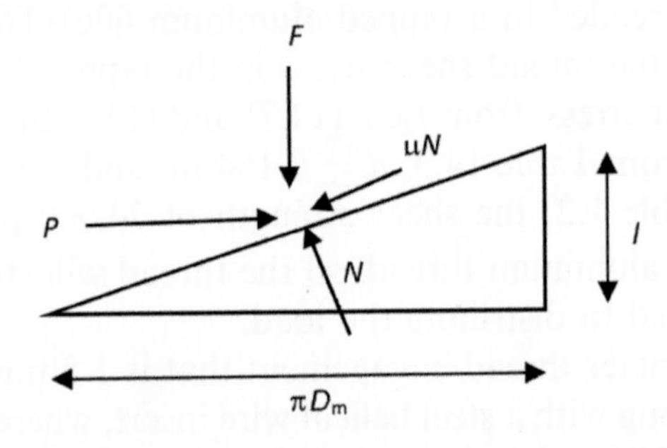

Figure 14.4 Thread friction under preload force [*N* depicts the normal (perpendicular) force exerted on the screw thread].

check is to attempt to tighten the bolt incrementally and note the screw motion. A further conclusion from the loosening equation [Eq. (14.13)] is that, for low friction, the screw may loosen on its own when

$$\mu < \tan\lambda. \tag{14.14}$$

For the UN series, where the lead angle λ is close to 3 deg, we find that a bolt having $\mu > 0.05$ is self-locking.

Typical friction values are $\mu = \mu_c = 0.15$ for "dry" screws and $\mu = \mu_c = 0.08$ for lubricated screws. Note that, in general, about 40% of torque goes to head/nut friction, 40% to thread friction, and only 20% to stretch the bolt.

During tightening torque, the bolt will twist. The torsional shear stress is given as

$$\tau = \frac{hTR}{J}, \tag{14.15}$$

where J is the polar moment of inertia, R is the bolt radius (equal to one-half of the minor diameter), and h is a fractional constant. While thread torque and thread friction twist the bolt, head-bearing (or nut-bearing) friction does not. Since 40% of torque goes into head (nut) friction, $h = \sim 0.6$ and

$$\tau = \frac{0.6TR}{J} = \frac{9.6T}{\pi D_{\min}{}^3}. \tag{14.16}$$

This stress is compared to allowable shear. When torque is removed, torsional stress is also removed, but preload and tensile stress remain in the bolt.

When friction values are substituted into Eq. (14.12), for UN series threads, we find that torque is (approximately)

$$\begin{aligned} T &= 0.2DP \text{ for a dry nut and dry thread;} \\ T &= 0.15DP \text{ for a dry nut and lubricated thread; and} \\ T &= 0.13DP \text{ for a lubricated nut and lubricated thread.} \end{aligned} \tag{14.17}$$

Due to friction uncertainties, a safety factor of 1.25 is applied in situations where preload is critical.

It is generally desirable to lubricate bolts. Typical dry lubrication for screws, nuts, or inserts is tungsten disulfide (WS_2) or molybdenum disulfide (MoS_2). Lubrication serves several purposes: it reduces thread and head or nut friction, reduces the uncertainty of the friction factor (making it more predictable), and prevents galling of similar materials.

Screw threads and heads are often lubricated simultaneously. Inserts are sometimes lubricated instead of screw threads. If screws are lubricated, inserts and nuts need not be lubricated. If nuts are not lubricated and the head is, there will be preload differences dependent on whether the nut (head) is torqued due to friction. Table 14.4 is a typical torque table based on Eq. (14.12) and includes recommended preload values.

14.5.2 Externally applied load

Screws and bolts are generally torqued to achieve 50% of their ultimate strength or 75% of their yield point, whichever is lowest. In general, the items being clamped together under this torque are much stiffer than the bolt itself. To this end, the bolt force is increased only marginally under external load since most of the external load goes toward relieving the compression force in the stiff, clamped parts, causing minimal bolt extension and hence minimal bolt force increase. Once the clamping force is overcome, the parts separate, and the bolt is then required to take the entire external load, its force increasing linearly with external load. It is therefore not good practice to exceed the bolt preload; doing so would eliminate the bolt-locking feature and could cause bolt bending as well as fatigue issues.

The above discussion is represented mathematically as

$$P_b = \frac{k_b P}{k_b + k_m}, \tag{14.18a}$$

$$F_b = P_b + F_i, \tag{14.18b}$$

$$P_m = \frac{k_m P}{k_b + k_m}, \tag{14.18c}$$

$$F_m = P_m - F_i, \tag{14.18d}$$

$$P = P_b + P_m, \tag{14.18e}$$

Table 14.4 Typical torque and preload requirements for select standard- and high-strength fasteners with and without lubrication.

		Lubrication							
		Dry Lubricant				None			
Screw	Material→	18-8	A 286	18-8	A 286	18-8	A 286	18-8	A 286
Size ↓		Torque (inch-pounds)		Preload (pounds)		Torque (inch-pounds)		Preload (pounds)	
4-40		3.5	8	200	480	4.5	11	200	480
8-32		12	28	470	1120	15	37	470	1120
10-32		19	46	670	1600	25	60	670	1600
1/4-20		40	95	1070	2540	55	130	1070	2540

where P is the applied load, P_b is the bolt load under applied load, P_m is the member load under applied load, F_i is the bolt preload, F_b is the total bolt load, F_m is the total member load, k_b is the bolt stiffness, and k_m is the member stiffness.

Combining Eqs. (14.18a) and (14.18b), the total bolt load F_b can also be given as

$$F_b = F_i + P\left(\frac{k_b}{k_b + k_m}\right). \tag{14.18f}$$

Of course, once the member compression is relieved, the members separate, and the entire load is carried by the bolt, so Eqs. (14.18a)–(14.18f) no longer apply.

To illustrate this, let us assume that the member stiffness is five times the bolt stiffness. Figure 14.5 shows bolt load versus applied load for a preload of 1000 lbs. When the applied load is zero, the bolt load is 1000 lbs. When the applied load is 1000 lbs, the bolt load is just 1180 lbs. When the applied load is 1200 lbs, the bolt load is 1200 lbs, and the joint separates. Above that, the bolt load equals the applied load.

Note in Fig. 14.5 that if no preload were applied, the total load would be lower than when preload is applied. This is misleading, as gapping would occur. High preload prevents gapping under external load. High preload also results in a low value of alternating stress reversals during cyclic loading, which greatly improves fatigue strength (Chapter 11). Fatigue is generally not an issue when bolts are properly preloaded—even when roots are not

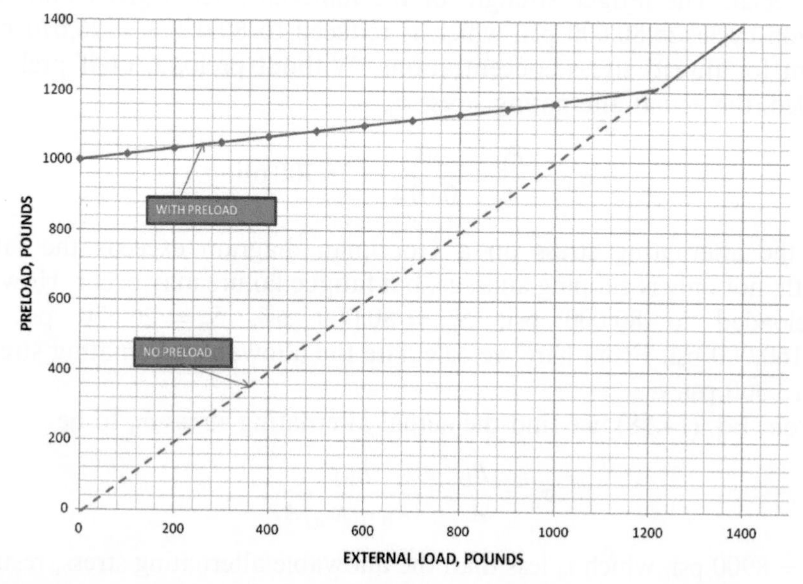

Figure 14.5 Total bolt load under external load application with and without preload.

radiused—because load fluctuation and reversal about the mean is small. However, due to machining, thread root notch, and the like, stress concentration factors must be accounted for in fatigue. Typically, a reduction by a factor of three in reported fatigue strength is used. Note that, due to local stress redistribution, stress concentrations do not affect static strength.

14.5.3 External load vibration: bolt fatigue

Modified Goodman fatigue diagrams (Chapter 11) are used to calculate allowable fatigue loads. Recall from Eqs. (11.1) and (11.2) that

$$\frac{S_a}{S_e} + \frac{S_m}{S_u} = 1,$$

or

$$S_a = S_e\left(1 - \frac{S_m}{S_u}\right),$$

where S_a is the allowable alternating stress, S_e is the fully reversed fatigue strength at a given number of cycles, S_u is the ultimate strength in tension, and S_m is the mean stress.

14.5.4 Example for consideration

Consider an A286 steel #8-32 screw preloaded to 1120 lbs (preload stress is 80,000 psi) that is joined to aluminum members which, in compression, are 3 times stiffer than the bolt. An external vibration load of 500 lbs is applied in full reversal. The fatigue strength of the material after a given number of vibration cycles is 80,000 psi, which is reduced to 80,000/3 = 26,670 psi to account for thread stress concentrations. Without preload, or if preload is exceeded, the actual alternating stress σ_a is

$$\sigma_a = \frac{P_b}{A_t} = \frac{500}{0.014} = 35{,}700 \text{ psi}.$$

Since the alternating stress on a Goodman diagram exceeds the fatigue strength, not only is gapping an issue, but fatigue failure may occur. However, if preloaded to 80,000 psi, $S_m = 80{,}000$ psi, $S_e = 26{,}670$ psi, and $S_u = 160{,}000$ psi. From here we calculate the allowable alternating stress as $S_a = 13{,}300$ psi.

From Eq. (14.18), we find the actual alternating stress σ_a to be

$$\sigma_a = \frac{P_b}{A_t} = \frac{k_b P}{(k_b + k_m)A_t},$$

so $\sigma_a = 8900$ psi, which is less than the allowable alternating stress, resulting in no fatigue issue. This is graphically depicted in Fig. 14.6.

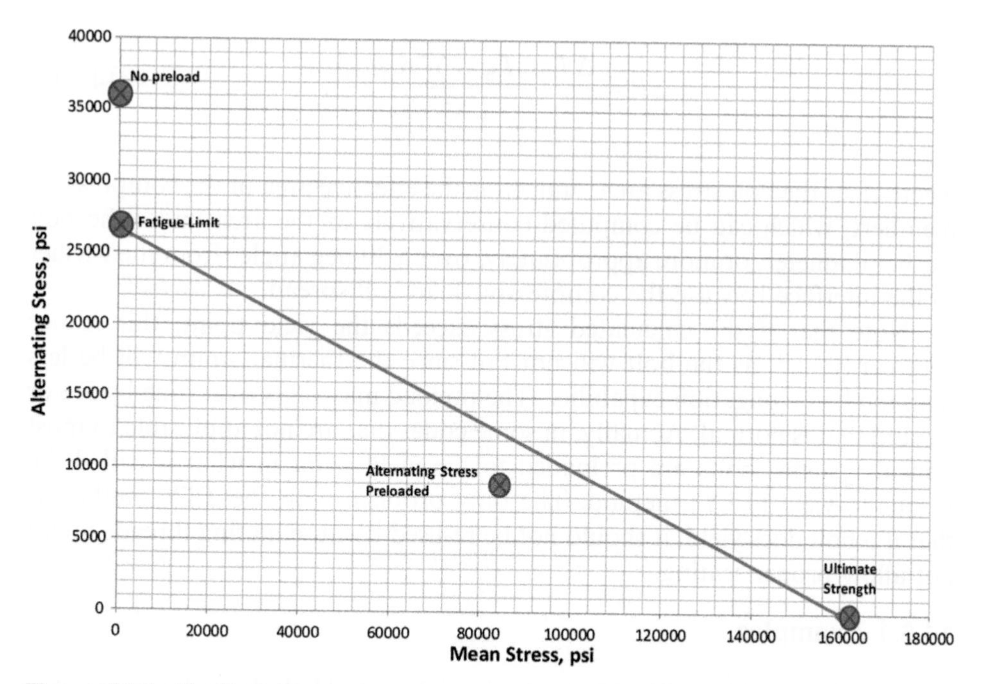

Figure 14.6 Modified Goodman diagram of bolt fatigue strength from Example 14.5.3 showing the benefit of preload.

14.6 Thermal Load

If the members being joined have a CTE that differs from that of the bolt, then the bolt will tighten or loosen under thermal soak condition. The bolt would normally tend to expand or contract freely (stress-free eigenstrain; see Chapter 1) relative to the members it joins, but the members resist that, causing a change in preload and stress.

Neglecting the small benefit of the joined members, which are generally much stiffer than the bolt, the additional load is given as [Eqs. (1.49) and (1.50)]

$$P = AE\Delta\alpha\Delta T \text{ (independent of bolt length)},$$

$$\sigma = E\Delta\alpha\Delta T \text{ (independent of bolt length } \textit{and} \text{ area)},$$

$$\text{where } \Delta\alpha = \alpha_b - \sum \frac{\alpha_m L_m}{L_b},$$

and AE is the bolt stress area × modulus, m denotes member, b denotes bolt, and L is length. This additional load can greatly increase or decrease preload.

If we do not neglect the benefit of members, the additional load can generally be computed as

$$\Delta T\left[\alpha_b L_b - \sum \alpha_m L_m\right] = \Delta P\left[\frac{L_b}{A_b E_b} + \sum \frac{L_m}{A_m E_m}\right], \tag{14.19a}$$

$$\Delta P = \frac{\Delta T[\alpha_b L_b - \sum \alpha_m L_m]}{\left[\frac{L_b}{A_b E_b} + \sum \frac{L_m}{A_m E_m}\right]}, \tag{14.19b}$$

where A_b is the bolt stress area, and A_m is the effective area of the compression members, which can be computed conservatively as the area under the bolt washer diameter, or more appropriately, over a 30-deg angle[1] from the washer to the interface because the compressive load will spread.

Note that if the calculated stress exceeds the yield point, this is not necessarily bad, as the bolt will not fail since the thermal strain will be less than the bolt elongation capability. The preload may relax over time by 5% or so, and that is normal. Actually, in some codes that are not applicable to most optical systems, bolts are preloaded to the yield point in what is called the torque-to-yield (TTY) method. The only problem with the TTY method is that if a bolt does yield, it should not be re-used because mating threads might not allow proper seating.

14.6.1 Examples

Example 1. Consider a #10-32 standard stainless-steel bolt through one member (as in Fig. 14.7) preloaded to 800 lbs, stressing the bolt to 40,000 psi. Compute the preload change under a soak from room temperature (20 °C) to −80 °C if the member CTE is 1.6 ppm/°C and the bolt CTE is 17 ppm/°C. The bolt area is 0.02 in.^{2} Let us conservatively assume that the member stiffness is high compared to the bolt. The bolt modulus is 29 Msi, and its yield point is 50,000 psi.

We again use Eqs. (1.49) and (1.50). The effective bolt length and member length are the same because the insert is embedded. We have

$$\Delta P_b = A_b E_b \Delta\alpha \Delta T = 0.02 \times 29 \times (17 - 1.6)[20 - (-80)] = 893\,\text{lbs},$$

$$\Delta\sigma = E_b \Delta\alpha \Delta T = 44{,}600\,\text{psi}.$$

The bolt will tighten by this amount, so the total load is 1693 lbs, or 84,600 psi, which is above the bolt's yield point. The bolt will not break, but if

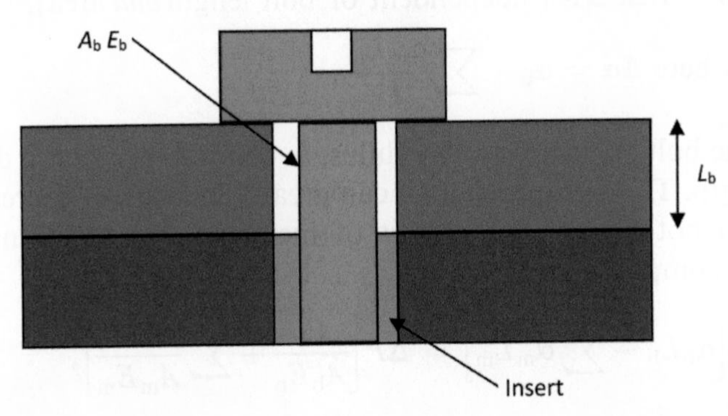

Figure 14.7 Screw and insert schematic for thermal load calculation.

it is to be reused, it might be best to instead select a high-strength bolt torqued to 1000 lbs, giving 40,000 psi preload against an ultimate strength of 160,000 psi and a yield strength of 120,000 psi. In this case, there is a positive margin against both yield strength and ultimate strength.

Example 2. Consider a #10-32 standard stainless-steel bolt through two members attached by a nut (as in Fig. 14.8) preloaded to 800 lbs, stressing it to 40,000 psi, or one-half of its ultimate strength. Including the stiffness of the members, compute the preload change under a soak from room temperature (20 °C) to −80 °C if the washer is stainless steel, the first member is Invar with a CTE of 1.6 ppm/°C, the second member is aluminum with a CTE of 22.5 ppm/°C, and the bolt and washer CTE is 17 ppm/°C. The bolt area is 0.02 in.^{2}. The bolt modulus is 29 Msi, the Invar modulus is 21 Msi, and the aluminum modulus is 10 Msi. The washer is 0.060 in. thick with outer dimension (OD) of 0.375 in. and inner dimension (ID) of 0.25 in. The members are each 0.125 in. thick.

We use Eq. (14.19b). To compute the area of the members, we approximate by conservatively taking the area under the washer, or, more appropriately, at a 30-deg angle from washer, as the compressive area will spread, as indicated in Fig. 14.8. We find that

$$A_{\text{washer}} = \frac{\pi}{4}(0.375^2 - 0.25^2) = 0.061\,\text{in.}^2,$$

$$A_{\text{members}} = \frac{\pi}{4}(0.519^2 - 0.25^2) = 0.162\,\text{in.}^2,$$

$$\Delta P = \frac{\Delta T[\alpha_b L_b - \sum \alpha_m L_m]}{\left[\frac{L_b}{A_b E_b} + \sum \frac{L_m}{A_m E_m}\right]},$$

$$\Delta P_b = \frac{100\,[17(0.31) - 17(0.06) - 1.6(0.125) - 22.5(0.125)]}{[0.31/[(0.02)(29)] + 0.06/[(0.061)(29)] + 0.125/[(0.162)(21)] + 0.125/[(0.162)(10)]]}$$

$$= 182\,\text{lbs}.$$

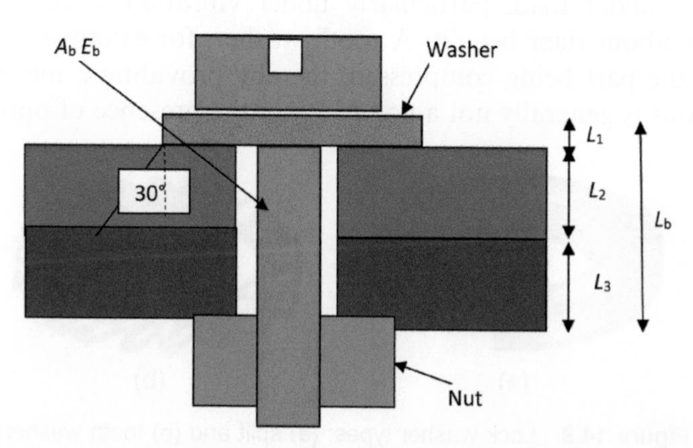

Figure 14.8 Bolt and nut schematic for thermal load calculation with multiple interfaces.

The bolt will tighten by 182 lbs, so the total load is 982 lbs, or 49,100 psi, which is below the ultimate strength and the yield strength, and is therefore okay. The high-expansion aluminum helps to negate the effect of the low-expansion Invar.

Note that if the thermal preload change is such that the bolt tightens or loosens too much, a Belleville® washer or washers can be inserted. These are significantly lower in stiffness than the bolt, and [according to Eqs. (14.19a) and (14.19b)] can greatly reduce preload change. Belleville washers must be carefully selected to apply proper preload, and stacking is subject to friction variability, so caution in their use is advised, as discussed further in Section 14.7.5.

14.7 Washers

Washers are used directly beneath a screw head or directly above a nut. They come in many varieties but in general belong to one of three classes: flat washers, lock washers, or spring washers.

14.7.1 Flat washers

Flat washers help to distribute load evenly under a bolt or nut. They also help to control friction for proper torque-induced preload. Finally, they provide a smooth surface, making the bolt less likely to loosen. Flat washers are a must in optical applications; there is nothing bad to say about them. To be effective, however, they need to encompass the area outside the clearance hole and not lie within it; otherwise, they will bend and compromise the desired effects. Washer stress is simply load divided by washer area. If properly used, flat washers will rarely pose a problem.

14.7.2 Lock washers

Lock washers come in several varieties, including split washers [Fig. 14.9(a)] and tooth washers [Fig. 14.9(b)]. While their intent is to prevent screws from backing out under load, particularly under vibration load, there is much controversy about their benefit. A tooth washer, for example, is designed to "dig into" the part being compressed, thereby providing a mechanical lock. However, this is generally not a good idea in the presence of optics, since the

Figure 14.9 Lock washer types: (a) split and (b) tooth washers.

small particles that are generated can cause contamination. Thus, their use is to be avoided.

Split washers provide a spring effect when applying preload torque to a bolt. However, when fully compressed, they become a flat washer, and the "springiness" benefits are rendered dubious. Furthermore, the split end [Fig. 14.9(a)], although it can provide an assist in locking by capturing the compressed part, can also pose contamination issues. Moreover, a series of well-documented tests[3] have shown that split washers do little, if anything, to prevent screw back-out in vibration. In fact, several codes do not allow their use.

14.7.3 Lock nuts

Lock nuts do provide the necessary locking features. Several varieties have "distorted" threads, which prevent back-out. However, this feature comes at the expense of high running torque and friction, which negate good preload control. Other varieties have wedged threads, as in a spiral lock nut. These have lower running torque but due to friction can also result in poor preload control.

14.7.4 Locking and staking

Use of epoxy to capture threads will prevent screw back-out. Of course, if screw removal is desired, this becomes problematic. Alternatively, epoxy can be used around a bolt head and washer, or around a pin, to "stake" it to the compressed member, as shown in Fig. 14.10, in which a small arc section of the screw head or pin is covered. Staking does not, in general, prevent loss of preload, as the epoxy strength is not high enough to withstand full preload torque, even for full coverage. Staking does, however, prevent screw back-out in the event that preload is lost, in which case the epoxy needs only to resist the screw mass times acceleration force.

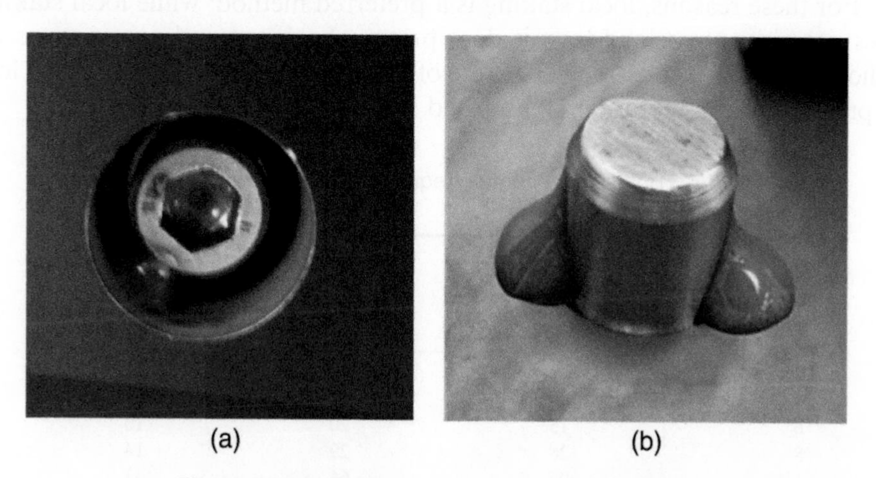

(a) (b)

Figure 14.10 (a) Bolt staking and (b) pin staking.

Consider, for example, a bolt head that is staked over its full circumference with epoxy of throat thickness t. Here we have the torque as

$$T = P'\pi DR = \frac{P\pi D^2}{2}, \qquad (14.20)$$

where P' is the shear flow in pounds per inch, D is the head diameter, and R is the head radius. The shear stress is thus

$$\tau = \frac{P'}{t} = \frac{2T}{\pi D^2 t}.$$

We can arrive at the same solution by consideration of the torque–stress relation [Eq. (1.31)] as

$$\tau = \frac{TR}{J} = \frac{TR}{2\pi R^3 t} = \frac{2T}{\pi D^2 t}.$$

Now, using the torque table for high-strength bolts (Table 14.4), and using the NAS standard of the bolt diameter being 1.5 times the bolt nominal diameter, we can determine the required loosening torque for a given epoxy shear strength, which is typically on the order of 2500 psi. We thus have

$$T = \pi D^2 t \frac{\tau}{2}. \qquad (14.21)$$

Table 14.5 presents the torque required to break the epoxy versus the preload. It shows that screw sizes greater than #8 for large bead thickness and greater than #2 for small beads of epoxy are insufficient to hold full torque. Additionally, full-coverage epoxy will require much work to remove if the screw needs replacement and will present contamination issues.

For these reasons, local staking is a preferred method; while local staking does not prevent preload loss, it does prevent back-out and can serve as an indication of screw loosening. The role of frictional forces[3] that may cause loss of preload during vibration is discussed in Section 14.8.

Table 14.5 Bolt-staking torque required to break epoxy for select epoxy bead sizes.

		Torque to Brake Stake (in-lb)	
Bolt Size	**Preload Torque**	**1/8-in. bead**	**1/16-in. bead**
2	4	8	4
4	8	13	7
6	15	20	10
8	28	29	14
1/4	95	66	33

In summary, it is a good idea to stake screws and washers—without the dubious lock washers.

14.7.5 Spring washers

Spring washers can take on several forms, such as a wave spring washer or a conical spring washer. Figure 14.11 is an example of a conical spring washer. Because spring washers are significantly less stiff than the bolt itself (by an order of magnitude), they can greatly reduce thermally induced force changes, as differential expansion is readily taken up by the spring. For a thermal soak change in which the bolt and mating part CTEs differ, ignoring the small member benefit, we can modify Eq. (14.19b) and add the spring washer stiffness effect k_w as

$$\frac{P}{k_b} = \Delta\alpha\Delta TL - \frac{P}{k_w}. \tag{14.22}$$

If $k_b = 10k_w$, then

$P = \frac{AE\Delta\alpha\Delta T}{11}$, which is greatly reduced from Eq. (1.49)!

The spring washer is not a panacea; if it is flattened by too much torque, its benefit is negated. Thus, careful control is required in its use. Furthermore, to obtain the desired preload, washers may need to be stacked on one another in parallel to increase load, or stacked upside down and back to back to decrease load, or a combination of the two. Friction between the stack results in uncertainties of preload, and the assembly is complicated. Spring washers are useful, but care is required in their proper use.

14.8 Friction Slip and Pins

14.8.1 Friction

As previously noted, bolts achieve their preload predominantly through friction. If friction is overcome, preload will drop. This is often the case during vibratory loading. Friction occurs between the bolt head, washer, and threads at the insert or nut, and at the mating interfaces as well. Friction coefficients

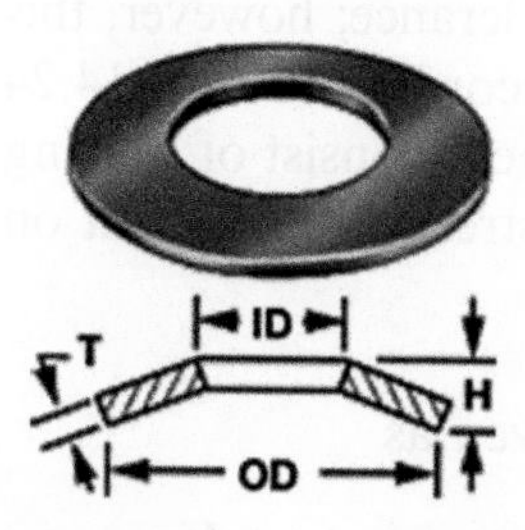

Figure 14.11 Belleville® conical spring washer (adapted from 1994 catalog of Associated Spring-Raymond, Barnes Group, Inc., Ohio).

are generally low (near 0.15) under bolt heads and are lower if lubricated (see Section 14.5.1). Lateral forces F, which tend to produce shear P, can slip the joint when

$$F > \mu P. \tag{14.23}$$

Any slip can move the threads due to male/female tolerance, further reducing preload. At the interface, slip is then possible as well, since preload or normal force is reduced. Generally, friction will be low here as well. One can count on joint slip in a high-acceleration vibration environment, such as that induced during spacecraft launch. Experience shows that slip is possible even when $F < \mu P$; as we have seen, efforts to lock the bolt are often fruitless or undesirable. This is not good in critical, high-acuity optical systems, where even small (sub-milli-inch) motions can cause misalignment as well as line-of-sight and wavefront errors.

14.8.2 Pins

The friction enigma can be solved by use of shear pins. Pins can take on several forms (dowel, taper, press fit, slip fit, etc.), but their intent is to prevent misalignment motion. When pins are used properly, bolt friction worry is moot.

Pin stress calculation is quite simple, as was indicated in the shear stress discussion from Chapter 1. We find that

$$\tau = \frac{V}{A} \text{ (for single shear)}, \tag{14.24a}$$

or

$$\tau = \frac{V}{2A} \text{ (for double shear)}, \tag{14.24b}$$

where V is the shear load force.

In the analysis, it must be assumed that for lateral loads, the pins carry all of the load in spite of bolt sharing, since the bolts are assumed to slip in their clearance holes. This is a conservative assumption as some bolts may be bearing against the surface due to tolerance; however, this is not always the case.

Once the pin is sized according to Eq. (14.24), the stresses in the mating part need to be assessed. These consist of bearing stresses (pin bearing on the surface) and shear tearout stresses (dependent on edge distance).

14.8.2.1 Bearing stress

The pin bearing stress is given as

$$\sigma_b = \frac{V}{Dt}, \tag{14.25}$$

where D is the pin diameter, and t is the surface engagement. This is a compressive stress; in general, parts in bearing will not fail in compression but are rather distortion limited. Many materials use an allowable bearing stress that is 1.5 times the parent material ultimate strength. Distortion energy theory (von Mises approach) indicates a failure in shear, resulting in an ultimate strength improvement of $\sqrt{3}$ (see Section 16.5.5 on Hertzian stress). We see that if full bearing strength is developed in the latter case, then

$$P = \sqrt{3}\sigma_{\mathrm{u}} Dt. \tag{14.26}$$

14.8.3 Shear tearout

With reference to Fig. 14.12, the tearout stress is given as

$$\tau = \frac{V}{2\left(e - \frac{D}{2}\right)t}, \tag{14.27}$$

where e is the distance from the pin hole center to the part edge. Noting that the shear strength for ductile von Mises materials is the ultimate strength divided by $3\sqrt{3}$, we can compute the e/D ratio required if full bearing strength is developed.

Here, using distortion theory, we see from Eqs. (14.26) and (14.27) that

$$P = \sqrt{3}\sigma_{\mathrm{u}} Dt = \frac{2\left(e - \frac{D}{2}\right)t\sigma_{\mathrm{u}}}{3};$$

therefore,

$$e/D = 2.0. \tag{14.28a}$$

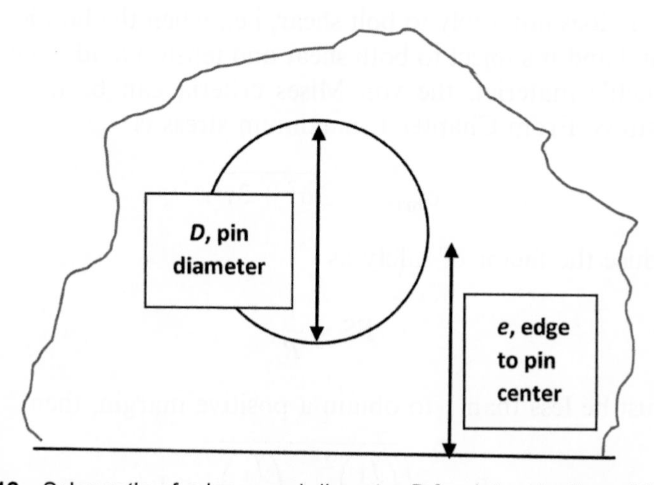

Figure 14.12 Schematic of edge e and diameter D for determination of the e/D ratio.

Several codes use this e/D ratio requirement; if we are less conservative on bearing strength and assume that $\sigma_b = \sigma_u$, then

$$e/D = 1.4. \tag{14.28b}$$

Several other codes use $e/D > 1.5$ as a criterion.

Of course, if the pin is oversized for its load, or if loads are lower than the maximum allowable, a significantly lower e/D ratio, even less than 1.0, would be acceptable, as would be indicated when the actual load is applied to Eq. (14.27). If pins are assumed to carry all shear load and bolts are assumed to carry all tension load, a stable optical system will result.

14.8.4 Example for consideration

Consider a steel pin of 3/16-in. diameter D in single shear under a load V of 500 lbs. The pin bears on an aluminum surface that is 3/16 in. in thickness t. Its edge distance e is given as 3/16 in. The pin shear yield strength is 40,000 psi; the aluminum body exhibits a shear yield strength of 20,000 psi and a bearing yield strength of 50,000 psi. Determine the adequacy of all hardware using a factor of safety (FS) of 1.25 against yield.

For the pin itself, from Eq. (14.24a), $\tau = \frac{(\mathrm{FS})V}{2A} = 22{,}650\,\mathrm{psi}$, which is greater than 40,000 psi, so is okay. For bearing against aluminum, from Eq. (14.25), $\sigma_b = \frac{(\mathrm{FS})\,V}{Dt} = 17{,}800\,\mathrm{psi}$, which is less than 50,000 psi, so is okay.

For shear tearout, we have $e/D = 1.0$, which is less than code recommendations, so we need to compute stress to ensure adequacy. From Eq. (14.27), $\tau = \frac{(\mathrm{FS})\,V}{2\left(e - \frac{D}{2}\right)t} = 18{,}000\,\mathrm{psi}$, which is less than 20,000 psi, so is okay.

14.9 Combined Bolt Loads

When friction does not apply to bolt shear, i.e., when the bolt is in bearing or is body bound and is subject to both shear and tension load, then, because the bolt is a ductile material, the von Mises criteria can be used to calculate maximum stress. From Chapter 1, maximum stress is

$$\sigma_{max} = \sqrt{\sigma^2 + 3\tau^2}. \tag{14.29}$$

If we introduce the factor of safety as

$$\mathrm{FS} = \frac{1}{R}, \tag{14.30}$$

where R must be less than 1 to obtain a positive margin, then

$$R = \sqrt{\left(\frac{f_t}{F_t}\right)^2 + \left(\frac{f_s}{F_s}\right)^2} \leq 1, \tag{14.31}$$

where f_t and f_s are the actual tensile stress and shear stress, respectively, and F_t and F_s are the allowable tensile stress and shear stress, respectively. Equation (14.31) is proven as shown below. We use the von Mises stress (MS) formulation to derive the bolt interaction equation:

$$\mathrm{MS} = \frac{1}{R} - 1,$$

where

$$R = \sqrt{\left(\frac{f_t}{F_t}\right)^2 + \left(\frac{f_s}{F_s}\right)^2},$$
$$\frac{1}{R} = \left[\left(\frac{f_t}{F_t}\right)^2 + \left(\frac{f_s}{F_s}\right)^2\right]^{-1/2}.$$

However,

$$\frac{1}{R} = \frac{F_t}{\sigma_{max}},$$
$$\sigma_{max} = \sigma_{\text{von Mises}} = \sqrt{f_t + 3f_s^2},$$
$$F_s = \frac{F_t}{\sqrt{3}}.$$

Therefore,

$$\frac{1}{R} = \left[\left(\frac{f_t}{F_t}\right)^2 + \left(\frac{f_s\sqrt{3}}{F_t}\right)^2\right]^{-\frac{1}{2}},$$
$$\frac{1}{R} = \left[\left(\frac{f_t}{F_t}\right)^2 + 3\left(\frac{f_s}{F_t}\right)^2\right]^{-\frac{1}{2}},$$
$$\frac{1}{R} = \left[\frac{f_t^2}{F_t^2} + \frac{3f_s^2}{F_t^2}\right]^{-\frac{1}{2}}, \tag{14.32}$$
$$\frac{1}{R} = \left[\frac{1}{F_t^2}(f_t^2 + 3f_s^2)\right]^{-\frac{1}{2}},$$
$$\frac{1}{R} = F_t(f_t^2 + 3f_s^2)^{-1/2} = \frac{F_t}{\sigma_{max}},$$
$$\sigma_{max} = \sqrt{(f_t^2 + 3f_s^2)}\ \text{Q.E.D.}$$

14.9.1 The bolt circle

Calculations are now made to determine the maximum bolt load under moment for any number of bolts located on a diameter. Consider a flanged connection in which two surfaces are attached by N bolts equally spaced on a

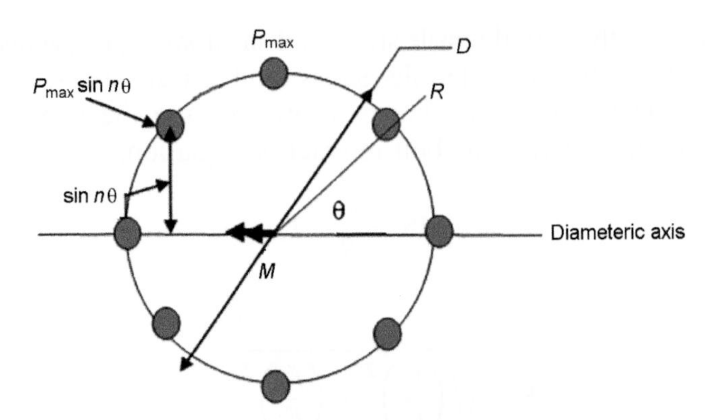

Figure 14.13 Bolt circle schematic for use in calculating bolt loads under moment loading.

bolt circle of diameter D (radius $R = D/2$) subjected to a moment load M about the diametric axis, as shown in Fig. 14.13.

We note that the maximum tensile bolt load P_{max} occurs at the most distant bolt ($\theta = 90$ deg). Furthermore, the load on the remaining bolts P_b is reduced as

$$P_b = P_{max} \sin(n\theta), \tag{14.33}$$

where $\theta = \frac{2\pi n}{N}$ (rad), and $n = 1 \ldots N$. Summing moments about the diametric axis, we have

$$M = \sum P_{max} R \sin^2 n\theta,$$
$$n = 1 \ldots N,$$

or

$$M = P_{max} R \sum \sin^2 n\theta. \tag{14.34}$$

Note that, from trigonometric identities,

$$2 \sin^2 n\theta = 1 - \cos 2n\theta,$$

so that

$$\sin^2 n\theta = \frac{(1 - \cos 2n\theta)}{2},$$
$$\sum \sin^2 n\theta = \sum \frac{(1 - \cos 2n\theta)}{2} = \frac{1}{2}\left(N - \sum \cos \frac{4\pi n}{N}\right).$$

From symmetry of the cosine function, we have

$$\sum \cos\frac{4\pi n}{N} = 0,$$

so that

$$\sum \sin^2 n\theta = \frac{N}{2}. \qquad (14.35)$$

Now substituting Eq. (14.35) into Eq. (14.34), we have

$$M = \frac{P_{max} RN}{2},$$

so that

$$P_{max} = \frac{2M}{RN}. \qquad (14.36)$$

Note that the value of Eq. (14.36) is not directly added to the bolt preload force P_i but is dependent on the member stiffness of the part being compressed, as given by Eq. (14.18f). We have, again due to the parallel load path,

$$P_b = P_{max}\left[\frac{k_b}{(k_b + k_m)}\right] + P_i, \qquad (14.37)$$

provided that the value of P_{max} is less than the preload, as is required by good design practice.

14.9.1.1 Example

An optical system assembly weighing 50 lbs (W) is bolted to an interface-mounting flange over a 20-in.-diameter bolt circle using eight ¼-20 bolts that are equally spaced. The system centroid e is 10 in. from the interface. The assembly undergoes a 35-g launch load (G) in all axes individually, and each bolt has been preloaded to 2500 lbs. The member stiffness at the interface is four times that of the bolt. Compute the new bolt load during launch.

For the axial direction, we have $P_{max} = WG/8 = 219$ lbs. For the lateral direction, we have, from Eq. (14.36), $P_{max} = \frac{2M}{RN}$, where $M = WGe$. $P_{max} = 438$ lbs. From Eq. (14.37), we calculate the bolt load as

$$P_b = P_{max}\left[\frac{k_b}{(k_b + k_m)}\right] + P_i = P_{max} = 438/5 + 2500 = 2588\,\text{lbs}.$$

References

1. J. Shigley and L. Mitchell, *Mechanical Engineering Design*, Fifth Edition, McGraw-Hill, New York (2000).
2. *Machinery's Handbook*, 27th Edition, Industrial Press, New York (2004).
3. G. H. Junker, "New criteria for self-loosening of fasteners under vibration," *Transactions of the Society of Automotive Engineers* **78**(1), 314–335 (1969).

Chapter 15
Linear Analysis of Nonlinear Properties

In Chapter 1 we found that both load-induced and thermally induced stress is readily calculated for linear elastic material properties using Hooke's law. In some cases, however, material behavior is nonlinear, and hence one needs to make use of nonlinear theory. Since this is now a departure from the usual Hookean approach, the use of plastic analysis may be required. However, we can avoid this complexity in some situations. In Chapter 4 we saw an example of nonlinear behavior of the CTE with temperature, which we handle quite nicely using secant CTE properties. We discuss this application and extend it to apply to a nonlinear elastic modulus. A theoretical approach is presented to calculate stresses for nonlinear, non-Hookean materials.

15.1 Linear Theory

Many materials, such as adhesives, exhibit nonlinear stress–strain behavior, particularly as temperature changes. Their elastic modulus increases as temperature decreases. Further, almost all materials exhibit nonlinear CTE properties as a function of temperature, often decreasing as temperature decreases and increasing as temperature increases. To understand how this might be addressed, we review Hooke's law from Chapter 1, since analysts tend to model things linearly. Set forth are analytical methods that consider such temperature-dependent behavior using the secant modulus of elasticity.

We start with linear elastic behavior. Consider a 1D beam with linear properties such that the displacement under axial load is given as [Eq. (1.1)] $F = kx$, where k is a constant that is often referred to as the spring constant of the material.

For the beam of Fig. 15.1 of length L with cross-sectional area A under load P, we have, by definition [Eq. (1.2)], $\sigma = P/A$ = stress, and [from Eq. (1.3)], $\varepsilon = x/L$ = strain (dimensionless).

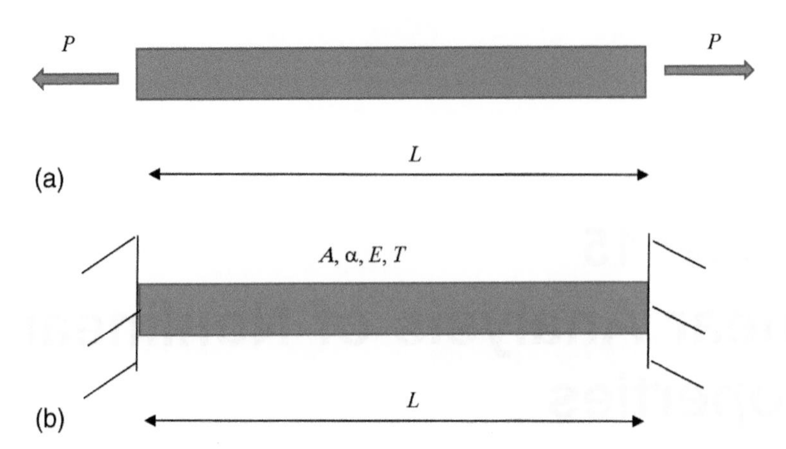

Figure 15.1 (a) Axially loaded beam in tension and (b) rigidly constrained beam under thermal soak.

Thus, as force is proportional to displacement, stress is proportional to strain, or [Eq. (1.4)—Hooke's law for stress)] $\sigma = E\varepsilon$, where E is Young's modulus, an inherent property of the material.

Substituting Eqs. (1.3) and (1.4) into Eq. (1.2) yields Eq. (1.5), which is repeated here:

$$x = \frac{PL}{AE}.$$

Under thermal stress, if the beam is restrained from expanding, we know [from Eq. (1.50)] that $\sigma = E\alpha\Delta T$.

Again, the above equations apply to a linear stress–strain relation. The value of k in Eq. (1.1) is simply the constant slope of the force displacement diagram as seen in Fig. 15.2, and the value of E in Eq. (1.4) is simply the constant slope of the stress–strain diagram, as seen in Fig. 15.3.

15.2 Nonlinear Systems: Secant and Tangent Properties

When the stress–strain curve of a material is nonlinear, Hooke's law *per se* no longer applies, and the material is said to be non-Hookean. Non-Hookean materials can tend to lose their full elastic recovery; i.e., when load is removed, the material might not return to zero strain as is the assumption in linear elastic theory. This now delves into the regime of plastic analysis, which has its own set of rules. However, in cases where nonlinear material returns to zero strain upon unloading, the material is said to be neo-Hookean. We can use the neo-Hookean model to avoid plasticity complications, and for simple cases, this may indeed serve as a good approximation for plastic behavior.

A few simple definitions are in order here. In geometry, a tangent is defined as a straight line that touches a curve or a curved surface at a single

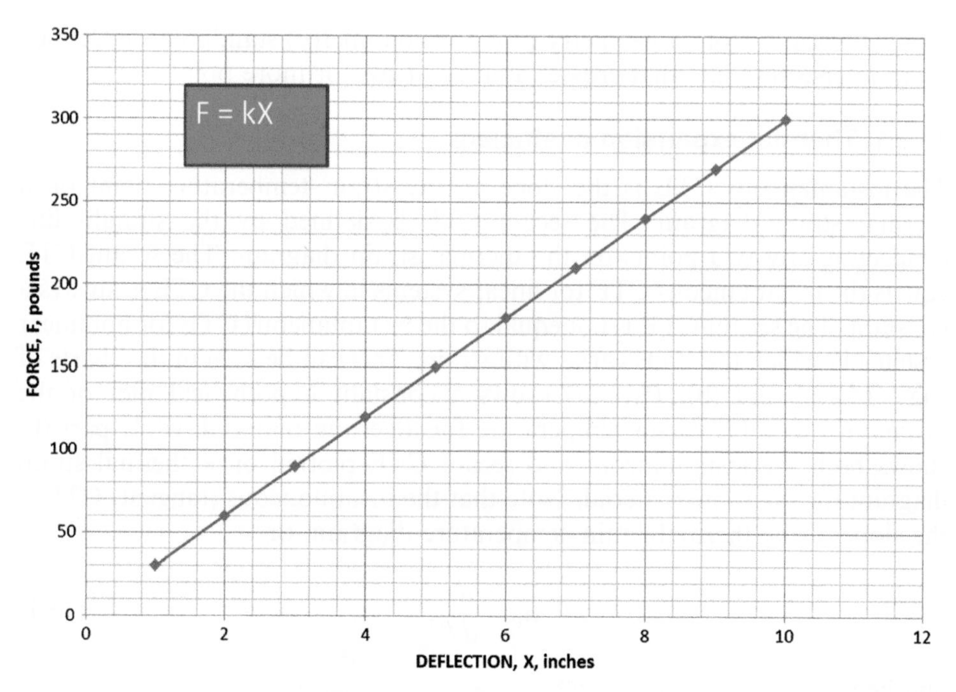

Figure 15.2 Hooke's law in one dimension.

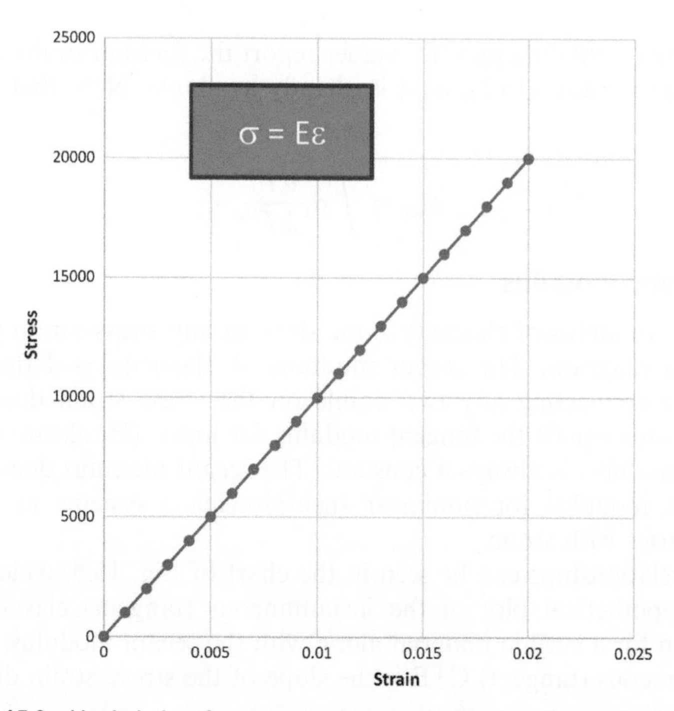

Figure 15.3 Hooke's law for stress and strain applied in a linear elastic region.

point. A secant, from a geometry (not a trigonometry) standpoint, is defined as an intersecting line that crosses a curve at two or more points.

15.2.1 Thermal expansion coefficient

The instantaneous CTE is the slope at any strain–temperature point on a thermal strain diagram. The secant CTE is defined by the straight line connecting any two points on the thermal strain diagram. The secant CTE equals the instantaneous CTE for linear systems in which the CTE is always a constant. The secant CTE is not equal to the instantaneous CTE for nonlinear systems in which the CTE varies with strain. This can be seen in the charts of Figs. 15.4(a) and (b). Part (a) shows a constant hypothetical plot of the instantaneous CTE versus temperature for a certain material, while part (b) shows both the instantaneous and secant CTE plotted on a thermal strain diagram for the same material. Note that the instantaneous (tangent) CTE is the slope of the thermal-strain–temperature diagram, or

$$\alpha_i = \frac{d\varepsilon}{dT}, \tag{15.1}$$

and hence,

$$\varepsilon = \int \alpha_i dT. \tag{15.2}$$

Since most material data on CTE values report the thermal strain, there is no need to integrate Eq. (15.2), as it is already available. Note that the secant CTE is

$$\alpha_{sec} = \int \alpha_i \frac{dT}{\Delta T}. \tag{15.3}$$

15.2.2 Elastic modulus

The tangent modulus of elasticity is the slope at any stress–strain point on a stress–strain diagram. The secant modulus of elasticity is defined by the straight line connecting any two points on the stress–strain diagram. The secant modulus equals the tangent modulus for linear (Hookean) systems in which the modulus is always a constant. The secant modulus does not equal the tangent modulus for nonlinear (neo-Hookean) systems in which the modulus varies with strain.

These relationships can be seen in the chart of Fig. 15.5, which shows a constant hypothetical plot of the instantaneous (tangent) elastic modulus versus strain for a certain material along with the secant modulus. Note that the instantaneous (tangent) CTE is the slope of the stress–strain diagram, or

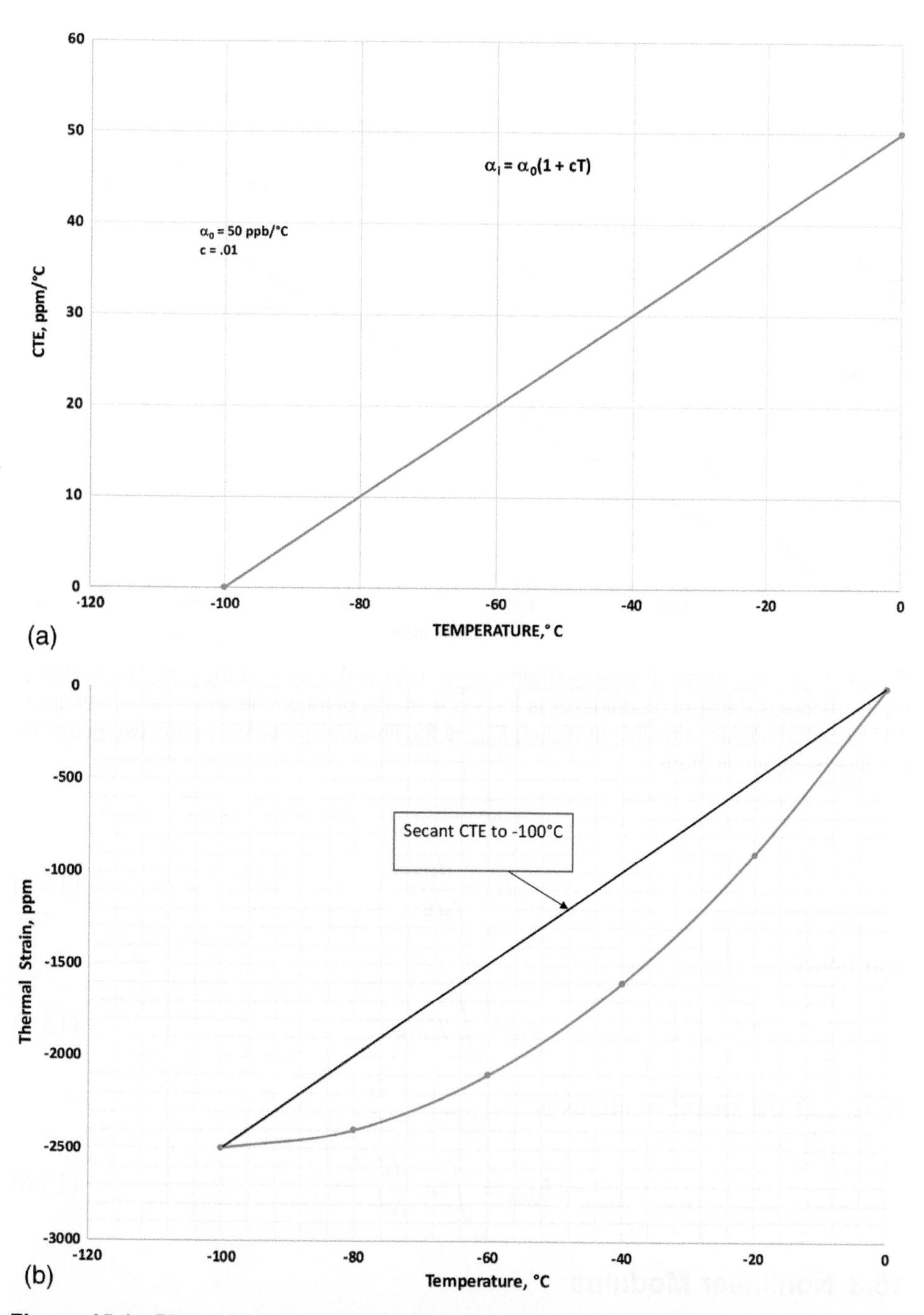

Figure 15.4 Plots of (a) instantaneous CTE versus temperature and (b) thermal strain versus temperature. The instantaneous CTE is the slope at any point on the thermal strain diagram, while the secant CTE is the effective CTE over a given range obtained by linearly connecting the range end points and determining the slope.

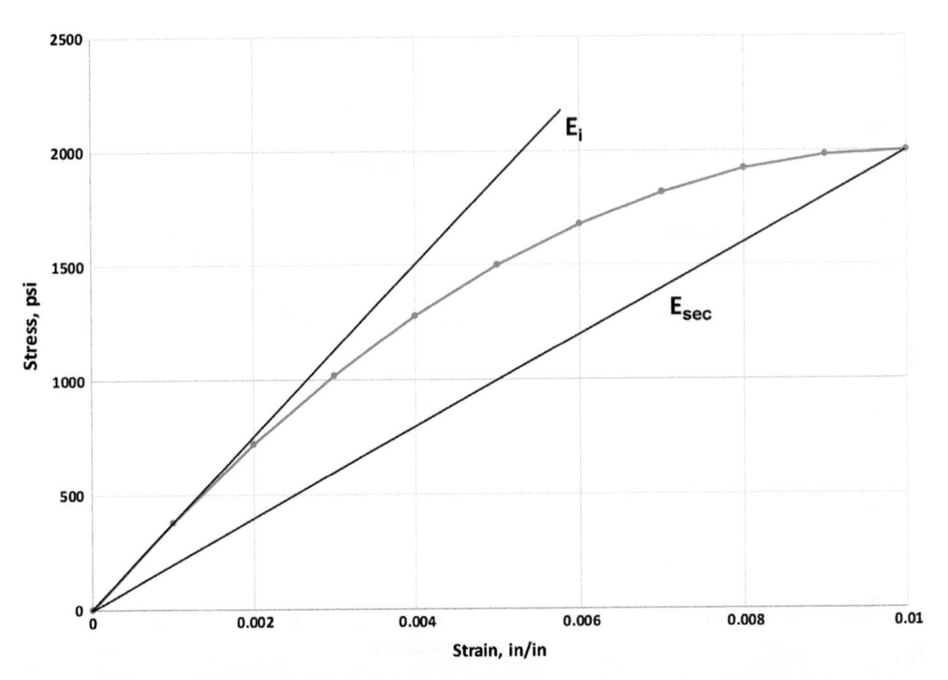

Figure 15.5 Hypothetical stress–strain diagram for a nonlinear adhesive. The instantaneous modulus (tangent modulus) E_i is the slope at any point on the stress–strain diagram. The secant modulus (effective modulus) E_{sec} is the linear slope between any two points on the stress–strain diagram.

$$E_{\mathrm{i}} = \frac{d\sigma}{d\varepsilon}, \tag{15.4}$$

and hence,

$$\sigma = \int E_{\mathrm{i}} d\varepsilon. \tag{15.5}$$

Note that the secant modulus is

$$E_{\mathrm{sec}} = \int E_{\mathrm{i}} \frac{d\varepsilon}{\Delta\varepsilon}. \tag{15.6}$$

15.3 Nonlinear Modulus

Figure 15.5 is a hypothetical stress–strain diagram for a nonlinear adhesive. While the stress–strain curve is linear for a small portion of strain, it quickly becomes nonlinear at larger strains. Here we can see from the slope of the curve that E is not constant, and at any point is given by Eq. (15.4).

Thus, from Eq. (15.5),

$$\sigma = \int \frac{d\sigma}{d\varepsilon} d\varepsilon, \tag{15.7}$$

and the resulting stress is readily computed.

For example, suppose that the adhesive material of Fig. 15.5 follows the relationship

$$\sigma = 2500 \sin(78.5\varepsilon), \tag{15.8}$$

where stress is in pounds per square inch, and strain is between 0 and 2% (angle in radians). Then,

$$E_i = \frac{d\sigma}{d\varepsilon} = 196{,}250 \cos(78.5\varepsilon). \tag{15.9}$$

The tangent modulus as a function of strain is shown in Table 15.1. Note that the modulus starts near 200,000 psi and drops to zero at the peak strain of 2%. Of course, the peak stress at peak strain is not zero but 2500 psi [from Eq. (15.8)].

We can arrive at the same result for any stress–strain combination on the curve of Fig. 15.5 by using the secant formulation of Eq. (15.6). For example, in this case, at the peak strain of 2%, the secant modulus is

$$E_{sec} = \frac{2500}{0.02} = 125{,}000 \text{ psi}.$$

15.4 Nonlinear Thermal Stress

We next review Hooke's law for thermal stress. Under thermal soak conditions, the conventional wisdom approach to nonlinear tangent modulus is to assume the highest value of tangent modulus in a linear analysis. This is based on the premise that strain is locked in and therefore stress is dependent only on the final modulus (is path independent). This premise is incorrect and is based on shoddy math. It does result in a conservative analysis, but the

Table 15.1 Tangent modulus and strain values for a select adhesive at room temperature.

Strain	Tangent Modulus (psi)
0	196,250
0.005	181,300
0.01	138,800
0.015	75,200
0.02	0

penalty is potential overdesign. On the other hand, the conventional approach to nonlinear CTE is to use the secant CTE in a linear analysis. This is correct!

Thermally induced stress and induced loads are conservative when both the modulus and the CTE vary, and the conventional approach is used. We have used the conventional approach because it is easier, but easier is not necessarily better. The following shows the correct way to use the full secant approach.

Consider, for example, the case when the stress–strain curve as a function of temperature is nonlinear (neo-Hookean). Here, the modulus varies as a function of temperature; also, the coefficient of expansion as a function of temperature may vary as well.

We can again make use of Eq. (15.4):

$$E_{\mathrm{i}} = \frac{d\sigma}{d\varepsilon},$$

and also note that the expansion coefficient may vary as well such that [Eq. (15.1)]

$$\alpha_{\mathrm{i}} = \frac{d\varepsilon}{dT}.$$

We now have

$$\sigma = -\int \frac{d\sigma}{d\varepsilon}\frac{d\varepsilon}{dT}dT. \tag{15.10}$$

Since $E_{\mathrm{i}} = \frac{d\sigma}{d\varepsilon}$ and $\alpha_{\mathrm{i}} = \frac{d\varepsilon}{dT}$, we write, substituting,

$$\sigma = -\int E_{\mathrm{i}}\alpha_{\mathrm{i}}dT, \tag{15.11}$$

which is the neo-Hookean relation for thermal stress. Since the integral is now a function of temperature, both the instantaneous (tangent) modulus and the instantaneous (tangent) CTE are expressed as a function of temperature.

15.5 Special Theory

15.5.1 Constant CTE

Consider an adhesive with a linearly varying modulus as a function of temperature (as in Fig. 15.6) and, for the moment, assume a constant CTE over the entire temperature range. The modulus–temperature relation in this example is

$$E_{\mathrm{i}} = E(T) = E_0 - bT,$$

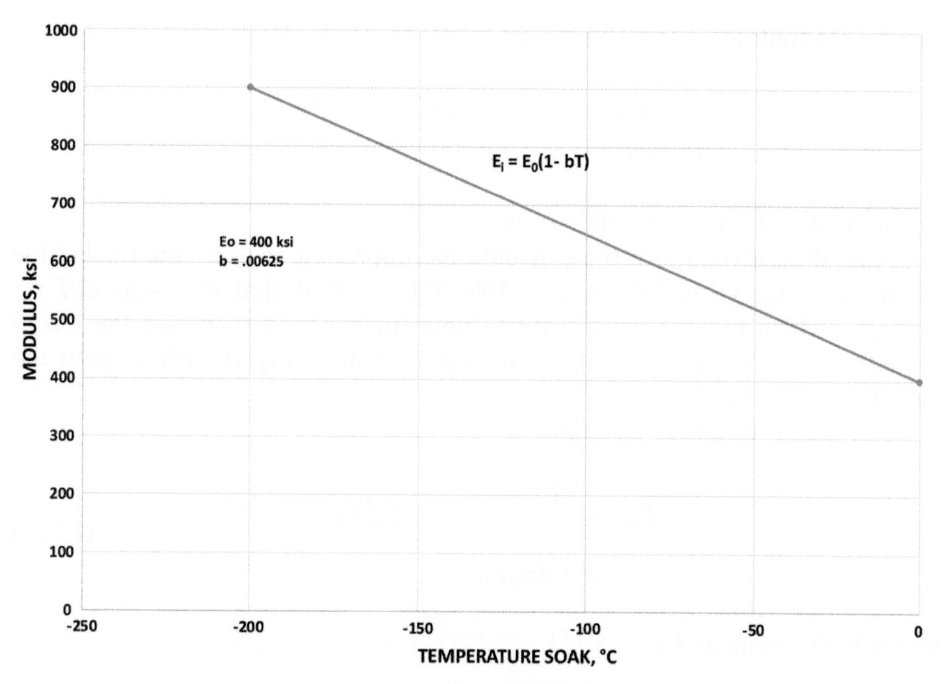

Figure 15.6 Modulus linearly varying with temperature.

where $b = 0.01E_0$, and the constant value of CTE is α_0. The temperature ranges from 0 to T in degrees Celsius.

From Eq. (15.11) we have

$$\sigma = -\int E_i\alpha_i dT = \alpha_0 \int E_i dT = \alpha_0 \int (E_0 - bT)dT, \tag{15.12}$$

$$\sigma = -\alpha_0\left(E_0T - \frac{bT^2}{2}\right). \tag{15.13}$$

Substituting for b, we have

$$\sigma = -\alpha_0\left(E_0T - \frac{0.01E_0T^2}{2}\right),$$

$$\sigma = -\left(\alpha_0E_0T - \frac{0.01\alpha_0E_0T^2}{2}\right), \tag{15.14}$$

$$\sigma = -\left(\alpha_0E_0T - \frac{0.01\alpha_0E_0T * T}{2}\right).$$

For a cold soak level of –100 °C, we have

$$\sigma = -(\alpha_0 E_0 T + 0.50\alpha_0 E_0 T) = -1.5\alpha_0 E_0 T, \qquad (15.15)$$
$$\sigma = 150 a_0 E_0.$$

For a positive CTE value, the stress is tensile.

Note that, if we adopt the conventional approach of using the final value of the modulus ($E = 2E_0$ at $T = 100$ °C), we find that $\sigma = -2\alpha_0 E_0 T = 200\alpha_0 E_0$, which is quite conservative. Note further that, if we use the *secant* modulus value, we arrive at the exact same result as Eq. (15.15) without the need for integration!

The secant modulus is simply

$$E_{\text{sec}} = \frac{(E_0 + 2E_0)}{2} = 1.5E_0, \qquad (15.16)$$
$$\sigma = 1.5\alpha_0 E_0 T,$$

which is the same as Eq. (15.15).

15.5.2 Constant modulus

Next, consider an adhesive with a linearly varying CTE over the temperature range as shown in Fig. 15.4(a). For the moment, assume a constant modulus over this entire temperature range. The modulus–temperature relation in this example is $E_i = E_0$ (constant) and $\alpha_i = \alpha_0 + 0.01\alpha_0 T$. The temperature ranges from 0 to T in degrees Celsius.

From Eq. (15.11), we have

$$\sigma = -\int E_i \alpha_i dT dT = -E_0 \int \alpha_i dT = -E_0 \int (\alpha_0 + 0.01\alpha_0 T) dT,$$
$$\sigma = -\left(\alpha_0 E_0 T + \frac{0.01\alpha_0 E_0 T^2}{2}\right), \qquad (15.17)$$
$$\sigma = -\left(\alpha_0 E_0 T + \frac{0.01\alpha_0 E_0 T * T}{2}\right).$$

For a cold soak level of –100 °C, we have

$$\sigma = -(\alpha_0 E_0 T - 0.50\alpha_0 E_0 T) = -0.5\alpha_0 E_0 T = 50\alpha_0 E_0 T. \qquad (15.18)$$

Note that if we adopt the conventional approach of using the final CTE value ($\alpha_i = 0$ at $T = -100$ °C), we have

$$\sigma = 0, \qquad (15.19)$$

which is blatantly wrong. Note further that if we use the *secant* CTE value, we arrive at the exact same result as Eq. (15.18) without the need for integration!

The secant CTE is simply

$$\alpha_{sec} = (\alpha_0 + 0)/2 = 0.5\alpha_0 E_0 T, \tag{15.20}$$

which is the same as Eq. (15.18).

The nice thing about the secant CTE is that it is obtained directly from thermal strain measurements. The thermal strain for the above example is shown in Fig. 15.4(b).

15.6 General Theory

In the general theory, we now assume that *both* the modulus and CTE vary with temperature. We consider the same examples as above for nonlinearity of the modulus and CTE, where

$$E_i = E(T) = E_0 - 0.01E_0 T,$$
$$\alpha_i = \alpha(T) = \alpha_0 + 0.01\alpha_0 T.$$

We saw in the previous section that we arrive at the correct result for stress with a constant CTE by use of the secant modulus:

$$\sigma = \alpha_0 E_{sec} T, \tag{15.21}$$

and the correct result for stress with a constant modulus by use of the secant CTE:

$$\sigma = \alpha_{sec} E_0 T. \tag{15.22}$$

We might assume that when both the modulus and CTE vary with temperature we can use the product of the secant properties as

$$\sigma = \alpha_{sec} E_{sec} T. \tag{15.23}$$

However, while this may be a useful approximation, it is incorrect when both terms vary in the integral. The integral of the product is not the product of the integrals. We need to carry out the integral fully by expanding it; if the integral is unwieldy, we need to use a modified form of the product rule for integration, as shown below.

Let $E_i = E(T)$ and $\alpha_i = \alpha(T)$, and integrate as before [using Eq. (15.11), repeated here, ignoring the sign in this case, for clarity]:

$$\sigma = \int E_i \alpha_i dT.$$

If the integration becomes complex, we can use the product rule:

$$uvdT = dT(uv) - \int \left(du/dT \int vdT \right) dT, \qquad (15.24a)$$

where u and v are the separate functions of temperature: the modulus and CTE functions, respectively. Applying the product rule to thermal stress, we have

$$\sigma = \int E(T)\alpha(T)dT = E(T) \int \alpha(T)dT - \frac{dE(T)}{dT}\left[\int \alpha(T)dT\right], \qquad (15.24b)$$

which simplifies the integration (although at first glance appears not to).

The midpoint rule of integration by summation can also assist. Here, we have

$$\sigma = \Sigma E_i \alpha_i T_i. \qquad (15.24c)$$

Returning to the example, we have

$$\sigma = -\int E_i \alpha_i dT = -\int (E_0 - 0.01E_0T)(\alpha_0 + 0.01\alpha_0 T)dT, \qquad (15.25)$$

$$\sigma = -\int (E_0\alpha_0 - 0.01E_0\alpha_0 T + 0.01E_0\alpha_0 T - 0.0001E_0\alpha_0 T^2)dT, \qquad (15.26)$$

$$\sigma = -(\alpha_0 E_0 T - 0.0001E_0\alpha_0 T^3/3).$$

For a cold soak level of –100 °C, we have

$$\sigma = -\left(\alpha_0 E_0 T - \frac{\alpha_0 E_0 T}{3}\right) = -0.667\alpha_0 E_0 T,$$
$$\sigma = 66.7\alpha_0 E_0. \qquad (15.27)$$

Note that, if we use the secant product, we have

$$\sigma = -(1.5)(0.5)E_0\alpha_0 T = 0.75E_0\alpha_0 T = 75\alpha_0 E_0, \qquad (15.28)$$

which is close but conservative. If we use the conventional approach of using the final modulus but with the secant CTE, we have

$$\sigma = -(2)(0.5)E_0\alpha_0 T = -E_0\alpha_0 T = 100\alpha_0 E_0, \qquad (15.29)$$

which is very conservative.

15.6.1 Example for consideration

Consider another hypothetical example in which both the modulus and CTE increase with cold soak. Assume that

$$E_{\mathrm{i}} = E(T) = E_0 - 0.01E_0T,$$
$$\alpha_{\mathrm{i}} = \alpha(T) = \alpha_0 - 0.01\alpha_0 T.$$

Performing the integration,

$$\sigma = -2.33\mathrm{E}_0\alpha_0 T = 233\alpha_0 E_0. \tag{15.30}$$

If we use the secant product,

$$\sigma = -2.25E_0\alpha_0 T = 225\alpha_0 E_0, \tag{15.31}$$

which is close. If we use the conventional approach of using the final modulus with the secant CTE,

$$\sigma = -3.0E_0\alpha_0 T = 300\alpha_0 E_0, \tag{15.32}$$

which is very conservative. If we (even more) erroneously use the conventional approach of using the final modulus with the *final* CTE,

$$\sigma = -4.0E_0\alpha_0 T = 400\alpha_0 E_0, \tag{15.33}$$

which is blatantly conservative and incorrect.

15.7 Using Secants

In review, we summarize the secant formulae for thermal stress:

1. For a constant CTE α_0 [Eq. (15.21)],

$$\sigma = \int E_{\mathrm{i}}\alpha_{\mathrm{i}} dT = \alpha_0 \int E_{\mathrm{i}} dT = \alpha_0 E_{\mathrm{sec}} \Delta T.$$

2. For a constant modulus E_0 [Eq. (15.22)],

$$\sigma = \int E_{\mathrm{i}}\alpha_{\mathrm{i}} dT = E_0 \int \alpha_{\mathrm{i}} dT = E_0 \alpha_{\mathrm{sec}} \Delta T.$$

3. When both modulus and CTE vary [Eq. (15.23)],

$$\sigma = \int E_{\mathrm{i}}\alpha_{\mathrm{i}} dT.$$

The secant modulus in Case 3 is not so simple, but from Eq. (15.3), repeated here,

$$\alpha_{\text{sec}} = \int \frac{\alpha_i dT}{\Delta T},$$

and from Eq. (15.6), repeated here,

$$E_{\text{sec}} = \int \frac{E_i d\varepsilon}{\Delta\varepsilon},$$

we have

$$E_{\text{sec}} = \frac{\int E_i \alpha_i dT}{\int \alpha_i dT} = \int \frac{E_i \alpha_i dT}{\alpha_{\text{sec}} \Delta T}. \tag{15.34}$$

15.8 Sample Problems

Example 1. A certain epoxy adhesive has a tangent modulus of 400,000 psi at 0 °C that increases linearly to 900,000 psi at −200 °C (see Fig. 15.6). The adhesive has an instantaneous CTE of 80 ppm/°C at 0 °C that decreases linearly to 30 ppm/°C at −200 °C (Fig. 15.7).

Compute the stress using the secant modulus method and final modulus method and compare the results. The epoxy strength at −200 °C is 12,000 psi in tension.

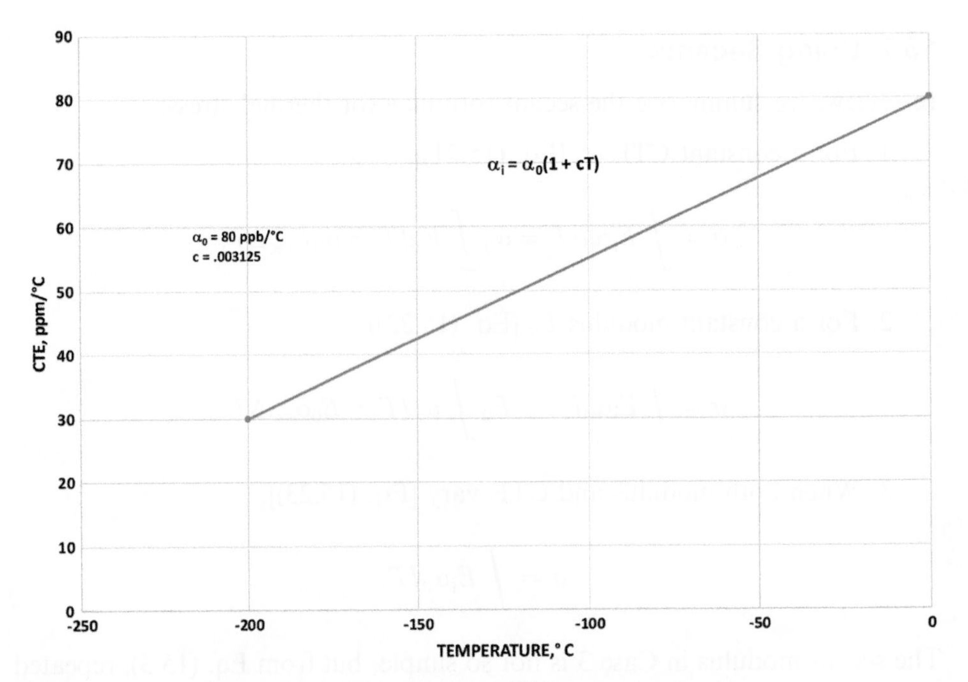

Figure 15.7 Instantaneous CTE versus temperature.

We have, where $\Delta T = 200\ °C$,

$$\sigma = \int E_i \alpha_i dT = \alpha_{sec} E_{sec} \Delta T,$$

$$E_{sec} = \frac{\int E_i \alpha_i dT}{\int \alpha_i dT} = \frac{\int E_i a_i dT}{\alpha_{sec} \Delta T},$$

$$\alpha_{sec} = \int \frac{\alpha_i dT}{DT},$$

$$\sigma = E_0 \alpha_0 \int (1 - bT)(1 + cT) = E_0 \alpha_0 \left(T - \frac{bT^2}{2} + \frac{cT^2}{2} - \frac{bcT^3}{3} \right) = 210 E_0 \alpha_0,$$

$$\sigma = 6720\ \text{psi},$$

$$\alpha_{sec} = \alpha_0 \int \frac{(1 + cT)}{DT} = 0.69 \alpha_0,$$

$$\alpha_{sec} = \frac{55\ \text{ppm}}{°C},$$

$$E_{sec} = \frac{\sigma}{\alpha_{sec} \Delta T},$$

$$E_{sec} = 1.52 E_0 = 608\ \text{psi},$$

$$E_0 = 400\ \text{ksi},$$

$$E_{final} = 900\ \text{ksi}.$$

Table 15.2 shows a comparison of these results to the results from the conventional final modulus method, the latter of which indicate a much higher stress.

For 2D analysis, a slight modification is made due to Poisson effects:

$$\sigma = \int E_i \alpha_i\, dT / (1 - \nu).$$

Using $\nu = 0.35$, we find that the secant method gives a stress of $\sigma = 10{,}300$ psi, while the final modulus method indicates 15,300 psi. This erroneous result exceeds the strength of this particular epoxy at the cold temperature, while the correct secant method does not.

Table 15.2 Conventional method and secant method comparison for silicone showing considerable difference in results.

	Method	
	Secant Modulus	**Final Modulus**
Stress, psi	6720	9950
Modulus, psi	608,000	900,000

Example 2. A silicone RTV adhesive has a modulus of 1000 psi between 293 K and 155 K, and a value of 1,000,000 psi between 155 K and 93 K as it passes through the glass transition temperature. The adhesive has a CTE of 220 ppm/K between 293 K and 15 5K, and a value of 60 ppm/K between 123 K and 93 K as it passes through the glass transition temperature. Compute the stress at 93 K using the secant modulus method and the final modulus method and compare the results. The silicone strength at 93 K is 9,000 psi. The modulus and CTE properties are indicated in Figs. 15.8(a) and (b), respectively.

Here, we use the summation technique and find that

$$\sigma = \sum E_i \alpha_i T_i,$$

$$\sigma = (0.001)(220)(138) + 1(60)(62),$$

$$\sigma = 3750 \text{ psi},$$

$$\alpha_{sec} = [(220)(138) + (60)(62)]/200,$$

$$\alpha_{sec} = 170 \text{ ppm/°C},$$

$$E_{sec} = \sigma/\alpha_{sec}\, \Delta T,$$

$$E_{sec} = 110{,}000 \text{ psi},$$

$$E_0 = 1{,}000 \text{ psi},$$

$$E_{final} = 1{,}000{,}000 \text{ psi}.$$

Table 15.3 shows a comparison of these results to the results of the conventional final modulus method, the latter of which indicates a much higher stress.

For 2D analysis, a slight modification is made due to Poisson effects:

$$\sigma = \int \frac{E_i \alpha_i dT}{1 - \nu}.$$

Using $\nu = 0.40$, the RTV strength at 93 K $\sim$ 9000 psi. The secant method gives a stress of 6250 psi, while the final modulus method would give a stress of 56,700 psi and erroneously show failure.

We have seen that when accounting for material nonlinearities in both modulus and expansion characteristics, thermal stress results are more meaningful than those calculated by a conventional linear model approach in which the final modulus property is assumed. It is cautioned, however, that, while these calculations are precise for performance at operational

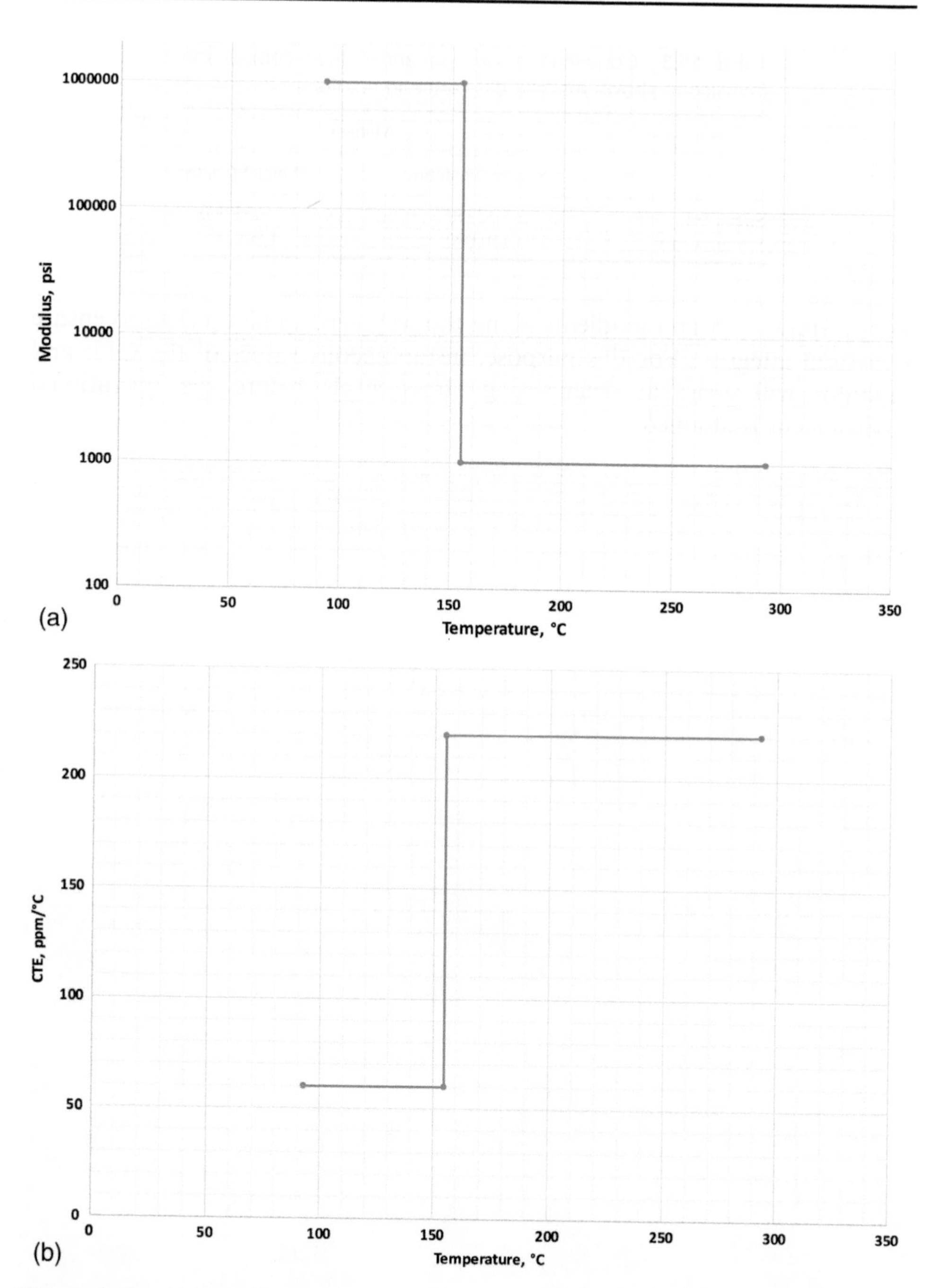

Figure 15.8 (a) RTV silicone tangent modulus versus temperature and (b) instantaneous CTE versus temperature.

Table 15.3 Conventional and secant method comparison for silicone shows marked difference in results.

	Method	
	Secant Modulus	**Final Modulus**
Stress, psi	3750	34,000
Modulus, psi	110,000	1,000,000

temperature, soak and gradients along the way need to be checked to ensure structural integrity. For this purpose, instantaneous values of the CTE and modulus will assist in determining stress levels before the operational environment is attained.

Chapter 16
Miscellaneous Analysis

This chapter presents some miscellaneous analyses and treatises that apply to optical systems and are critical to the successful performance of high-acuity optical systems.

16.1 Venting

When systems are to be operated in vacuum or in space, it is important to provide proper venting to release trapped air in enclosed volumes. Otherwise, differential pressure (up to 14.7 psi) could cause undue distortion of parts, or, worse yet, catastrophic failure due to high stresses, particularly in brittle materials. Figure 6.6, for example, shows a closed-back glass optic with each hexagonal cell pocket containing a hole for venting. Without vent holes, the glass could shatter in vacuum under the trapped internal pressure.

The question arises as to how large the vent hole needs to be in order to minimize pressure variation differential either during launch vehicle ascent or in vacuum chamber testing. If such differential pressures can be kept to below 0.2 psi, analyses generally show positive safety margins in most cases.

Consider the launch-scenario pressure rate, which is generally much more severe than slow vacuum-chamber testing would dictate. Pressure rate variations prior to orbital insertion are dependent on vehicle conditions. These generally range from 1 to 10 kPa/sec (0.15 to 1.5 psi/sec). [Nominal values for the space shuttle are near 5 kPa/sec (0.76 psi) maximum.] At any rate, these values give rise to variations in pressure between enclosed volumes and the surrounding atmosphere, where pressure reduces to near-vacuum conditions over a duration of about of 100 sec.

Flow rate analysis by Mironer and Regan[1] indicates that the pressure differential is limited to 0.5 psi for a ratio of one cubic foot volume per ¼-in.-diameter hole area [ratio of volume to hole area (V/A)] equal to 35,000 in. From the work of Mironer and Regan, if we size the hole for one cubic foot volume per 0.25-square-in. area, the resulting V/A ratio of 6,912 in. will limit the differential to only 0.2 psi. If we reduce the V/A ratio to 4000 in., the pressure differential is kept to less than 0.1 psi, a rather safe value.

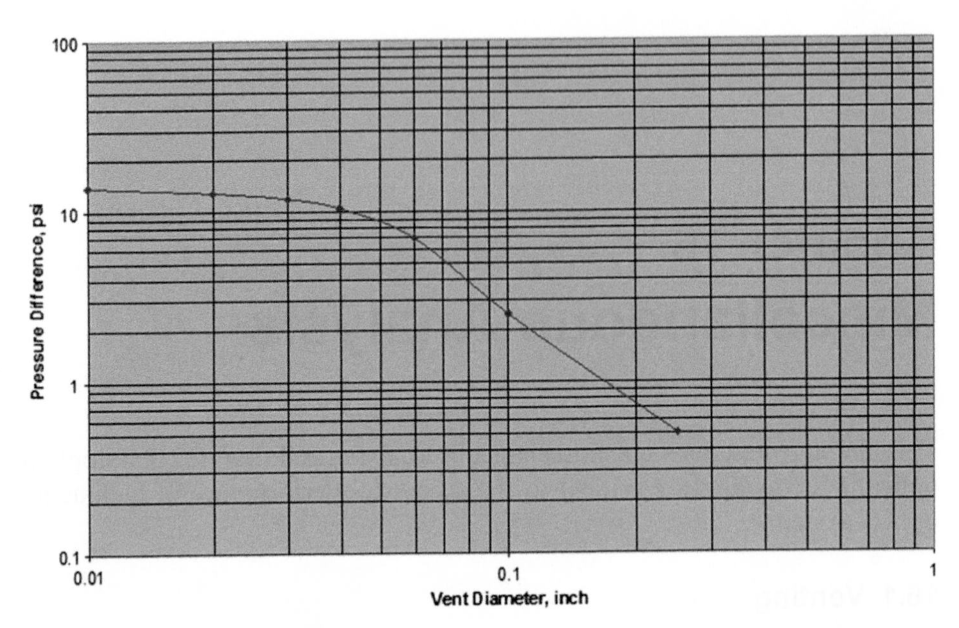

Figure 16.1 Pressure differential versus vent-hole diameter per cubic foot volume containment (adapted from Ref. 1).

A plot of pressure differential versus vent-hole diameter is given in Fig. 16.1 as extrapolated from the data of Ref. 1. Using a *V*/*A* ratio of 4000 in. for venting requirements is a conservative approach to ensure minimum pressure differential and hence minimal stresses and loads for enclosed but vented volumes within a structure. The vent area should include obstructions in flow due to filter screens or the like.

16.1.1 Contaminants

Another reason for venting is to provide an escape path by which contaminants can exit prior to precision cleaning. In stringent applications, some customers may desire venting of screws placed in blind holes. This is accomplished by drilling (or purchasing) a bolt with a small hole through its center. The pressure differential in a blind hole is far from a level that could cause concern of overload to a bolt or its cavity; it is contamination that is of concern here.

16.2 Stress Birefringence

We now assess the wavefront error of a coated optical glass window due to stress birefringence. Error is dependent on residual stress uniformity, stress-optical coefficients, and the thickness of the coating and optic.

Residual stresses in uncoated glass arise during fabrication processes. These stresses vary depending on such processes, resulting in a low value of stress for precision-annealed processes and a higher value for coarser annealing. Applied stress will increase or decrease the residual stresses depending on direction. Wavefront error in transmission is produced when the polarized index of refraction varies with stress in two orthogonal directions.

The peak wavefront error is given as

$$\Delta \mathrm{OPD} = (n_e - n_o)t = \Delta n t, \tag{16.1}$$

where n is the index of refraction in the subscripted (e for extraordinary and o for ordinary) orthogonal polarized directions, and t is the material thickness.

Every material has a stress optic coefficient K in transmission such that

$$\Delta n = K\Delta\sigma = K(\sigma_1 - \sigma_2) = \text{stress birefringence}, \tag{16.2}$$

where $\Delta\sigma$ is the principal stress difference in the orthogonal axes.

Thus, with increased stress difference, we have increased index difference, and the OPD is therefore [substituting Eq. (16.2) into Eq. (16.1)]

$$\Delta \mathrm{OPD} = K(\sigma_1 - \sigma_2)t. \tag{16.3}$$

Note that the value of $\Delta\sigma$ is the integrated average through the material thickness. Therefore, if a material is free of residual stress and an external bending stress is applied, there is no resultant birefringence because the average stress (compression on one side, zero at the neutral axis, and tension on the opposite side) is zero. Note further that differential stress must be nonuniform to produce wavefront error. If principal stresses are equal, there is no wavefront error.

The units of K are the inverse of stress. For many glasses, the value of K lies near $K = 3 \times 10^{-12}$/Pa, and residual stresses in manufacturing vary between 10 psi and 100 psi (68,950 Pa to 689,500 Pa). Thus, the stress birefringence[2] is approximately $\Delta n = K\Delta\sigma = 2$–$20$ nm/cm due to residual stress alone.

Therefore, for a 1-cm-thick window, we have $\Delta \mathrm{OPD} = 2$–20 nm, which, for a wavelength of 1000 nm, results in $\lambda = 0.002$ to 0.02 peak-peak waves. For a 0.5-cm window and a 1550-nm wavelength, we have $\lambda = 0.0006$ to 0.006 peak-peak waves.

16.2.1 Coating-induced birefringence

Coatings applied to the transmissive optic usually have some degree of residual stress caused by thermal expansion characteristics and internal stresses induced in the deposition process. These stresses can be quite high. Noting that it is the average stress through the optic that induces birefringence, we have

$$\Delta\text{OPD} = \sum K_i[(\sigma_1 - \sigma_2)_i]t_i, \tag{16.4}$$

where i denotes each coating or glass layer.

In this case, the average stress through the total part thickness is still zero, independent of the magnitude of the coating residual stress, as the stress is self-equilibrating. For example, in a part whose top and bottom are coated with a material having high residual compressive stress, there will exist a low tensile stress in the glass, and forces must be directly in balance, producing an integrated average zero stress. Thus, provided that the coating stress optic coefficient K equals the glass stress optic coefficient, we have no wavefront error. This is true even if the bottom-coating stress is different from the top-coating stress since all stresses are still self-equilibrating, as is readily shown.[3]

Note, however, that if the value of K differs between the coating[4] and the glass, we do get birefringence and therefore OPD as well, if both the stress differential and magnitude vary with distance. In general, since the coating thickness is usually miniscule relative to the glass thickness, this situation would require high coating stress, a high coating stress optic coefficient, and a large thickness of the coating.

16.2.2 Residual stress

In the same way that we compute birefringent wavefront error (or retardation) from stress, we can also reversely compute stress if we measure retardation. This can be done with polarimetry, as long as the material is transparent. Having already measured retardation, we use Eq. (16.2) to compute the principal stress difference.

For example, consider a circular optic with a residual hoop stress at the edge due to edge cooling during the manufacturing process. The principal hoop stress ($\sigma 1$) occurs along the tangential axis, while there is no principal stress ($\sigma 2$) along the radial axis. The stress at the optic center is significantly lower; here, radial stress equals tangential stress.

Note from Eq. (16.1) that it is the principal stress difference that is measured; therefore, at the center of the optic no retardation can be measured, while at the edge the retardation will be evident. Here, since the radial stress is zero, we find the principal tangential stress directly.

In cases where both radial and tangential principal stresses exist, we can only determine the principal stress difference when using polarimetry normal to the surface. However, if we assume that the third principal stress ($\sigma 3$) normal to the optic is zero, we can view the surface through an angle about which the polarimeter device is rotated. In this case we can measure retardation, which will differ from and complement the initial normal-incidence value. With two equations and two unknowns, knowing the view angle, we can compute the individual stress magnitudes. A well-documented and more-thorough discussion on stress birefringence for nonuniform stress is given by Doyle et al.[5] for readers desiring more-advanced analyses.

16.3 Bonded Tubes and Grooves

Calculations can be made to determine epoxy stresses for a tube bonded into a groove and, similarly, for a tube bonded over a boss. The bonded tube is subjected to loading in all degrees of freedom, as shown in Fig. 16.2.

16.3.1 Bending moment

First we review bending moment only. Consider a thin-walled circular tube bonded into a groove on both the inner and outer diameters. The reactive forces under a total moment M are resisted in both shear in the bond and in tension/compression due to socket action, as shown in Fig. 16.2. A portion of the load is resisted in shear, and a portion is resisted in tension/compression (bearing) due to load sharing in the dual-stiffness path. We determine the proportion of each below. Note that peel is not an issue due to the bearing load path.

To determine the shear stress, recall in Chapter 14 that we developed equations for calculating bolt loads on a circular flange under bending moment. There, we saw [Eq. (13.36)] that

$$P_{\max} = \frac{2M}{RN},$$

where N is the number of bolts. We can expand this for a continuous, adhesive-circle-bonded groove by dividing the circumference into N parts, where each part is a unit inch in width; i.e, $N = \pi D = 2\pi R$; therefore, Eq. (13.36) becomes

$$P = \frac{2M}{RN} = \frac{M}{\pi R^2}. \tag{16.5}$$

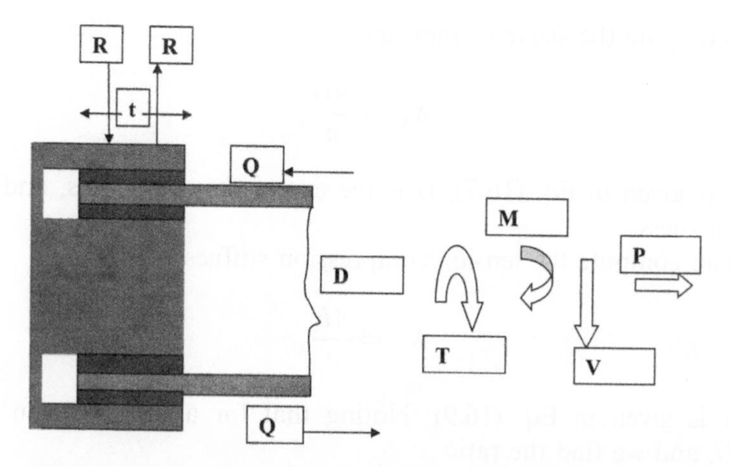

Figure 16.2 Bonded groove reaction to tensile, shear, moment, and twist loads.

The bond shear stress is

$$\tau = \frac{P}{A},$$

where A is the bond engagement $t \times 1$ in. $= t$ square in. Thus,

$$\tau = \frac{M}{\pi R^2 t} = \frac{4M}{\pi D^2 t} \tag{16.6a}$$

for a bond in single shear, and

$$\tau = \frac{M}{\pi R^2 t} = \frac{2M}{\pi D^2 t} \tag{16.6b}$$

for a bonded groove in double shear. In this case, $M = QD$ (see Fig. 6.2), the portion carried in shear, where D is the tube mean diameter, and t is the bond engagement. Since $\tau = \frac{Q}{A}$, we can set the resulting value (in double shear) equal to Eq. (16.6b) and compute the effective area as

$$A = \frac{\pi D \tau}{2}. \tag{16.7}$$

We next determine the bearing stress as

$$\sigma = \frac{P}{2Dt} = \frac{M}{Dt^2}, \tag{16.8}$$

where, in this case, $M = \frac{Pt}{2}$, the portion carried in tension/compression. The effective area is

$$A = 2Dt. \tag{16.9}$$

We can compute the shear stiffness as

$$K_Q = \frac{AG}{n}, \tag{16.10}$$

where A is given in Eq. (16.7), G is the epoxy shear modulus, and n is the epoxy thickness.

We can compute the tension/compression stiffness as

$$K_P = \frac{AE}{n}, \tag{16.11}$$

where A is given in Eq. (16.9). Noting that for a high Poisson ratio μ, $E = {\sim}3G$, and we find the ratio

$$\frac{K_P}{K_Q} \sim 4. \tag{16.12}$$

The shear rotational stiffness is

$$\begin{aligned} K_{Q\Theta} &= \frac{M}{\Theta}, \\ \Theta &= \frac{M}{D}, \\ \Theta &= \frac{Q}{K_Q D} = \frac{M}{K_Q D^2}, \end{aligned} \tag{16.13}$$

$$K_{Q\Theta} = K_Q D^2, \tag{16.14}$$

$$\begin{aligned} K_{P\Theta} &= \frac{M}{\Theta}, \\ \Theta &= \frac{2P}{K_P t} = \frac{4M}{K_P t^2}, \end{aligned} \tag{16.15}$$

$$K_{P\Theta} = \frac{K_P t^2}{4}. \tag{16.16}$$

However, from Eq. (16.12),

$$K_P = 4K_Q,$$

so

$$K_{P\Theta} = K_{Q\Theta}\left(\frac{t}{D}\right)^2. \tag{16.17}$$

It is seen that the moment is thus shared in shear and tension/compression by the square of the engagement-to-diameter ratio $(t/D)^2$. For small engagement and large diameter, almost all of the load is in shear. For large engagement and small diameter, almost all the load is in tension/compression.

The stress is thus computed from Eqs. (16.6) and (16.8) as

$$\tau = \frac{\frac{2M}{\pi D^2 t}}{1 + \left(\frac{t}{D}\right)^2}, \tag{16.18}$$

$$\sigma = \frac{\frac{M}{Dt^2}\left(\frac{t}{D}\right)^2}{1 + \left(\frac{t}{D}\right)^2}. \tag{16.19}$$

The stresses can be combined using the von Mises criterion [Eq. (1.44)] as

$$\sigma = \sqrt{(\sigma^2 + 3\tau^2)}.$$

Von Mises stress can be compared to tension ultimate strength, and shear stress can be compared to shear ultimate strength with an appropriate factor of safety.

16.3.2 Axial load

Under axial load P, we have

$$\tau = \frac{P}{A} = \frac{P}{2\pi Dt} \tag{16.20}$$

since the inner and outer bonds of the groove share the load.

16.3.3 Torsion

Under torque load T, we have [from considerations of shear flow shown in Chapter 14, specifically Eq. (14.21), repeated here]

$$\tau = \frac{2T}{\pi D^2 t}$$

computed for the groove in single shear; therefore,

$$\tau = \frac{T}{\pi D^2 t} \tag{16.21}$$

computed for the groove in double shear.

16.3.4 Shear

Here, the load is reacted in tension and compression (bearing) at the tube-to-groove interface. The stress is

$$\sigma = \frac{V}{A} = \frac{V}{2Dt} \tag{16.22}$$

since there are two bonds (inner and outer) and two sides (tension and compression) over one-half of the engagement depth. We can combine the shear and normal stresses from all sources by use of the von Mises criterion.

16.3.5 Tube over boss

If a tube is bonded over a boss, as in Fig. 16.3, Eqs. (16.19) through (16.22) are identical except that the bond is in single shear and we therefore multiply all stresses by a factor of 2. A summary of these stresses is given in Table 16.1

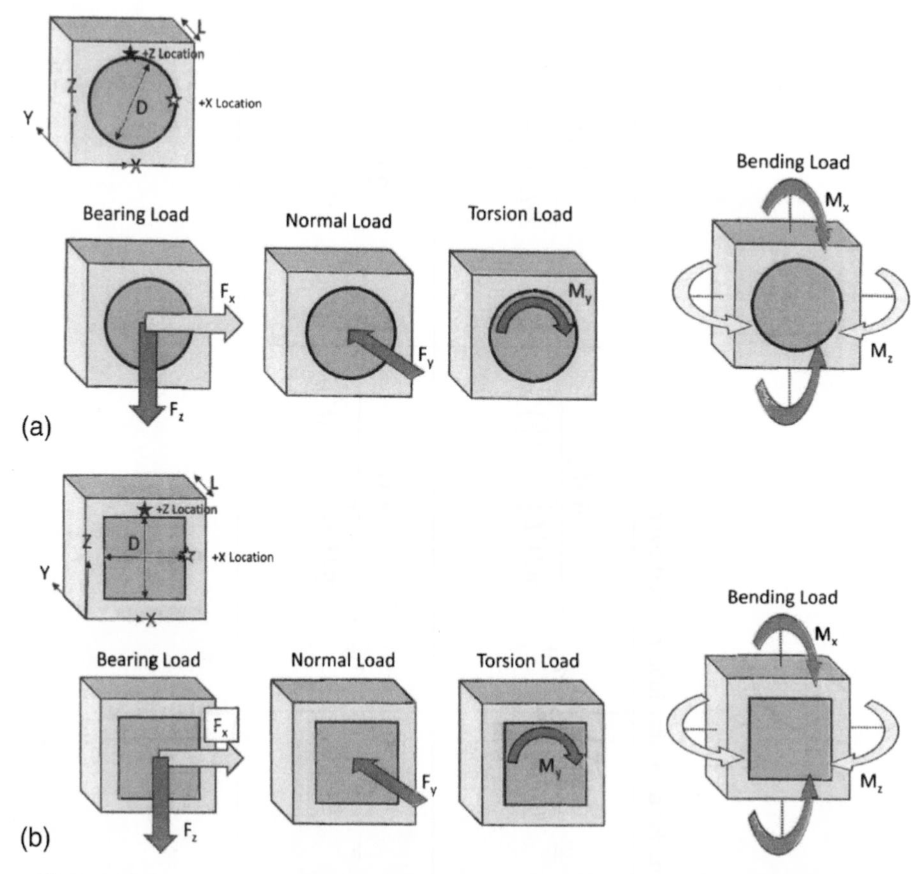

Figure 16.3 (a) Circular and (b) square tubes bonded over a boss in single shear.

for a circular tube bonded over a boss. These stresses are then simply divided by 2 to obtain the bonded-groove cases.

16.3.6 Square boss

If a boss is made square instead of round, an added benefit occurs under torsion, or twist shear, as a bearing load path also exists. Equations (16.16) through (16.22) can be modified to result in those presented in Table 16.2.

16.4 Bonded Flexures

Chapter 3 discusses a quasi-kinematic optic support that uses blade flexures. As noted in Section 3.3, rigidly bonding these flexures directly to an optic (without a pivot) does not release the required degrees of freedom about a radial line. Thus, if a flexure is bonded and inertial loads are applied, the bond sees not only direct shear load, but moment load as well. It is important to calculate the bond stresses induced by moment to preclude failure of the adhesive.

Table 16.1 Adhesive stress equations for a circular tube bonded over a boss in single shear.

+Z Location	$\sigma_z = \dfrac{F_z}{DL}$	$\tau_{xy} = \dfrac{F_y}{\pi DL}$	$\tau_{zx} = \dfrac{2M_y}{\pi D^2 L}$	$\tau_{xy} = \dfrac{4M_x}{\pi D^2 L\left[1 + \left(\frac{L}{D}\right)^2\right]}$	$\sigma_z = \dfrac{2M_x\left(\frac{L}{D}\right)^2}{DL^2\left[1 + \left(\frac{L}{D}\right)^2\right]}$
+X Location	$\sigma_x = \dfrac{F_x}{DL}$	$\tau_{yz} = \dfrac{F_y}{\pi DL}$	$\tau_{xy} = \dfrac{2M_y}{\pi D^2 L}$	$\tau_{yz} = \dfrac{4M_z}{\pi D^2 L\left[1 + \left(\frac{L}{D}\right)^2\right]}$	$\sigma_x = \dfrac{2M_z\left(\frac{L}{D}\right)^2}{DL^2\left[1 + \left(\frac{L}{D}\right)^2\right]}$

Table 16.2 Adhesive stress equations for a square tube bonded over a boss in single shear.

+*Z* Location	$\sigma_z = \frac{F_z}{DL}$	$\tau_{xy} = \frac{F_y}{4DL}$	$\tau_{zx} = \frac{M_y}{2D^2L\left[1+\left(\frac{L}{D}\right)^2\right]}$	$\tau_{xy} = \frac{M_x}{1.5D^2L\left[1+\left(\frac{L}{D}\right)^2\right]}$	$\sigma_z = \frac{2M_x\left(\frac{L}{D}\right)^2}{DL^2\left[1+\left(\frac{L}{D}\right)^2\right]}$
+*X* Location	$\sigma_x = \frac{F_x}{DL}$	$\tau_{yz} = \frac{F_y}{4DL}$	$\tau_{xy} = \frac{M_y}{2D^2L\left[1+\left(\frac{L}{D}\right)^2\right]}$	$\tau_{yz} = \frac{M_z}{1.5D^2L\left[1+\left(\frac{L}{D}\right)^2\right]}$	$\sigma_x = \frac{2M_z\left(\frac{L}{D}\right)^2}{DL^2\left[1+\left(\frac{L}{D}\right)^2\right]}$

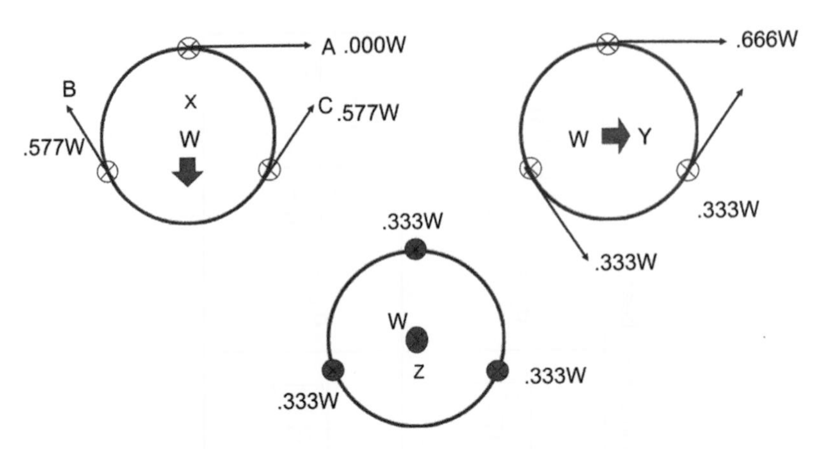

Figure 16.4 Idealized tangent flexure reactions assuming pure kinematics (pinned joints).

Consider the reactive flexure loads in all axes for a blade flexure, as is shown in Fig. 16.4 for a purely kinematic mount; i.e., all required degrees of freedom are released.

However, if the flexures are rigidly bonded, a torsional moment exists in the *Z* (vertical) direction of load (see Fig. 16.5), which will put both torsion and direct shear stress on the bond. This combination will stress the bond according to

$$\tau = \frac{kM}{bt^2} + \frac{W}{3bt}, \tag{16.23}$$

where b is the bonded area width, and t is its height. Here, $3 < k < 5$, depending on the rectangular shape (b/t ratio), as discussed in Section 1.3.5. The value of the moment can be calculated as that of a guided cantilever (see Chapter 1 and Table 1.1), or

$$M = \frac{WL}{6}. \tag{16.24}$$

Comparing this stress to the adhesive shear strength, the former is conservative at room temperature, as the stress is maximum at the outer edge

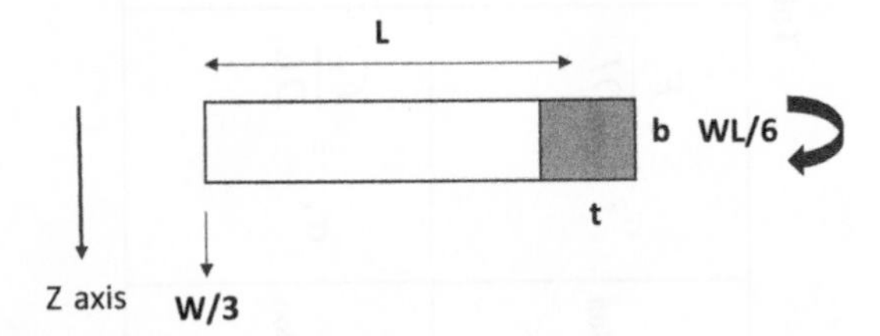

Figure 16.5 Rigidly bonded tangent flexures create a moment about the optic center, resulting in high bond stress.

but minimal at the center. The adhesive will plastically flow and not fail until reaching an average strength value, typically about one-half of the outside edge stress. However, counting on full plastic flow is not a good idea, as the bond may prematurely fail on repeated cycling or loading. Thus, reducing the k factor by no more than 1.5 is suggested.

For lateral (X, Y) loading, no moment exists for straight flexures, but if flexures are curved, bent, or otherwise nonrigidly attached to their mounting bezel, a peel, or cleavage, moment will exist about the Z axis. This will stress the bond approximately as

$$\sigma = \frac{V}{A} = \frac{V}{bt} + \frac{6Mz}{tb^2}, \tag{16.25}$$

where the moment is determined from more-detailed analyses. In this case, the epoxy cleavage strength is generally less than the shear strength, perhaps one-half of the value, so caution is advised.

16.4.1 Example for consideration

A certain tangent flexure 1 in. long (L) and 3/8 in. wide is adhesively bonded to an optic over a 3/8 in. square area, as schematically shown in Fig. 16.6. Due to envelope constraint, the eccentric bend is 1/8 in. For a vertical (individual) flexure launch load of 20 lbs (P) and a lateral load of 40 lbs (V), determine the adequacy of the bond if the allowable shear stress is 1250 psi and the allowable cleavage stress is 625 psi.

Here the bond area is 0.14 square in.; the bending moment for the vertical load as a guided cantilever is $M = PL/2 = 10$ in.-lbs. From Eq. (16.23),

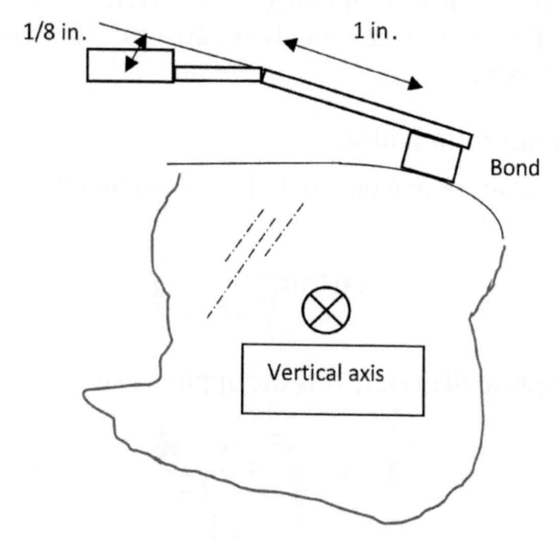

Figure 16.6 Bonded tangent flexure adhesively bonded to an optics, as in Example 16.4.1.

using $k = 4.8$ from Figs. 1.7 and 1.8 as well as Eq. (1.29), we calculate stress as $\tau = 1050$ psi, which is less than 1250 psi, so is okay.

For the lateral load, using Eq. (16.25), we find that $\sigma = 500$ psi, which is less than 625 psi, so is okay. We have not used any plastic reduction factor k in these calculations.

16.5 Contact Stress

Contact stresses arise when two bodies are loaded in contact with one another over a small area. These stresses are quite localized in nature; a few characteristic distances away, they are computed in the usual fashion. For ductile materials, contact stresses are forgiving, as yield may occur, along with stress redistribution. However, these contact stresses cannot be ignored; permanent set will occur, and localized imperfections could be sources for fatigue failure after repeated load. For brittle materials, contact stress is not forgiving, and fracture can readily occur. Thus, contact stress analysis is important for design.

Chapter 3 discusses kinematics and explains that an ideal kinematic mount resists load in only six stable degrees of freedom. Ideally, this mount consists of a three-point mount support: ball on flat (one degree of freedom), ball on cone (three degrees of freedom), and ball in groove (two degrees of freedom) for the mandated six restraints. Again, near-kinematic, or quasi-kinematic, mounting is preferred over 'ideal' kinematic mounting, which is never ideal due to friction, stiction, locking, and moving potential. However, in many designs (e.g., a pivot design for a fast-steering mirror, a ball bearing design, a hard stop design, etc.), contact stress is unavoidable in a design.

Many sources are available that provide detailed contact stress calculations for a variety of cases. It is not the intent to go into the details of the theory of elasticity required to solve such cases. Here, we point out the nature of contact stress for the most commonly occurring conditions: ball on flat and ball in cone or groove.

16.5.1 Ball-on-flat formulation

For a spherical surface resting on a flat plate, the contact[6] stress is computed as

$$\sigma = 0.918\sqrt[3]{\left(\frac{P}{K^2\lambda^2}\right)}, \tag{16.26}$$

where K is the sphere diameter, P is the applied load,

$$\lambda = \frac{1-\nu_1^2}{E_1} + \frac{1-\nu_2^2}{E_2}, \tag{16.27}$$

and the contact displacement y is

$$y = 1.04\sqrt[3]{\left(\frac{P^2\lambda^2}{K}\right)}.$$

The contact radius a is given as

$$a = 0.721\sqrt[3]{(PK\lambda)}. \tag{16.28}$$

16.5.2 Ball-in-cone formulation

To understand the case of a spherical surface resting in a cone, we first look at a cylinder on a flat with line contact. The contact stress is computed[7] as

$$\sigma = 0.798\sqrt{\frac{p}{D\lambda}}, \tag{16.29}$$

where $p = P/L$. The contact rectangular width b is given as

$$b = 1.6\sqrt{(pD\lambda)}, \tag{16.30}$$

and the contact displacement y is given as

$$y = \frac{2p\lambda}{\pi}\left[\frac{1}{3} + \ln\left(\frac{2D}{b}\right)\right], \tag{16.31}$$

where L is the length of the cylinder.

16.5.2.1 Examples for consideration

Example 1. Consider a stainless-steel spherical ball of diameter 0.50 in. resting on a flat plate of the same material under a compressive load of 25 lbs. Compute the contact radius, deformation, and resulting compressive stress.

We have (from Table 2.2) $E_1 = E_2 = 29$ Msi, and $\nu_1 = \nu_2 = 0.30$. We find upon substitution into Eqs. (16.26) through (16.28) that $\sigma = 266{,}000$ psi, $y = 0.00018$ in., and $a = 0.0067$ in.

Example 2. Consider a 1-in.-long stainless-steel cylinder with a diameter of 0.50 in. resting on a flat plate of the same material under a compressive load of 25 lbs. Compute the contact radius, deformation, and resulting compressive stress. Compare the results to the results of Example 1.

We find upon substitution that $\sigma = 22{,}500$ psi, $y = 0.0000075$ in., and $b = 0.0015$ in. Note that both stress and deflection are reduced by an order of magnitude compared to the spherical case in Example 1.

16.5.3 Ball-in-cone analysis

For a ball in cone, we compute the developed length of the cone contact, which results in the case of a cylinder on a flat with line contact. With reference to Fig. 16.7(a), for the vertical load P, the developed length of contact is

$$L = \pi D \cos \Theta. \tag{16.32}$$

The reactive force is related to $P/(\sin \Theta)$, and the deflection is related to $P/(\sin^2 \Theta)$. We can thus rewrite Eq. (16.29) as

$$\sigma = 0.798\sqrt{\frac{q}{D\lambda}},$$

where $q = \frac{P}{\pi D \cos \Theta}$. The contact rectangular width b is given as

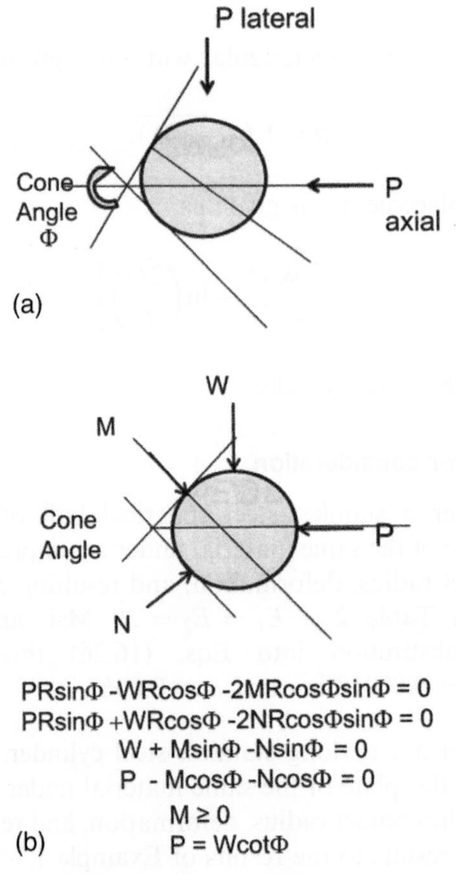

Figure 16.7 (a) Ball-in-cone arrangement and (b) ball in cone loaded laterally plus the equations of equilibrium.

$$b = 1.6\sqrt{\frac{q}{\sin\Theta}D\lambda},$$

and the contact displacement is given as

$$y = \frac{\frac{2q\lambda}{\pi}\left[\frac{1}{3} + \ln\left(\frac{2D}{b}\right)\right]}{\sin^2\Theta}. \tag{16.33}$$

With reference to Fig. 16.7(b) for the lateral load W, the developed length L of contact (with only one-half contact across the circumference) is

$$L = \frac{\pi D\cos\Theta}{2}.$$

The reactive force is related to $W/(\cos\Theta)$, and the deflection is related to $P/(\cos^2\Theta)$. The force is sinusoidally distributed, and we can rewrite Eq. (16.29) as

$$\sigma = 0.798\sqrt{\frac{2q}{D\lambda}}.$$

The contact rectangular width b is given as

$$b = 1.6\sqrt{\frac{2q}{\cos\Theta}D\lambda^2},$$

and the contact displacement y is given as

$$y = \frac{\frac{4q\lambda}{\pi}\left[\frac{1}{3} + \ln\left(\frac{2D}{b}\right)\right]}{\cos^2\Theta}. \tag{16.34}$$

The contact stiffness under axial preload K_a and under lateral in-plane gravity load K_l can be computed as a developed cylinder on a flat plate for a ball on cone by alternative techniques.[8,9] The equations are basically for a cylinder housed between two flat plates, modified for the sinusoidal and cosinusoidal distribution of load and resulting displacement.

Stiffness is computed[7] [from Eqs. (16.33 and 16.34)] as

$$K_a = \frac{\pi^2 D\sin^2\theta\cos\theta}{2\lambda\left[\frac{1}{3} + \ln\left(\frac{2D}{b}\right)\right]}, \tag{16.35}$$

$$K_l = \frac{\pi^2 D \cos^3\theta}{4\lambda\left[\frac{1}{3} + \ln\left(\frac{2D}{b}\right)\right]}. \tag{16.36}$$

Alternatively,[8,9]

$$K_a = \frac{2\pi^2 E_c R \cos\alpha \sin^2\alpha}{\ln\left(\frac{4d_1}{b_{max}}\right) + \ln\left(\frac{4d_2}{b_{min}}\right) - 2}, \tag{16.37}$$

$$K_l = \frac{\pi^2 E_c R \cos^3\alpha}{\ln\left(\frac{4d_1}{b_{max}}\right) + \ln\left(\frac{4d_2}{b_{min}}\right) - 2}. \tag{16.38}$$

Here,

$$b_{max} = \sqrt{\frac{4P_{max}R}{\pi E_c}},$$

$$b_{min} = \sqrt{\frac{4P_{min}R}{\pi E_c}},$$

$$P_{max} = \frac{1}{2\pi R\cos\alpha}\left(\frac{F_a}{\sin\alpha} + \frac{2F_r}{\cos\alpha}\right),$$

$$P_{min} = \frac{1}{2\pi R\cos\alpha}\left(\frac{F_a}{\sin\alpha} - \frac{2F_r}{\cos\alpha}\right),$$

$\alpha = 90$ deg (cone angle)/2 (deg) $= 45$ deg; the cone angle (Fig. 16.7) is 90 deg; E is Young's modulus $= 44{,}000{,}000$ psi (alumina–silicon nitride); $F_r = 10$ lbs (weight of the optic-moving mass); $F_a = 13$ lbs (preload); $E_c = E/2$; b is the half-width of the contact; $d_1 = R$; $d_2 = 10R$; and R is the ball radius $= 0.25$ in.

Note that Eqs. (16.35) and (16.36) give results that are nearly identical to those of Eqs. (16.37) and (16.38), respectively.

16.5.4 Kinematic coupling

A three-tooth kinematic coupling[10,11] is reviewed for the resulting contact stresses when clamping forces are applied. Results show that high preload and g-forces are acceptable for many applications using line contact in contrast to devices that use spherical contacts, which produce significantly higher stress.

Use of a kinematic coupling is attractive when precision registration of optics is required for optical systems. Unlike traditional ball-socket-groove kinematic mounts, a coupling makes use of cylindrical line contacts that reduce local contact stresses. A schematic of the device, which maintains six degrees of freedom, is shown in Fig. 16.8(b). The coupling consists of two halves: a diametrical body with an inner recessed cavity that has cylindrical teeth on one half and flat planes on the other. These provide the necessary six kinematic degrees of freedom without overconstraint, except for frictional forces, which are not critical, as independent mount systems will accommodate these forces. The kinematics provide for repeatable registration.

For precision systems that may be subjected to high g-forces, the coupling halves must be clamped together to resist such forces. These clamping forces induce contact stresses between the cylinder and the flat plane that must be kept to allowable values to prevent fretting, wear, or damage. To this end, we review these stresses as a function of clamping force and contact area in the following subsections.

16.5.4.1 Contact stress

Consider in Fig. 16.8 the couplings under clamping load in which the load is shared by three line contacts of length L. The length L equals the outer diameter of the body D minus the diameter of the recessed cavity. To provide registration, ensure repeatability, and minimize stresses, the cylinder contact radius is suggested to lie between 0.5 to 2.0 times the outer body radius with a preferred[7] value of 1.0. The flat contact planes are suggested to be at a 45-deg angle to the cylinder with a height of 1/15 of the body radius.

From the standard theory of Section 16.5.2, the compressive contact stress σ is again given as [Eq. (16.29)]

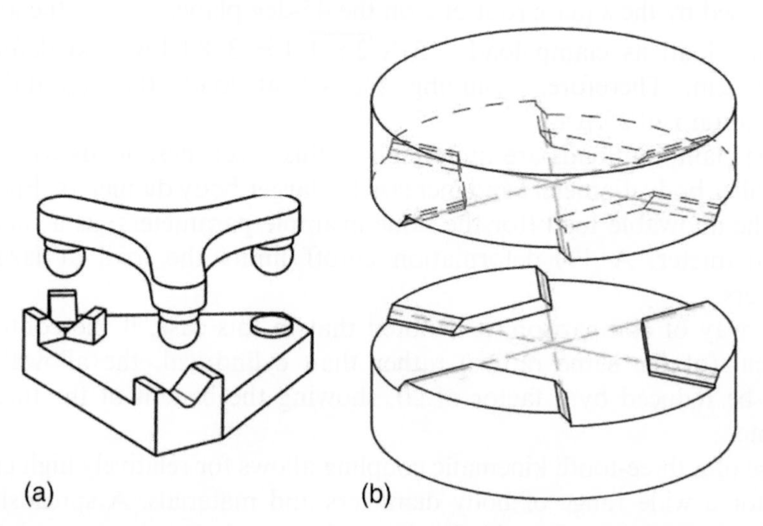

Figure 16.8 (a) Standard kinematic mount. (b) Kinematic coupling reduces contact stress (figures adapted from Ref. 11).

$$\sigma = 0.798\sqrt{\frac{q}{K\gamma}},$$

where q is the contact load P divided by the contact length L; K is the cylindrical contact diameter, where $R < K < 4R$, and R is the body radius, which is $D/2$; and $\gamma = 2(1 - \mu^2)/E$, where μ is the material's Poisson ratio, and E is the material elastic modulus.

The surface deformation is given as

$$y = \frac{2q\gamma}{\pi}\left[\frac{1}{3} + \ln\left(\frac{2D}{b}\right)\right],$$

where the contact width $b = 1.6\sqrt{qK\gamma}$. We compute both stress and deflection under load, and compare the results to allowable values, or else we compute the allowable load under a given allowable stress or deformation.

16.5.4.2 Example

Consider an Invar 36 coupling having an outer diameter of 1 in. and an inner-cavity recess diameter of 0.5 in., resulting in a contact length of 0.5 in. Consider the cylindrical contact radius to equal $2R = D$, the outer diameter. The elastic modulus of Invar is 2.05×10^7 psi, and its Poisson ratio is 0.3. The ultimate strength of Invar is 71,000 psi, so its Hertzian strength is 1.732 (71,000) = 123,000 psi. Using a safety factor of 1.40, and noting that the clamping load is shared by 3 line contacts, and that the resultant normal load is increased by the square root of 2 on the 45-deg plane, we find the allowable clamping load as clamp load $= 3P\sqrt{2 \times 1.4} = 3780$ lbs, and deformation $y = 25$ μm. Therefore, assuming 1% set at load, the repeatability of deformation is 0.25 μm.

The clamping loads are quite high in this case; these loads will decrease for smaller body diameters and increase for larger body diameters. Figure 16.9 plots the allowable load (for the same example parameters) as a function of body diameter. A 1% deformation cutoff limits the load at large body diameters.

By way of comparison, it is noted that in this case, if the contact were spherical (of the same radius) rather than cylindrical, the allowable load would be reduced by a factor of 20, showing the benefit of the three-tooth coupling.

Use of a three-tooth kinematic coupling allows for relatively high clamping loads for a wide range of body diameters and materials. A spreadsheet can assist in determining allowable load and deformation with appropriate safety factors applied.

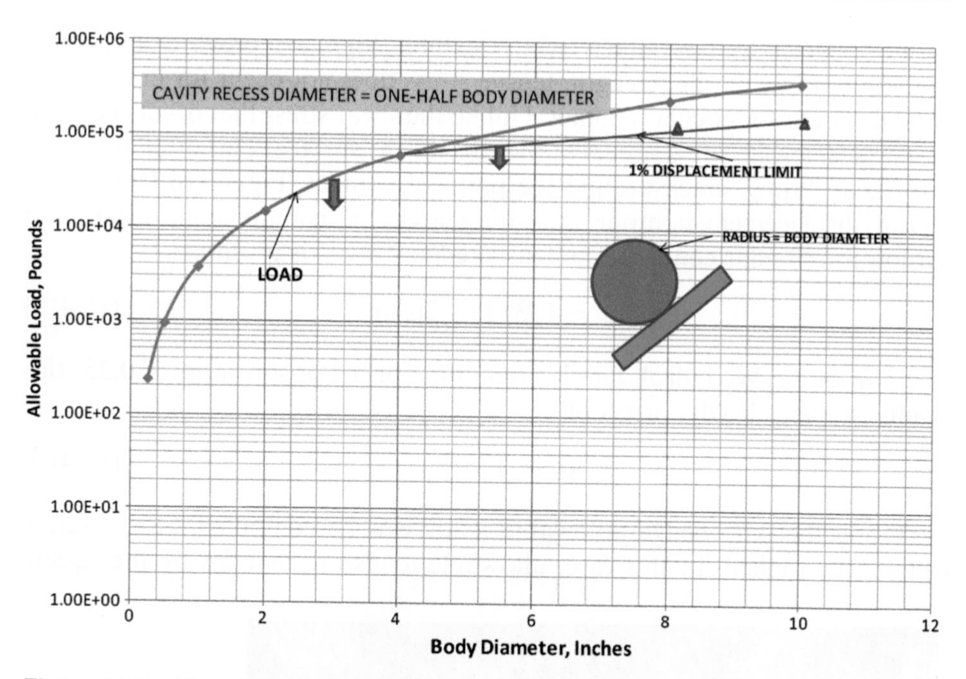

Figure 16.9 Allowable coupling load versus body diameter for use in Example 16.5.4.2.

16.5.5 Allowable load: Hertzian stress

For Hertzian-type contact stresses, allowable stresses are higher than the usual ultimate material strength due to the local and compressive nature of Hertzian stress. Various sources show Hertzian strength to be 1.6 to 4 times the material ultimate compressive strength. Almost all sources indicate that failure occurs not in compression but in shear and occurs somewhat below the contact surface, and that Hertzian shear stress values are one-third of the calculated compressive stress values.

Using this assumption, and noting that (using strain energy theory) for most metals, the shear strength is $1/\sqrt{3}$ of the compressive strength, we find an allowable compressive strength of

$$S = \frac{3\sigma}{\sqrt{3}} = 1.732\sigma, \tag{16.39}$$

which lies at the lower end of the range of values found in the literature and is thus conservative.

For materials such as beryllium, which exhibits a shear strength of 80% of its compression strength, we find that

$$S = 2.4\sigma. \tag{16.40}$$

For glasses and ceramics, which exhibit high compressive and shear strength but low tensile strength, tensile stress does not occur for line contact but does occur for circular contact. Unlike the shear stress that occurs beneath the surface, tensile stress occurs at the edge of the surface. This often results in a Hertzian "ring" crack around the point of contact. Figure 16.10 shows a typical displacement contour indicating areas of compression, shear, and tension for circular contact. The tensile edge stress[12] is given as

$$\sigma_t = 1/3\,(1 - 2\mu)\,\sigma_c, \tag{16.41a}$$

where σ_t is the Weibull A tensile strength. For a Poisson ratio of 0.25, this results in a tensile edge stress of

$$\sigma_t = 1/6\,(\sigma_c). \tag{16.41b}$$

Using this knowledge, allowable loads can be computed with a safety factor also applied. In this case, allowable loads are reduced by the square

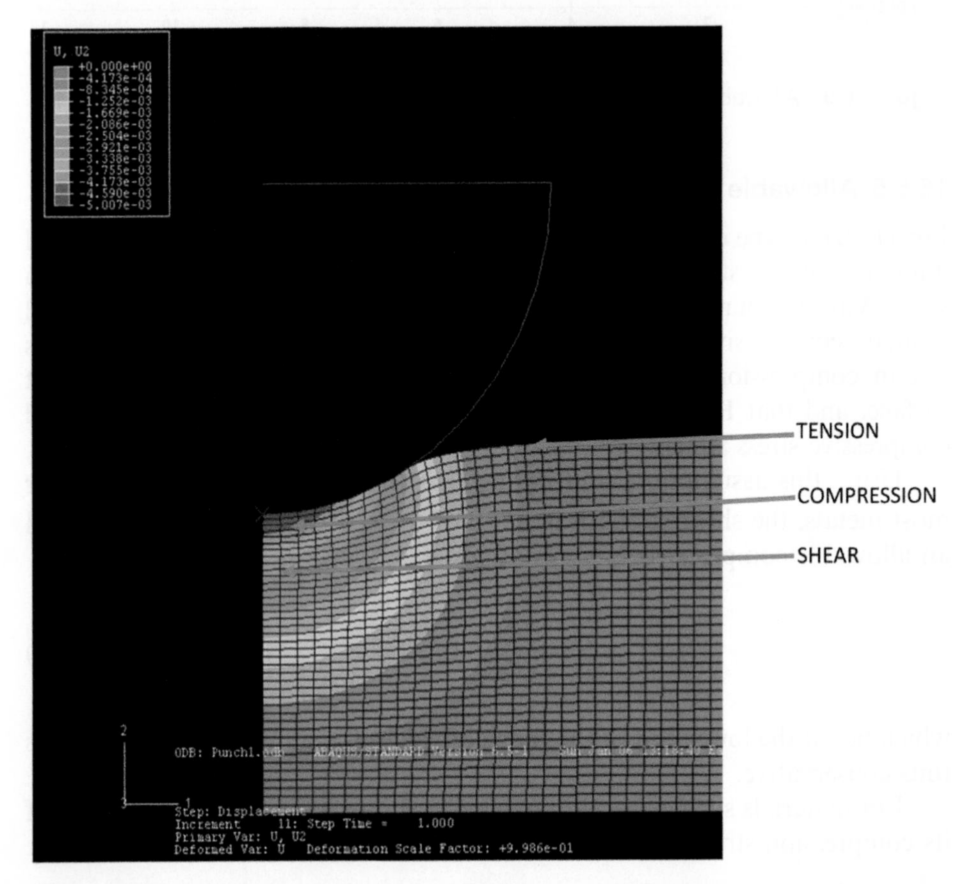

Figure 16.10 Typical stress pattern for a sphere on a flat, indicating areas of maximum compression, shear, and strain.

root of the safety factor, since load is not linear but square-root proportional to stress. Finally, deformation is computed; in general, stress will be the driver as long as the deformation is elastic; however, as load and deformation increase, one may consider using a fractional percentage permanent set that may affect repeatability.

16.6 Friction

Friction is a resistant force encountered when two surfaces move over one another. This force is independent[13] of the apparent area of contact. It is, however, dependent on the finish of the surfaces.

The coefficient of friction μ between two materials is the relationship between the normal force N acting on an object and the lateral force F required to move it, as

$$F = \mu N. \tag{16.42}$$

Consider the object in Fig. 16.11 resting on an inclined plane in the presence of gravity. From equilibrium considerations, we have

$$\begin{aligned} \Sigma Fxx' &= 0, \\ F &= W\sin\Theta, \\ \Sigma Fyy' &= 0, \\ N &= W\cos\Theta. \end{aligned} \tag{16.43}$$

Here F is the friction force, W is the object weight, and Θ is the angle of incline. Substituting Eq. (16.42) into Eq. (16.43) gives

$$\mu = \tan\Theta. \tag{16.44}$$

Thus, for a general indication of the friction coefficient, we can rest an object on a horizontal plane and pivot the plane through an angle, measuring

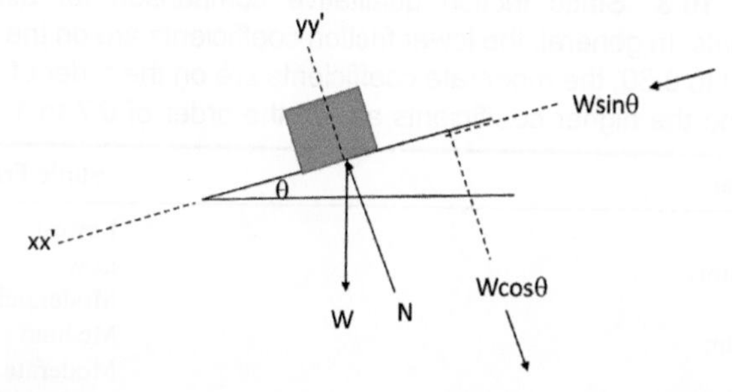

Figure 16.11 Diagram of an object resting on an inclined plan in the presence of gravity showing friction equilibrium.

the angle where the object first starts to slip. The coefficient of friction is the tangent of that angle.

While this is a simplistic view, friction is anything but simple. Over the years, many theories[14] have been proposed to explain it. Depending on the surface, there are intermolecular and adhesive forces at play, as well as plowing action and surface roughness, or asperities (points of contact) with which to contend. As such, the coefficient of friction depends on the true contact area and can widely vary.

Most theories indicate that the coefficient of friction depends on the true area of contact, not the apparent area of contact. Thus, when two optical surfaces are perfectly lapped and polished, the very high contact results in a large surface area. Intermolecular actions govern here, and the coefficient of friction can be quite high. With such optical contact, the surfaces are difficult to separate.

If two surfaces are smooth but not lap polished, the asperities tend to be fewer because the object rolls only at several points, resulting in low friction. If the surface is roughened, the asperities and true area increase, resulting in higher friction. Here, plowing action can be developed, as the asperities are plastically deformed[15] to produce motion.

The asperity contact radius can be determined with Hertzian theory, assuming that this radius is much smaller than the radius of curvature of the asperity. The theory[16] also shows that the higher the material strength the lower the friction. Still other theories show that friction is dependent on elastic asperity deformation and load magnitude. Others show the friction coefficient decreasing as the material tensile strength increases. Finally, if asperities are high and load is light, friction tends to be lower again.

Because friction is so highly dependent on many factors, no tabular data is presented here. In general, the static friction coefficient can range from about 0.05 (on a highly lubricated surface) to more than 1.0, with a typical range being between 0.2 and 0.8 for many materials, again dependent on the finish. Table 16.3 gives a qualitative summary of static friction for various materials.

Table 16.3 Static friction qualitative comparison for different materials. In general, the lower friction coefficients are on the order of 0.03 to 0.30, the moderate coefficients are on the order of 0.3 to 0.7, and the higher coefficients are on the order of 0.7 to 1.2.

Material	Static Friction
Ice	Lowest
Lubricants	Low
Steel	Moderately low
Titanium	Medium
Glass	Moderate
Aluminum	Moderately high
Silver	High
Rubber	Highest

16.6.1 Surface roughness

Based on the above discussion, the literature indicates that surface roughness is important to reducing friction: The greater the roughness the greater the friction. This relationship is due to contact points of the roughness, called asperities, which are the peaks of the roughness. Roughness is defined as R_a or s, is measured as mean or RMS, and is reported in either microns or micro-inches.

When a spherical ball is preloaded to a flat, a contact stress and a contact area are created. If the asperity is high, and if the preload is light, contact is only made on the high points, and the contact area is greatly reduced from the Hertzian calculation. The influence on surface roughness is negligible when the parameter

$$\alpha = \frac{\sigma R}{a^2} < 0.05. \tag{16.45}$$

Here R is the sphere radius, and a is the contact area radius; σ is the roughness dimension in inches, and the value of α is nondimensional.

Note from Eq. (16.45) that the smaller the contact radius a the higher the value of α. For high preloads, the value of a will increase, resulting in a lower α value; this means that the roughness concern applies particulary to light loads.

For example, consider a contact ball with a spherical radius of 0.25 in. and contact load of 13 lbs. For a ball (sphere) on flat we compute from Eq. (16.28) a contact area radius a of 0.0048 in. Substituting into Eq. (16.45), we find a required roughness of $\sigma = a^2\alpha/R = 4.6\,\mu\text{in}$. This will keep friction to a minimum.

16.7 Large Displacements

When displacements exceed one-half of the material thickness, the small-displacement theory we have been assuming does not apply. In this case, membrane action will prevent excessive bending stress from developing. For example, if a pressure load is applied to a drumhead, most of the load will be membrane induced and not bending induced. This has application to optical supports, e.g., as in a diaphragm mount designed to be radially stiff but axially soft.

The equations for combined bending and membrane loads are transcendental and nonlinear,[17] and require significant calculations and tables to solve, or, alternatively, require complex nonlinear finite element model approaches. However, assuming that membrane stresses alone exist, we can approximate the drumhead as a tension-carrying membrane only[18] by computing displacement under pressure as

$$y = \sqrt[3]{\frac{qa^4}{K_2 Et}}, \tag{16.46}$$

where $K_2 = 3.44$, E is the plate elastic modulus, and t is the plate thickness.

We then compute stress as

$$\sigma = \frac{K_4 y^2 E}{a^2}, \tag{16.47}$$

where $K_4 = 0.748$ at the edge, and $K_4 = 0.965$ at the center.

For other edge conditions and for rectangular plates, where uniform pressure is applied, stress–strain handbooks[19] can be consulted.

16.8 Windows

Windows are required for many optical systems to protect the critical components while allowing light to be properly transmitted. Although distortion of a flat window is not as critical as distortion of a powered optic, wavefront aberrations in the window do need to be addressed.

16.8.1 Bending

Windows can deform under a variety of conditions, including pressure differential (bending) and axial (through-thickness) temperature gradients, as seen in Chapter 4. Here we introduce an optical path difference given by Barnes[20] as

$$\mathrm{OPD} = \left[(n-1)\frac{h}{2n}\right]\left[\left(\frac{\partial Y}{\partial x}\right)^2 + \left(\frac{\partial Y}{\partial y}\right)^2\right], \tag{16.48}$$

where n is the window index of refraction, and h is its thickness. Knowing the deflection of the bent surface, we can then compute the OPD. Fortunately, for most applications, this effect is small. For example, substituting the axial thermal gradient displacement from Eq. (4.28) into Eq. (16.48) gives

$$\mathrm{OPD}_{\mathrm{ax}} = \frac{h(n-1)}{2n}\left(\frac{\alpha \Delta T_{\mathrm{ax}} r}{h}\right)^2. \tag{16.49}$$

Note that the squared term will assist in keeping the OPD small. An example of keeping the OPD small is provided in the next subsection.

16.8.2 Lateral thermal gradient

Lateral temperature gradients will not deform a flat window to a large extent but will produce undesirable effects due to refractive index n and, most notably, its change with respect to temperature, dn/dT. Here, the optical path difference is given as[20]

$$\mathrm{OPD_r} = h\left[\alpha(n-1)(1+v) + \frac{dn}{dT}\right]\Delta T_{\mathrm{r}}. \tag{16.50}$$

We see a stronger dependence on the CTE than in the bending case, and a dominant dependence of the index of refraction change with respect to temperature. To this end, Table 16.4 gives these properties for a low-expansion glass with poor *dn/dT* and a high-expansion glass with excellent *dn/dT*. We can illustrate a comparison in the following example.

Example. A 6-in.-diameter glass window that is 0.6-in. thick is subjected to a 0.1 °C radial gradient and 1 °C axial gradient. Compute and compare the OPD for fused silica and BK-7 glass candidates.

Under the lateral gradient, we have [with reference to the data of Table 16.4 and Eq. (16.50)], OPD = 7.32×10^{-7} in. (0.029 λ peak visible waves) for fused silica and OPD = 3.64×10^{-7} in. (0.0145 λ peak visible waves) for BK-7. The BK-7 window outperforms fused silica by a factor of 2 despite its high thermal expansion, due to its excellent *dn/dT*.

Under the axial gradient, we have [with reference to the data of Table 16.4 and Eq. (16.49)], OPD = 3×10^{-8} peak visible waves for fused silica and OPD = 5×10^{-6} peak visible waves for BK-7, both of which are quite negligible despite the fact that the axial gradient is tenfold that of the radial gradient. Results are shown in Table 16.5.

16.9 Dimensional Instability

Dimensional instability refers to the property of all materials by which a change of dimension takes place due to internal processes and/or an external environment. As we know from previous sections, materials will change dimension under a load environment (strain) or temperature environment (eigenstrain). In the former case, strain is accompanied by stress; in the latter

Table 16.4 CTE and index of refraction for select glasses.

Material		Modulus (Msi)	Poisson's Ratio	CTE PPM/°C (α)	Index (n)	*dn/dT* (PPM/°C)
Glass	Fused Silica	10.6	0.17	0.56	1.47	11.9
Glass	BK-7	11.8	0.21	7.1	1.52	1.6

Table 16.5 Low *dn/dT* improves performance more than low CTE under lateral gradient. Axial gradients have minimal effect on performance.

Glass Material	Thickness (inch)	*dn/dT* (PPM/°C)	Radial Gradient (°C)	Axial Gradient (°C)	OPD Error λ p-p surface	
			ΔT		Radial	Axial
Fused Silica	0.6	11.9	0.1	1	0.029	3.00E-08
BK-7	0.6	1.6	0.1	1	0.015	5.00E-06

case, stress may also occur if the material is constrained from movement. While we expect such stress and strain to be stable, dimensional instability can occur in the absence of a change in external environment, or under the influence of an internal process.

For example, if a material is strained at a fixed temperature or fixed load yet continues to change dimension, then instability is occurring. The sources of instability are numerous and all involve internal changes at the molecular level. Some common examples are creep, hysteresis, residual (internal) stress, external stress relaxation, and glass transition. By definition, all instability is temporal.

16.9.1 Glass transition temperature

The effects of glass transition temperature are fully discussed in Chapter 9, where a one-time strain change is noted when certain materials, most notably polymers, pass through a brittle (glassy) and more ductile (rubbery) stage under load. This results in a change in both modulus and CTE (methods to account for these property changes in analysis are discussed in Chapter 15). Of prime interest is a phase change that occurs at the glass transition temperature, resulting in a creep effect that causes a permanent change in dimension that is not reversible if the glass transition has not been previously passed through. Subsequent cycling beyond the glass transition does not extend the creep process. This change can be on the order of 3%. The glass transition creep must be accounted for if the epoxy is used in a critical, optical-alignment-sensitive region. This creep occurs as a function of stress level and could be higher at higher stresses. Typically, at low stress (under 200 psi), the change is not necessarily high enough to produce misalignment (a 3% change in a 0.005-in.-thick epoxy bond line results in a set of 0.00015 in.). On the other hand, if occurring during bonding to a finished optic, such a change could change the figure error. Often, it is desirable to bond to optics prior to finishing to avoid this costly concern.

16.9.2 Hysteresis

Chapter 2 presents the resulting small changes in both CTE and hysteresis for a particular glass ceramic after cycling to warm temperatures. Hysteresis is dependent on the cooling rate and can be controlled. Hysteresis (permanent strain change) can also occur in other materials (after thermal cycling) due to various processes. For example, due to grain redistribution, early forms of vacuum hot-pressed (VHP) beryllium did not return to zero strain after cycling to cold temperatures and strained permanently even farther after multiple cycling. This anomaly has largely been solved with hot isostatic pressing (HIP) processes and, in particular, is absent with fine-powder beryllium productions, such as optical-grade O30.

As another example, graphite and fiberglass composites (as well as metal matrix composites) can also exhibit hysteresis and a change in CTE after cycling to cold temperatures. This is due to the large differences in thermal expansion between the glass (or graphite) fibers and the epoxy resin system. The cold soak to cryogenic extremes results in localized stresses producing translaminar stress relief, also known as micro-cracking. These micro-cracks cause the CTE to shift permanently toward the lower-CTE fibers. The micro-cracks also produce strength degradation on the order of 10% to 20% and are not catastrophic if a sufficient safety margin is provided. Figure 16.12 shows the CTE shift after thermal cycling (and is the same as Fig. 13.5). The micro-crack phenomenon can be greatly reduced by appropriate fiber layups and the use of ultra-thin lamina. Advances in resin matrix technology and, in particular, the use of cyanate esters have significantly reduced concerns.

16.9.3 External stress relation

Chapter 7 shows the effects of molecular rearrangement of ion groups (alkali oxides) for certain glass ceramics caused by stress relaxation (even at room temperature) after application of external stress. (This so-called Pepi effect was first discovered by Benjamin Franklin, placing the author in good company.) Such changes occur both at a fixed load (delayed strain creep) and after removal of the fixed load (delayed strain recovery). These delayed effects are more pronounced at warmer temperatures. The alkali oxides causing this phenomenon are cesium oxide, rubidium oxide, potassium oxide, sodium oxide, and lithium oxide. While all of these oxides produce the delayed effect, the highest contributor is lithium oxide.

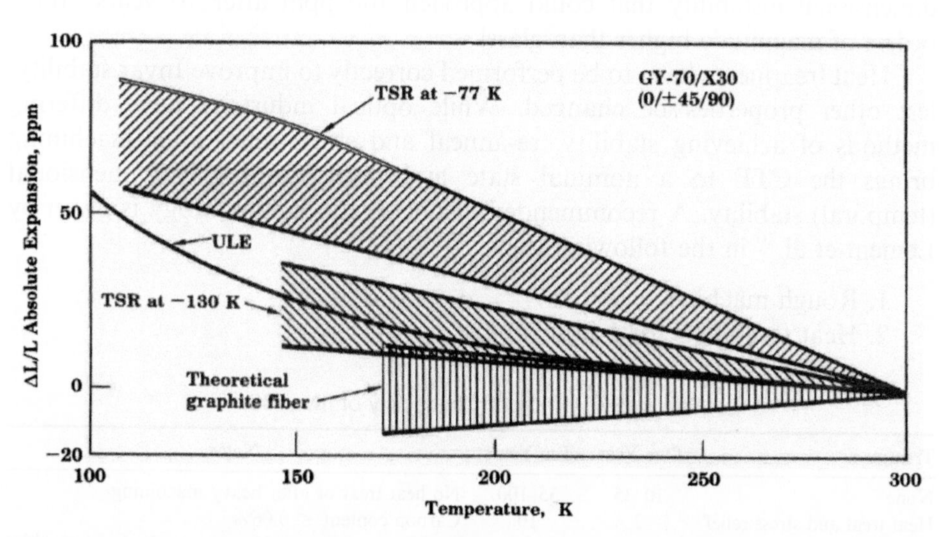

Figure 16.12 Translaminar stress relief (TSR) of a thick-ply composite causing a negative shift in CTE over cold soak as fibers dominate (reprinted from Ref. 21).

16.9.4 Creep

Creep is another form of dimensional instability in a fixed-temperature environment, particularly on the warm side of room temperature. While creep is pronounced at extremely high temperatures for many metals, it is particularly pronounced for epoxy adhesives at relatively cool temperatures: on the order of 25 °C to 70 °C and increasing with temperature. Chapter 9 discusses this phenomenon in relation to adhesives.

16.9.5 Glass and ceramics

Glass and glass ceramics (such as ZERODUR®) have excellent temporal stability. Left to themselves, changes on the order of only 0.1 ppm/year[22] have been noted. Thus, after 10 years, a strain change on the order of 1 ppm can be expected. This is good news for glass optics, where stringent performance is indicated. Note that the excellent stability is true for Zerodur as well as for fused-silica glass, the former of which is subject to delayed elastic[23] effects (Chapter 7) under stress and hysteresis effects in warm[24] and cold[25] environments.

16.9.6 Invar 36

Chapter 2 points out some excellent (if not peculiar) properties of Invar 36, one of these being Invar's excellent CTE. However, Invar's temporal stability (expansion with time often in the absence of temperature change) is highly dependent on heat treatment. Variations in carbon content and heat treatments have an effect on temporal stability, as seen in Table 16.6. Note that not including a heat treatment gives high dimensional instability, while reducing carbon content has a positive effect. Note also that untreated Invar has a dimensional instability that could approach 100 ppm after 10 years—three orders of magnitude higher than glass!

Heat treatments have to be performed correctly to improve Invar stability, lest other properties be changed. While optical industries have differing methods of achieving stability, re-anneal and stress relief after machining brings the CTE to a nominal state and results in higher dimensional (temporal) stability. A recommended, if not costly, methodology is given by Lement et al.[26] in the following steps:

1. Rough machine.
2. Heat to 840 °C for 1 hour in salt.

Table 16.6 Temporal instability of Invar 36.

Treatment	One Year	Ten Years	Notes
None	10–35	35–100	No heat treat or after heavy machining
Heat treat and stress relief	2	10	Carbon content < 0.06%
Special low carbon	1	3	Carbon content < 0.02%; *minimal* other impurities

Temporal instability values in ppm

3. Water quench.
4. Semi-finish machine.
5. Stress relieve at 315 °C for 1 hour.
6. Air cool.
7. Stress relieve at 93 °C for 24 hours.

While this heat treatment will result in temporal stability, the quench operation will lower the CTE to an uncertain value, so caution is advised. Note that eliminating carbon and other impurities[26] will result in a very stable Invar; however, such a material is essentially unavailable without great cost. The usual heat treatment methods are preferred unless absolute stability is required.

16.9.7 Internal (residual) stress

Perhaps the most common form of instability in optical systems is due to residual stress. Residual stress is a material's inherent internal stress due to various processes. Residual stress in a particular member results in self-equilibration; i.e., there are no net forces on the member. Thus, if residual stress is highly compressive over a particular layer, it is balanced by tensile stresses over the remaining layer. If the compressive layer is relatively thin, the tensile stresses will be considerably lower in magnitude due to force equilibrium.

16.9.7.1 Deposition residual stress

Chapter 4 discusses the bimetallic effect under thermal soak due to CTE differences between a thinly coated or clad optic relative to its substrate. A similar effect will occur in the presence of residual stress in the coating or cladding. These stresses are induced in the manufacturing deposition and are dependent on the process. An example of deposition residual stress for a clad optic is illustrated at the end of this subsection.

Because the stress is residual, it will be subject to stress relaxation; i.e., the stress will change with time and cause dimensional change. This is particularly evident when the temperature is raised above room temperature as molecular motions increase and cause redistribution of the stresses. The higher the temperature the more the relaxation. Thus, it is important to cycle critical optical components to just above the maximum temperature they are to experience in an operational or non-operational (survival) environment. The stress relaxation at temperatures below the deposition temperature can be very small but might be too large for the stringent performance criteria demanded of high-acuity optical systems.

The processes involved in coating optics (usually submicron thickness) and cladding optics (up to 100-μm thickness) give rise to internal residual stress. This stress is due in part to cooling from high temperatures (CTE differential) and in part to the deposition process itself. Depending on the process, residual stresses of 4000 to 40,000 psi can exist. The equations of

Chapter 4 show how these stresses can deform optics; the thicker the deposition the higher the deformation.

After coating an optic, one must simply accept this deformation. After cladding, on the other hand, the induced deformation can be polished out prior to coating. However, the residual stress still exists. Accordingly, stress relaxation must be accommodated.

Residual stress relaxation in claddings can be relieved by thermal cycling to very high temperatures;[27] it can be partially relieved by thermal cycling to just above and just below survival temperatures, and can certainly be relieved by thermal cycling to above coating temperature. Experience has shown that it will take three or more thermal cycles to slow the relaxation rate to a reasonable level before final polish and coating.

For example, Fig. 16.13 shows the effects of residual stresses on relaxation for a particular clad optic. Note that the stress reduction after cycling to 175 °C is on the order of only 2%; however, note from Eq. (4.48) that this small reduction can degrade performance for high-aspect-ratio optics. By cycling the optic several times to above (and below, for good measure) the maximum expected extremes, relaxation is stabilized as long as those extremes are not exceeded. It is then that the optic can be polished. Of course, if such changes (due to high coating residual stress) are found to occur after polish during the coating process, one has to live with those changes or otherwise use a form of active control, such as by focus or actuating methods. Fortunately, most coatings are thin (less than 1 μm) and inconsequential to lower-aspect-ratio optics.

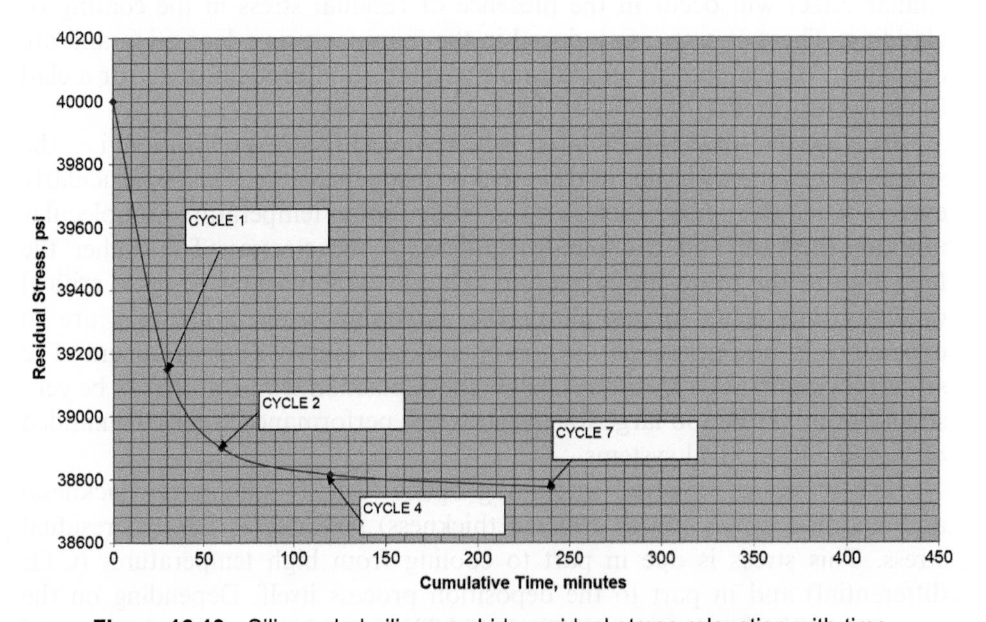

Figure 16.13 Silicon-clad silicon carbide residual stress relaxation with time.

16.9.7.2 Machining residual stress

Machining of surfaces produces residual stress. The machining process produces small subsurface flaws in the material beneath the removed surface, creating, in general, a very high internal compressive stress near the surface. The flaw depth depends on the machining cut depth. Similar compressive stresses are produced in grinding or abrasive processes, with flaw size dictated by grit size. At any rate, again, residual stress needs to be stabilized to avoid stress relaxation if the residual stress is high enough or is not subsequently removed by control grinding techniques.[28] Stabilization of residual stress from machining is achieved by stress reversal and redistribution of locally high stress values, which often exceed the material yield point.

16.9.8 Metal optics

Machining of metal optics such as aluminum or beryllium induces residual stresses that can be removed to some extent by heat treatment. In the case of beryllium, however, to be fully successful, such a heat treatment needs to approach the anneal point, which is much higher than one would care to go for precision optics.

Residual stress in machining beryllium causes basal plane re-orientation of the crystals[29] at the surface, resulting in a higher CTE near the surface. Thus, a change in temperature will show a change in deformation if the residual stress is not eliminated. To this end, isothermal exposure or thermal cycling can mitigate the effects.

A classic example[29,30] is shown in Fig. 16.14. Here, a beryllium component exhibiting a high degree of residual stress is thermally cycled over various temperature extremes. Note that the higher temperatures produce significant stress reduction, but even lower temperatures produce some relief. Again, as a minimum, it is important to cycle to just above and just below the specification temperature extremes. Several cycles should stabilize the material.

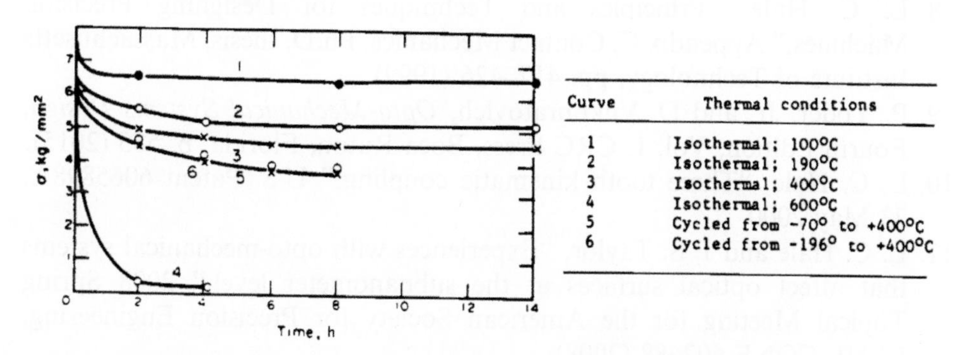

Curve	Thermal conditions
1	Isothermal; 100°C
2	Isothermal; 190°C
3	Isothermal; 400°C
4	Isothermal; 600°C
5	Cycled from -70° to +400°C
6	Cycled from -196° to +400°C

Figure 16.14 Beryllium residual stress relief (reprinted from Ref. 29).

It should be pointed out again that in the case of beryllium, whose grain structure is anisotropic, a "double whammy" effect occurs. In addition to residual stress relaxation, the surface grain structure in the presence of the residual stress has a different CTE compared to the undisturbed material, perhaps on the order of 1 ppm/°C. Thus, a bimetallic effect occurs with temperature, compounding the relaxation phenomenon. Therefore, many processes cycle to significantly higher-than-survival extremes to remove as much stress as possible and minimize both relaxation and CTE concerns.

A similar phenomenon occurs in machining aluminum and other metals; the reader should adhere to recommended guidelines of stress relief temperature cycling for these materials.

References

1. A. Mironer and F. Regan, "Venting of space shuttle payloads," Shuttle Environment and Operations Meeting, American Institute of Aeronautics and Astronautics, Washington, D.C. (1983).
2. K. Schwertz and J. Burge, *Field Guide to Optomechanical Design and Analysis*, SPIE Press, Bellingham, Washington (2012) [doi: 10.1117/3.934930].
3. S. Timoshenko, "Analysis of bi-metal thermostats," *J. Optical Society of America* **11**(3), 233–255 (1925).
4. T.-C. Chen, C.-J. Chu, C.-H. Ho, C.-C. Wu, and C.-C. Lee, "Determination of stress-optical and thermal-optical coefficients of Nb_2O_5 thin film material," *J. Applied Physics* **101**, 043513 (2007).
5. K. B. Doyle, V. L. Genberg, and G. J. Michels, *Integrated Optomechanical Analysis*, Second Edition, SPIE Press, Bellingham, Washington, p. 265 (2012) [doi: 10.1117/3.974624].
6. R. J. Roark and W. C. Young, *Formulas for Stress and Strain*, Fourth Edition, McGraw-Hill, New York, p. 319 (1965).
7. A. Foppl, *Technische Mechanik*, Vol. **5**, p. 350, Kessinger Legacy Reprints, Germany (1905).
8. L. C. Hale, "Principles and Techniques for Designing Precision Machines," Appendix C, Contact Mechanics, Ph.D. thesis, Massachusetts Institute of Technology, pp. 417–426 (1999).
9. P. Yoder, Jr. and D. Vukobratovich, *Opto-Mechanical Systems Design*, Fourth Edition, Vol. **1**, CRC Press, Boca Raton, Florida, p. 585 (2015).
10. L. C. Hale, "Three tooth kinematic coupling," U.S. Patent 6065898A, 23 May 2000.
11. L. C. Hale and J. S. Taylor, "Experiences with opto-mechanical systems that affect optical surfaces at the subnanometer level," 2008 Spring Topical Meeting for the American Society for Precision Engineering, LLNL-CONF-402688 (2008).

12. S. Timoshenko and J. Goodier, *Theory of Elasticity*, Second Edition, McGraw-Hill, New York, p. 376 (1951).
13. G. Amontons, "On the resistance caused in machines, both by the rubbing of the parts that compose them, and by the stiffness of the cords that one uses in them, & the way of calculating both," *Histoire de l'Académie royale des sciences*, Paris, pp. 206–222 (1699).
14. J. F. Archard, "Contact and rubbing of flat surfaces," *J. Applied Physics* **24**(8), 981–988 (1953).
15. F. P. Bowden and D. Tabor, *The Friction and Lubrication of Solids*, Oxford University Press, Oxford, pp. 87–89 (1950).
16. K. Miyoshi and D. H. Buckley, "Relationship between the ideal tensile strength and the friction properties of metals in contact with nonmetals and themselves," NASA Technical Paper 1883 (1981).
17. S. P. Timoshenko and S. Woinowsky-Krieger, *Theory of Plates and Shells*, Second Edition, McGraw-Hill, New York, pp. 4–32 (1959).
18. H. H. Stevens, "Behavior of circular membranes stretched above the elastic limit by air pressure," *Proc. Soc. Experimental Stress Analysis* **2**(1) (1944).
19. R. Roark and W. Young, *Formulas for Stress and Strain*, Fifth Edition, McGraw-Hill, New York, p. 408 (1975).
20. W. P. Barnes, "Optical windows," *Proc. SPIE* **10265**, *Optomechanical Design: A Critical Review*, 102650B (1992) [doi: 10.1117/12.61108].
21. J. W. Pepi, M. A. Kahan, W. H. Barnes, and R. J. Zielinski, "Teal Ruby: design, manufacture, and test," *Proc. SPIE* **0216**, pp. 160–173 (1980) [doi: 10.1117/12.958459].
22. S. F. Jacobs, "Variable invariables: dimensional instability with time and temperature," *Proc. SPIE* **10265**, *Optomechanical Design: A Critical Review*, 102650I (1992) [doi: 10.1117/12.61115].
23. J. W. Pepi and D. Golini, "Delayed elasticity in Zerodur® at room temperature," *Proc. SPIE* **1533**, pp. 212–221 (1991) [doi: 10.1117/12.48857].
24. S. F. Jacobs, S. C. Johnston, and G. A. Hansen, "Expansion hysteresis upon thermal cycling of Zerodur," *Applied Optics* **23**(17), 3014–3016 (1984).
25. S. C. Wilkins, D. N. Coon, and J. S. Epstein, "Elastic hysteresis phenomena in ULE and Zerodur optical glasses at elevated temperatures," *Proc. SPIE* **0970**, pp. 40–46 (1989) [doi: 10.1117/12.948176].
26. B. S. Lement, B. L. Averbhach, and M. Cohen, "The dimensional behavior of Invar," *Transactions American Society for Metals* **43**, pp. 1072–1097 (1951).
27. J. Mullin, "Viscous Flow and Structural Relaxation in Amorphous Silicon Thin Films," Ph.D. thesis, Harvard University, Cambridge (2000).
28. J. W. Pepi, *Strength Properties of Glass and Ceramics*, SPIE Press, Bellingham, Washington, p. 133 (2014) [doi: 10.1117/3.1002530].

29. R. A. Paquin, "Dimensional instability of materials: How critical is it in the design of optical instruments?" *Proc. SPIE* **10265**, *Optomechanical Design: A Critical Review*, 1026509 (1992) [doi: 10.1117/12.61106].
30. I. Kh. Lokshin, "Heat treatment to reduce internal stresses in beryllium," [translated from Russian] *Metal Science and Heat Treatment* **12**(5), pp. 426–427 (1970).

Epilogue

As stated in the Preface, the intent of this text was to provide a basic understanding of opto-structural design principles using hand analyses. This is critical in a proposal or early program phase, enabling a first-order system design in short time prior to diving in to the more detailed finite element models, the latter of which will be necessary for fine tuning the performance requirements.

Even in later design phases in which finite element analyses are critical, the hand analysis basics are of equal or greater importance. For it is these basics that validate the models, and not the models that validate the basics.

While some of the analysis reported herein can be found in other publications, much more is not; at any rate, this text gathers the structural analysis that is critical to optical instruments. And although some of the first-order analyses are basic, others are more esoteric in nature and, indeed, will *not* be found elsewhere.

While finite element tools are of absolute necessity, the techniques in this text allow for the development of improved models as well as validation of models, which is essential. There is a process from which one can start with hand calculations, go to simple "stick" models, and then move to models of increasing fidelity in a "crawl, walk, run" strategy. The hand calculations fit nicely into this process.

It is hoped that the preceding chapters will mitigate the risk associated with the use of detailed modeling prior to the application of the basics of engineering analysis principles.

Perhaps someday we will look back fondly on these things.

Index

John W. Pepi received his undergraduate education from Tufts University, graduating with a Bachelor of Science degree in Civil Engineering in 1967. He obtained a Master of Science degree in Structural Engineering in 1968 from Northwestern University.

He has been employed with L-3 Technologies SSG for the past fifteen years as a lead staff mechanical engineer for precision, lightweight, space-based optical systems. Prior to his employment at L-3 he was employed at Lockheed Martin and was extensively involved as lead engineer for the successfully launched AIRS satellite program for JPL/NASA Goddard Space Flight Center. Prior to that assignment, he held the positions of lead engineer, chief engineer, department manager, and program manager for large optical systems at ITEK Optical Systems, where he worked for 22 years directing the work of up to 55 individuals. His last assignment for the Keck Telescope led to the discovery of delayed elasticity effects in certain ceramics.

John is an internationally recognized authority on lightweight mirror design and a member of several international ground-based large-telescope oversight committees. He is an SPIE Fellow and is the author of more than one dozen publications on lightweight optics and mirror design principles, including the book *Strength Properties of Glass and Ceramics*, SPIE Press (2014), and has been an instructor for SPIE at its international meetings.